Lasers and Mass Spectrometry

OXFORD SERIES ON OPTICAL SCIENCES

EDITORS
MARSHALL LAPP
HENRY STARK

Lasers and
Mass Spectrometry

Edited by
DAVID M. LUBMAN
The University of Michigan

New York   Oxford
OXFORD UNIVERSITY PRESS
1990

Oxford University Press

Oxford   New York   Toronto
Delhi   Bombay   Calcutta   Madras   Karachi
Petaling Jaya   Singapore   Hong Kong   Tokyo
Nairobi   Dar es Salaam   Cape Town
Melbourne   Auckland

and associated companies in
Berlin Ibadan

Library of Congress Cataloging-in-Publication Data
Lasers and mass spectrometry / edited by David M. Lubman.
p. cm.—(Oxford series on optical sciences ; 1)
Bibliography: p.   Includes index.
ISBN 0-19-505929-8
1. Laser spectroscopy. 2. Mass spectrometry.
I. Lubman, David M. II. Series.
QD96.L3L365   1989
543′.0858—dc19   88-39793   CIP

9 8 7 6 5 4 3 2 1
Printed in the United States of America
on acid-free paper

Preface

This volume deals with one of the most exciting and rapidly expanding fields of chemical analysis, that is, the use of lasers in mass spectrometry. This field encompasses a range of different methodologies applied to a wide variety of problems in detection and identification of atoms and molecules. The growth of this field has been in part the result of rapid advances in laser technology including the advent of high-powered, narrow-bandwidth pulsed tunable dye lasers that can produce wavelengths ranging from the infrared to the vacuum ultraviolet. These lasers have been used in a number of different configurations when interfaced to various mass spectrometer designs to solve problems not easily solved by conventional sources.

The use of lasers in mass spectrometry has been applied to problems as diverse as single atom detection and the sequencing of peptides. Laser methodology has also been used to solve problems in surface analysis, semiconductor analysis and depth profiling, organic mass spectrometry, biomedical and polymer analysis, supersonic jet spectroscopy and conformational structure analysis, and atmospheric pressure detection. These have used a range of laser-based methods such as laser desorption/ablation for evaporation of atoms and molecules and laser multiphoton ionization for selective and sensitive detection in a mass spectrometer. Thus, the uses of lasers and mass spectrometry constitute a field of broad interest for solving a wide range of problems, not only to the analytical chemistry community, but to a number of peer groups in materials science, engineering, medicine, pharmacy, biophysics, synthetic chemistry, and spectroscopy.

This volume attempts to address the interests of both the chemical analysis group and the broader community that might use laser-based analytical techniques. It also attempts to address the graduate student community to introduce them to this dynamic field. The only reasonable way in which this could be accomplished was by an edited volume consisting of a number of investigators who could best write about their own work in their specialized field of research. Thus, this volume consists of a collection of 23 articles contributed by 57 authors working in numerous different laboratories around the country. Each of these authors is an expert in his field and the articles represent the state of the art in laser-based methods. Although a diverse number of authors is represented, it should be noted that these authors do not in any way represent a definitive list of the important groups active in this field. The particular selection of authors was based most notably on the desire to obtain a broad coverage of the field. The work of many excellent investigators could not be included because of space limitations. In addition, the contributors selected are from the United States and Canada to facilitate the assembling of this collection of manuscripts. However, this volume should not be considered the final work in the field. Hopefully, as a result of the increasing demand for the use of the techniques described and the growth of new methodologies, other edited collections will soon follow.

I would like to thank the many contributors for the great effort invested in each article and for their timely response. I am very much indebted to several of my students, Ho Ming Pang, Michael Tierney, Liang Li, David Lustig, Lori McCaig, Louis Grace and Ton: Barstis for a critical reading and aid in editing many of these manuscripts. I would like to thank Dr. Jacqueline Hartt, the Oxford University Press Editor, for her enduring patience and the Oxford University Press staff for their fine copyediting. I would also like to thank the National Science Foundation, The U. S. Army Research Office, the Petroleum Research Fund 1, Shell Development Co., and Dow Chemical Co. for generous support of my research in this field.

Ann Arbor   D.M.L.
October 1988

Contents

Contributors

Steven L. Allman   Health and Safety Research Division, Oak Ridge National Laboratory, Oak Ridge, TN 37831

Peter Arrowsmith   IBM General Products Division, San Jose, CA 95193

Christopher H. Becker   Chemical Physics Laboratory, SRI International, Menlo Park, CA 94025

Asgeir Bjarnason   Science Institute, University of Iceland, Dunhaga 3, 107 Reykjavik, Iceland

Chung H. Chen   Health and Safety Research Division, Oak Ridge National Laboratory, Oak Ridge, TN 37831

M. Paul Chiarelli   Midwest Center for Mass Spectrometry, Department of Chemistry, University of Nebraska, Lincoln, NE 68588

Robert B. Cody   Nicolet Analytical Instruments, 6416 Schroeder Rd., Madison, WI 53711

Steven M. Colby   Department of Chemistry, Indiana University, Bloomington, IN 47405.

R. Graham Cooks   Department of Chemistry, Purdue University, W. Lafayette, IN 47907

Terrill A. Cool   School of Applied and Engineering Physics, Cornell University, Ithaca, NY 14853

Roy L. M. Dobson   Procter & Gamble Company, Cincinnati, OH

Arthur P. D'Silva   Ames Laboratory, Iowa State University, Ames, IA 50011

W. Bart Emary   Department of Chemistry, Purdue University, W. Lafayette, IN 47907

Ron C. Estler   Chemistry Department, Fort Lewis College, Durango, CO 81301

Ben S. Freiser   Department of Chemistry, Purdue University, West Lafayette, IN 47907

Michael L. Gross   Midwest Center for Mass Spectrometry, Department of Chemistry, University of Nebraska, Lincoln, NE 68588

James W. Hager   SCIEX, 55 Glen Cameron Rd., Thornhill, Ontario L3T 1A2 Canada

Owen W. Hand   Department of Chemistry, Purdue University, W. Lafayette, IN 47907

Willard W. Harrison   Department of Chemistry, University of Florida, Gainesville, FL 32611

John C. Hemminger   Department of Chemistry, University of California, Irvine, CA 92717

Kenneth R. Hess   Department of Chemistry, Franklin and Marshall College, Lancaster, PA 17604

Donald F. Hunt   Department of Chemistry, University of Virginia, Charlottesville, VA 22901

Heinrich E. Hunziker   IBM Research Division, Almaden Research Center, 650 Hary Road, San Jose, CA 95120

G. Samuel Hurst   Institute of Resonance Ionization Spectroscopy, University of Tennessee, Knoxville, TN 37932

Steven D. Kramer   Health and Safety Research Division, Oak Ridge National Laboratory, Oak Ridge, TN 37831

Donald P. Land   Department of Chemistry, University of California, Irvine, CA 92717

Donald H. Levy   The James Frank Institute and Department of Chemistry, University of Chicago, Chicago, IL 60637

Liang Li   Department of Chemistry, The University of Michigan, Ann Arbor, MI 48109

David M. Lubman   Department of Chemistry, The University of Michigan, Ann Arbor, MI 48109

Timothy J. MacMahon   Department of Chemistry, Purdue University, West Lafayette, IN 47907

Robert T. McIver, Jr.   Department of Chemistry, University of California, Irvine, CA 92717

John J. Morelli   McDonnell Douglas Research Laboratories, Dept. 221, Building 110, St. Louis, MO 63166

Nicholas S. Nogar   Chemical and Laser Sciences Division, MS G738, Los Alamos National Laboratory, Los Alamos, NM 87545

Lydia M. Nuwaysir   Department of Chemistry, University of California–Riverside, Riverside, CA 92521

Robert W. Odom   Charles Evans & Associates, 301 Chesapeake Dr., Redwood City, CA 94063

Richard B. Opsal   Department of Chemistry, Indiana University, Bloomington, IN 47405

James E. Parks   Atom Sciences, Inc., 114 Ridgeway Center, Oak Ridge, TN 37830

Marvin G. Payne Health and Safety Research Division, Oak Ridge National Laboratory, Oak Ridge, TN 37831

Ronald C. Phillips Health and Safety Research Division, Oak Ridge National Laboratory, Oak Ridge, TN 37831

James P. Reilly Department of Chemistry, Indiana University, Bloomington, IN 47405

Thomas R. Rizzo Department of Chemistry, University of Rochester, Rochester, NY 14627

Bruno Schueler Charles Evans & Associates, 301 Chesapeake Dr., Redwood City, CA 94063

Jeffrey Shabanowitz Department of Chemistry, University of Virginia, Charlottesville, VA 22901

Michael G. Sherman Department of Chemistry, University of California, Irvine, CA 92717

Jack A. Syage The Aerospace Corp., Aerophysics Laboratory, P. O. Box 92957, Los Angeles, CA 90009

Tsong-Lin Tai Department of Chemistry, University of California, Irvine, CA 92717

Mattanjah S. de Vries IBM Research Division, Almaden Research Center, 650 Hary Road, San Jose, CA 95120

Stephen C. Wallace Department of Chemistry, University of Toronto, Toronto, Ontario M5S 1A2 Canada

Diana T. S. Wang Department of Chemistry, University of California, Irvine, CA 92717

Stephen Weeks Ames Laboratory, Iowa State University, Ames, IA 50011

David A. Weil Nicolet Analytical Instruments, 6416 Schroeder Rd., Madison, WI 53711

Robert R. Weller Department of Chemistry, Purdue University, West Lafayette, IN 47907

H. Russel Wendt IBM Research Division, Almaden Research Center, 650 Hary Road, San Jose, CA 95120

Charles W. Wilkerson, Jr. Department of Chemistry, Indiana University, Bloomington, IN 47405

Charles L. Wilkins Department of Chemistry, University of California—Riverside, Riverside, CA 92521

Lasers and Mass Spectrometry

1

Isotopically Selective Counting of Atoms and Molecules Using Resonance Ionization Spectroscopy

CHUNG H. CHEN, MARVIN G. PAYNE, G. SAMUEL HURST,
STEVEN D. KRAMER, STEVEN L. ALLMAN, and
RONALD C. PHILLIPS

Analytical chemists have dreamt of having an instrument with the following properties: (1) detection sensitivity that reaches the ultimate limit: single atom or single molecule detection, (2) high discrimination between various isotopes, and (3) very broad dynamic range. When the concentration of the desired species is low in a sample, the species can be individually counted. Further, a sample containing a high level of selected atoms or molecules can be measured in a very short period of time in an analog mode. In addition, (4) the time for analysis is short and (5) the instrument is inexpensive and easy to operate. At this time there are no instruments available that meet these five conditions. However, during the last decade, resonance ionization spectroscopy $(RIS)^{1-4}$ was developed and has been used to demonstrate single atom detection$^{5-8}$ and isotopically selective rare gas counting$^{9-13}$ at the Oak Ridge National Laboratory (ORNL). This may possibly be an important step toward having a super analytical tool that meets all five of the requirements listed.

In general, it is quite difficult to detect the occurrence of rare events in nature. However, the resolution of these rare events often leads to the discovery of very fundamental principles of physics or chemistry. For example, the flux of solar neutrinos is strongly related to the model of our sun.^{14} The cross section of capturing a solar neutrino is so low that the detection of solar neutrino flux needs 100,000 gallons of perchloroethylene sample placed underground to produce a few ^{37}Ar atoms to be counted.^{15} However, the measurement of the solar neutrino flux may have significant impact on the understanding of the solar model.^{16} Another example of the need to count low levels of isotopically selective rare gas atoms is the dating of the polar ice cap^{17} and ocean current studies.^{18} For polar ice cap dating, a 1-kg block of modern ice contains only about 1400 atoms of ^{81}Kr. However, there are about 2.5×10^{14} atoms of ^{82}Kr in the same sample. Only the accurate counting of ^{81}Kr can lead to the accurate determination of the age of the ice. It is obvious that isotopically selective detection of molecules will have a very large impact on biological, medical, and chemical research as well. Since radioactive doping is still among the most popular methods used in checking mechanisms of chemical and biochemical reactions, the ability to detect isotopically selected atoms or molecules is critical.

RESONANCE IONIZATION SPECTROSCOPY (RIS)

The concept of RIS is quite straightforward. An atom or a molecule is excited by photons that promote an electron into a selected discrete state. The atom with an excited electron is then subsequently promoted to the continuum to form an ion by the absorption of another photon or by other methods such as field ionization or collisional ionization. RIS is a very selective ionization method when compared to nonresonance ionization caused by electron impact, X rays, or high-velocity charged particles. As a result of the rapid progress in the development of high-power lasers, a multiphoton ionization process that does not use the resonant excitation of an excited state becomes possible.[19-21] However, RIS is not the same as nonresonance multiphoton ionization procesess that usually have very little selectivity. The other distinct feature of RIS is the capability of achieving complete ionization for all the atoms or molecules from a selected state. In contrast, it is unusual to completely ionize any atomic or molecular species within a laser beam with nonresonant multiphoton ionization. Since RIS uses one- or two-photon resonance excitation, the saturated transition of the ground state to the selected excited state can be achieved with commercially available laser beams. Thus, measurements of the selected atoms or molecules can be determined quantitatively. On the other hand, it is almost impossible to measure selected atoms or molecules quantitatively with a nonresonant multiphoton ionization process without resorting to very tedious calibrations of a known sample.

Figure 1.1 schematically shows what is involved in the most elementary of all RIS processes. A pulsed laser beam produces photons of the resonant energy, $\hbar\omega$, to excite an atom initially in its ground state 1, to an excited level 2. Another photon can ionize state 2, but not state 1. Typically within 10^{-9} sec, the rate of stim-

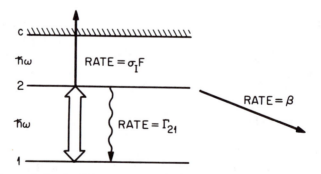

Fig. 1.1. Schematic for the RIS process in which the final state is reached by photoionization. An atom in ground state (1) can be ionized by first exciting to another bound state (2), driving ionization through a resonance process.

ulated emission from the excited state will be about equal to its rate of production. Thus, an equilibrium persists between states 1 and 2 as long as the laser intensity is kept sufficiently high during a pulse. However, such an equilibrium is actually a quasi-steady state since the intermediate state can be ionized slowly. If the photon flux, F (number of photons per cm^2/sec) is higher than the decay rate β, and the photon fluence (number of photons per cm^2) is large enough to have $\sigma_I F \tau$ much larger than 1, the RIS process is said to be saturated (τ is the pulse duration of the laser; σ_I is the ionization cross section). Then, all the atoms in the laser beam should be ionized.

To show saturation of the RIS process in a quantitative manner, a rate formulation is given in the following:

$$\frac{dn_1}{dt} = -n_1 \int d\nu \sigma_a(\nu) I(\nu) + \Gamma_{21} n_2 + n_2 \int d\nu \sigma_s(\nu) I(\nu)$$

$$\frac{dn_2}{dt} = n_1 \int d\nu \sigma_a(\nu) I(\nu) - n_2 \int d\nu \sigma_s(\nu) I(\nu) - \Gamma_{21} n_2 - \beta n_2 - \sigma_I F n_2$$

$$\frac{dn_I}{dt} = \sigma_I F n_2$$

In the above equations, n_1, n_2, and n_I represent the number of atoms in state 1, state 2, and continuum, respectively. $\sigma_a(\nu)$ is the cross section for photon absorption from state 1 to state 2, $\sigma_s(\nu)$ is the stimulated emission cross section, and $\sigma_I(\nu_0)$ is the cross section for photoionization of state 2 at the frequency ν_0. A laser beam is characterized by intensity $I(\nu)$ where $I(\nu) \, d\nu$ equals the number of photons per cm^2/sec and in the frequency interval between ν and $\nu + d\nu$, where a Gaussian lineshape is assumed.

$$I(\nu) = \frac{F}{(2\pi \Delta \nu^2)^{1/2}} \exp \left[-\frac{(\nu - \nu_0)^2}{2 \Delta \nu^2} \right]$$

F is the photon flux and $\Delta \nu$ is the laser linewidth.

By solving the rate equations above with the condition of $\sigma_I F \gg \beta$ and both σ_a and σ_s are much larger than σ_I, the production rate of ions is

$$\frac{dn_I}{dt} = \frac{g_2}{g_1 + g_2} \sigma_I F n_1(0) \exp\{-[g_2/(g_1 + g_2)] \sigma_I F t\}$$

Integrating over the length of a laser pulse gives

$$n_I = n_I(0) \left[1 - \exp \left(\frac{-g_2}{g_1 + g_2} \sigma_I F \tau \right) \right]$$

where g_1 and g_2 are the statistical weights of level 1 and level 2, $n_1(0)$ is the initial ground state population, and τ is the laser pulse duration assuming a square pulse. If $\sigma_I F \tau \gg 1$, all the atoms initially in the ground state will be ionized during the laser pulse.

A wide variety of laser schemes can be employed to carry out a resonance ionization process.[22,23] Most of the schemes can easily lead to complete ionization of the resonant species utilizing commercially available lasers. The common RIS schemes are shown in Fig. 1.2. The notations used here are explained by the following examples:

1. $Cs[\omega_1, \omega_1 e^-]Cs^+$ [7] means that Cs can be excited resonantly by a photon of frequency ω_1, the excited Cs is subsequently ionized by the photon with the same frequency to produce a Cs^+.
2. $Kr[\omega_1\omega_2, \omega_1 e^-]Kr^+$, a two-photon excitation of Kr followed by an one-photon ionization process[24,25] (see Fig. 1.2, scheme 5).
3. $Xe[\omega_1\omega_1, \omega_1 e^-]Xe^+$, a two-photon excitation of Xe followed by ionization by a laser photon.[9]
4. $NO[2\omega_1, 2\omega_1 e^-]NO^+$, ground state NO molecules are excited by a photon that is obtained by a frequency doubling process. The excited NO molecules get ionized by another photon with the same frequency.[26]

The notation $\omega\omega$ represents a two-photon process whereas 2ω represents a single photon produced by a frequency doubling process. Each of the basic photoionization schemes can be modified so that the ionization process can go through associative ionization, chemionization, field ionization, etc. A possible laser scheme for each element in the periodic table (except for He and Ne) is also shown in Fig. 1.2. The schemes shown in Fig. 1.2, in general, can be used for saturating the ionization process (i.e., ionizing all of the selected species within the laser beam) by commercially available lasers. However, that does not imply that the proposed schemes are always the most efficient process for RIS since the cross sections for most two-photon processes are not known and most of the ionization cross sections of an excited atom are not determined either. If the saturation of the RIS process is not of great concern, RIS of He and Ne can also be pursued since coherent vacuum ultraviolet (VUV) beams with wavelengths shorter than 40 nm have been produced by a high order harmonic generation process in the pulsed molecular beam.[27]

Since the energy level of the first excited state of all the stable molecules is below 12 eV, RIS can be applied to any molecule. However, the Franck-Condon factor and dissociation or predissociation process can preclude an efficient RIS process for many molecules.

DETECTION OF A SINGLE ATOM IN A LASER BEAM BY RIS

It is well known that a proportional counter[28] can be used to detect a single electron as a result of gas amplification initiated by a free electron. Because RIS can be used to completely ionize all atoms of a selected type within the laser beam, the combination of RIS and a proportional counter can provide the capability of single atom detection. The first demonstration was carried out with a proportional counter to show that a single Cs atom could be picked out in a gas containing an enormously large number of atoms or molecules of another type.[5] In Fig. 1.3 a detailed diagram of spectral wavelengths, transition rates, and lifetimes for the lower levels of the cesium atoms is shown. A flashlamp pumped dye laser with wavelength at 455.5 nm was used to promote Cs atoms from the ground state to $Cs(7^2P_{3/2})$. Another photon with the same wavelength ionizes the atom. Only a very

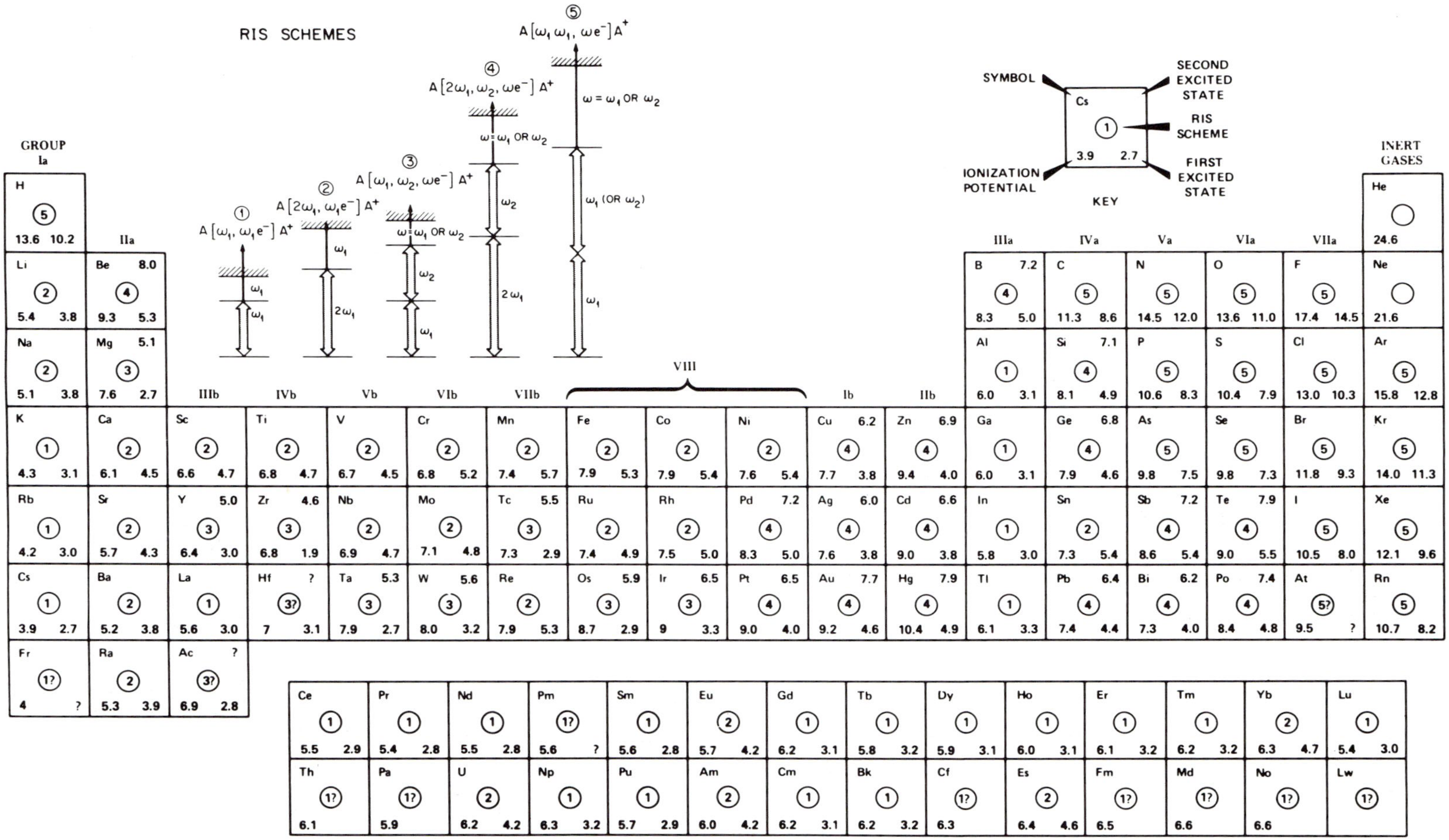

Fig. 1.2. Classification of RIS scheme for detection of the elements.

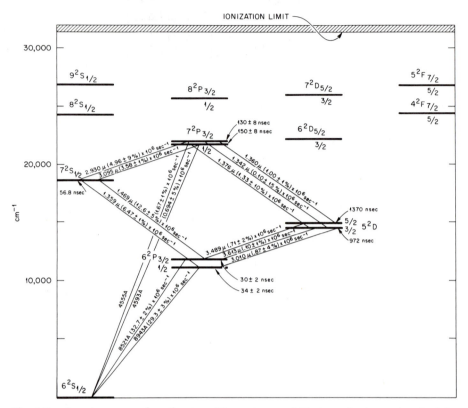

Fig. 1.3. Spectrum wavelengths, transition rates, and lifetimes for lower level of the cesium atom.

modest laser energy (~ 100 mJ/cm^2) was necessary to achieve saturation of the ionization.

The apparatus used for the detection of Cs atoms is shown in Fig. 1.4. With p-10 counting gas (90% Ar + 10% CH$_4$) at 100 torr, the gas amplification was on the order of 10^4 when the wire was at 1000 V. The laser beam was focused below the center of the counter wire using a 25-cm focal length lens. The laser beam volume in the active region of the counter was estimated to be $\sim 5 \times 10^{-2}$ cm^3. Under these conditions and with the field tubes decoupled from the wire to suppress photoelectrons from the quartz windows, there was no measurable photoelectron background within a few microseconds of the laser pulses. The concentration of Cs vapor in the active laser volume was varied by changing the distance of the cesium source to the laser beam or by controlling the temperature of the cesium source.

Under conditions of saturation, every Cs atom in the laser beam was photoionized and detected. Figure 1.5 shows a pulse-height distribution for a large number of Cs atoms with the condition that the laser always saturates the ionization. From the known fact that each 6.4-keV X ray produces 250 electrons in p-10 counting gas at high pressure, the Cs peak in Fig. 1.5 was found to correspond to 10^4 Cs atoms.

One electron in a proportional counter produces an exponential-like pulse-height distribution. Figure 1.6 was obtained by using a mercury lamp to release single photoelectrons from the inner walls of the counter at a low rate. With the Cs sample in its lowest position and a gas pressure of 200 torr, a nearly identical distribution due to ionization of Cs atoms was found. At all population levels, including one atom within the beam, it was observed that the ionization signal vanished when the wavelength of the laser beam was detuned from the Cs resonance transition. Thus, the demonstration of the detection of a single Cs atom by RIS is clearly demonstrated. In addition to the detection of a single atom with a proportional counter by RIS, it can easily be visualized that the detection of a single atom in a laser beam under ultrahigh vacuum conditions can be achieved by using an electron multiplier or a channeltron to detect a single electron or ion. Detection of a single atom in a laser beam was also demonstrated by some laser fluorescence methods. Greenless et al.[29] used the photon burst method to show this capability. Similar fluorescence methods with certain improvements have been used by Kaufman[30] as well as She et al.[31]

Although the detection of a single atom in a laser beam reaches the optimum

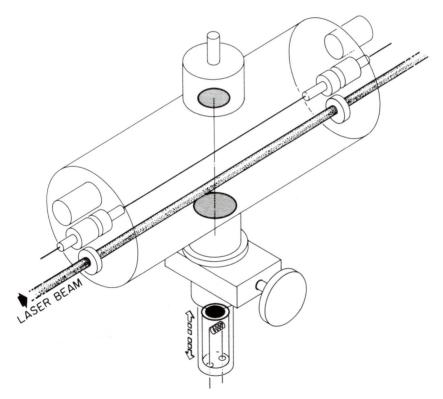

Fig. 1.4. Schematic of apparatus used for the detection of cesium atoms. A pulsed laser beam is used to remove one electron from the selected atom. The electron produced is subsequently detected with a proportional counter.

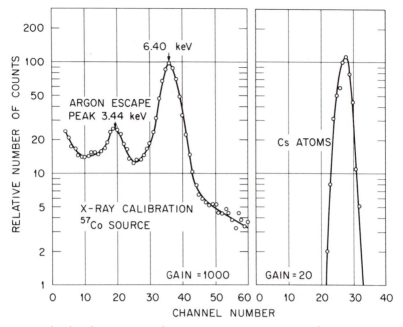

Fig. 1.5. Pulse-height spectrum due to resonance ionization of Cs atoms. From the gain ratio and the fact that the ^{57}Co X ray produces 250 electrons, the most probable number of Cs atoms detected is 10^4.

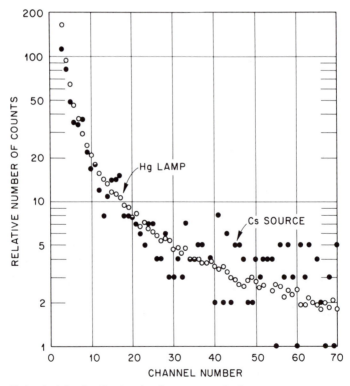

Fig. 1.6. Pulse-height distribution for the case in which one Cs atom was counted per 20 laser pulses, compared with a one-electron distribution.

detection sensitivity limit, this does not imply that all atoms of interest can be detected because, in a given circumstance, many desired atoms may simply not be made to reside in the laser beam. Thus, we tried to develop a rare gas atom counter[11,32,33] that could count every single rare gas atom in the sample. By this method, as little as a single atom in a sample can be detected.

ISOTOPICALLY SELECTIVE COUNTING OF NOBLE GAS ATOMS

The first isotopically selective detection of rare gases by RIS was pursued by two-photon excitation and three-photon ionization of Xe.[9] In this experiment, an Nd-YAG laser beam was frequency doubled and used to pump a red dye (DCM) to obtain tunable light between 645 and 675 nm[34] with an output of 40 mJ/pulse. The dye laser output was doubled in a KDP crystal, and this light was mixed with 1.06-μm radiation from the Nd:YAG laser to yield tunable light from 246 to 258 nm. At a two-photon resonance in Xe (i.e., 252.6 or 249.6 nm), the energy per pulse was ~ 1 mJ and the pulse duration was 4 nsec. During the experiment, the dye laser beam was focused with a 13.5-cm focal length lens into the center of the ionizer in a quadrupole mass spectrometer. The ionization lineshape for three-photon ionization of Xe near the two-photon resonance with the $\mathrm{Xe}[\frac{1}{2}]_0(5p^56p)$ and $\mathrm{Xe}[\frac{3}{2}]_2(5p^56p)$ state is shown in Fig. 1.7. The mass number of the quadrupole filter was set at 131.

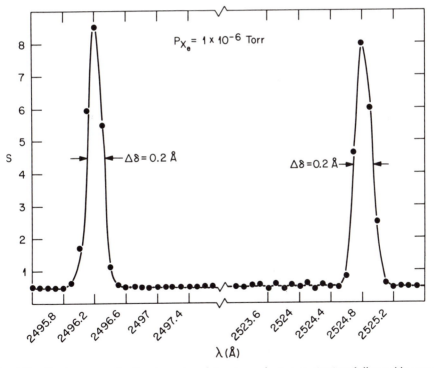

Fig. 1.7. Tuning curve for Xe atoms involving two-photon excitation followed by one-photon ionization.

The background due to the multiphoton ionization process was not observable, because most of the background due to multiphoton ionization could not pass through the mass filter.

We now want to consider the RIS of Xe by the two-photon excitation and the three-photon ionization process in more detail since a two-photon excitation process is broadly used in RIS of molecules. When the power density is less than 10^{10} W/cm^2 and the laser linewidth is greater than 0.3 Å, the RIS process of Xe can be described by the following rate equations:

$$\frac{d\rho_{00}}{dt} = -R(\rho_{00} - \rho_{11})$$

$$\frac{d\rho_{11}}{dt} = R(\rho_{00} - \rho_{11}) - \Gamma_I \rho_{11}$$

with $n_I = \sigma_I F$, where ρ_{00} and ρ_{11} represent the probabilities of a Xe atom being in ground state and excited state, respectively. R, Γ_I, σ_I, and F are two-photon excitation rates, ionization rate of the excited state, ionization cross section from the excited state, and photon flux, respectively.

In the approximation of a square pulse envelope for the power density, the ionization probability (P_I) of an atom can be obtained.

$$P_I = 1 - \frac{F_I R}{2J} \left[\frac{\exp(-\beta_1 \tau)}{\beta_1} - \frac{\exp(-\beta_2 \tau)}{\beta_2} \right]$$

where $\beta_1 = R + \Gamma_I/2 - J$, $\beta_2 = R + \Gamma_I/2 + J$, $J = \sqrt{R^2 + \Gamma_I^2/4}$, and τ is the pulse duration of the laser. The ionization probability depends only on the values of the saturation parameters $R\tau$ and $\Gamma_I \tau$ at the location. Figure 1.8 shows the ion signals as a function of the square of the energy per pulse. We have analyzed the power dependence of the signal in order to infer experimental values for Γ_I and R as a function of power density. In the analysis, a square pulse for the time dependence was used. We assume the power density at a distance x from the focal point and a distance ρ from beam center is

$$I(\rho,x) = (E/\tau)A(x) \exp\{-[\rho - \rho_0(x)]^2/d^2(x)\}$$

where E is the energy per pulse, $\rho_0(x) = \rho_{00}x/f$, ρ_{00} is the distance from beam center to the peak intensity before focusing, f is the lens focal length, $d(x) = [(F\theta_{1/2})^2 + (d_0 x/F)^2]^{1/2}$, where d_0 is the half-width for e^{-1} drop in the intensity of the doughnut-shaped beam before focusing, $\theta_{1/2}$ is the half-angle beam divergence of the initial beam before focusing, and $A(x)$ is determined by

$$A(x) = \left[2\pi \int_0^\infty \rho d\rho \exp\{-[\rho - \rho_0(x)]\}^2/d^2(x) \right]^{-1}$$

The values of $\theta_{1/2}$, d_0, and ρ_{00} were determined experimentally to make the fit to observation as close as possible. Thus, if P_I is regarded as a function of power density, the number of ions produced per pulse is

$$N_I = n \int_{-L/2}^{L/2} dx \int_0^\infty 2n\rho P_I[I(\rho, x)] d\rho$$

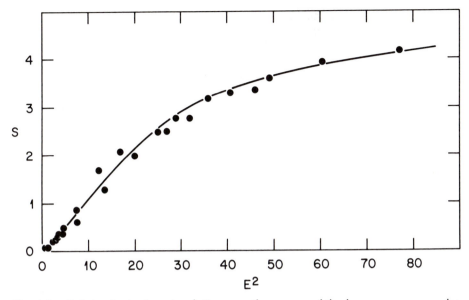

Fig. 1.8. Relative ionization signal, S, versus the square of the laser energy per pulse at 252.6 nm.

where n is the Xe density in atom/cm^3 and $-L/2 \leq x \leq L/2$ is the region over which Xe$^+$ is seen by the detector. Data on Fig. 1.8 were fitted very well by the choice $\Gamma_I = 3I$ and $R = 2 \times 10^{-9} I^2$. The unit of I is W/cm^2. With a laser energy at 1 mJ/pulse, we find the effective volume in which Xe can be fully ionized is about 10^{-4} cm^3. The time required is too long to be practical to count every selected Xe atom in a mass spectrometer. Thus, a second generation experimental setup was designed to make more efficient counting of isotopically selected rare gas atoms.

General Concept of the Isotopically Selective Counting of Rare Gas Atoms

The experimental schematic for efficient isotopically selective counting of rare gas atoms is shown in Fig. 1.9. Lasers are used for the selective ionization of noble gas atoms in the RIS process. Selected rare gas atoms are ionized and a fraction of these ions passes through a mass filter. For example, a quadrupole mass spectrometer would select the atom according to its mass number. The transmitted ions are then accelerated to 10 keV and implanted into a target that emits a burst of secondary electrons that can be detected to record each implanted atom. Any ion not transmitted will be neutralized by collision with the wall and returned to the gaseous sample in the static system. After a large number of shots, all of the mass-selected rare gas atoms are implanted and the count rate approaches zero. In principle, the total number of counts is just equal to the number of the selected rare gas isotope atoms. However, even a very high-resolution quadrupole mass spectrometer cannot have an abundance sensitivity better than 10^6 (i.e., the transmitted fraction of $A \pm 1$ is about 10^{-6} of A when the mass number is set at A). This limitation can be overcome by pumping the system to remove the other isotopes after all those

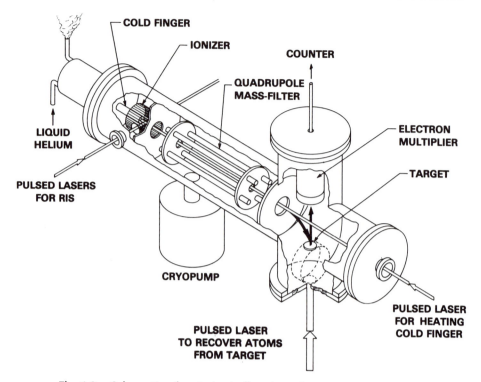

Fig. 1.9. Schematic of an isotopically selected rare gas atom counter.

isotopically selected atoms have been implanted. Then the atoms are released from the implanted target back into the enclosure for another cycle through the mass filter. After several such cycles, the undesired isotopes can totally be eliminated.

To describe more clearly how the rare gas atom counter (Fig. 1.9) should work, an assumption of counting $\sim$1000 ^{81}Kr atoms in a sample of 10^{12} atoms of ^{82}Kr is used as an example. The krypton atoms are introduced into the vacuum chamber through a needle valve and condensed onto a cold finger at 25 K. When the light of a pulsed visible laser heats the cold finger, krypton atoms are quickly released and travel a few millimeters in a few microseconds; then, another pulsed laser is fired just above the cold finger to resonantly ionize a significant fraction of the krypton atoms. The Kr$^+$ ions enter a quadrupole mass filter tuned to mass 81. After exiting the mass filter, Kr$^+$ ions are implanted at about 10 keV into a silicon target. If the abundance sensitivity is about 10^4 at mass 81, 10^8 atoms of ^{82}Kr are also implanted into the target. After all of the ^{81}Kr atoms are implanted, all other Kr isotope atoms remaining in the space will be pumped out. After the remaining gases are removed, the silicon target can be annealed with a pulsed laser to release the implanted atoms. This process completes a cycle that returns the $\sim$1000 atoms of ^{81}Kr and 10^8 atoms of ^{82}Kr to the static vacuum system. By repeating the above process two more times, the final count of ^{81}Kr atoms can be made. Then, the target is changed to a Be–Cu disk for the final counting since the secondary electron yield of Be–Cu is high. A Johnston electron multiplier with high gain ($\sim$$10^6$) is used to do the final counting of the ^{81}Kr atoms.

The following problems need to be resolved to count the desired rare gas atoms accurately and efficiently: (1) A coherent VUV beam needs to be developed for the excitation of rare gas atoms by a one-photon process. Although two-photon excitation of Xe has been demonstrated, any two-photon process will have a small effective excitation volume and requires very high-power density. Thus, it is more likely to produce background due to multiphoton ionization processes. The ionization cross section due to a multiphoton process can be roughly estimated as

$$\sigma_n \simeq 3 \times 10^{-18} \left| \frac{3n \times 10^{-7}}{E_1} \right|^{2(n-1)} I^{n-1}$$

in which E_1 is the energy of the first excited state in units of eV, I is the laser power density in units of W/cm^2, and n is the number of photons required to ionize the atoms. If the laser power density is $\sim 1 \times 10^9$ W/cm^2 at 250 nm, the probability for a multiphoton ionization of a molecule with ionization potential of 10 eV is about 10^{-6}. In most vacuum systems ($\sim 10^{-7}$ torr), a few ions can be produced from the ionization of residual gas. (2) The time for counting isotopically selected noble gas atoms needs to be reasonably short. If the excitation volume by laser beams is $\sim 10^{-3}$ cm^3, more than 100 hr is needed to complete the counting of rare gas atoms in a 4-L volume. Thus, an atom buncher was developed and tested to reduce the time needed for counting. (3) The implantation probability should be close to 1 to prevent any double counting. (4) The release of desired rare gas atoms should be close to 100% during a laser annealing process. (5) The outgassing of rare gases from the chamber wall should be kept low enough to permit the enrichment process.

Laser Scheme for Resonance Ionization of Rare Gas Atoms

Because of the requirement of high power densities and narrow bandwidths by two-photon excitation process, a more efficient way to ionize rare gas atoms is to generate VUV light and promote rare gas atoms from the ground state to an excited state by one-photon resonant absorption process. We use the four-wave mixing process[35-37] to produce a coherent VUV beam for rare gas atom excitation. The schematic of the four-wave mixing process to produce a coherent VUV beam is shown in Fig. 1.10. We consider a situation in which unfocused, plane polarized light from two lasers passes through a gas cell containing xenon as an active medium and Ar as a buffer gas. The two laser beams are arranged to overlap in time and space and have frequency, ω_{L1} and ω_{L2}. ω_{L1} is tuned to a two-photon resonance in Xe and ω_{L2} chosen so that $2\omega_{L1} + \omega_{L2}$ is just on the high energy side of a three-photon resonance in Xe. The efficiency of the four-wave mixing scheme depends primarily on the following factors: (1) the nonlinear susceptibility. This quantity determines the amplitude of the polarizability generated at $2\omega_{L1} + \omega_{L2}$; (2) the condition of phase matching. The ratio of Ar to Xe has to meet the condition of $\kappa(2\omega_{L1} + \omega_{L2}) = 2\kappa(\omega_{L1}) + \kappa(\omega_{L2})$ where $\kappa = 2\pi/\lambda$, λ and ω are the wavelength and angular frequency of a photon; and (3) the absorption of the medium. In the absence of absorption by dimers, the number of photons at $2\omega_{L1} + \omega_{L2}$ emerging from the cell should be proportional to the square of the number of Xe provided the ratio of Ar density to Xe is set at optimum phase matching conditions. In general, the Ar/Xe ratio remains constant for optimum phase matching at a fixed VUV

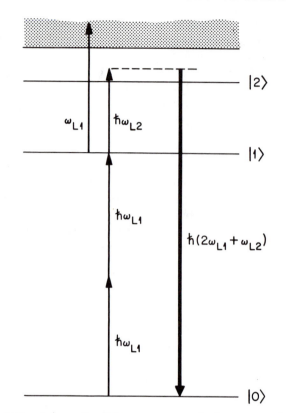

Fig. 1.10.   Schematic of the sum process in four-wave mixing.

wavelength for various Xe pressures. However, the increase of Xe_2 and ArXe is about proportional to the square of Xe pressure as the pressure of Xe increases.[38,39] Thus, saturation of the production of the coherent VUV beam is expected.

The number of VUV photons generated can be expressed as follows, if the absorption of the medium can be neglected and the laser beam profile has a Gaussian distribution:

$$N_{2\omega_{L1}+\omega_{L2}} = \frac{\pi N}{3} K\tau \left| \frac{\Omega_2 \Omega_1 LR}{\Delta_m \Gamma_L} \right|^2$$

where $\Delta_m = 2\omega_{L1} + \omega_{L2} - \omega_r$, N is the number density of the Xe atom, τ is the laser pulse duration, Ω_1 is the one-photon Rabi frequency for $|1> \rightarrow |2>$, Ω_2 is the two-photon Rabi frequency for $|0> \rightarrow |1>$, $K = \pi(e^2/mc)F_{02}$, Γ_L is the laser bandwidth, R is the laser beam radius, F_{02} is the oscillator strength for the transition of $|0> \rightarrow |2>$, and L is the length of the nonlinear medium.

We take the estimate of 116.5 nm for excitation of Kr as an example. The schematic of four-wave mixing to produce a VUV beam for the excitation of Ar and Xe is shown in Fig. 1.11. Consider a xenon–argon mixture in which the active gas is xenon having a two-photon resonance at 252.5 nm and a two-photon Rabi frequency of $\Omega_2 = 3I_1$, where I_1 is the power density at ω_{L1} in W/cm², and Ω_2 is in rad

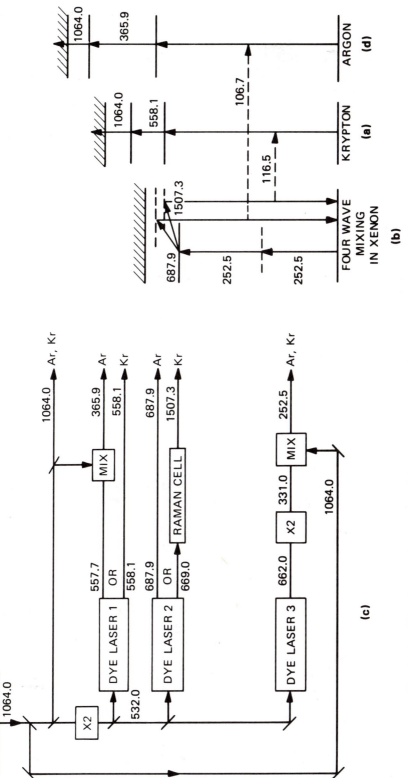

Fig. 1.11. RIS schemes for argon and krypton using one-photon excitation. Wavelengths are shown in nm.

$\sec^{-1}$. When ω_{L1} corresponds to a vacuum wavelength of 252.5 nm, 116.5 nm can be generated by the $2\omega_{L1} + \omega_{L2}$ process if ω_{L2} corresponds to 1506 nm. An estimate of the one-photon Rabi frequency is $|\Omega_1| = 10^8(I_2)^{1/2}$. The detuning from three-photon resonance with $5p^2(^2p_{3/2})7s$ $(J = 1)$ is $\Delta_m = 7.75 \times 10^{13}\ \sec^{-1}$. Assuming 1 mJ of 252.5-nm and 1506-nm laser beams is generated with $R = 0.7$ mm, $L = 5$ cm, $\tau = 5$ nsec at both ω_{L1} and ω_{L2}, optimum phase matching is observed to occur at 116.5 nm when the ratio of Ar pressure to Xe pressure is 9. With the following conditions:

$$I_1 = I_2 = 1.2 \times 10^7\ \text{W/cm}^2$$

$$\Gamma_L = 2.7 \times 10^{12} \quad \text{and} \quad F_{02} = 0.09$$

the number of 116.5-nm photons can be estimated to be 5×10^{11}.

Figure 1.11a shows a two-step resonant, three-photon ionization for krypton. To excite the $5s'$ transition, VUV radiation at 116.5 nm is required. The s' excitation is a strongly allowed transition that has an oscillator strength of 0.18. A power density of 700 W/cm² is enough to saturate the transition. It is very much smaller than the 10^{10} W/cm² needed to saturate a two-photon allowed transition.

Although it is possible to directly ionize the krypton $5s'$ excited state by a 252.5-nm laser beam, the photoionization cross section at threshold is very small (2×10^{-19} cm²). It is more efficient to ionize an electron from a p state. By using another laser tuned to 558.1 nm, krypton in the $4p^5 5s'[\frac{1}{2}]$ can be excited to the $4p^5 6p[\frac{1}{2}]_0$ state. This is a strongly allowed transition that can easily be saturated with the output of a simple dye laser. The $6p$ state then can be completely ionized using a 1.06-μm beam of a Nd:YAG laser. Since the photoionization cross section from the p state is about 5×10^{-17} cm², a modest fluence of 0.03 J/cm² will saturate the ionization step.

Figure 1.11b shows the relevant energy levels in the xenon four-wave mixing scheme and Fig. 1.11c shows a simplified diagram of the experimental apparatus used for the generation of the radiation at the wavelengths needed for resonance ionization of krypton according to the scheme in Fig. 1.11a. A laser beam with 200 mJ/pulse at 532 nm was split equally into two beams and used to pump dye lasers 2 and 3. The wavelength, 252.5 nm, which was tuned to excite the $5p^5 6p[\frac{3}{2}]_2$ two-photon allowed transition in xenon was generated by mixing the doubled output of dye laser 3 with the residual 1.06-μm pump laser output. The output of dye laser 2 was used to pump a high-pressure hydrogen Raman cell whose second Stokes' shift produced radiation at 1506 nm when dye laser 2 was tuned to 669.0 nm. Both dye lasers used DCM dye. To shift the tuning curve of the dye to the spectral region needed, dimethylsulfoxide (DMSO) was used as a solvent.[34]

The linearly polarized beams at 252.5 nm and near 1.5 μm were separated from their respective generating wavelengths by using Pellin–Broca prisms. To optimize the light transmission through the prism, a polarization rotator was used. Both lasers 2 and 3 produced visible light with a bandwidth of 0.3 cm⁻¹. The Nd:YAG fundamental output had a bandwidth of about 1 cm⁻¹. Under these conditions, the bandwidth of VUV produced would be about 1.5 cm⁻¹. The light pulses at 252.5 nm (1 mJ/pulse) and near 1506 nm (1 mJ/pulse) were focused with separate lenses and made coaxial by use of a dichroic beam splitter before entering the xenon VUV generation cell. About 10% of the second harmonic output of the Nd:YAG laser pumped a small home-made dye laser (1) to generate 3 mJ/pulse of 558.1 nm radiation. Approximately 70 mJ of a 1.06-μm laser beam from the fundamental of the Nd:YAG laser was used to complete the RIS process.

Since xenon is negatively dispersive at 116.5 nm, the VUV output can be increased by phase matching with Ar, which is positively dispersive in the region of interest. At 116.5 nm, the optimum phase matching conditions required an argon–xenon ratio of 9 $\pm$ 0.1. The pulse energy at this wavelength was 0.5 μJ, which was measured using a nitric oxide calibration chamber.[40]

An LiF exit window was used on the xenon–argon VUV generation cell to maximize the transmission at 116.5 nm. In the actual krypton detection region, the diameter of the VUV beam was about 1 mm. The other two beams, 558.1 nm and 1.06 μm, were focused to 3-mm beam diameter and were made coaxial with the generated VUV beam. With a 2-mJ/pulse beam at 558.1 nm and a 70-mJ/pulse beam at 1.06 μm, the $5s'$ to $6s$ transition and the final photoionization step were saturated. Experiments demonstrated that with this laser system, the volume for effective ionization of Kr is about 3×10^{-4} cm^3.

To excite Ar by a one-photon absorption process, a coherent VUV beam at 106.7 nm or 104.8 nm must be generated. The laser scheme is shown in Fig. 1.11a and d. A similar scheme for Kr can be used for Ar except the wavelength of the second laser beam is 687.9 nm. It is relatively easy to carry out RIS of xenon. The VUV beam needed for the first excitation can be obtained by a four-wave mixing process in a mercury cell. Figure 1.12 shows a possible ionization scheme. A 125.02-nm beam can promote Xe atoms from $5p^6$ to $5p^55d[\frac{1}{2}]_1$. A 532.0-nm beam from the second harmonic of the Nd:YAG laser can saturate the ionization process.

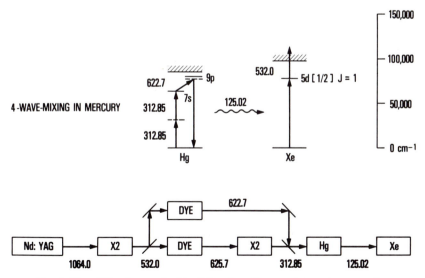

Fig. 1.12. RIS scheme for Xe. Wavelengths are shown in nm units.

The Atom Buncher

Since the effective ionization volume of rare gases by the above laser scheme is usually less than 10^{-3} cm^3, it should take more than 10^7 laser shots to count all the rare gas atoms in a 4-L chamber that contains a mass filter and other detection devices. The time required to count all the atoms is unacceptably long for a laser with a 10 Hz repetition rate. Thus, we conceived an atom buncher[41] that would put a large fraction of the target atoms into the laser beam at the time of the laser pulse. The basic idea of the atom buncher is to use a cold surface to condense the atoms of interest and a suitable pulse of laser light to momentarily heat a thin layer of the cold surface to release inert gas atoms. Thus, there is a much higher probability that the atoms will be in the laser volume at the desired time. Long before atoms can return by random walk, the surface is again cold enough to condense the heavy noble gas atoms.

Physical adsorption of rare gases on cooled surfaces can be characterized by the concept of mean stay time, that is, $\tau = \tau_0 e^{\epsilon/\kappa T}$. Values for τ_0 for xenon and krypton on a copper surface range from 10^{15} to 10^{20} sec whereas ϵ values range from 0.22 to 0.28 eV, depending on the surface treatment.[42] It is estimated that for Kr to stay on a nickel surface for longer than 1 μsec the temperature must be about 110 K. However, if the temperature of the surface is lower than 60 K, Kr atoms will for all practical purposes stay on that surface forever.

Recurrence times to a small area A in an open system of volume V are approximately $t_R = 4V/A\bar{v}\xi$, where $\bar{v}$ is the mean speed of the free atoms and ξ is the sticking probability. This recurrence time describes the number of atoms, N, in the free state at time t according to $N = N_0 \exp(-t/t_R)$, where N_0 is the number of free atoms at $t_0 = 0$. For a 4-L volume with A as small as a few mm^2, t_R can be about 1 min provided that the sticking probability is close to 1. Thus, rare gas atom samples can be recycled frequently.

With an atom buncher, the rare gas sample is introduced by a well-controlled leak-valve to allow as low as 10^{-9} cm^3 into the chamber. All of these heavy rare gas atoms, such as Kr, will condense on the cold surface which is kept at $\sim$25 K. A pulsed laser is used to evaporate rare gas atoms. If the effective ionization volume ($\sim 3 \times 10^{-4}$ cm^3) is centered about 1 mm above the cold tip, it takes about 10 μsec for Kr atoms to fill it if the atoms have a mean velocity distribution characteristic of release at $T = 120$ K. The RIS lasers are delayed accordingly ($\sim$10 μsec) to achieve the optimum resonance ionization process. In this case, the ionization probability of the desorbed Kr atoms can reach as high as 10%. Since the recurrence time is estimated as 50 sec and the laser repetition rate is 10, the overall ionization percentage is $\sim 10^{-4}$ per pulse. Thus, with a small number of atoms ($<10^3$) the time required to adequately count all the Kr atoms in the sample is less than 1 hr. If the sample contains a large number of Kr atoms ($>10^4$), analog detection in which more than 10 atoms are detected during each pulse is more appropriate.

The atom buncher was constructed for the use of liquid helium as a coolant, with dimensions for use inside the ionization region of a quadrupole mass spectrometer. The design is shown in Fig. 1.13. Coupling the liquid helium to the cryotip was accomplished by using very high-purity (99.998%) nickel. Stainless steel (304) was chosen for the thin disk material of the cryotip. This type of stainless

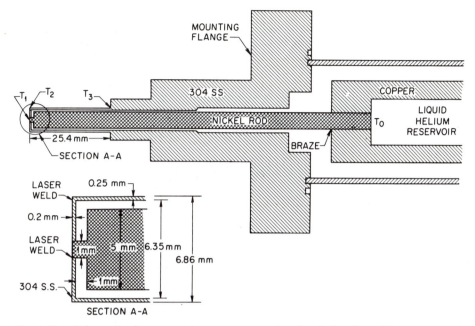

Fig. 1.13. Schematic diagram and some construction details for the cold tip portion of the atom buncher.

steel is easily rolled to the desired thickness and laser welded to nickel. The resistive tube, the body, and the mounting flange were constructed from one piece of 304 stainless steel. The low thermal conductivity of the thin tube lowers the overall heat load of the system. Figure 1.14 illustrates data from the test of the atom buncher using ^{85}Kr gas. A G-M tube was positioned as close to the cryotip as possible to monitor the count rate on the surface. The top graph shows that the atoms do come off of the surface during each individual laser pulse (50 mJ/pulse) and return to the surface with a recurrence time on the order of 10 sec. The center graph illustrates what happens when the laser irradiated the surface at a 1 Hz repetition rate. These results also show that all of the krypton atoms are on the surface in an area that is heated by the laser. If the entire cold surface was not heated to above 120 K, the count rate would steadily increase because all of the atoms would eventually collect on the unheated area.

Concern of the Possible Loss of the Sample and Contamination Due to Outgassing

After rare gas atoms are desorbed from the cold finger, in general, only a fraction of the desorbed atoms is ionized in one laser pulse. Thus, the stay time of rare gas atoms on the wall of the chamber needs to be very short to prevent the loss of the desired rare gas atoms. At room temperature, the stay time for rare gas atoms on a clean metal surface is shorter than 1 nsec.[42] Thus, every rare gas atom introduced into the chamber will get counted. However, if the wall of the chamber becomes contaminated by oil, grease, or water, the stay time of rare gas atoms on these con-

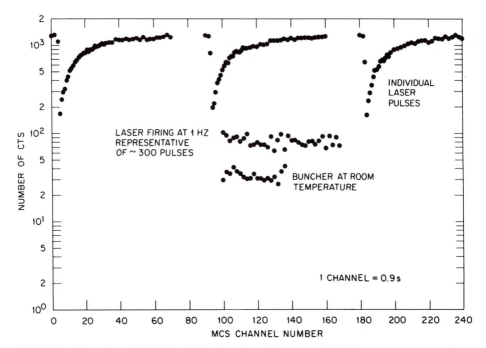

Fig. 1.14. Experimental tests of the atom buncher, a small G-M counter was used to view the ^{85}Kr atoms remaining on the cool tip as a function of time. Each laser pulse causes nearly all of the atoms to leave the cold surface, as shown in the top portion of the figure.

taminated walls can increase drastically. It is necessary to bake the whole chamber for a couple of days under ultrahigh vacuum conditions before performing isotopically selective rare gas counting.

A known quantity of ^{85}Kr was introduced into the chamber and tested several days for the loss of Kr. No significant reduction of ^{85}Kr was observed. Most stainless steel material is produced under high Ar pressure. It is not surprising that some Ar gas is trapped in the stainless steel. Since a minor amount of Kr and Xe always exists in Ar gas, the outgas of Kr is a concern when doing ultralow level rare gas counting. The outgas rate of Kr was measured at $\sim 10^3$/min for a 4-L chamber.[43] With the abundance sensitivity of $\sim 10^4$ for a quadrupole mass spectrometer, the upper limit of the time needed to count the selective atoms is 10^4 min. Indeed, we cannot be positive that the increase of the Kr level is the result of the outgas or the leak because the level of increase of Kr is equivalent to 10^{-13} cm^3/sec of air, which is very difficult to detect by any available leak detector.

Implantation and Gas Release

There is a sticking fraction η for the process of an energetic ion impact with a surface. The sticking coefficient is a function of the type of ion and its energy as well as the type and temperature of solid material and the number of particles previously implanted. For rare gas atoms impinging on tungsten, η approaches unity for ion energies about 5 keV.[44] Other authors have observed that for most metal, η approaches unity for ion energies greater than 5 keV.[45] The average implantation depth for 10 keV Kr ions on Si is about 100 Å. In this work, a simple experimental facility was developed to study implantation of ions and recovery of rare gas atoms from targets. This facility consisted of an electron source, a target at 10 keV, and a G-M tube mounted in a vacuum enclosure with suitable traps and getter pumps. Electron impact ionization of ^{85}Kr was followed by ion implantation into a Be–Cu target at 10 keV. As ions were implanted, the count rate of the G-M tube decreased because it was positioned to detect atoms in the gas phase. From the emission current and ^{85}Kr density, the implantation probability, η, can be measured. The probability for a 10 keV Kr$^+$ to implant into an activated Be–Cu or polished Si wafer is 0.9 ± 0.2. For the isotopically selective counting discussed in this work, the rare gas ions were accelerated to 10 keV and implanted into the Be–Cu target to ensure the implantation and prevent any possible double counting.

It is desirable to recall nearly all of the implanted noble gas atoms for further isotopic enrichment or for repeated counting. If the release of the implanted atoms does not reach 100%, part of the sample is lost. We discovered that more than 90% of the implanted rare gas atoms with average depth of 100 Å in silicon can be released by a laser annealing process.[46] The laser beam used in this process is a doubling of the Nd:YAG laser with 1 J/cm^2 pulse. In Fig. 1.15, the arrangement of an ion target and a Johnston multiplier similar to a Daly configuration[47] is shown. The rare gas ions after passing through quadrupole mass filter are accelerated to 10 keV to implant on an Si target. Then, a few secondary electrons are ejected and

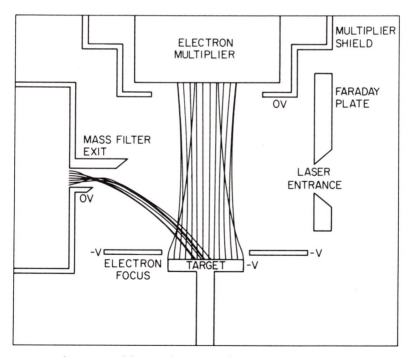

Fig. 1.15. Configuration of the ion detector and associated ion and electron trajectory.

detected by the multiplier. The accuracy of the final counting of rare gas atoms depends on having a target with a high electron ejection coefficient.[48] For an activated Be–Cu target this coefficient is 2.5 for Kr^+ at 3 keV, which ensures that a Johnston electron multiplier can perform digital counting.

Demonstration of Counting [81]Kr Atoms

The concept for counting small numbers of rare gas atoms required the development of several basic components of the complete apparatus before a crucial demonstration could be made. With the progress on (1) an efficient laser scheme for resonant ionization, (2) the atom buncher to put desired atoms in the laser beam, (3) efficiency of implantation and desorption reaches to 100%, and (4) efficient digital counting, it was then possible to prove the feasibility of isotopically selective rare gas counting. An experiment for the demonstration of counting ~1000 [81]Kr atoms was pursued. A vacuum chamber containing a quadrupole mass filter, an atom buncher, implantation target, Johnston electron multiplier, and other necessary hardware as shown in Fig. 1.9 was baked at 300°C under ultrahigh vacuum conditions for a few days before the sample was introduced. The extensive pumping and baking were to prevent any significant outgas of Kr and the possibility of Kr atoms sticking to the wall of the chamber.[43] After mixing helium with a sample of enriched [81]Kr, which was obtained from the National Bureau of Standards (NBS), ~1000 atoms of [81]Kr with 2×10^5 other krypton atoms and ~10^{10} atoms of He were introduced. Krypton atoms were expected to freeze on the surface of the atom buncher, which was set at 25 K. A four-wave mixing scheme similar to Fig. 1.11 was used to generate a 116.5-nm laser beam with energy of 500 nJ/pulse and a bandwidth of 1.5 cm^{-1}. The laser energy of a 558.1-nm beam was obtained at 3 mJ/pulse. Thus, krypton atoms could be efficiently excited stepwise by resonance radiation and subsequently ionized. A flash lamp pumped dye laser beam with 20 mJ/pulse was used to desorb krypton atoms on the surface of the atom buncher. The RIS laser beams were aimed 0.2 cm from the tip of the atom buncher and synchronized with the laser that was used to desorb krypton atoms from the buncher. The delay between the RIS laser beams and the laser beam for desorption is set at about 10 μsec to compensate for the time needed for the Kr atoms to travel from the tip to the RIS beams. The quadrupole mass filter was set at mass 81 with the abundance ratio at ~2×10^5. After the krypton atoms were ionized and isotopically selected by the mass filter, the ions were accelerated to 10 keV onto a Be–Cu target where they were implanted. An electron multiplier with gain higher than 10^6 was used to count each implanted ion by detecting the secondary electrons emitted by the target. Since the implanted krypton atoms stayed in the beryllium–copper target indefinitely, nearly all of the isotopically selected krypton atoms were eventually ionized and counted. Thus, the number of isotopically selected krypton atoms was determined. Experimental data for counting ~1000 atoms of [81]Kr are shown in Fig. 1.16.

During this experiment, the buncher laser was started before the RIS lasers to make sure that only a very small fraction of Kr atoms was sitting on the cold spot at any given time. Thus, the probability of ionizing and counting one atom of [81]Kr by a single laser pulse is less than 1%. This procedure ensured digital counting in

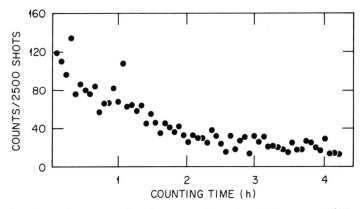

Fig. 1.16. Atom counting demonstration using 1000 atoms of ^{81}Kr.

which each atom was counted one at a time. The sum of the count in Fig. 1.16 is 2100 atoms of ^{81}Kr. However, it was later found that the Be–Cu used as a target was not activated. The sticking coefficient of this Be–Cu target was measured as 0.6. Taking the correlation factor of the sticking probability, about 1300 atoms were actually counted. This rather good agreement between our RIS counting and decay counting by NBS proves the capability of isotopically selective counting of heavy rare gas atoms.

The isotopically selective rare gas counting technique was also used to count ^{81}Kr from a groundwater sample that was taken from a sandstone aquifer near Zürich, Switzerland. The sample contains 5×10^{15} Kr atoms with less than 1700 ^{81}Kr atoms. Since the overall ionization probability for each laser pulse is $\sim 10^{-5}$, the number of ions produced by one RIS laser pulse is about 5×10^{11}. The transmission of the produced ions through mass filter will be a very small fraction because of the serious space charge condition. Thus, a preenrichment system was used to enrich the concentration of ^{81}Kr in the sample.

Preenrichment of ^{81}Kr was performed with a separate quadrupole mass spectrometer that used its conventional electron-impact ionization source shown in Fig. 1.17.[49,50] The vacuum chamber was pumped below 10^{-8} torr by a helium-compressed cryopump. This pump was shut off from the mass spectrometer before the gas sample was introduced. The vacuum was maintained using a ST 101 zirconium/aluminum getter pump (which does not remove rare gases). The mass spectrometer had a throughput of about 4 mA/torr with abundance sensitivity of 4×10^3. Once inside the mass spectrometer enrichment chamber, the krypton gas was ionized by electron bombardment. The krypton ions then entered the quadrupole mass spectrometer that was set to pass ^{81}Kr. After leaving the mass filter, the ^{81}Kr ions and any residual krypton ions of other masses were accelerated by a potential of -10 kV and focused onto a target. The target was a Kapton film that had a 600 nanometer-thick aluminum film. The implantation efficiency was measured to be near 100%. The krypton implanted in the foil was isotopically enriched for ^{81}Kr compared to ^{82}Kr by a factor of 4000. After 1 hr of running the mass spectrometer, the remaining gas was pumped out. The krypton was freed from the foil by vaporizing the aluminum film by the second harmonic of an Nd:YAG beam at 100 mJ/

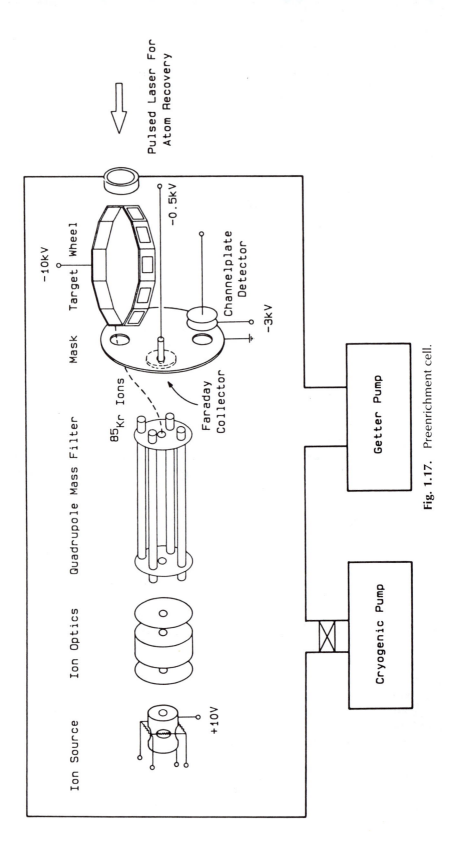

Fig. 1.17. Preenrichment cell.

pulse. Previous studies had indicated that the procedure yielded essentially 100% recovery of krypton into the gas phase. The enrichment cycle was repeated three times. The target used for the last enrichment process was a silicon wafer that was estimated to contain 10^5 ^{82}Kr atoms and most of the ^{81}Kr atoms. After implantation, the silicon wafer was removed and placed in the ultrahigh vacuum RIS final counting chamber. Then the krypton atoms implanted in the silicon wafer were released by a laser annealing process using a second harmonic of an Nd:YAG laser. Because of the very good abundance sensitivity (2×10^5) of the quadrupole system when operated with the laser and buncher source, the number of ^{81}Kr atoms was measured as 1200 ± 300 atoms.

Applications of Isotopically Selective Counting of Rare Gases

Groundwater Dating

The age of groundwater represents the length of time that the water has been isolated from the atmosphere. This age is of fundamental importance to hydrology and to the problem of the disposal of radioactive waste.[51] ^{81}Kr has generated great interest among hydrologists as a potential dating tool. The age of a groundwater sample is determined by comparing the measured $^{81}Kr/Kr$ ratio with the modern $^{81}Kr/Kr$ ratio and applying the known half-life. However, the radioactive decay counting of ^{81}Kr needs much larger quantities of ^{81}Kr since the half-life of ^{81}Kr is very long (2×10^5 years). The capability of measuring small numbers of ^{81}Kr provides a convenient method for groundwater dating.

Solar Neutrino Flux Measurements

Solar energy is believed to be produced by the fusion of light atoms into heavier elements. During these nuclear processes, several steps involve the emission of solar neutrinos that can carry important information on the interior of the sun. The only solar neutrino experiment ever undertaken was based on the reaction $^{37}Cl(\nu,e^-)^{37}Ar$ using a volume of 380 m^3 of C_2Cl_4. This experiment was primarily sensitive to the weak ^8B neutrino source.[15,16] The capture rate predicted by the standard solar model was 7.6 solar neutrino units (SNU) (1 SNU equals 10^{-36} captures per second per target atom), which disagreed with the value from the chlorine experiment of 1.9 SNU. Use of the reaction $^{81}Br(\nu, e^-)^{81}Kr$, which is primarily sensitive to the flux of ^7Be neutrino, is now feasible because of the development of isotopically selective counting of ^{81}Kr. The radioactive decay counting is not practical because of the long lifetime of ^{81}Kr.

Oceanography

Atmospheric ^{81}Kr is a by-product of plutonium production and nuclear power production and, consequently, first appeared in the atmosphere in the early 1950s. The source function of ^{85}Kr in the oceans is well known and the penetration of ^{85}Kr into deep waters with time can be used to estimate the volume rate of the formation of new deep water.[18] Thus, ^{85}Kr would be a particularly useful tracer in studying mixing processes in the near-surface waters. The present concentration of ^{85}Kr in ocean water is such that a 200-L sample of modern seawater will yield 600 counts per day by a decay counter method. The additional sensitivity offered by the RIS method in this work allows reduction in the sample size and could extend the dating range of the ^{85}Kr isotope.

Waste Isolation and Pu in Soil

It has been shown that the RIS technique for detection of Xe can be used to identify materials that contain ^{239}Pu.[52] For instance, a thermal neutron exposure of 10^8 neutrons/cm^2 will produce 3×10^4 atoms of Xe in 1 kg of waste material containing Pu at the level of 10 nCi/g. Detection of xenon could produce a rapid scan procedure to determine whether a given batch of material must be stored as hazardous waste. Similarly, irradiation of a 100-kg soil sample with 10^8 thermal neutrons/cm^2 would enable the detection of even the lowest levels (4 pCi/g) thus far recommended for Pu in soil.

Diagnosis of Bone Diseases

It has been shown that the exposure of an individual to a very modest level of fast neutrons (1 mrad) will produce measurable quantities of ^{37}Ar in the bone via the reaction[53] ^{40}Ca$(n, \alpha)^{37}$Ar. In fact, an average person exhales $\sim$2300 atoms of ^{37}Ar per minute shortly after this small exposure. Thus, an ultrasensitive detection of isotopically selected Ar can be used to determine the loss of Ca in bone, and could possibly be an indicator of certain types of bone disease. A few possible applications are listed in Table 1.1.

RIS Molecules from a Supersonic Nozzle Beam

The use of lasers to drive electronic transitions of molecules as a sensitive detection method is complicated by the high degree of rotation excitation. Even when only one rovibrational level is occupied, the absorption spectrum of a molecule is far more dense than that of an atom because of the degree of freedom of the nuclei. When the population is spread over many rotational levels, the percentage of molecules excited by a narrow bandwidth is relatively small. Thus, ultrasensitive detection of molecules by RIS could not be achieved. However, a supersonic beam with molecules in question being seeded into a rare gas carrier can efficiently reduce the rotational temperature.[57-60] Another great concern about resonance ionization of molecules is the Franck–Condon factor for each transition. A low Franck–Condon

Table 1.1. Some possible applications of noble gas detectors, assuming a few atoms can be counted directly

1. Baryon conservation[54,55] (e.g., ^{37}Ar or ^{38}Ar from decay of ^{39}K)
2. Solar neutrino flux and neutrino oscillations [^{81}Kr from ^{81}Br$(\nu e)^{81}$Kr]
3. $\beta^-\beta^-$ decay[56] (e.g., ^{82}Kr from $\beta^-\beta^-$ decay of ^{82}Se)
4. Oceanic circulation (naturally occurring ^{39}Ar)
5. Polar ice caps (naturally occurring ^{81}Kr)
6. Aquifers (naturally occurring ^{81}Kr)
7. Waste isolation (Xe from neutron or photofission of transuranic elements)
8. Pu in soil (^{136}Xe or ^{86}Kr from neutron fission of Pu)
9. Diagnoses of bone diseases [^{37}Ar from ^{40}Ca$(n,\alpha)^{37}$Ar]
10. Fast neutron dosimetry [^{37}Ar from ^{40}Ca$(n,\alpha)^{37}$Ar]

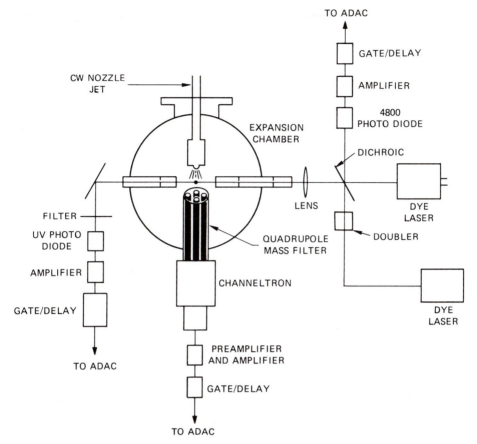

Fig. 1.18. Experimental scheme for RIS of CO.

coefficient for a transition will demand higher laser energy. If the excited molecules are in a repulsive potential or a predissociative state, it is more difficult to obtain the parent ions by an RIS process.

A mass-selective RIS method to study CO was pursued.[61] The spectroscopy of the $A^1\Pi$ and $B^1\Sigma$ states was studied by the four-photon, two-resonant process:

$$X^1\Sigma^+(n = 0, J) + 2\hbar\omega \rightarrow A^1\Pi(n = 1, J')$$

$$A^1\Pi(n = 1, J') + \hbar\omega \rightarrow B^1\Sigma^+(n = 0, J'')$$

$$B^1\Sigma(n = 0, J'') + \hbar\omega \rightarrow CO^+ + e$$

The Franck–Condon factors for these two processes are favorable. Both the two-photon ($X^1\Sigma^+ \rightarrow A^1\Pi$) and ionization steps were driven by the doubled output of one laser with a resulting wavelength region of 300–303 nm. A second laser of wavelength region, 470–474 nm, was the source of the second excitation step ($A^1\Pi \rightarrow B^1\Sigma^+$). The experimental apparatus is shown in Fig. 1.18. The two laser beams were focused collinearly by a 25-cm lens to the center of the ionization region of a quadrupole mass spectrometer. With calibration of the transmission of the mass spectrometer and the detection efficiency of the channeltron, the two-photon absorption rate from $X \rightarrow A$ transition was measured as 5×10^{-11} cm^4 W^{-2}/sec and the ionization cross section of CO(A) was obtained as 1.4×10^{-20} cm^2. By fixing the wavelength of one dye laser and tuning the wavelength of the other laser, the absorption spectra of the two-photon ($X \rightarrow A$) and the one-photon process ($A \rightarrow B$) were obtained. A typical spectrum is shown in Fig. 1.19. The resulting excitation energy is then given by the wave number expression in the following for both $X_{v=0} \rightarrow A_{v=1}$ and $A_{v=1} \rightarrow B_{v=0}$ transitions.

$$\Delta E(J, J') = 66227.9 - 0.346J(J + 1) + 1.577J'(J' + 1)$$

for

$$CO(X^1\Sigma)_{v=0} \rightarrow CO(A^1\Pi)_{v=1}$$

$$\Delta E(J'J'') = 20688 + 1.948J''(J'' + 1) - 1.577J'(J' + 1)$$

for

$$CO(A^1\Pi)_{v=1} \rightarrow CO(X^1E)_{v=0}$$

the unit of ΔE is cm^{-1}.

Isotopically Selective Detection of CO

We want to show an example of the RIS of isotopically selective CO as an illustration of the capability of detecting low levels of molecules with isotopic selection. The goal of the study of ^{14}CO is to see if ^{14}C detection using the RIS technique can be competitive with accelerator techniques.[62] To be competitive, a laser technique should be capable of detecting the presence of $\sim 10^5$ of ^{14}C^{16}O molecules in 0.05 g

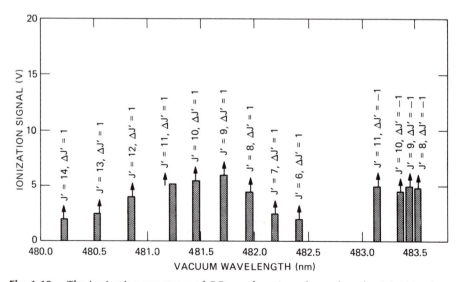

Fig. 1.19. The ionization spectrum of CO as a function of wavelength of the blue laser for $A \rightarrow B$ transition. The wavelength for $X \rightarrow A$ transition was fixed at 603.65 nm.

of CO. Thus the technique should be able to discriminate against 10^{16} times as many $^{12}C^{16}O$ and $\sim 10^{13}$ times as many $^{12}C^{18}O$ molecules. The isotopic effects on the energy levels in CO can be approximately described for small v and J by the following equation:

$$E_{v,J} = T_e + \omega_{e0}(\mu_0/\mu)^{1/2} (v + \tfrac{1}{2}) - x_{e0}\omega_{e0}(\mu_0/\mu) (v + \tfrac{1}{2})^2$$
$$+ B(\mu_0/\mu)J(J + 1)$$

where T_e is the term energy of the electronic state in question, ω_{e0} is the vibrational constant for $^{12}C^{16}O$, $x_{e0}\omega_{e0}$ is a harmonic vibrational constant for $^{12}C^{16}O$, B is the rotational constant for $^{12}C^{16}O$, μ is the reduced mass of the isotope in question, μ_0 is the reduced mass of $^{12}C^{16}O$, J is the rotational quantum number, and v is the vibrational quantum number.

Resonance steps for the RIS of $^{14}C^{16}O$ starting from $X^1\Sigma^-(0, 0)$ are given in the following:

$$X^1\Sigma^- + \hbar\omega_1 \, (\lambda = 4858.5 \text{ nm}) \rightarrow X^1\Sigma^+(1, 1)$$

$$X^1\Sigma^-(1, 1) + \hbar\omega_2 \, (\lambda = 159.48 \text{ nm}) \rightarrow A^1\Pi(0, 0)$$

$$A^1\Pi(0, 0) + \hbar\omega_3 \, (\lambda = 413.98 \text{ nm}) \rightarrow B^1\Sigma^-(1, 1)$$

$$B^1\Sigma^-(1, 1) + \hbar\omega_4 \, (\lambda < 450 \text{ nm}) \rightarrow X^2\Sigma^-(0, 0) + e$$

where the numbers in parentheses indicate the values of v and J, respectively.

The laser beam with wavelength at 4858 nm can be obtained by frequency doubling a tunable CO_2 laser with ~ 500 mJ/pulse and bandwidth ~ 0.02 cm^{-1}. It is very easy to obtain a few millijoules of laser energy at 413.98 nm with a commercially available dye laser. Thus, all of the light source except a coherent VUV beam at 159.48 nm is available with characteristics that should allow $\sim 10\%$ ionization of the $^{14}C^{16}O$ molecules in a pulsed nozzle beam. Figure 1.20 shows the potential energy curves for $^{14}C^{16}O$. It is clear that Franck–Condon factor is favorable for $X \rightarrow A$ transition. Since the lifetime of $A^1\Pi(0,0)$ is $\sim 10^{-8}$ sec, a peak power of a few watts with a linewidth $\sim 4 \times 10^{-4}$ Å of 5 nsec duration would be sufficient to provide a scheme for ionizing $\sim 5\%$ of all of the $^{14}C^{16}O$ present in a nozzle jet pulse. A four-wave mixing process in Xe with $2\omega \, (\lambda = 249.6 \text{ nm}) - \omega_1 \, (\lambda = 570 \text{ nm})$ should give 159.48 nm radiation, which has an energy estimated at about 500 nJ. Thus, the only requirement for a powerful light is in the ionizing laser. By carefully choosing its wavelength to stay away from any near resonances and keeping power densities as low as possible, we estimate that the ionization probability of $^{14}C^{16}O$ can be made 0.5. Table 1.2 gives energies of various molecular states of CO for the various isotopes. A close examination, together with the fact that the power densities of all lasers are to be kept as small as possible and still be consistent with ionizing several percent of $^{14}C^{16}O$, indicates why the selectivity is excellent. If the wavelengths for vibrational excitation and first electronic excitation ($X \rightarrow A$) are fixed for the resonance excitation of $^{14}C^{16}O$, the background contribution by $^{12}C^{16}O$ and $^{12}C^{18}O$ through nonresonant multiphoton excitation is estimated to be about $\sim 10^{12}$ lower. However, this estimate can be off a couple orders of magnitude. A decent quadrupole mass spectrometer can distinguish between $^{12}C^{16}O$ and $^{14}C^{16}O$ to

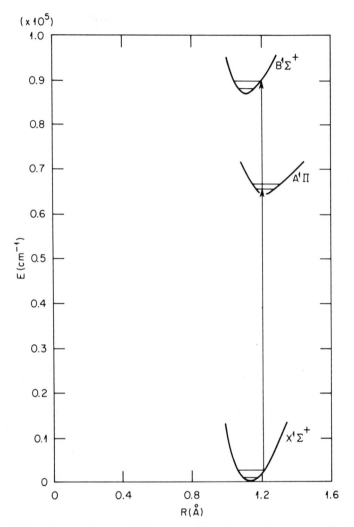

Fig. 1.20. Potential curves for isotopically selective detection of $^{14}C^{16}O$.

at least a factor of 1×10^5. Under this condition, the background due to $^{12}C^{16}O$ can be totally eliminated. Since the abundance of $^{12}C^{18}O$ is about three orders of magnitude lower than $^{12}C^{16}O$, background due to $^{12}C^{18}O$ can also be neglected even if the mass spectrometer has the discrimination capability of only a factor of 10 between $^{14}C^{16}O$ and $^{12}C^{18}O$. The experimental schematic shown in Fig. 1.21 is a minor modification of the facility of Fig. 1.18. A pulsed nozzle is used instead of a continuous nozzle to reduce the critical demand of pumping power and to provide a coincidence between the laser beam and the release of CO. With CO samples seeded in He (10% CO in He) at 10 atm behind the nozzle, each single pulse contains $\sim 10^{17}$ $^{12}C^{16}O$ and ~ 10 $^{14}C^{16}O$. Thus, the detection of $^{14}C^{16}O$ can possibly be achieved.

Table 1.2. Energy levels in different isotopic species of CO

Isotopic species	n	J	$X^1\Sigma^+$ of CO	$A^1\Pi$ of CO	$B^1\Sigma^+$ of CO
				$E_{n,J}$ (cm^{-01})	
$^{12}C^{16}O$	0	0	1081.61	65830.08	87997.80
	0	1	1085.47	65833.28	88001.70
	0	2		65839.72	
	1	0	3224.88	67309.49	90080.11
	1	1	3228.74	67312.71	
$^{14}C^{16}O$	0	0	1036.65	65798.82	87954.04
	0	1	1040.20	65801.77	87957.65
	0	2		65807.69	87964.85
	1	0	3091.65	67218.12	89950.80
	1	1	3095.19	67221.08	89954.40
	1	2		67226.99	89961.60
	2	0	5122.24		91919.63
	2	1	5125.78		
	2	2	5132.88		
$^{12}C^{18}O$	0	0	1055.62	65812.00	87972.49
	0	1	1059.29	65815.07	87976.23
	0	2		65821.20	
	1	0	3147.87	67256.69	90005.36
	1	1	3151.54	67759.75	90009.10
	1	2			90016.57
	2	0			92009.28
	2	1			92013.01

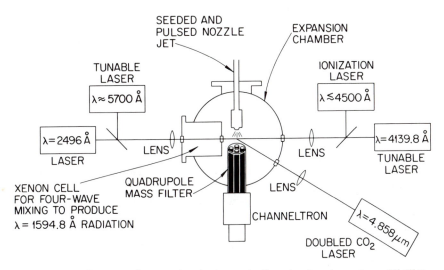

Fig. 1.21. Schematic diagram for the isotopically selective detection of $^{14}C^{16}O$.

CONCLUSION

The apparatus used to do isotopically selective counting of atoms and molecules is still relatively complicated and reasonably expensive at the present time. However, the fast progress of the extension of diode array lasers to the UV region and the improvement of laser energy can possibly reduce the cost and complication significantly in a few years. By that time, the method presented in this work may turn the dream of most analytical chemists into reality.

ACKNOWLEDGMENTS

The authors would like to thank W. R. Garrett and J. P. Judish for valuable discussions and D. C. Crowell for preparation of this manuscript.

Research was sponsored by the Office of Health and Environmental Research, U.S. Department of Energy under contract DE-AC95-84OR21400 with Martin Marietta Energy Systems, Inc.

REFERENCES

1. G. S. Hurst, M. G. Payne, M. H. Nayfeh, J. P. Judish, and E. B. Wagner, *Phys. Rev. Lett.* **35**, 82 (1975).

2. G. S. Hurst, M. G. Payne, and E. B. Wagner, United States Patent No. 3,987,302 (1976).

3. J. P. Young, G. S. Hurst, S. D. Kramer, and M. G. Payne, *Anal. Chem.* **51**, 1050A (1979).

4. G. S. Hurst and M. G. Payne, *Rev. Mod. Phys.* **51**, 767 (1979).

5. G. S. Hurst, M. H. Nayfeh, and J. P. Young, *Appl. Phys. Lett.* **30**, 229 (1977).

6. G. S. Hurst, M. H. Nayfeh, and J. P. Young, *Phys. Rev. A* **15**, 2283 (1977).

7. S. D. Kramer, C. E. Bemis, Jr., J. P. Young, and G. S. Hurst, *Op. Lett.* **3**, 16 (1978).

8. J. Iturbe, S. L. Allman, G. S. Hurst, and M. G. Payne, *Chem. Phys. Lett.* **93**, 460 (1982).

9. C. H. Chen, G. S. Hurst, and M. G. Payne, *Chem. Phys. Lett.* **75**, 473 (1980).

10. C. H. Chen, S. D. Kramer, S. L. Allman, and G. S. Hurst, *Appl. Phys. Lett.* **44**, 640 (1984).

11. C. H. Chen, G. S. Hurst, and M. G. Payne, in *Progress in Atomic Spectroscopy, Part C,* edited by H. F. Beyer and H. Kleinpoppen. Plenum, New York, 1984, pp. 115–150.

12. M. G. Payne, G. S. Hurst, C. H. Chen, and G. W. Foltz, *Advances in Atomic and Molecular Physics,* Vol 17, edited by D. R. Bates and B. Bederson. Academic Press, New York, 1981, pp. 229–274.

13. G. S. Hurst, M. G. Payne, S. D. Kramer, and C. H. Chen, *Phys. Today* **33**, 24 (September 1980).

14. R. Davis, Jr., D. S. Harmer, and K. C. Hoffman, *Phys. Rev. Lett.* **20**, 1205 (1968).

15. J. K. Rowley, B. G. Cleveland, R. Davis, and J. C. Evans, in *Neutrino 77 (Proceedings of the International Conference on Neutrino Physics, Baksan Valley, USSR, 1977).* Brookhaven National Laboratory Report BNL-23418.

16. J. N. Bahcall, *Rev. Mod. Phys.* **50,** 88 (1978).

17. W. Dansgaard, H. B. Clausen, N. Gundesrtrup, C. U. Hammer, S. F. Johnsen, P. M. Kristindottir, and N. Reek, *Science* **218,** 1273 (1982).

18. H. H. Loosli and H. Oeschger, *Earth Planet Sci. Lett.* **7,** 67 (1969).

19. P. Lambropoulos, *Adv. At. Mol. Phys.* **12,** 87 (1976).

20. D. Zakheim and P. Johnson, *J. Chem. Phys.* **68,** 3644 (1978).

21. D. L. Feldman, R. K. Lengel, and R. N. Zare, *Chem. Phys. Lett.* **52,** 413 (1977).

22. M. G. Payne, C. H. Chen, G. S. Hurst, S. D. Kramer, W. R. Garrett, and M. Pindzola, *Chem. Phys. Lett.* **79,** 142 (1982).

23. S. D. Kramer, J. P. Young, G. S. Hurst, and M. G. Payne, *Opt. Commun.* **30,** 47 (1979).

24. M. P. McCann, C. H. Chen, and M. G. Payne, *Appl. Spectrosc.* **41,** 399 (1987).

25. M. P. McCann, C. H. Chen, and M. G. Payne, *Chem. Phys. Lett.* **138,** 250 (1987).

26. J. C. Miller and W. C. Cheng, *J. Phys. Chem.* **89,** 1647 (1985).

27. P. H. Buckshaum, J. Bokor, R. H. Storz, and J. C. White, *Opt. Lett.* **7,** 399 (1982).

28. H. Geiger and W. Mueller, *Z. Phys.* **29,** 839 (1928).

29. G. W. Greenless, D. L. Clark, S. L. Kaufman, D. A. Lewis, J. F. Tomm, and J. H. Broadhurst, *Opt. Commun.* **23,** 236 (1977).

30. S. L. Kaufman, *Opt. Commun.* **17,** 309 (1976).

31. C. Y. She, W. M. Fairbank, Jr., and K. W. Billman, *Opt. Lett.* **2,** 30 (1978).

32. G. S. Hurst, M. G. Payne, S. D. Kramer, C. H. Chen, R. C. Phillips, S. L. Allman, G. D. Alton, J.W.T. Dabbs, R. D. Willis, and B. E. Lehmann, *Rep. Prog. Phys.* **48,** 1333 (1985).

33. G. S. Hurst, M. G. Payne, C. H. Chen, R. D. Willis, B. E. Lehmann, and S. D. Kramer, in *Laser Spectroscopy V,* edited by A.R.W. McKellar, T. Oka, and B. P. Stoichoff. Springer-Verlag, New York, 1981, pp. 59–66.

34. C. H. Chen and S. D. Kramer, *Appl. Opt.* **23,** 526 (1984).

35. F. S. Tomkins and R. Mahon, *Opt. Lett.* **6,** 179 (1981).

36. S. D. Kramer, C. H. Chen, M. G. Payne, G. S. Hurst, and B. E. Lehmann, *Appl. Opt.* **22,** 3271 (1983).

37. S. D. Kramer, C. H. Chen, and M. G. Payne, *Opt. Lett.* **9,** 347 (1984).

38. W. R. Garrett, W. R. Ferrell, M. G. Payne, and J. C. Miller, *Phys. Rev. A* **34,** 1165 (1986).

39. W. R. Ferrell, M. G. Payne, and W. R. Garrett, *Phys. Rev. A* **35,** 5020 (1987).

40. K. Watanake, F. M. Matsunaga, and H. Sakai, *Appl. Opt.* **6,** 391 (1967).

41. G. S. Hurst, M. G. Payne, R. C. Phillips, J.W.T. Dabbs, and B. E. Lehmann, *J. Appl. Phys.* **55,** 1278 (1984).

42. R. G. Wilmoth and S. S. Fisher, *Surf. Sci.* **72,** 693 (1978).

43. R. D. Willis, S. L. Allman, C. H. Chen, G. D. Alton, and G. S. Hurst, *J. Vac. Sci. Technol.* **A2,** 57 (1984).

44. D. Lal, W. F. Libby, G. Wetherill, J. Leventhal, and G. D. Alton, *J. Appl. Phys.* **40,** 3257 (1969).

45. I. B. Khaibullin, E. I. Shtyrkov, and M. M. Zaripov, *J. Phys. Soc. Jpn. Suppl.* **A49,** 1281 (1980).

46. R. F. White, *Appl. Phys. Lett.* **38,** 357 (1981).

47. N. R. Daly, *Rev. Sci. Instrum.* **31,** 720 (1960).

48. R. Baumhakel, *Z. Phys.* **199,** 41 (1967).

49. C. H. Chen, R. D. Willis, and G. S. Hurst, *Vacuum* **34,** 5811 (1984).

50. B. E. Lehmann, H. Oeschger, H. Loosli, G. S. Hurst, S. L. Allman, C. H. Chen, S. D. Kramer, M. G. Payne, R. C. Phillips, R. D. Willis, and N. Thonnard, *J. Geophys. Res.* **90,** 11547 (1985).

51. H. H. Loosli and H. Oeschger, *Earth Planet Sci. Lett.* **7,** 67 (1967).

52. L. A. Franks, H. M. Borella, M. R. Cates, G. S. Hurst, and M. G. Payne, *Nucl. Instrum. Methods* **173,** 317 (1980).

53. R. E. Bigle, J. S. Laughlin, R. Davis, Jr., and J. C. Evans, *Radiat. Res.* **67,** 266 (1976).

54. S. P. Rosen, *Phys. Rev. Lett.* **34,** 774 (1975).

55. F. Reines and M. F. Crouch, *Phys. Rev. Lett.* **32,** 493 (1974).

56. H. Primakoff and S. P. Rosen, *Phys. Rev.* **184,** 1925 (1969).

57. R. E. Smalley, L. Wharton, and D. H. Levy, *J. Chem. Phys.* **63,** 4977 (1975).

58. R. Tembreull and D. M. Lubman, *Anal. Chem.* **58,** 1299 (1986).

59. R. Tembreull and D. M. Lubman, *Appl. Spectrosc.* **41,** 431 (1987).

60. C. H. Chen, S. D. Kramer, D. W. Clark, and M. G. Payne, *Chem. Phys. Lett.* **65,** 419 (1979).

61. W. R. Ferrell, C. H. Chen, M. G. Payne, and R. D. Willis, *Chem. Phys. Lett.* **97,** 460 (1983).

62. K. H. Purser, A. E. Litherland, and H. E. Gove, *Nucl. Instrum. Methods* **162,** 637 (1979).

2

Surface Analysis Using Resonance Ionization Spectroscopy

JAMES E. PARKS

The concept of resonance ionization spectroscopy (RIS) was recognized from the very beginning to "have exciting potential for ultrasensitive detection of elemental substances and other materials."[1] Since then, this potential has been proven by the development of several ultrasensitive analysis techniques for practical analytical applications. One of these techniques, the subject of this chapter, is sputter initiated resonance ionization spectroscopy (SIRIS). SIRIS combines lasers, energetic ion beams, and mass spectrometers in a unique way so that solid material can be analyzed quantitatively for its elemental constituents with sensitivities in the parts per billion to parts per trillion range.

In RIS, atoms of a preselected element are resonantly excited and ionized using tunable lasers. Since atoms must be free to be ionized by RIS, SIRIS uses an energetic ion beam to sputter solid samples and atomize the constituents. Generally, the RIS technique does not provide isotopic information, so that a mass spectrometer is introduced to separate the ions by mass. The mass spectrometer also is used to discriminate against unwanted background signals that may be the result of secondary ionization produced by the sputtering process and/or nonresonant ionization of other atoms and molecules.

The ideal technique for analyzing materials for elemental impurities would (1) require no sample preparation, (2) be selective to the element of interest, (3) be independent of the matrix material, (4) have no interferences, (5) be applicable to small sample sizes, (6) be sensitive to the ultimate one atom level, (7) be equally applicable to all elements, (8) be quantitative and linear in response, and (9) survey all elements at the same time. The SIRIS technique begins to approach these ideals with the exception of the last one. SIRIS is not inherently a survey technique, however, multielement analysis using laser wavelength sequencing becomes increasingly feasible with the development of laser technology.

SIRIS was conceived in early 1981 by Atom Sciences, Inc.[2] and since then has been developed to the point of being used routinely for the analysis of electronic and biological materials.[3-10] In this chapter the basic concepts of RIS are briefly reviewed and illustrated with data obtained with the SIRIS technique and apparatus. The important features of the RIS process are emphasized for their significance in the practical application of SIRIS analyses. The basic concept of the SIRIS technique is presented with particular attention given to the factors that affect the sensitivity of the technique. The SIRIS instrument developed by Atom Sciences, Inc., and used for most of the measurements presented here, is described along with instruments developed at other laboratories. The types of analyses that can be made with SIRIS are bulk analyses, surface analyses, and depth profiles, and examples of these are presented. Quantification of measurements is a distinct advantage that SIRIS has over some other existing techniques and the various means for quantification of results are presented along with supporting data.

SIRIS has been used for practical applications in the fields of semiconductors, electro-optics, biology and medicine, health physics, basic physics, and others. Selected results from these fields are presented to illustrate the significance of the technique.

REVIEW OF RESONANCE IONIZATION SPECTROSCOPY

The RIS technique and theory has been described in an excellent treatise by Hurst et al. in the *Reviews of Modern Physics*[11] and in other sections of this book. Here, the RIS process is briefly reviewed with particular attention to some of the practical aspects of the application of RIS. In RIS, an atom, usually in a state in or near the ground state, is excited to a higher lying state by the absorption of one or more photons of the correct wavelength and energy. The excited atom is then photoionized by the absorption of a second or third photon of the same or different wavelength(s). The basics of RIS are summarized in Fig. 2.1. In the RIS process, gaseous atoms of the element to be measured are ionized in a two- (or three)-step process. In the first step (Fig. 2.1b), light from a tunable laser is used to excite the atoms resonantly to a bound state lower in energy than the ionization potential. The state is chosen to be one for which the transition from the ground state has high probability. In the second step (Fig. 2.1c), each excited atom absorbs a second quantum of light (from the same or from a different laser), and is thereby ionized. For some atoms it is necessary or desirable to use two steps of resonant excitation before the final ionization (Fig. 2.1d). The ions so produced can then be counted, usually after a mass separation.

The excitation step in RIS involves resonant excitation and is very important in practice. This is the step that gives RIS its selectivity. Since the excitation is resonant, the transition will be easily induced by weak light, and, in the practice of RIS, particular attention needs to be given to this step. Frequently, the excitation is accomplished with ultraviolet light, and if the intensity of the light is too great, this light can directly ionize some atoms and molecules with a finite probability. This nonresonant ionization is then a source of background noise that can limit the detection limit of the measurement. Therefore in the practical application of RIS, the intensity of the laser light used for excitation should be just large enough to nearly saturate the excitation process and no more. Frequently, it is necessary to attenuate the light to avoid this problem.

The photoionization step in RIS can sometimes involve a resonant transition to an autoionizing state (Fig. 2.1e). In such cases RIS is more selective and more

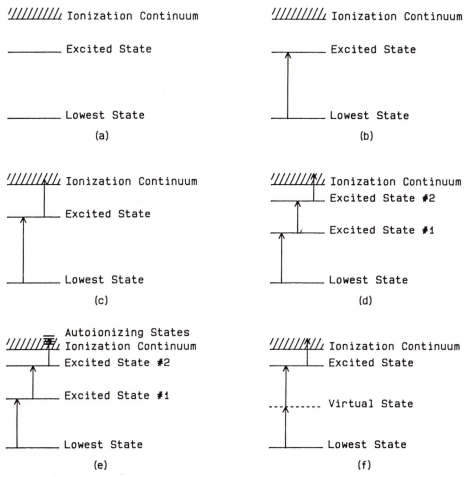

Fig. 2.1. (a) The steps of the RIS process can be explained by considering Grotrian diagrams representing the discrete energy levels of the selected atom to be ionized. (b) The selected atom is excited from its lowest ground state to a higher energy state in an allowed, resonant transition by the absorption of a photon having the correct wavelength or energy. (c) The excited atom is then ionized by the absorption of a second photon from the same laser or a different one. (d) Sometimes it is necessary or desirable to excite the atom to a second, higher level before photoionization. (e) Sometimes the photoionization step is a transition to an autoionizing state, which is also resonant. (f) For some elements the first excited state is too energetic to be excited by a single photon and the atom is excited by using a two-photon allowed transition.

efficient. Most elements can be excited with a single photon, however, a few of the elements require high two-photon allowed transitions (Fig. 2.1f). As a result, all elements of the periodic chart except helium and neon can be ionized in RIS processes using commercially available lasers. Those schemes involving two-photon excitation usually require a more intense source of light and the chances of non-resonant ionization occurring are greater.

The key features of RIS are a high degree of elemental selectivity and very high sensitivity. The selectivity arises at the excitation step, because it is possible to find levels in each element that are not matched by those of any other element, and because tunable lasers are available with the wavelength resolution necessary to differentiate between such different levels. The high sensitivity arises from the high probability of the resonant transitions used and from the high power available today in tunable lasers. For most elements, it is fairly easy to saturate a suitable transition, including the ionization step.

THE SIRIS CONCEPT

SIRIS combines two established and well-understood technologies, ion beam sputtering and resonance ionization spectroscopy, to make analyses of impurities in solids at concentrations of a few parts per billion or lower. The SIRIS technique was first conceived by workers at Atom Sciences in early 1981, which led to a patent being filed in January 1982.[2] The commercial development of SIRIS was then begun by Atom Sciences. Since then the concept has been described and the results of SIRIS measurements have been reported by Atom Sciences in various publications.[3-10,12] Others, notably researchers at Pennsylvania State University and Argonne National Laboratory, have R&D programs to develop the SIRIS technique, and they have also reported their progress.[13-22] The basic concept of SIRIS is illustrated in Fig. 2.2. An inherent condition of the RIS technique is that the atoms must be free in the gas phase. In a few cases, a sample is already a gas, but most samples are either solid or liquid and must first be atomized. SIRIS uses sputtering to accomplish this. Sputtering is a well-established technique for vaporizing solids in a controlled manner and it is very reproducible. Therefore, in SIRIS, an energetic, pulsed ion beam (usually argon) is focused onto a solid sample, thus producing a cloud of vapor immediately above the target, the contents of which are representative of the constituents of the sample. RIS lasers then selectively ionize atoms in the vapor cloud of the chosen element that are subsequently extracted and directed to an ion detector to be measured or counted. In a more practical form a mass spectrometer is added to the detection system to confirm the elemental identity of the ionized atom and/or to add isotopic identification. The mass spectrometer may be one of several types, a magnetic sector, a time-of-flight, or an rf quadrupole mass spectrometer. Secondary ions, produced by the impact of the ion beam, can be rejected by electrostatic fields, electrostatic energy analysis, the relative timing between the ion beam pulse and RIS laser pulse, or a combination of these.

For many years after 1950, secondary ion mass spectrometry (SIMS)[23-26] was considered the most sensitive, practical technique for the analysis of trace elements in solids and on the surface of materials. In SIMS an energetic ion beam sputters a solid sample and in the process creates secondary ions. Although the use of SIMS is currently experiencing an exponential growth,[26] there are problems that limit the effectiveness of the SIMS technique and the interpretation of results. The most serious of these problems is the strong dependence of the secondary ion yield from a given matrix on the element being sputtered as well as the matrix material. In addition, the secondary ion yield is very small compared with the yield of neutral

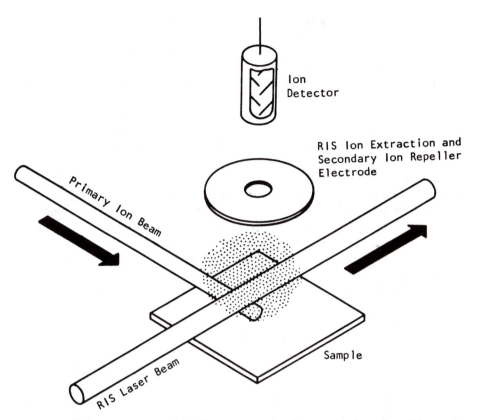

Fig. 2.2.   The basic concept of SIRIS uses sputtering for atomization of a solid sample and RIS to ionize a selected element.

atoms. SIMS has been limited by the fact that most of the sputtered particles are neutral atoms that are not detected. In addition, the ionization that takes place is not selective and a mass spectrometer is required for the elemental identification. There are also isobaric interferences from atomic or molecular ions of the same nominal mass that cause difficulties in the interpretation and quantification of data. The SIRIS technique is ideally suited to avoiding some of these problems. SIMS and SIRIS are complementary techniques; where one has deficiencies, the other excels.

ELEMENTARY THEORY OF THE SIRIS TECHNOLOGY

The efficiency and sensitivity of the SIRIS technique and apparatus can be modeled quite well using RIS, sputtering, and ion optics theory. When a pulse of argon ions sputters a particular matrix material, the yield of sputtered neutral atoms can be predicted from prior measurements made as a function of energy of the primary ion and the angle it makes with the sample.[27] The sensitivity of the SIRIS technique depends primarily on how many atoms can be sputtered in a single pulse. To achieve the best sensitivity, the efficiency of ionizing and of detecting the RIS ions must be maximized. The efficiency of ionizing the selected element from the sputtered cloud and then counting the ionization in a detector depends on the temporal and spatial overlap of the beam of laser light and the atomized cloud of atoms, the angular and energy distribution of the sputtered particles, and the efficiency of the mass spectrometer.

The sensitivity of the SIRIS technique and apparatus is determined by the number of atoms sputtered by the primary ion pulse and the fraction of those sputtered atoms ionized by the RIS laser process and then extracted and detected in the mass spectrometer. The number of atoms sputtered by the primary ion beam depends on the sputtering yield and the number of primary ions. The sputtering yield depends on the energy of the ion beam, the type of primary ion, the target matrix material, and the angle of incidence between the primary ion and the target. The sputtering yield increases at least as fast as the inverse cosine of the angle of incidence.[27] The number of primary ions is directly proportional to the ion beam current and the pulse length of the ion pulse. For an 18.5-keV argon ion beam normally incident on silicon, the sputtering yield[27] is approximately 2 and for a 60° angle of incidence the sputtering yield is about $2 \times 1/\cos 60° = 4$. For an argon ion beam of 1 μA and an ion beam pulse length of 1 μsec, the number of primary ions per pulse would be about $[(1 \times 10^{-6} \mu A) \times (1 \times 10^{-6} \sec)]/(1.6 \times 10^{-19}$ C/ion$) = 6.25 \times 10^6$ ions/pulse. If the primary beam is incident on silicon at 60°, the yield of sputtered atoms will be about 2.5×10^7 sputtered atoms.

The fraction of sputtered atoms ionized by the lasers and then detected depends on the temporal and spatial overlap of the laser beams and the sputtered ions, the fraction of sputtered neutrals in the lowest energy state to be excited, the efficiency of extracting the ions, and the transmission of the mass spectrometer. The fraction of sputtered atoms that spatially overlap the laser beams is a geometrical factor, F_G, that is the fraction of the total number of atoms sputtered within the solid angle defined by the laser beam and the acceptance region of the extraction electrodes. The solid angle subtended by a laser beam 1 mm in diameter and 5 mm long located with its center 3 mm above the sample is approximately $\frac{5}{9}$ sr and the fraction of the total solid angle 2π sr is roughly 10%.

The temporal overlap of the pulse of secondary neutrals and the pulse of laser light is determined mostly by the energy distribution of the sputtered particles. The fraction of sputtered particles located in the region of the laser beam at the time the laser is on is the time factor F_T. Since the pulse length of the laser is short compared to the time the sputtered particles take to transverse the diameter of the laser beam (8 nsec compared to about 200 nsec), the laser ionization can be considered to be instantaneous. The fraction of sputtered particles in the sensitive volume when the laser beam is turned on is those particles that are fast enough to reach the beam in time, but slow enough not to have passed through it. This time fraction, F_T, can be determined by using the energy distribution[28]

$$f(E) = CE/(E + U)^3$$

where C is a constant independent of E, and U is the binding energy. This energy distribution can be transformed to a time distribution, $T_0(t)$ for particles traveling between a distance x_1 to reach the near edge of the laser beam and a distance x_2 to reach the far edge of the laser beam. If the primary ion beam consists of a pulse of ions given by $S(t)$, then the fraction of sputtered particles in the sensitive volume at some time T will be given by the convolution integral

$$F_T(T) = \int_0^T S(t)T_0(T - t)\, dt$$

Assume that gallium-69 atoms are being sputtered from a silicon matrix and that the binding energy is 4.5 eV. Assume, also, that a 1-μsec ion beam pulse is used to sputter a sample and that the sensitive volume is 1 mm across and centerd 2.5 mm from the sample. Then the fraction of sputtered atoms in the sensitive volume can be computed as a function of time. Figure 2.3 is a graph of this fraction, and it indicates that a maximum fraction of about 17% of the atoms will be in the sensitive volume if the laser is turned on at the optimum time of about 1.15 μsec. Shorter ion beam pulses usually allow larger maximum fractions, but this is offset by fewer atoms being sputtered.

The useful fraction of sputtered analyte atoms in the electronic state being probed by RIS can be denoted by F_R. For gallium, Atom Sciences has found this fraction to be about 50%, and Kimock et al.[15] have found that about 90% of sputtered neutral indium atoms are in the lowest electronic state. Kimock et al. have also reported that in some cases the excited state populations near the lowest state

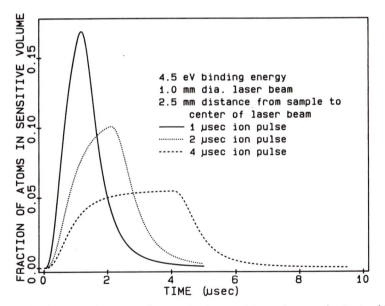

Fig. 2.3. The fraction of sputtered atoms in the sensitive volume to be ionized by RIS is a maximum of 17% for a 1-μsec ion pulse and a laser beam of 1.0 mm diameter centered 2.5 mm from the sample. For longer pulses, more of the sputtered atoms are unused and the fraction is reduced.

depend on the energy of the primary ion beam energy. One of the advantages of the SIRIS technique is this ability to investigate the excited states created by sputtering. It is reasonable that F_R is about 50% in estimating the sensitivity of the SIRIS process.

In estimating the fraction of atoms ionized by RIS and then detected by the mass spectrometer it is useful to divide the fraction into two parts: one is the fraction of RIS atoms, F_E, extracted and focused into the entrance aperture and acceptance angle of the mass spectrometer and another is the fraction of these ions, F_M, transmitted through the mass spectrometer and finally detected. In the magnetic sector instrument of Atom Sciences, the extraction efficiency is estimated conservatively to be at least 50% and the transmission of the mass spectrometer has been demonstrated to be 80%. The total fraction of sputtered atoms detected by SIRIS, Y, is the product of these separate fractions,

$$Y = F_G \times F_T \times F_R \times F_E \times F_M$$

and is generally referred to as the "useful yield" by practitioners of SIMS. Using the estimates and examples just stated, the useful yield of the magnetic sector instrument of Atom Sciences is estimated to be

$$Y = 10\% \times 17\% \times 50\% \times 50\% \times 80\% = 0.34\%$$

If the analyte concentration is C_A, then the number of analyte atoms, N_A, out of N_T sputtered atoms that could be detected by SIRIS in N_L laser pulses is

$$N_A = Y \times C_A \times N_T \times N_L$$

In 9000 laser pulses (5 min at 30 Hz) with N_A equal to 1 and a signal-to-noise ratio of 1, the sensitivity of SIRIS is estimated to be

$$C_A = 1/(0.0034 \times 2.5 \times 10^7 \times 9000) = 1 \text{ ppb}$$

by using the estimates for N_T of 2.5×10^7 atoms/pulse, and Y of 0.34%. Longer integration times, higher beam currents, and higher detection efficiencies could lead to even better sensitivities. A sensitivity of 2 ppb has been reported by Atom Sciences, Inc.[5] for a 5-min measurement for gallium in silicon, consistent with these estimates.

THE SIRIS APPARATUS

The SIRIS apparatus designed and constructed at Atom Sciences to obtain some of the results presented here is illustrated in the schematic diagram in Fig. 2.4. Sputtering is accomplished with a pulsed argon ion beam generated by a commercially available duoplasmatron microbeam ion source that is pulsed synchronously with the 30-Hz laser. The argon ions are given an energy of approximately 10 keV. The ion beam intersects the sample plane at 60° with respect to the normal, and typically with argon, ion beams of 5 μA can be generated and brought to a focus in a spot size from 5 to 100 μm diameter. The ion beam pulses are typically 1 μsec in duration and are generated to coincide with the pulse repetition rate of the laser.

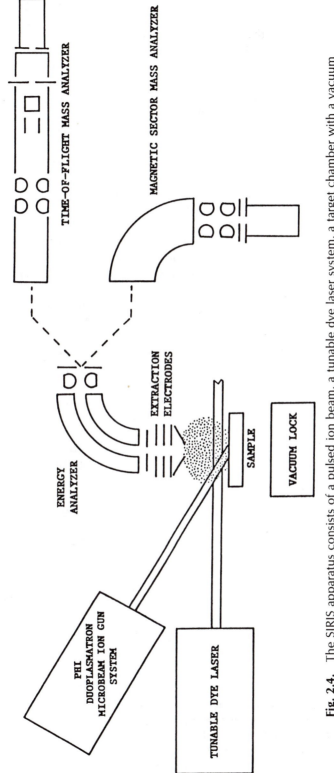

Fig. 2.4. The SIRIS apparatus consists of a pulsed ion beam, a tunable dye laser system, a target chamber with a vacuum lock, extraction electrodes, an energy analyzer, and an option for a time-of-flight or magnetic sector mass spectrometer.

   The dye laser system used in the SIRIS apparatus at Atom Sciences is illustrated schematically in Fig. 2.5. The system is pumped by a frequency doubled Quanta-Ray DCR-2A neodymium-YAG laser that can produce approximately 600 mJ of light in a 8-nsec pulse. The infrared light of fundamental wavelength of 1.06 μm is frequency doubled to produce 532-nm green light that is used to pump one or two dye lasers. Typically, green light of about 200 mJ/pulse is used to produce approximately 60 mJ/pulse of visible light that is tunable from blue to red light by the proper choice of dyes. The visible light in turn can also be frequency doubled to produce UV light ranging from about 270 to 340 nm with energies as high as 10 mJ/pulse. Wavelengths down to 220 nm can be generated by mixing the doubled visible frequencies with the fundamental YAG light. This system, with two dye lasers, is a versatile RIS laser system and allows the use of a variety of different RIS schemes using both fixed and variable wavelengths.

   Once the selected element has been ionized, the ions are electrostatically accelerated and focused into the entrance of a double focusing magnetic mass spectrometer that consists of an electrostatic sector for energy analysis and a magnetic sector for mass analysis. The mass spectrometer serves three purposes. First, the mass spectrometer is used for verification of the RIS process. Second, the mass spectrometer is for isotopic identification. With the laser systems currently being used with the RIS technique, the bandwidths are large compared to the natural linewidths of the atoms, and no isotopic selectivity is achieved. The third use of the mass spectrometer involves the rejection of SIMS ions. The energy analyzer is used primarily to reject the secondary ions, and it accomplishes this by making use of the fact that the secondary ions and RIS ions are formed in different locations in an electric field and hence are accelerated differently and are caused to have different energy distributions. Of course, mass analysis in the magnetic sector also discriminates against the SIMS ions. Since the lasers are pulsed, time discrimination can also be employed to reject the SIMS ions. The creation of the SIMS ions and the RIS ions

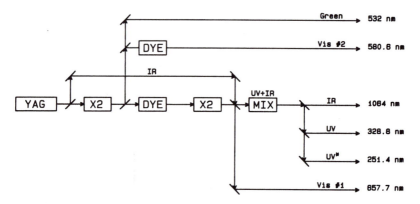

Fig. 2.5.   The tunable dye laser system for the SIRIS instrument consists of a frequency doubled neodymium-YAG laser that is used to pump two dye lasers. Light from the dye lasers is sometimes frequency doubled and the doubled light is sometimes mixed with the fundamental IR light from the YAG pump laser to generate light with wavelengths as low as 220 nm.

can be controlled to occur at different times so that time gating the electronics discriminates against the unwanted ions. The potentials of the extraction electrodes and sample bias can also be pulsed to help reject SIMS ions and improve the signal-to-noise ratio.

As an alternative, a time-of-flight mass spectrometer can be substituted for the magnetic sector instrument. This is particularly advantageous when making isotopic ratio measurements since the dependence on the fluctuations in the shot to shot signals is removed from the measurements.

The SIRIS ions are detected with an electron multiplier with a conversion electrode, operating either as an analog detector or as a single ion counter. Data collection is accomplished with a computer-based data acquisition system and the instrument is also automated and controlled with microprocessors and a host computer.

Other researchers have also developed SIRIS instruments and have measured and reported results. Winograd and his coworkers at Penn State University have developed SIRIS instruments and are using them primarily for the investigation of the fundamentals of sputtering.[13-15] Recently, they have designed a clever SIRIS type instrument for measuring energy- and angle-resolved neutral particle distributions.[16,17] This instrument uses RIS [or multiphoton resonance ionization (MPRI)], a position-sensitive detector, and a computer-based data digitization system for measuring the energy- and angle-resolved sputtered neutral-particle distributions. The RIS process is used to identify the element and to pick out the various excited states of the sputtered neutrals. The position-sensitive detector provides the necessary angular and distance information whereas the short pulse length of the dye laser system provides excellent timing information for the velocity measurement. Winograd has used both quadrupole and time-of-flight mass spectrometers in his instruments.

Pellin and co-workers[18-22] have produced several SIRIS instruments with which to investigate the SIRIS technique that they refer to as SARISA, surface analysis by resonance ionization of sputtered atoms. Most of their configurations make use of hemispherical energy analyzers of their own proprietary design. The mass spectrometers in their instruments operate on the time-of-flight principle, and they have been able to achieve sensitivities in the parts per trillion range.

One of the basic operational differences between the SIMS and SIRIS techniques is that SIMS uses a dc ion beam for sputtering during data acquisition whereas SIRIS uses a pulsed ion beam. As a result SIRIS typically does not remove more than an equivalent monolayer of material from the sample during the measurement period. This feature permits the use of very small samples for analysis with SIRIS since very little sample is consumed in the measurement. Basically, therefore, SIRIS is a surface technique. When an analysis is needed as a function of depth into the bulk of the material, SIRIS can alternate between using a pulsed ion beam for analysis and a dc beam to mill away upper layers of the sample to a depth at which the measurement is desired. The ion beam size on target typically is approximately 0.1 mm and can be rastered over an area approximately 1×2 mm during an ion milling step. Typically raster areas need to be 10 times the dimensions of the ion beam, and this is important in depth profiling to avoid edge effects.

The advantages and disadvantages of SIRIS are summarized in Table 2.1. The

Table 2.1.   Advantages and disadvantages of SIRIS

Advantages
   1. Has high sensitivity; less than 1 part per billion
   2. Has selective ionization of the elements
   3. Has efficient ionization of the elements
   4. Measures sputtered neutrals
   5. Is generally applicable to all the elements of the periodic chart (although helium and neon are
         currently excluded for practical reasons)
   6. Accommodates small sample size
   7. Is free from isobaric interferences
   8. Is a surface measurement
   9. Can be used for depth profiles
   10. Is less dependent on chemical and matrix composition

Disadvantages
   1. Is not well suited to survey analyses (many elements at one time)

selectivity and efficiency of the RIS process are the main advantages that give SIRIS its high sensitivity. SIRIS measures sputtered neutrals away from the surface, free of chemical and other surface effects, is generally applicable to all the elements of the periodic chart, although helium and neon are currently excluded for practical reasons, and is a surface measurement that can be used for depth profiles and is accommodating to small sample size.

Perhaps, most importantly, SIRIS is free from isobaric interferences. Mass spectra of RIS-produced ions are easily interpreted because of the selectivity of the RIS process. Because only atoms of a given element are ionized, a mass spectrum generated by RIS contains only the isotopes of the element being measured, with no isobaric interferences. Identification of masses is straightforward and simple. Figure 2.6 shows a mass spectrum of an analysis of steel in which the apparatus has been tuned to ionize silicon. Only the isotopes of silicon, masses 28, 29, and 30, appear in the SIRIS spectrum.

The most outstanding disadvantage of SIRIS is that it is not well suited to survey analysis when many elements are measured at one time. This is contrary to the very nature of SIRIS, which discriminates against all other elements except the one that has been selected. Inherently, SIRIS is a single element technique, but it does provide isotopic information.

QUANTIFICATION OF SIRIS

In theory, SIRIS measurements can be calibrated from first principles using known or measured parameters that describe the laser beam, the RIS process, the sputtering process, and the ion optics and transport that extract and detect the ions. However, as a practical matter, SIRIS uses standards for calibration. SIRIS measurements have been shown to be linear over a wide range of concentrations by using Standard Reference Materials (SRMs) from the National Bureau of Standards. A number of elements, including Al, V, B, Cu, Mo, and Si, have been measured in steel SRMs with SIRIS, and the measurements have correlated very well with the

NBS certified values. A linear relationship has been shown between the measurements and the SRM values. As an example, Fig. 2.7 shows a comparison of SIRIS measurements for vanadium in steel and the NBS reference values. The concentration of vanadium ranges from a few parts per million to pure vanadium, and linearity is maintained over the entire range.

The technique of isotope dilution is an excellent method with which to calibrate SIRIS measurements, and this technique has been successfully used and reported for SIRIS measurements of copper in blood serum.[10] Using this technique, an appropriate amount of isotopically pure copper (99.61% ^{65}Cu) was added to the sample in solution, and then the copper from the spiked sample was electrodeposited on a high-purity (99.9999%) gold substrate. The altered isotope ratio was then determined with the SIRIS instrument. With the known concentration and volume of the spike solution added to the sample, the volume of the sample solution before spiking, the natural isotopic abundance, and the altered isotope ratio, the concentration of the unknown sample can then be computed. A system calibration was performed by making a series of ^{63}Cu/^{65}Cu isotope ratio solutions prepared by mixing accurately weighed portions of a ^{65}Cu solution and a natural copper solution. The natural copper solutions were prepared by dissolving high-purity copper wire in nitric acid and diluting with laboratory distilled water. The isotope ratios determined with SIRIS are compared with the prepared gravimetric values in Fig. 2.8. The results show a linear relationship between the measured and prepared values for ratios varying more than a factor of 500. This capability coupled with the pulsed feature that utilizes small samples allows the determination of analytes in the picogram range using samples as small as 10 μL.

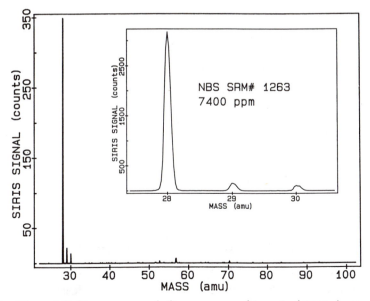

Fig. 2.6. Time-of-flight mass scan of silicon measured in a steel SRM shows no interferences by the presence of iron and other impurities.

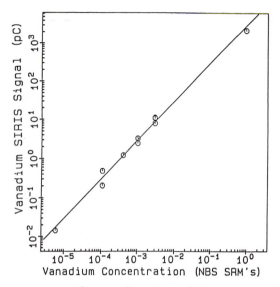

Fig. 2.7. SIRIS measurements for vanadium in steel SRMs compared to the standard reference value shows good linearity over a range of 10^5.

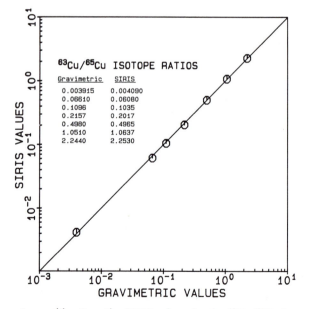

Fig. 2.8. In a system calibration, the SIRIS values for the $^{63}Cu/^{65}Cu$ isotopic ratio were in excellent agreement with the prepared gravimetric values.

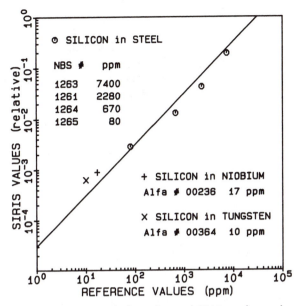

Fig. 2.9. An early measurement of silicon in steel SRMs and samples of niobium and tungsten indicated little matrix dependence for SIRIS measurements.

An early measurement of silicon in steel SRMs and two samples of Nb and W indicated little matrix dependence for the SIRIS measurements. These results are shown in Fig. 2.9. The agreement of the SIRIS values with the concentrations obtained by other methods is good in spite of the fact that Nb, W, and steel are different matrices. This evidence suggested a more careful study in which the SIRIS signal was corrected for differences in sputtering yield and concentration. The results of those measurements are shown in Table 2.2 in which the SIRIS signal per unit concentration and per unit sputtering yield has been normalized to the value for the Si matrix. Except for the $TiSi_2$ matrix, the SIRIS response to silicon in the different matrices was nearly the same. Boron, also, was measured in several

Table 2.2. Relative SIRIS response for silicon and boron in various matrices

Matrix	Element	Relative SIRIS response
Si	Si	1.00
Al	Si	0.96
$PdSi_2$	^{28}Si	0.96
$PdSi_2$	^{30}Si	0.96
$TiSi_2$	Si	3.60
GaAs	Si	3.65
Si	B	1.00
GaAs	B	0.77
BPSG	B	0.23

matrices and the SIRIS response was normalized to the sputtering yield and known concentration of boron. As with the silicon analyses, the boron response in the different matrices was normalized to a single matrix (silicon in this case). The relative response of the SIRIS determination of boron in the different matrices is nearly the same, well within an order of magnitude. This is a significant improvement over what is normally observed in SIMS with regard to matrix effects.

APPLICATIONS FOR SIRIS

Semiconductor Analysis

The semiconductor industry has motivated the development of many analysis techniques and has been responsible for the progress that has been made in the SIMS technique and instrumentation. The need for better semiconductor measurements and for improvements in SIMS analysis has encouraged the development of the SIRIS technique and instrumentation. Semiconductor measurements have required that better sensitivities and detection limits be attained. Most semiconductor materials can be made with a high degree of purity and are ideally suited for ultrasensitive measurements. They are very amenable to ultrahigh vacuum systems. The sputtering technique provides the necessary atomization of the solid samples consistent with the operating criteria of a mass spectrometer environment. Semiconductor devices are very small and a microbeam ion source allows the analysis to be well defined spatially in a small area of micron dimensions.

The semiconductor community has a wealth of well-characterized samples and can make customized samples for testing purposes. SIRIS has been used to analyze these semiconducting materials for bulk impurities and for measuring depth profiles of implanted samples in which the concentration varies with depth.

Silicon-doped samples of gallium arsenide were prepared by molecular beam epitaxy (MBE) by the Avionics Laboratory of Wright-Patterson Air Force Base and were used to demonstrate SIRIS measurements of the bulk silicon concentration. A typical time-of-flight mass spectrum of silicon in one of the samples is shown in Fig. 2.10. This sample was prepared to contain 8 ppm ($4 \times 10^{17}/cm^3$) of silicon in the gallium arsenide epilayer. Integration of the mass peaks yielded results very near the expected isotopic abundances, shown in parentheses. The correct isotopic ratios and the clean spectrum in the other mass areas indicate that there are no isobaric or molecular interferences.

There is great interest in the semiconductor industry in depth profiles in which the concentrations of both dopants and impurities are measured as a function of depth. Samples can be implanted with a dopant at known energies and doses so that the concentration distribution can be computed from well known theory. Wafers of gallium arsenide implanted with 200 keV ^{29}Si ions with peak concentrations of $5 \times 10^{18}/cm^3$ and $5 \times 10^{17}/cm^3$ were supplied by the Avionics Laboratory of Wright-Patterson Air Force Base. A plot of ^{29}Si concentration as a function of depth in one of the samples with the higher concentration is shown in Fig. 2.11. The SIRIS signal was calibrated by normalizing the intensity to the known peak

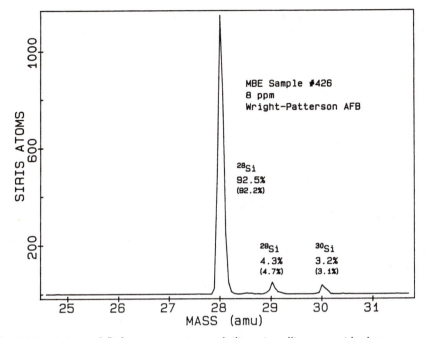

Fig. 2.10. A time-of-flight mass spectrum of silicon in gallium arsenide that was grown by MBE and doped to have a concentration of 8 ppm was measured with SIRIS.

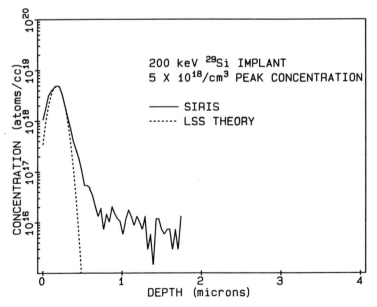

Fig. 2.11. A depth profile of 200-keV ^{29}Si ions implanted into gallium arsenide was measured with SIRIS and compared with LSS theory.

concentration, and the depth scale was calibrated by measuring the depth of the crater with a Dek tak profilometer. The data show a background limit near $1 \times 10^{15}/cm^3$ (20 ppb), which probably comes from natural silicon in the sample matrix or from contamination within the apparatus. For silicon in gallium arsenide, this measurement is state-of-the-art.

A theoretical distribution for silicon implanted in gallium arsenide with an implant energy of 200 keV and $5 \times 10^{18}/cm^3$ peak concentration was calculated with LSS theory[29] and compared with the SIRIS results. That comparison is also shown in Fig. 2.11. The SIRIS results agree with the theoretical distribution at the smaller depths, nearer the surface than the depth of the peak concentration, but at the larger depths, the agreement diminishes. These deviations of the measured profile from the theoretical have been explained by Shepherd et al.[30] to arise from channeling of Si^+ during implantation. Shephard et al. demonstrated that if the samples were amorphized before implantation to avoid channeling, the sample profile becomes Gaussian and fits the LSS theory. The sample shown in Fig. 2.11 was not amorphized before implantation and channeling was possible.

Two additional depth profiles of silicon implanted into gallium arsenide are shown in Fig. 2.12. In one sample, the implant dose gave a peak concentration of $5 \times 10^{18}/cm^3$ and in the other sample the peak concentration was a factor of 10 less, $5 \times 10^{17}/cm^3$. In each case, the SIRIS measurements were normalized to the given peak concentration.

The technologies of MBE and metal organic chemical vapor deposition (MOCVD) have made it possible to construct devices composed of many layers of

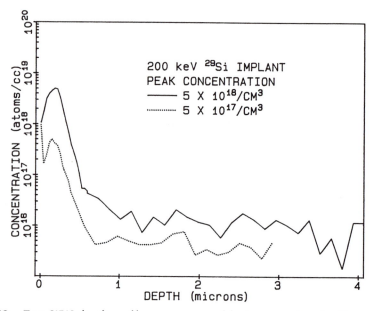

Fig. 2.12. Two SIRIS depth profiles are compared for two samples of gallium arsenide implanted with 200-keV ^{29}Si ions with peak concentrations of $5 \times 10^{18}/cm^3$ and $5 \times 10^{17}/cm^3$.

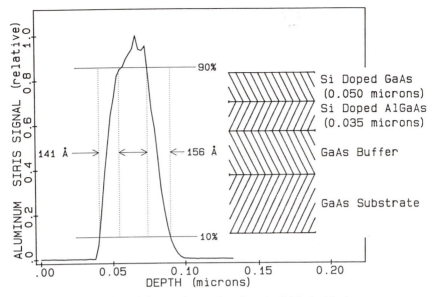

Fig. 2.13. A silicon-doped, layered sample of GaAs/AlGaAs/GaAs was constructed and the aluminum concentration profile measured with SIRIS.

different materials suitably doped to create the desired characteristics. Impurities, coming from the growth process or through diffusion from layer to layer, can degrade the operation of the devices. This has created a need for an analytical procedure that can measure an impurity or dopant across an interface without being subject to unwanted matrix effects and SIRIS is ideally suited for this kind of measurement. A problem encountered in SIMS analysis is that an enhanced signal is often observed near the interface of two materials when a depth profile is performed. This enhanced signal can often be attributed to an oxygen buildup at the interface during a pause in the deposition while changing the material for the next layer. It is well known that in many cases oxygen enhances the SIMS response. A technique that analyzes sputtered neutrals, such as SIRIS, should be less susceptible to differences in matrices and the chemical effects arising in the SIMS technique.

A layered sample of GaAs and AlGaAs was depth profiled using the SIRIS technique to investigate the capability of SIRIS for depth profiling layered materials. The sample, supplied by Hanscom Air Force Base, had been grown by MBE and electrically characterized. The structure of the sample is illustrated in Fig. 2.13. The layers were grown on a substrate of undoped LEC GaAs. A buffer layer of GaAs (approximately 1 μm thick) was first grown, followed by a layer of Al$_{.30}$GaAs doped with silicon. This AlGaAs layer was approximately 350 Å thick and the concentration of the silicon was determined electrically to be approximately 7×10^{17}/cm^3. A final layer of GaAs that was also doped with Si was grown on top of the AlGaAs. This layer was approximately 500 Å thick and the silicon concentration was determined electrically to be approximately 2×10^{18}/cm^3. The physical concentrations and the electrical concentrations are different in that the electrical concentration is

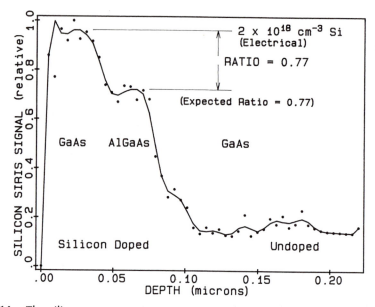

Fig. 2.14. The silicon concentration was measured as a function of depth with SIRIS in a layered sample of GaAs/AlGaAs/GaAs.

based on the sum of N and P type carriers. The physical concentration of the silicon in each of the layers is unknown; however, the ratio of the physical concentrations in the two layers is known from the parameters of the MBE growth process to be 0.77.

Because silicon was the dopant, it was first measured as a function of depth. The results are shown in Fig. 2.14 with the indicated data points and a smooth line fit to guide the eye. The silicon concentration, based on electrical carrier measurements, in the top layer is in the $2 \times 10^{18}/\text{cm}^3$ range (40 ppm) and the SIRIS signal was high. The SIRIS signal for silicon in the AlGaAs layer decreased to 77% of the value found in the GaAs layer when the top layer was sputtered through and the measurements were made in the second layer. This agrees exactly with the expected ratio of silicon concentrations in the layers of GaAs and AlGaAs based on the MBE growth rate parameters. The SIRIS measurements have a precision of approximately 10%, therefore this exact agreement is somewhat fortuitous. The ratio based on the MBE growth parameters is significantly different from the ratio determined by the electrical carrier measurements. The free silicon carrier concentration in the AlGaAs layer was determined electrically to be approximately $7 \times 10^{17}/\text{cm}^3$ so that the ratio based on electrical measurements is 35%. However, if the ratio based on the MBE parameters, 77%, is correct, then the SIRIS measurement indicates that there is no significant matrix effect for Si between the two matrices, GaAs and AlGaAs.

The concentration of silicon in the undoped buffer layer was found to be surprisingly high, about $4 \times 10^{17}/\text{cm}^3$ (when normalized to the $2 \times 10^{18}/\text{cm}^3$ electrical measurement of the top layer). There should be no problem at this level ($4 \times 10^{17}/\text{cm}^3$) with background or detection limit, since in the depth profiles the detection limit for Si in GaAs has been shown to be about $5 \times 10^{15}/cm^3$ (see Figs. 2.11 and 2.12).

The depth scale was calibrated by measuring the depth of the sputtered crater with a Dek tak profilometer. According to the data supplied with the sample, the interfaces between the first and second layers (from the top), and between the second and third layers, should occur at depths of approximately 0.050 and 0.085 μm. Figure 2.14 shows that the SIRIS measurements are in good agreement with these values.

The aluminum concentration was also measured as a function of depth with the results shown in Fig. 2.13.This profile indicates the depth resolution of the measurements. The aluminum signal rises from 10 to 90% of its peak value in a depth of 140–150 Å, which is typical of measurements with argon ions with energies of 10 keV. This result is encouraging since depth resolutions of 30 Å can be achieved usually by reducing the primary beam energy from 10 to 3 keV. Presently, the source of the SIRIS apparatus can be operated as low as 0.5 keV and depth resolutions of 30 Å or less should be feasible.

Electro-optics Application

The analysis of materials by SIRIS has been shown to be useful for the electro-optics industry. Titanium-diffused lithium niobate is an electro-optic material being developed for waveguides and high-speed modulation of light in fiber optic communication. The ability to couple light to the waveguide and to efficiently couple the modulating electric field to the waveguide depends on the effective mode index, which in turn depends on the distribution of the index of refraction and hence of the titanium. SIRIS measurements of the titanium distribution have been made and found to be different from some SIMS measurements in that only a single gaussian profile was observed as opposed to two reported by others.[31]

SIRIS was used to measure the concentration distribution of titanium that had been diffused into a lithium niobate crystal for 3.0 hr at 1000°C from a 420-Å surface layer of titanium. The initial thickness, time, and temperature were chosen to give a gaussian distribution. In that case the concentration gradient should follow the equation $C(x) = C_0 \exp(-x^2/4Dt)$ where C_0 is the concentration at $x = 0$, D is the diffusion coefficient, and t is the diffusion time. The data, shown in Fig. 2.15 and plotted as log concentration versus depth squared, produces a straight line, consistent with diffusion theory. Some diffusion studies with SIMS have reported a peak in the concentration at low depths. The SIRIS data shown here display a higher number of data points taken in the first 1.5 μm of depth, and a low-depth peak is not apparent. There is a fundamental difference between the SIMS and SIRIS techniques and this difference should not be disregarded. Although the measurement is not ultrasensitive, there are problems with this type of measurement because lithium niobate is an oxide and an insulator material.

The depth scales of the profiles were accurately determined by measuring the rastered craters with a profilometer and the concentrations were calibrated by comparison with a standard, by integrating the distribution, and equating the integral to the initial amount of titanium. The results indicated that the standard value was a factor of two too large, and this correction had to be applied before the diffusion data became compatible with the optical data.

The diffusion coefficient is related to temperature, T, by $D(T) = D_0 \exp(-E_0/kT)$ where E_0 is the activation energy. The values obtained from SIRIS, plotted in Fig. 2.16, agree within experimental error with those reported by Holmes[32,33] for congruent lithium niobate.

Biological Applications

The sensitivity and selectivity of RIS and the small sample utilization of SIRIS provide the bases for the extension of SIRIS into the biological area for elemental and isotopic analysis. The ability to ionize an element selectively in the presence of many atoms of other elements permits the simplification of chemical separations and the commensurate reduction of the analytical blank that would otherwise limit much of conventional technology. As a result, SIRIS is being used to measure trace elements in blood serum for nutrition studies in which small sample size is particularly important, as in cases such as blood samples from premature infants.

Nanogram quantities of copper in microliter samples of human and reference bovine serum were accurately measured by SIRIS using the technique of isotope dilution. A simplified two-step sample preparation was developed by utilizing wet ashing followed by simultaneous separation and electrodeposition of copper (together with other metals) onto a high-purity gold substrate. The pulsed ion beam of the SIRIS technique typically removes less than a fraction of an equivalent

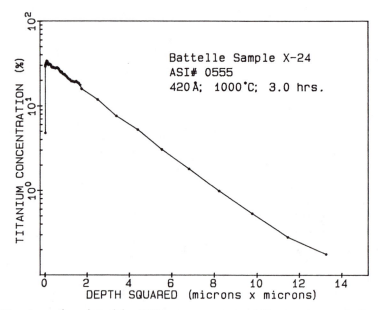

Fig. 2.15. A semilog plot of the SIRIS measurements of titanium concentration versus depth squared shows a linear relationship for a sample of titanium-diffused lithium niobate and good agreement with diffusion theory.

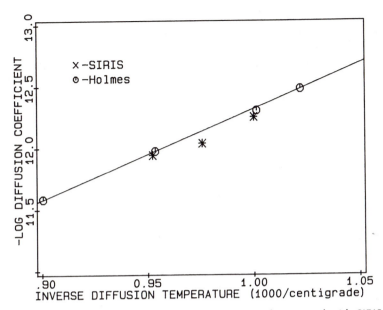

Fig. 2.16. A plot of $-\log_{10}$ of the diffusion coefficients determined with SIRIS measurements indicates a linear relationship with the inverse diffusion temperature. The data show good agreement with diffusion theory and the data of Holmes.[32,33]

monolayer of material during measurement, so that only a few monolayers of material need be deposited on the substrate for measurement. Sensitivities of a few picograms have been observed with molybdenum and sample sizes as small as 1 μL should be possible.

Data for the isotope dilution determination of copper in bovine serum, NBS Reference Material 8419, and human serum samples are tabulated in Table 2.3. The SIRIS isotope dilution value for bovine serum is in good agreement with the suggested value derived from several atomic spectroscopic techniques utilizing 2- to 3-mL sample sizes.[10]

The sample size used for SIRIS was 250 μL. The reproducibility of the SIRIS measurements is illustrated by the replicate analyses of sample F. A 1-g sample of

Table 2.3. Copper in blood serum (μg/g)

Sample	Suggested value	SIRIS value[a]
Bovine serum RM #8419	0.73±0.1[b]	0.763
Human serum sample F (1 g)		1.31[c]
		1.30[c]
		1.41[c]
Human serum sample (0.1 g)		1.18

[a]Estimated uncertainty = ±15%.

[b]The value from Veillon et al.[34] was listed originally in mg Cu/L; the value here is converted to a weight basis for comparison using a density of 1.03 g/mL.

[c]Aliquots from the same sample.

F was processed and split into three aliquots. Isotope dilution analyses of the aliquots produced concentrations of 1.308 ± 0.029 (SD, internal to an analysis), 1.304 ± 0.047 and 1.408 ± 0.057 μg Cu/g. A separate analysis, 1 week later with a 100-μg sample, produced a concentration of 1.18 μg Cu/g, in good agreement. The uncertainty of each determination is estimated to be $\pm 15\%$, which includes contributions from chemical blank ($<10\%$ of the total Cu per sample), stoichiometry and impurities of the ^{65}CuO, and isotope ratio accuracy and precision. Typical isotope ratio precisions for copper, internal to an isotopic analysis, range from 0.9 to 4.0%, relative standard deviation (RSD), with errors of the mean down to 0.25%. Thus, the methodology described here would also be useful to determine isotopic enrichments in small samples for metabolism studies.

Other Applications

SIRIS has been suggested and tested for a number of other applications, primarily because of its sensitivity, its lack of interferences, and the fact that complicated sample preparation procedures are not necessary. SIRIS, many times, can utilize a sample as it naturally occurs or with little modification. Frequently, sample size is limited, and the pulsed operation that utilizes very little of the sample makes SIRIS very amenable to various applications. In a geological application, SIRIS has been used to measure isotopic ratios that can be used for studying and dating geological materials. SIRIS's freedom from isobaric interferences makes it the preferred technique for the measurement of accurate isotopic ratios. Often the geological samples are small, and it is desirable to measure an element and its isotopes in a particular portion of the sample that has been identified by some other means, for example, with an optical microscope. The small, probing ion beam, is well suited for such an application.

Similar to the geological application, SIRIS has been suggested for the study of the composition of certain biological tissues such as brain tissue. A slice of tissue can be studied with an optical microscope and correlated with the findings of SIRIS using its small ion probe. By rastering the ion beam across the sample and monitoring the secondary electrons emitted by the sputtering, an electronic image of the sample can be formed and compared with the optical image. The RIS process can then be used to determine the concentration of the element of interest for the correlation studies with the optical observations.

The potential use of SIRIS for bioassays and radiological assays is of particular interest for two reasons. One, the sample preparation procedures can be greatly simplified, and two, SIRIS offers a viable alternative to radioactive decay counting techniques, particularly for the the long-lived radioisotopes. A bioassay study for the feasibility of using SIRIS to determine the uranium contamination in urine has been conducted and reported.[35] This study was motivated by SIRIS's potential for greater sensitivity and simplified sample preparation. A detection limit of 1 μg/L was demonstrated with SIRIS, however, this was somewhat short of the goal of 0.05 μg/L. In the case of uranium, many states other than the ground state are populated with the sputtering process, and various molecular oxides of uranium are easily formed and emitted in the atomization process. The net result is a lower sensitivity for uranium, although sample preparation can be reduced. A study has been conducted by Christie and Goeringer[36] with similar results.

A study of the determination of uranium in soil samples has been conducted and reported.[37] Soil samples of NBS standard reference materials were compressed and inserted into the analysis chamber of SIRIS. The technique of isotope dilution was used for calibration purposes and various mixtures of graphite were added to the samples to overcome charging problems. The results were mixed; however, the predominant problem still pertained to the lower sensitivity of SIRIS to the detection of uranium as was found in the bioassay measurements.

A very important application of SIRIS is its use in the investigation of basic science. The ultrasensitivity that SIRIS offers has led Fairbank and his collaborators[38] to use this technique to search for exotic species and rare events such as quarks and superheavy atoms. Theoretical predictions for the correct spectroscopy for these investigations are possible and allow for systematic searches and the means for correctly identifying the unknown species. This work demands the ultimate sensitivity and pushes the development effort to lower the detection limit.

In other basic science applications, SIRIS is being used to study isotope fractionation effects, both of the sputtering process and the RIS process. Isotopic fractionation in SIMS has been documented[39] and is well known in the case of thermal atomization. The possibility of such an effect not being present in SIRIS would be an important revelation, however, recent unpublished work has indicated the possibility of the existence of isotope fractionation as a result of a difference in RIS ionization between even and odd isotopes of an element. Further work on this problem is important to future development of the SIRIS technique and close attention will need to be given to these investigations.

SUMMARY

Among the techniques available for measurement down to trace levels of almost any element in almost any material, sputter initiated resonance ionization spectroscopy offers a number of useful characteristics. Most importantly, it is highly selective, so that only the element selected is measured, even in the presence of much larger amounts of any other element. It is sensitive at the present stage of development to the low part per million range for almost every element, to the low part per billion range for many, and to the part per trillion range for a few. In most cases, these limits are not intrinsic in the method, but arise from factors such as lack of time or incentive for developing higher sensitivity in a particular case, or from unavailability of standards at the lowest levels. Where incentives exist for sensitivities in the ppb range and below, sensitivities can be attained for most of the metallic elements. A third characteristic, of particular importance to those who grow crystals for electronic and optical devices, is that SIRIS can make these selective and sensitive measurements as a function of lateral position and depth. The 5- to 100-μm ion beam gives good lateral resolution, and the very low rate of erosion of the solid when the beam is in the pulsed mode for measurement ensures that the concentration measured pertains to the surface of the material. At the same time, the intensity of the ion beam in continuous mode is sufficient to remove material rapidly for producing a depth profile. The absence of large matrix effects on the sensitivity is important for work on layered devices, because it allows the concentration of a desired trace element to be followed through regions of different bulk composition without the problems that a depth-sensitive calibration would introduce.

ACKNOWLEDGMENTS

The following persons at Atom Sciences have made significant contributions to the development of SIRIS and the author greatly appreciates their collaboration: D. W. Beekman, W. M. Fairbank, Jr., L. J. Moore, R. Sangsingkeow, H. W. Schmitt, M. T. Spaar, E. H. Taylor, and N. Thonnard. Part of the work presented here was supported by Air Force Wright Aeronautical Laboratories, Aeronautical Systems Division (AFSC), United States Air Force, Wright-Patterson AFB, Ohio 45433 and Electronic Systems Division, Air Force Systems Command USAF, Hanscom AFB, MA 01731.

REFERENCES

1. G. S. Hurst, J. P. Judish, M. H. Nayfeh, J. E. Parks, M. G. Payne, and E. B. Wagner, *Proceedings of the Third Conference on Applications of Small Accelerators, Volume I, The Use of Small Accelerators in Research and Teaching.* United States Energy Research and Development Administration Report CONF-741040-P1, 97 (October 21–23, 1974).

2. G. S. Hurst, J. E. Parks, and H. W. Schmitt, U.S. Patent 4,442,354 (April 10, 1984).

3. J. E. Parks, H. W. Schmitt, G. S. Hurst, and W. M. Fairbank, Jr., *Thin Solid Films* **108,** 69 (1983).

4. J. E. Parks, H. W. Schmitt, G. S. Hurst, and W. M. Fairbank, Jr., in *Laser-Based Ultrasensitive Spectroscopy and Detection V,* edited by Richard A. Keller. Proc. SPIE 426, 32 (1983).

5. J. E. Parks, H. W. Schmitt, G. S. Hurst, and W. M. Fairbank, Jr., in *Resonance Ionization Spectroscopy 1984, Invited Papers from the Second International Symposium on Resonance Ionization Spectroscopy and Its Applications,* (Conference Series Number 71), edited by G. S. Hurst and M. G. Payne. The Institute of Physics, Bristol and Boston, 1984, p. 167.

6. J. E. Parks, D. W. Beekman, H. W. Schmitt, and E. H. Taylor, *Nucl. Instrum. Methods Phys. Res.* **B10,** 280 (1985).

7. J. E. Parks, D. W. Beekman, H. W. Schmitt, and M. T. Spaar, in *Applied Materials Characterization, Materials Research Society Symposia Proceedings,* Vol. 48, edited by W. Katz and P. Williams. Materials Research Society, Pittsburgh, Pennsylvania, 1985, p. 309.

8. J. E. Parks, *Optics News* **12,** 22 (October 1986).

9. J. E. Parks, D. W. Beekman, L. J. Moore, H. W. Schmitt, M. T. Spaar, E. H. Taylor, J.M.R. Hutchinson, and W. M. Fairbank, Jr., in *Resonance Ionization Spectroscopy 1986, Proceedings of the Third International Symposium on Resonance Ionization Spectroscopy and Its Applications held at the University College of Swansea, Wales, on 7–12 September 1986* (Conference Series Number 84), edited by G. S. Hurst and C. Grey Morgan. The Institute of Physics, Bristol, 1987, p. 157.

10. L. J. Moore, J. E. Parks, E. H. Taylor, D. W. Beekman, and M. T. Spaar, in *Resonance Ionization Spectroscopy 1986, Proceedings of the Third International Symposium on Reso-

nance Ionization Spectroscopy and Its Applications held at the University College of Swansea, Wales, on 7–12 September 1986 (Conference Series Number 84), edited by G. S. Hurst and C. Grey Morgan. The Institute of Physics, Bristol, 1987, p. 239.

11. G. S. Hurst, M. G. Payne, S. D. Kramer, and J. P. Young, *Rev. Mod. Phys.* **51**, 767 (1979).

12. J.M.R. Hutchinson, K.G.W. Inn, J. E. Parks, D. W. Beekman, M. T. Spaar, and W. M. Fairbank, Jr., *Nucl. Instrum. Methods* **B26**, 578 (1987).

13. N. Winograd, J. P. Baxter, and F. M. Kimock, *Chem. Phys. Lett.* **88**, 581 (1982).

14. F. M. Kimock, J. P. Baxter, D. L. Pappas, P. H. Kobrin, and N. Winograd, *Anal. Chem.* **56**, 2782 (1984).

15. F. M. Kimock, J. P. Baxter, and N. Winograd, *Surf. Sci.* **124**, L41 (1983).

16. J. P. Baxter, G. A. Schick, J. Singh, P. H. Kobrin, and N. Winograd, *J. Vac. Sci. Technol.* **A4**, 1218 (1986).

17. P. H. Kobrin, G. A. Schick, J. P. Baxter, and N. Winograd, *Rev. Sci. Instrum.* **57**, 1354 (1986).

18. M. J. Pellin, C. E. Young, W. F. Calaway, and D. M. Gruen, *Surf. Sci.* **144**, 619 (1984).

19. D. M. Gruen, M. J. Pellin, C. E. Young, and W. F. Calaway, *J. Vac. Sci. Technol.* **A4**, 1779 (1985).

20. M. J. Pellin, C. E. Young, W. F. Calaway, and D. M. Gruen, *Nucl. Instrum. Methods* **B13**, 653 (1986).

21. C. E. Young, M. J. Pellin, W. F. Calaway, B. Jorgensen, E. L. Schweitzer, and D. M. Gruen, in *Resonance Ionization Spectroscopy 1986, Proceedings of the Third International Symposium on Resonance Ionization Spectroscopy and Its Applications held at the University College of Swansea, Wales, on 7–12 September 1986* (Conference Series Number 84), edited by G. S. Hurst and C. Grey Morgan. The Institute of Physics, Bristol, 1987, p. 163.

22. M. J. Pellin, C. E. Young, W. F. Calaway, J. W. Burnett, B. Jørgensen, E. L. Schweitzer, and D. M. Gruen, *Nucl. Instrum. Methods* **B18**, 446 (1987).

23. A. Benninghoven, C. A. Evans, Jr., R. A. Powell, R. Shimizu, and H. A. Storms, eds., *Secondary Ion Mass Spectrometry, SIMS II, Proceedings of the Second International Conference on Secondary Ion Mass Spectrometry*, Vol. 9. Springer-Verlag, New York, 1979.

24. A. Benninghoven, J. Giber, J. Laszlo, M. Riedel, and H. W. Werner, eds., *Secondary Ion Mass Spectrometry, SIMS III, Proceedings of the Third International Conference*, Vol. 19. Springer-Verlag, New York, 1982.

25. A. Benninghoven, J. Okano, R. Shimizu, and H. W. Werner, eds., *Secondary Ion Mass Spectrometry, SIMS IV, Proceedings of the Fourth International Conference*, Vol. 36. Springer-Verlag, New York, 1984.

26. R. E. Honig, in *Secondary Ion Mass Spectrometry, SIMS V, Proceedings of the Fifth International Conference*, edited by A. Benninghoven, R. J. Colton, D. S. Simons, and H. W. Werner. Springer-Verlag, New York, 1986, p. 2.

27. M. W. Thompson, *Phys. Lett.* **69**, 335 (1981).

28. M. W. Thompson, *Phil. Mag.* **18**, 377 (1968).

29. J. Lindhard, M. Scharff, and H. E. Schiott, *Mat. Fys. Medd. Dan. Vid. Selsk.* **33**(14) (1963).

30. F. R. Shepherd, W. Vandervorst, W. M. Lau, W. H. Robinson, and A. J. Spring-Thorpe, in *Secondary Ion Mass Spectrometry, SIMS V, Proceedings of the Fifth International Conference*, edited by A. Benninghoven, R. J. Colton, D. S. Simons, and H. W. Werner. Springer, New York, 1986, p. 350.

31. W. K. Burns, P. H. Klein, E. J. West, and L. E. Plew, *J. Appl. Phys.* **50**, 6175 (1979).

32. R. J. Holmes, Diffusion of Titanium into Lithium Niobate. Ph.D. Thesis, Lehigh University, Bethlehem, Pennsylvania, 1982.

33. R. J. Holmes, *J. Appl. Phys.* **55,** 3531 (1984).

34. C. Veillon, S. A. Lewis, K. Y. Patterson, W. R. Wolf, J. M. Harnly, J. Versieck, L. Vanballengerghe, R. Cornelis, and T. C. O'Haver, *Anal. Chem.* **57,** 2106 (1985).

35. J. E. Parks, E. H. Taylor, D. W. Beekman, and M. T. Spaar, U.S. Nuclear Regulatory Commission Report NUREG/CR-4419 (January 1986).

36. W. H. Christie and D. E. Goeringer, *Anal. Chem.* **60,** 345 (1988).

37. J.M.R. Hutchinson, K.G.W. Inn, J. E. Parks, D. W. Beekman, M. T. Spaar, and W. M. Fairbank, Jr., *Nucl. Instrum. Methods* **B26,** 578 (1987).

38. W. M. Fairbank, Jr., W. F. Perger, and E. Riis, G. S. Hurst, and J. E. Parks, in *Laser Spectroscopy IV, Proceedings of the Seventh International Conference, Hawaii, June 24–28, 1985,* edited by T. W. Hansch and Y. R. Shen. Springer-Verlag, New York, 1985, p. 53.

39. N. Shimizu and S. R. Hart, *J. Appl. Phys.* **53,** 1303 (1982).

3

Laser Desorption/Laser Ablation with Detection by Resonance Ionization Mass Spectrometry

NICHOLAS S. NOGAR and RON C. ESTLER

Resonance ionization mass spectrometry (RIMS) is becoming an accepted and widely used tool for chemical analysis. This technique combines multistep photoionization with mass spectral sorting and detection of the resultant ions. High spectral brightness pulsed lasers are typically used to effect ionization. When used in conjunction with high detection efficiency mass spectrometers, this combination offers the possibility of unparalleled sensitivity and selectivity.

Initial applications[1-3] often emphasized the use of RIMS to reduce or eliminate isobaric interferences in isotopic or elemental analysis. As a result, the sample preparation method was typically the same as that used for conventional isotope-ratio mass spectrometry (i.e., Langmuir evaporation[4]) from a metal sample filament or ribbon. These sources have a number of significant attributes, including reproducibility, stability, and a substantial literature of work in surface ionization on which to draw. A significant drawback stems from the common use of pulsed lasers for ionization. The short pulses and relatively low repetition rates of most pulsed lasers, coupled with the constant evaporation of sample from thermal sources, result in a low effective duty cycle, and substantial loss of analyte. For a laser beam focused to $\omega = 2.5$ mm (radius), and an atomic velocity $v = 5 \times 10^4$ cm/sec, the rate of sample turnover will be $\approx 10^5$ sec^{-1}. Since most pulsed lasers operate at repetition rates of 10–100 Hz, this yields an effective duty cycle of 10^{-2} to 10^{-3}, and a substantial loss (nonuse) of analyte. This inefficiency can be a substantial burden when the sample is difficult to obtain or when sample size must be minimized, as for radioactive materials.

Several alternatives have been explored to overcome this problem. The use of CW lasers,[5] alternate source geometries,[6] and pulsed thermal sources[7] have all been reported. More recently, a variety of sputter sources have been utilized, including particle bombardment,[8-10] covered elsewhere in this volume, and laser ablation or desorption.

The use of laser ablation or desorption coupled with RIMS detection of the sputtered neutrals has a number of interesting applications and advantages. The duty cycle argument previously mentioned is extremely important for conventional "wet" analytical samples. In addition, the use of laser desorption can allow high spatial resolution (≈ 1 μm) and the evaporation of difficult refractory or tightly bound samples. Last, this method can be used to study the fundamentals of laser–material interactions by diagnosing the evolution and speciation of evaporated material, as well as the kinetic and internal energy distributions of the evaporated atoms, molecules, and fragments.

The remainder of this chapter will be divided into several sections. Initially we will present a theoretical framework with which to evaluate the results of these experiments. Second, we will briefly describe the apparatus typically used for these investigations. Last, we will present the results and analysis of a number of investigations in which we have applied laser desorption/ablation in combination with RIMS diagnostics.

THEORETICAL DEVELOPMENT

For concreteness, we will assume the geometry shown in Fig. 3.1, a configuration commonly used in RIMS experiments. A number of simplifying assumptions will be made in considering the evolution of material from the surface:

- A point source for emission. This is a reasonable assumption, since the incident laser can be focused to $\omega = 0.5$ μm whereas the other characteristic dimensions of the apparatus are $\approx$ cm.

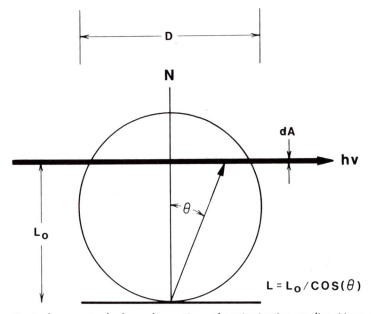

Fig. 3.1. Typical geometry for laser desorption–photoionization studies. Here, L_0 is the distance from the surface to the interrogating laser beam, and D represents the linear distance over which collection of the resultant ions takes place.

- All atoms (molecules) are emitted at the same surface "temperature." This is a reasonable distribution to model, and can provide baseline information from which to interpret experimental results.
- The thermal temperature transient is short compared to the flight time. A two-dimensional calculation of the temperature transient from a metal surface suggests that the duration of the evaporation pulse is ≤ 100 nsec, whereas other characteristic times in the system are $\approx \mu$sec.
- The ionization laser beam has constant diameter in the ionization region. For long confocal parameters, this accurately reflects the actual behavior of the beam. Even for cases in which the beam may be slightly converging/diverging in the interaction region, previous work[11] suggests that the error in calculation will be minimal.
- Collection efficiency for the ion optics is constant over the ionization region. Although there may be small fringing effects, model calculations[12] suggest that this effect will be minimal.

We begin[13-15] by calculating the flux density of atoms (or molecules), $\Gamma(v,\mathbf{R},t)$, with velocities in the range (v to $v + dv$) per unit area per unit time from the surface:

$$\Gamma(v,\mathbf{R},t_0) = v n_0 f(v,t_0) \cos(\theta)\, dv\, d\theta \tag{3.1}$$

where θ is defined as the polar angle with respect to the surface normal (Fig. 3.1), and $n_0 f(v,t_0)$ is the distribution of molecules per unit volume in the appropriate velocity range at the time of evolution from the surface, and we have implicitly assumed an isotropic distribution. Next, it is convenient to introduce the speed distribution in terms of spherical polar coordinates, $dv = v^2 \sin(\theta)\, d\theta\, d\Phi\, dv$,

$$\Gamma(v,\mathbf{R}) = n_0 v^3 (m/2\pi kT)^{3/2}\, e^{-(mv^2/2kT)} \cos(\theta) \sin(\theta)\, d\theta\, d\Phi\, dv \tag{3.2}$$

where we have assumed a thermal (Boltzmann) temperature distribution, $f(v)$. Here, Φ is the azimuthal angle, m is the atomic or molecular mass, k is the Boltzmann constant, and T is the absolute temperature. Next, we substitute for the solid angle element, $d\Omega = \sin(\theta)\, d\theta d\Phi$, to yield

$$\Gamma(v,\mathbf{R}) = n_0 v^3 (m/2\pi kT)^{3/2}\, e^{-(mv^2/2kT)} \cos(\theta)\, d\Omega\, dv \tag{3.3}$$

Since the interrogation laser pulse is extremely short, $\tau \leq 15$ nsec, the molecules are essentially frozen in space for the duration of the pulse. Thus, number density, N, rather than flux is detected,

$$N(v,\mathbf{R}) = \Gamma(v,\mathbf{R})/v = n_0 v^2 (m/2\pi kT)^{3/2}\, e^{-(mv^2/2kT)} \cos(\theta)\, d\Omega\, dv \tag{3.4}$$

Since the laser beam ionizes atoms or molecules at a fixed linear distance from the surface, we transform coordinates, $v = L/t$, and $dv = Ldt/t^2 + dL/t \approx Ldt/t^2$, where t is the time of detection, so that

$$N(L,t,\theta) = n_0 (m/2\pi kT)^{3/2} (L^3/t^4)\, e^{-[m(L/t)^2/2kT]} \cos(\theta)\, d\Omega\, dt \tag{3.5}$$

For the geometry depicted in Fig. 3.1, $L = L_0/\cos(\theta)$. In addition, the substitution $C'(T) = n_0(m/2\pi kT)^{3/2}$ is made, to yield

$$N(L_0,t,\theta) = C'(T)\{L_0^3/[\cos^2(\theta)\, t^4]\}\, e^{-\{m[L_0/t\,\cos(\theta)]^2/2kT\}}\, d\Omega\, dt \tag{3.6}$$

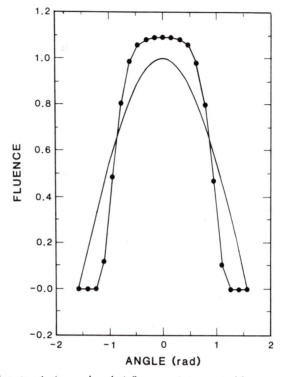

Fig. 3.2. Angular atomic (or molecular) fluences intercepted by a constant radial distance detector [solid line—cos(θ) distribution] and similar calculation for a constant linear distance detector (circles) such as the laser photoionization source.

Last, the detected signal, S, can be related to the number density of evolved atoms or molecules by $S(L_0,t,\theta) = \kappa N(L_0,t,\theta) \, dA/d\Omega$, or

$$S(L_0,t,\theta) = C(T)\{L_0^3/[\cos^2(\theta) \, t^4]\} \, e^{-\{m[L_0/t \, \cos(\theta)]^2/2kT\}} \, dA \, dt \qquad (3.7)$$

where $C(T) = \kappa C'(T)$, and κ is a measure of signal collection efficiency. In general, this parameter will be a function of many variables, including laser intensity and detector efficiency, but is assumed to be independent of L and θ. Here we have also assumed that $dA/d\Omega$ is not a function of θ, where dA is the area of the laser beam. Although this is not rigorously correct, it does not introduce a significant error into the calculations over the range of θ involved in our experiments.

The observed signal calculated from (3.7) will clearly show a dramatically different behavior than the cos(θ) angular dependence normally observed for an isotropic expansion. This is due to the fact that we are detecting at a constant linear distance from the surface, rather than a constant radial distance. The angular dependence for (3.7) can be calculated by $S(L_0,\theta) = \int S(L_0,t,\theta) \, dt$. This function is displayed in Fig. 3.2, along with a plot of cos(θ) for comparison. In performing this (and subsequent) calculations, $L_0 = D = 1$ cm was chosen as a reasonable model of existing apparatuses, with $T = 2000$ K and $M = 100$ au. Calculations were performed both by direct evaluation of (3.7) and by evaluating Monte Carlo (MC) trajectories. We have assumed a laser beam diameter $dA = 1$ mm. The results were in agreement to within the accuracy of the MC calculations.

We are also interested in the temporal evolution of the signal at various angles from the normal. Figure 3.3 shows the temporal distributions $S(L_0,t,\theta = 0)$ and $S(L_0,t,\theta = 45°)$ evaluated for the condition listed earlier. Note that the signal observed normal to the surface is both more sharply peaked, and peaked at a shorter flight time than for the $\theta = 45°$ signal. This is a manifestation of the difference in flight paths: at $\theta = 45°$ the radial distance from the origin to the interrogation zone is longer than at $\theta = 0°$. It thus requires more time for the atoms (or molecules) to reach the interrogation zone, and allows more time for the packet spread out due to velocity dispersion.

This result raises the question of whether a measured signal, integrated over the entire angular distribution, will yield a temperature, and whether this temperature will be characteristic of the actual surface temperature. To explore this question, (3.7) was integrated over the angular range appropriate for the apparatus of Fig. 3.1, $S(L_0,t) = \int S(L_0,t,\theta) \, d\theta$. This result is shown in Fig. 3.4 and is remarkably similar to the thermal distribution normal to the surface (Fig. 3.3), due in part to the strong on-axis weighting, as indicated by Fig. 3.2. This suggests that the measured signal, integrated over the appropriate angular range, can be well represented by a single temperature.

Last, spatial distributions, integrated over the experimental angles, are calcu-

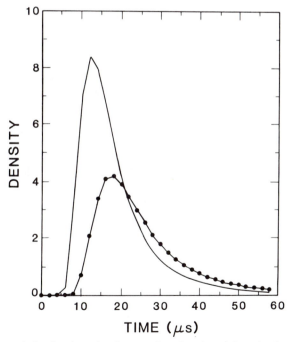

Fig. 3.3. Temporal distributions for the number density of desorbed species at normal incidence (solid line) and 45° off axis (circles).

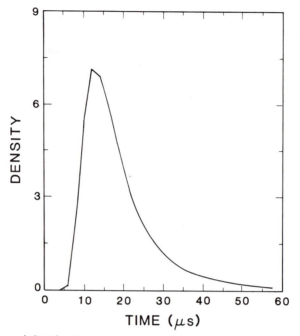

Fig. 3.4. Temporal distribution integrated over the entire range of appropriate angles for the geometry of Fig. 3.1 for comparison with the solid line of Fig. 3.3.

lated for three different delay times. In this case, $dv = Ldt/t^2 + dL/t \approx dL/t$. By analogy with (5)–(7):

$$S(L',t_i,\theta) = C(T)\{L'^2/[\cos^2(\theta)\ t_i^3]\}\ e^{-\{m[L_0/t\ \cos(\theta)]^2/2kT\}}\ dA\ dL \qquad (3.8)$$

where L' and t_i are, respectively, the perpendicular distance from the surface and the time at which the distribution is evaluated. For Fig. 3.5 we have evaluated $S(L',12) = \int S(L',12,\theta)\ d\theta$, and similarly for $S(L',18)$ and $S(L',36)$. Several features should be noted. For maximum sensitivity, the desorbed pulses should be interrogated with the ionization laser as close to the surface, and as soon after impact of the desorption laser, as possible. For realistic laser beam diameters and delay times, it may be possible to achieve an effective duty cycle $\approx 10\%$. This sensitivity is achieved at a cost, however. As larger and larger fractions of the atomic pulse are evaluated, less and less resolution is possible for the temporal distribution (temperature). With this background, we can now discuss the results of several recent experiments performed in this laboratory.

APPARATUS

Mass Spectrometry

Mass spectral studies of laser desorption processes are facilitated by multiplex detection, that is, the ability to obtain a mass spectrum (or selected mass intensities) quickly, and ideally, on a single-scan (or single laser-shot) basis. Ion cyclotron resonance Fourier transform mass spectroscopy (ICR-FTMS) and time-of-flight mass spectroscopy (TOF-MS) are therefore the mass spectral techniques of choice. In addition, Fourier transform time-of-flight (FT-TOF) techniques have recently been investigated and show promise with respect to such experiments.[16]

Regardless of the specific technique involved, high sensitivity is paramount, since desorbed particle concentrations may be extremely low. This consideration alone places severe constraints on the method of ionization for the ablated material as well as ion-optic design.[12] As previously discussed, further constraints may also include the desire to maintain both the spatial and velocity distributions of the nascent ejecta. For such experiments, the ion collection optics must provide a high throughput for high sensitivity, while avoiding distortions to the spatial and velocity distributions. It should also be noted that some dynamic information is more difficult to attain with some of the techniques previously mentioned (e.g., ablation velocity distributions have not yet been measured with ICR-FTMS).

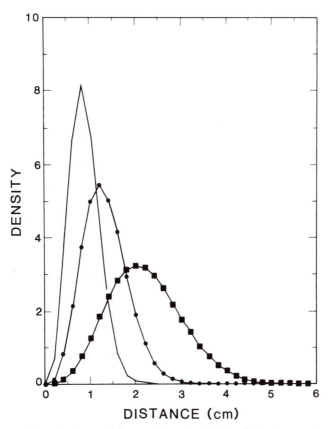

Fig. 3.5. Spatial distributions of the number density for delay times of 12 μsec (solid line), 18 μsec (circles), and 30 μsec (squares).

Laser Hardware

To maximize the dynamic information content of the laser desorption experiment, it is important to both temporally and spatially separate the ablation and ionization events. By the judicious selection of the variable parameters of both the ablation laser and the resonantly tuned ionization laser (energy, wavelength, pulse width, respective orientation) both sensitivity and dynamic information are maximized.

Ideally, the desorption laser has a short pulse duration (≤ 100 nsec). Short desorption laser pulses maximize detection probability by maximizing the spatial overlap of the ablation plume and the ionizing laser field. In addition, the short desorption pulse can effectively be treated as an "impulse" gate function (Delta function) in any time-of-flight analysis, thereby considerably simplifying the ablation velocity distribution analysis.[15] Long duration pulses require extensive deconvolution analysis in which the temporal characteristics (pulse shape) of the desorption laser must be accurately known. This complication becomes increasingly important as the pulse duration approaches an appreciable fraction of the desorbed species flight time between desorption and ionization. Variable wavelength is also a desirable feature, since desorption of adsorbed species and bulk ablation of substrate material are sensitive to incident laser frequency and absorption bands.

Many of the features desired in the desorption laser are appropriate for the ionization laser as well. Broad tunability ensures a wide range of detectable species through various ionization schemes, whereas short pulse durations facilitate velocity distribution analysis and multiphoton ionization. The requirements of both laser systems are not stringent and are easily found in many commercial systems.

Use of the "2 + 1" Ionization Processes

A significant fraction of the published work in RIMS analysis describes the use of conventional thermal sources to vaporize and atomize the sample. Multistep laser photoionization is then followed by mass spectral separation and detection. Although numerous ionization processes have been proposed[17] and demonstrated,[18,19] most analytical work to date has utilized simple 1 + 1 (photons to resonance + photons to ionize) ionization schemes, using the fundamental or frequency-doubled output from a single dye laser. This bias is due in part to two factors: the simplicity and ease of modeling for this ionization process and a commonly held belief that n-photon resonances ($n \geq 2$) are difficult to drive efficiently. The potential advantages derived from the use of n-photon resonances include minimal laser hardware since the fundamental from a single dye laser is normally sufficient to effect ionization, and the potential for Doppler-free excitation.[20] On the other hand, dye lasers of reasonably narrow bandwidth (<0.5 cm^{-1}) and good beam quality are typically needed for efficient n-photon excitation.

To examine the sensitivity of 2 + 1 ionization, we consider details of the two-photon excitation process.[21] In general, a calculation of a two-photon transition rate must be summed over all intermediate states. For the case in which a single intermediate state clearly dominates the resonant enhancement,[22] this summation can be reduced to

$$P_{m_i} = (1/\Gamma)[(3r_0\lambda_{ir}\lambda_{rf})/(\pi\hbar c)]^2(I)^2[(\omega_{ir}\omega_{rf})/(\Delta\omega)^2]$$
$$\times(f_{ir}f_{rf})|\langle C^{J_f,1,m_f,0,J_r,m_r}\rangle\langle C^{J_r,1,m_r,0,J_i,m_i}\rangle|^2 \qquad (3.9)$$

where P_{m_i} is the two-photon transition rate associated with the particular initial state m_i; Γ is the larger of the laser linewidth or the spontaneous transition rate from the final state; λ_{ir}, λ_{rf}, and f_{ir}, f_{rf} are, respectively, the wavelengths and oscillator strengths for transitions from the initial state to the resonantly enhancing state, and from the resonantly enhancing state to the final state; ω_{ir} and ω_{rf} are the angular frequencies associated with λ_{ir} and λ_{rf}; $\Delta\omega$ is the detuning of the laser frequency from ω_{ir}; the $C^{J,1,m,0,J,m}$ are the Clebsch–Gordon coefficients for the excitation of the appropriate transitions with linearly polarized light; r_0, π, $\hbar$, and c have their usual meanings; and I is the intensity of the laser beam. In our experiments, the laser bandwidth is normally large enough that all magnetic or hyperfine (m_j or l_j) components are excited simultaneously. Hence the total rate of transfer for the initial state to the final state is the result of the sum

$$P_{if} = (1/\Gamma)[(3r_0\lambda_{ir}\lambda_{rf})/(\pi\hbar c)]^2(I)^2[(\omega_{ir}\omega_{rf})/(\Delta\omega)^2]$$
$$\times(f_{ir}f_{rf})(\Sigma_j|\langle C^{J_f,1,m_f,0,J_r,m_r}\rangle\langle C^{J_r,1,m_r,0,J_i,m_i}\rangle|^2) \qquad (3.10)$$

where the sum is taken over the initial m_j states. Once the two-photon excitation rate is determined, an effective two-photon cross section can be defined by $\sigma = P_{if}/I^2$, where σ has the units of cm^4 sec and the intensity I is given in units of photons cm^{-2} sec^{-1}.

For a numerical example, consider lead.[23] In this case, the two-photon excitation takes place from the $(6p^2)$ $^3P_0$ ground state to the $(6p7p)$ $^3P_0$ state at 44401 cm^{-1}, with the $(6p7s)$ $^3P_1$ state at 35287 cm^{-1} as the enhancing state. Using published oscillator strengths, calculated Clebsch–Gordon coefficients, and a laser bandwidth of 7.5×10^9 Hz, we calculate a two-photon cross section of 1×10^{-44} cm^4 sec. Note that for typical pulsed-laser parameters (5 mJ, 10 nsec, 10 GHz, 1-mm-diameter spot) the two-photon excitation will be saturated, $P_{if} \approx 1.8 \times 10^8 > 1/\tau_f$, $1/\tau_{laser}$ where τ_f is the fluorescence lifetime, and τ_{laser} is the laser pulse duration. Figure 3.6 shows the result of a rate equations calculation[11] for the energy dependence in this system, assuming a constant beam diameter. Three distinct regions are evident: a second-order dependence at low pulse energies prior to the saturation of the two-photon transition, a linear dependence after saturation due to the ionization step, and finally, a turnover at the highest energies due to the saturation of the ionization step.

Los Alamos Apparatus

The experiments that will be described took place in the source region of a time-of-flight (TOF) mass spectrometer.[24,25] Figure 3.7a is a schematic of this region, while Fig. 3.7b describes the typical timing sequence used. Briefly, the Q-switch synch-out from the Nd:YAG laser (Quanta Ray/Spectra Physics Model DCR 1A) was used to master the timing sequence. This laser was used as the ablation/desorption source and operated at either the fundamental frequency or one of the harmonics, 3ω or 4ω. For most studies, the laser was equipped with beam-filling optics and produced a near-Gaussian profile. The output pulse was ≈ 10 nsec in duration and smooth on the time scale of our detection electronics (≈ 2 nsec). The laser output was focused to varying spot sizes on the ablation samples with several different lenses. The residual fundamental and/or other harmonics were removed from the ablation beam by using dichroic mirrors.

For some experiments, the samples were prepared on wire ribbons (Ta, Re, or W) as slurries, and were suspended within the source region after drying. In other experiments, optical substrates were cemented to a 15-cm stainless-steel rod for mounting with the source region. The targets could be translated and rotated with a high-vacuum feedthrough (Varian Model 1371). The rotation axis of the substrate was parallel but not coincident with the ablation laser beam axis. A displacement of 5 mm between these two axes permitted the interrogation of different substrate sites through substrate rotation. Typical pressures within the source for all experiments were maintained at 10^{-7} torr by an L-N$_2$ trapped oil diffusion pump.

Pulses from an excimer-pumped dye laser (Lambda Physik Model 101/2002), propagating perpendicular to the Nd:YAG laser at approximately 1.6 cm below filament sources and 2.9 cm below optical substrate surfaces, were used to interrogate the ablated neutral species. Ions produced directly by the ablation laser were strongly discriminated against by the ion optics.[12] Ionization of the ablated neutrals

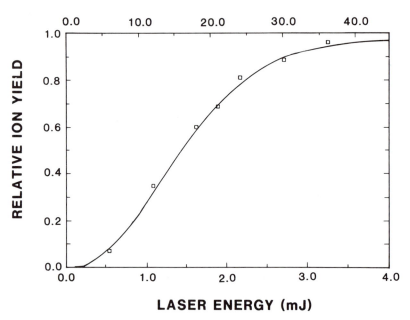

Fig. 3.6. Pulse-energy dependence of the lead "2 + 1" photoionization process, utilizing a two-photon excitation to the $^3P_0$ state at 44,401 cm^{-1}.

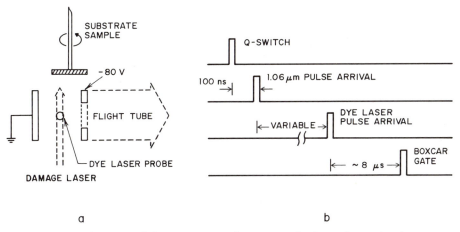

Fig. 3.7.   (a) Schematic of the experimental apparatus for laser desorption/resonance ionization experiments. (b) Timing sequence used in these experiments. The signal is captured either with a gated integrator or transient recorder.

was effected by a variety of multiphoton transition sequences, each of which is described for the particular application. Dye-laser pulses (≈ 2 mJ, 10 nsec) produced at a variable delay relative to the ablation laser were spatially filtered and loosely focused through the ionization region. The extraction field was approximately 110 V/cm, perpendicular to the paths of both laser beams. This was followed by a drift tube at -1000 V; the flight path was 0.4 m. A pair of deflection plates between the extractor and the flight tube maximized the transmission of ions to the detector and minimized any transmission variation that was due to ion velocity components perpendicular to the flight tube. Detection electronics consisted of a channel electron multiplier, a preamplifier, and a boxcar integrator (PAR 162/164) whose gate delay was set to the flight time of the ion of interest. The boxcar width was adjusted so as to encompass only a few mass units about that of the detected species. For studies involving transient-species ablation, the ion signal was processed with a high-speed waveform recorder (Tektronix 2430).

EXPERIMENTAL RESULTS

Analytical Applications

As previously outlined, there is a significant drawback in using pulsed lasers for RIMS analyses. The short pulses and relatively low repetition rates of most pulsed lasers, coupled with the constant evaporation of sample from thermal sources, result in a low (typical $\leq 10^{-3}$) effective duty cycle. The alternative explored within this article, pulsed sample evaporation via laser ablation, has been explored in several laboratories.[26-28]

    In an initial experiment at Los Alamos, infrared pulsed desorption from a metal ribbon was investigated.[29] In these experiments, a tantalum filament (0.60×0.075

$\times$ 0.0025 cm) was maintained by resistive heating at a base temperature slightly below that required to produce a detectable resonance ionization signal without laser desorption ($\approx$1200°C). Pulses of 1.06-μm laser radiation were focussed to 0.12 cm diameter at the filament, producing intensities $\leq$ 3 $\times$ 10^8 W cm^{-2} (35 mJ, 10 nsec). Ground-state desorbed Ta atoms were detected via a two-photon transition to the $^4F_{3/2}$ (43,964 cm^{-1}) excited state[19] followed by ionization using a third photon of the same color (2 + 1).

The pulsed desorption observed in these experiments appears to be a quasi-thermal rather than a plasma-driven process. Initial evidence of this fact is provided by the absence of primary ions produced solely by laser impact as opposed to the presence of such ions in the well-known laser microprobe.[30] In addition, the estimated intensity used for desorption in these experiments, 10^8 W/cm^2, falls within the range observed for thermal desorption in previous experiments. Primary ions could be observed at somewhat higher intensities, $\approx$5 $\times$ 10^8 W/cm^2, by pulsing the extraction field on several microseconds after the arrival of the infrared laser pulse.

Velocity distributions were recorded by measuring the dependence of the detected Ta$^+$ signal on the time delay between the desorption and interrogation pulses. These spectra are fitted to a shifted Maxwellian velocity distribution using the mass flow velocity (hydrodynamic, center of mass) and thermal velocity (kinetic, thermal width) as parameters in the fit. For the Ta desorption, the hydrodynamic and thermal velocities are far from equilibrium, the former corresponding to a temperature (defined by $v^2\pi m/8k$) of $\approx$8500 K, the latter to a temperature of $\approx$450 K. At varying desorption pulse energies, the general trend observed is one of decreasing v_{cm} and increasing v_{therm} with decreasing 1.06-μm intensities.

Although it is not yet possible to unequivocally determine the cause of the nonthermal velocity distributions, several alternatives appear possible: nonequilibrium interactions on the surface, collisional interactions subsequent to desorption, and potential energy barriers to tantalum atom desorption. Further study is needed to determine which mechanism(s) are responsible for the observed velocity spreads in these experiments. Such studies are hindered, however, by the difficulty of optimizing the desorption intensity so as to prevent primary ion formation while ensuring the detection of laser-desorbed neutrals (rather than those thermally evaporated due to the resistive heating).

The net analytical result of this nonthermal velocity distribution is that the atom pulse is temporally narrow, so that the effective sampling duty cycle is greatly increased relative to continuous thermal desorption. A rough calculation of the total ionization probability for the desorption/RIMS detection process yields $\approx$2 $\times$ 10^{-4}, with a geometric overlap of 5 $\times$ 10^{-2}, a temporal overlap of 10^{-1}, a partition function of 0.5, and an ionization/detection probability for this three-photon process of 10^{-1}. This estimate in turn suggests that the total number of atoms removed from the surface is $\approx$10^5 per pulse. For sampling small deposits, this allows both high sensitivity and reasonable statistical precision through repetitive sampling.

For small amounts of analyte material in particular, the higher duty factor in laser desorption experiments is desirable. Initial experiments on both calcium and magnesium show that materials applied to substrates are easily desorbed and detected via 2 + 1 ionization schemes. For these experiments, 1 μg of material was placed on tungsten or rhenium filaments in the form of an aqueous solution and allowed to dry. After placement into the apparatus, the analytes were desorbed with 1.06-μm radiation, and ionized via multiphoton ionization. For calcium and magnesium, the ionization occurred via a 2 + 1 ionization process. In all cases, the analytes were removed from the substrates with a few pulses of the 1.06-μm laser. The difficulties encountered here in maintaining a stable analyte signal can be easily overcome through the use of the transient detection.

Interesting results were obtained for the analysis of Al/Fe mixtures. There is an accidental coincidence between a 1 + 1 ionization process for aluminum, using the $(4s)\,^2S_{1/2} \leftarrow (3p)\,^2P_{1/2}$ transition (394.4 nm), and a variety of 2 + 1 ionization processes in iron. As can be seen in Fig. 3.8, the 1 + 1 ionization in aluminum is power broadened, whereas the 2 + 1 features in iron remain sharp. This allows the opportunity to "tune" the ratio of ionization efficiencies, thus greatly increasing the potential dynamic range of the measurement. In combined laser desorption/RIMS experiments, we were able to detect Al in an iron sample at the 1 ppt (thousand)

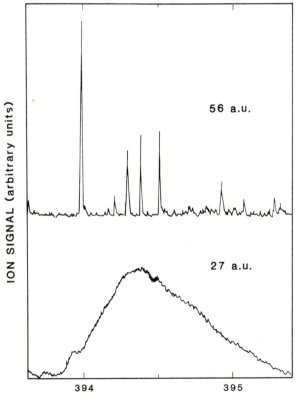

Fig. 3.8. Wavelength dependence of the Al "1 + 1" ionization process, utilizing the $(4s)^2S_{1/2} \leftarrow (3p)^2P_{1/2}$ transition (394.4 nm), and a variety of "2 + 1" ionization processes in iron.

level by tuning the laser off any iron resonances, and on to the peak in the aluminum RIMS spectrum. Conversely, Fe was easily detected in aluminum at the 10 ppm level by tuning to a sharp iron feature in the tail of the aluminum spectrum.

Optical Damage Studies

Pulsed lasers have shown great promise in material-processing applications including the ablation (etching) of resist materials and photodeposition of semiconductors inherent to the fabrication of integrated circuits. Consequently, over the past few years, numerous mechanistic studies of the laser–solid interaction have appeared in the literature. Few investigations have, however, coupled the sensitive and specific laser-based schemes now available for the detection of desorbed material to the study of the phenomena of optical damage.

The work described here investigates[25] the interaction of relatively high-fluence pulses (≥ 10 J cm^{-2}) with a (nominally) transparent medium: calcium fluoride, CaF_2. Calcium fluoride was chosen for initial studies as it is a common dielectric used in excimer laser optics and has ablation products (Ca, CaF) with visible absorptions and low ionization energies. Information concerning the laser–solid interaction was obtained through measurement of the velocity distributions and internal energies of the particles ejected from the surface during breakdown (damage).

Detection of Ca atoms by RIMS involved a 2 + 1 ionization scheme. A two-photon transition from the ground state ($4s^2\ ^1S_0$) to the excited resonance state ($4s4d\ ^1D_2$) is followed by another photon of the same frequency (18,649 cm^{-1}) to effect ionization. The ionization potential of Ca is 6.111 eV. The two-photon transition is saturated, and thus small fluctuations in the laser intensity will have little effect on the signal. This becomes important when the TOF data are considered.

The detection of CaF radicals parallels that of the Ca atom detection. Ionization occurred through a two-photon transition to the $F^2\Pi$ state ($T_0 = 37,548$ cm^{-1}) from the $X^2\Sigma^+$ ground state, followed by the absorption of a third photon of the same color. The ionization energy of CaF is approximately 6.0 eV, very similar to that of Ca atoms.[31] Although other ionization schemes are possible, the transition through the $F^2\Pi$ state proved to be convenient, since both Ca and CaF could be detected by using the same laser dye.

Even though the detection sensitivity for Ca is high, neither Ca atoms nor CaF radicals could be detected before breakdown. In separate experiments, the ratio of neutrals to ions[12] produced during breakdown was determined to be $\geq 100:1$. Again, TOF spectra were recorded for the ejected species and fitted to a shifted Maxwellian velocity distribution.

For the Ca TOF signal recorded at 1.06 μm and high fluence (25 J/cm^2), the data are adequately fit using a purely thermal distribution (mass flow velocity of zero) characterized by a kinetic temperature of 834 $\pm$ 30 K. At an ablation wavelength of 266 nm and the same fluence, the data are significantly different.[25] In particular, the data cannot be fitted by a single thermal distribution or by a single shifted Maxwellian distribution with a nonzero mass flow velocity. The best fit obtained corresponded to a model incorporating the sum of two thermal distributions: one fast distribution and one slow distribution, where the temperature and the mixing coefficients were varied to produce the best fit. Here, the best fit was obtained by a fast component characterized by a kinetic temperature of 4970 ± 90 K and the slow component by a temperature of 830 ± 90 K. The parameter fit also indicates that 33% of the Ca atoms are accounted for by the fast component: the remaining 67% are contained in the slow component. The distributions are a function of fluence: the fast-component percentage increases with fluence. Spectra recorded at 355 nm produced results similar to those of the 266-nm data.

TOF signals were also recorded for CaF. The results obtained for this radical, however, are less reliable due to the weaker signals (sensitivity loss due to oscillator strength, partition function, etc.). The spectra here also are best fit with a two-velocity distribution model having fast and slow components characterized by temperatures of 8250 and 1940 K, respectively. In this case, the fast component contributes 44% to the overall distribution; the slow component contributes 56%.

In addition to the TOF spectra, laser excitation spectra of the multiphoton ionization spectra were recorded for the CaF radicals to investigate the partitioning of energy among its various degrees of freedom. Although direct inversion of spectral intensities to state populations is not currently possible (as in much laser-induced fluorescence work) because of the lack of knowledge of the two-line strengths and spectroscopic constants of the intermediate $F^2\Pi$ state, general trends are observed. Through computer simulations of the vibrational and rotational structure using the best available spectroscopic constants, it is clear that the internal excitation of the CaF radical decreases with decreasing wavelength. Internal rotational temperatures appear to be of the same order of magnitude as that of the kinetic temperatures for the 1.06-μm ablation, whereas they appear to be far below both the fast- and the slow-component kinetic temperatures of the shorter wavelength ablations.

Since the early work of Yablonovitch[32] showing a close correlation between laser-induced breakdown thresholds in the infrared and dc breakdown thresholds for alkali-halide crystals, avalanche ionization has often been indicated as the mechanism responsible for laser-induced damage. However, the results summarized here are not consistent with this model. Other models introduced in the consideration of laser sputtering seem to be more applicable, and some of these mechanisms can lead to thermal velocity distributions.[33] The results summarized here do not point to one single mechanism in the initiation of the breakdown event but are significant from the standpoint that a change in mechanism (i.e., an opening of a second channel) is indicated. The apparent bimodal velocity distribution coupled with the disparity between the kinetic- and internal-energy estimates observed for the ejected particles at shorter wavelengths indicate a direct photophysical interaction. Such an interaction might include rapid-energy-deposition models, in which excitation of individual atoms is followed by nonadiabatic transitions of the atoms to antibonding states.[34] Such a model could also account for the wavelength threshold of this second channel, even though the multiphoton ($n = 2$) absorption coefficient for CaF_2 at 266 nm is known to be small.[35]

In subsequent studies of optical damage, commercial Sc_2O_3/SiO_2 multilayer coated 7940 optics were studied using RIMS hoping to characterize the dynamics of macroscopic damage events.[36] It is laser-induced damage lore that the presence of near-surface contaminants and defects correlate with damage initiation. Direct evidence for this correlation is rare.

A variety of species were observed in the low-fluence ($1-3$ J/cm^2) 1.06-μm irradiation of these optics. Several transient species were observed by RIMS at a single site prior to any macroscopic damage detected by the naked eye. Figure 3.9 shows a typical TOF mass spectra, with the lower horizontal axis indicating flight time, and the upper corresponding to mass. The lower trace is the (null) mass spectrum observed when the YAG laser fires, but no photons impinge on the sample. This display is the result of a four-shot average, and displays the mass range of $\approx 4-250$ au. The deflection immediately following $t = 0$ is the result of rfi due to firing the ionization (excimer-pumped dye) laser. The time between firing the desorption and excimer lasers (≈ 14 μsec) was adjusted to address particles having a velocity normal to the surface of $\approx 7 \times 10^4$ cm/sec. This time delay was found empirically to maximize the observed signal.

The middle trace (b) of Fig. 3.9 is typical of the mass spectra observed when the desorption laser is just sufficient (100 mJ/cm^2) to produce any detectable signal. In this case, the only observable signal (equivalent to one or two detected ions per laser shot) occurs at mass 56. Since the ionization laser is tuned to $^5G_5$ (50,703.9 cm^{-1}) $\leftarrow$ $^5G_4$ (0.000 cm^{-1}) two-photon transition in iron (394.4 nm), and since the observed signal is sharply peaked at this wavelength, both the mass and the optical spectral signatures suggest that iron is being desorbed from the surface. It is important to note that for a particular location on the sample, this signal could only be

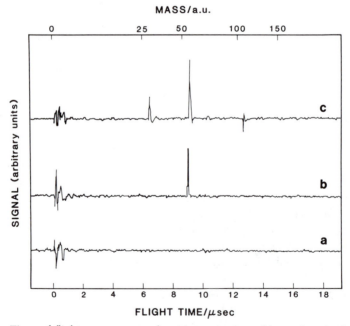

Fig. 3.9. Time-of-flight mass spectra from interrogation of laser-desorbed atoms and molecular fragments from a commercial coated optic: (a) the null signal in the absence of laser heating of the surface; (b) the transient iron signal due to gentle irradiation (≤ 100 mJ/pulse); (c) additional desorbed material removed at high fluence irradiation (≈ 1 J/pulse).

observed for two to six shots at a given fluence of the desorption laser. To observe subsequent signal, either the location of irradiation had to be changed or the fluence raised. The uppermost trace, (c) of Fig. 3.9 shows the signal generated at signficantly higher fluence (6.4 J/cm^2). In addition to the iron signal observed in trace (b), the signal is now also present at 28 au, corresponding to silicon, and ≈ 106 au, corresponding to Sc_2O. Note that although the 28 and 106 au signals are smaller than that observed for 56 au, this does not necessarily indicate that more iron is being removed from the surface than silicon or scandium. Since the ionization laser is tuned to the iron resonance, its ionization efficiency is expected to be much greater than for the nonresonant silicon- or scandium-containing fragments.

These results can be profitably compared to Nomarski microscope photographs of the samples. The appearance of a number of 10-μm-diameter round pits within the footprint of irradiation, with the concomitant detection of iron emission from the surface, strongly suggests a correlation. The evidence suggests that the observed signals are due to the presence of small, iron-containing microinclusions in the optical coatings. Low fluence irradiation removes near-surface contaminants with minimal damage, whereas higher fluence irradiation may remove more deeply imbedded, or lower susceptibility, contaminants, with concomitant removal of surrounding coating material. This interpretation is consistent with a number of previous observations[37,38] including a laser-annealing effect. That is, a gradual increase in the fluence with which a sample is irradiated results in a higher measured damage threshold than for immediate irradiation at high fluences. This can be rationalized by low fluence removal of included impurities below the threshold for macroscopic damage. In addition, the association, in the experiments described here, of the circular pits with the macroscopic damage sites suggests that at high fluences, absorption by impurity inclusions can result in a high local temperature or field to cause macroscopic damage.

CONCLUDING REMARKS

RIMS is a powerful diagnostic tool for elemental and transient analysis in studies of laser desorption and laser ablation. The multistep laser photoionization process, when coupled with conventional mass analysis, can provide exceptional detectivity, dynamic range, and discrimination against interfering species. These properties can be used to great advantage both in the analysis of conventional materials and in the interrogation of interfacial phenomena.

In the examination of routine analytical samples, the effective duty cycle can be greatly improved relative to continuous sample evaporation. In addition, these methods allow direct analysis without extensive sample preparation.

For the study of mechanisms of laser–material interactions, RIMS detection provides a means to determine the internal and velocity distributions of the desorbed species. Even though it provides good spatial and temporal resolution, its application is somewhat limited. For resonant ionization, the identity of the ablation species must be known or guessed beforehand or the less efficient nonresonant ionization schemes must be used at high laser intensities.

It is anticipated that RIMS will play an increasingly important role in the study of the laser desorption processes, particularly as surface imaging capabilities become coupled to the ionization detection.

REFERENCES

1. C. M. Miller, N. S. Nogar, A. J. Gancarz, and W. R. Shields, *Anal. Chem.* **54**, 2377 (1982).

2. D. L. Donahue, J. P. Young, and D. H. Smith, *Int. J. Mass Spectrom. Ion Phys.* **43**, 293 (1982).

3. J. D. Fassett, J. C. Travis, L. J. Moore, and F. E. Lytle, *Anal. Chem.* **55**, 765 (1983).

4. J. D. Fassett, L. J. Moore, J. C. Travis, and F. E. Lytle, *Int. J. Mass Spectrom. Ion Process.* **54**, 201 (1983).

5. C. M. Miller and N. S. Nogar, *Anal Chem.* **55**, 1606 (1983).

6. S. V. Andreev, V. I. Mishin, and V. S. Letokhov, *Opt. Commun.* **57**, 317 (1986).

7. J. D. Fassett, L. J. Moore, R. W. Shideler, and J. C. Travis, *Anal. Chem.* **56**, 204 (1984).

8. F. M. Kimock, J. P. Baxter, D. L. Pappas, P. H. Kobrin, and N. Winograd, *Anal. Chem.* **56**, 2782 (1984).

9. C. H. Becker and K. T. Gillen, *Appl. Phys. Lett.* **45**, 1063 (1984).

10. M. J. Pellin, C. E. Young, W. F. Calaway, and D. M. Gruen, *Surf. Sci.* **144**, 619 (1984).

11. C. M. Miller and N. S. Nogar, *Anal. Chem.* **55**, 481 (1983).

12. The authors would like to thank D. A. Dahl and J. E. Delmore, EG&G Idaho, for providing an IBM-compatible copy of SIMION, an ion optics program that we have used to estimate and optimize transmission efficiencies in our apparatus.

13. R. S. Berry, S. A. Rice, and J. Ross, *Physical Chemistry*, Chap. 28. Wiley, New York, 1980.

14. M. J. Pellin, R. B. Wright, and D. M. Gruen, *J. Chem. Phys.* **74**, 6448 (1981).

15. R. A. Olstad and D. R. Olander, *J. Appl. Phys.* **46**, 1499 (1975).

16. F. J. Knorr, M. Ajami, and D. A. Chatfield, *Anal. Chem.* **54**, 690 (1986).

17. G. S. Hurst, M. G. Payne, S. D. Kramer, and J. P. Young, *Rev. Mod. Phys.* **51**, 767 (1979).

18. J. P. Young, D. L. Donahue, and D. H. Smith, *Int. J. Mass Spectrom. Ion Process.* **56**, 307 (1984).

19. N. S. Nogar, S. W. Downey, and C. M. Miller, *Anal. Chem.* **57**, 1144 (1985).

20. T. B. Lucatorto, C. W. Clark, and L. J. Moore, *Opt. Commun.* **48**, 406 (1984).

21. E. C. Apel, J. E. Anderson, R. C. Estler, N. S. Nogar, and C. M. Miller, *Appl. Opt.* **26**, 1045 (1987).

22. G. Grynberg and B. Cagnac, *Rep. Prog. Phys.* **40**, 791 (1977).

23. B. L. Fearey, M. W. Rowe, J. E. Anderson, C. M. Miller, and N. S. Nogar, *Anal. Chem.,* **60**, 1786 (1988).

24. S. W. Downey, N. S. Nogar, and C. M. Miller, *Int. J. Mass Spectom. Ion Process.* **61**, 337 (1984).

25. R. C. Estler, E. C. Apel, and N. S. Nogar, *J. Opt. Soc. Am. B* **4**, 281 (1987).

26. D. W. Beekman, T. A. Callcott, S. D. Kramer, E. T. Arakawa, G. S. Hurst, and E. Nussbaum, *Int. J. Mass Spectom. Ion Phys.* **34**, 89 (1980).

27. S. Mayo, T. B. Lucatorto, and G. G. Luther, *Anal. Chem.* **54**, 553 (1982).

28. M. W. Williams, D. W. Beekman, J. B. Swan, and E. T. Arakawa, *Anal. Chem.* **56**, 1348 (1984).

29. N. S. Nogar, R. C. Estler, and C. M. Miller, *Anal. Chem.* **57**, 2441 (1985).

30. R. J. Cotter, *Anal. Chem.* **56,** 485A (1984).

31. D. L. Hildenbrand and E. Murad, *J. Chem. Phys.* **43,** 1400 (1965).

32. E. Yablonovitch, *Appl. Phys. Lett.* **19,** 495 (1971).

33. N. G. Stoffel, R. Riedel, E. Colavits, G. Margaritondo, R. F. Haglund, E. Taglaner, and N. H. Tolk, *Phys. Rev. B* **32,** 6805 (1985).

34. B. Jost, B. Schueler, and F. R. Krueger, *Z. Naturforsch. Teil A* **37,** 18 (1982).

35. P. Liu, W. Lee Smith, H. Lotem, J. H. Bechtel, N. Bloembergen, and R. S. Adhav, *Phys. Rev. B* **17,** 4620 (1978).

36. R. C. Estler and N. S. Nogar, *Appl. Phys. Lett.,* **52,** 2205 (1988).

37. M. E. Frink, J. W. Arenberg, D. W. Mordaunt, S. C. Seitel, M. T. Babb, and E. A. Teppo, *Appl. Phys. Lett.* **51,** 415 (1987).

38. J. B. Franck, and M. J. Soileau, *Proc. 15th Ann. Symp. Laser Induced Damage in Opt. Mat.,* 1981. NBS Special Publication 638, p. 114.

4

The Analysis of Inorganic and Organic Surfaces by Ionization of Desorbed Neutrals with Untuned Ultraviolet (UV) and Vacuum Ultraviolet (VUV) Lasers

CHRISTOPHER H. BECKER

This chapter is concerned with surface chemical analysis and material analysis with both lateral and in depth spatial resolution. The approach in this work, as with all of the work described in this book, is based on mass spectrometry. Mass spectrometry has played a very honored role in the history of modern science. It is a tremendously powerful and general tool, still playing a vital role in fields from atomic physics to biochemical analysis to coal pyrolysis to the thermodynamics of inorganic substances. Mass spectrometry, fundamentally, is a vacuum-based technique performed on gaseous ions. When considering material surfaces, it is necessary to address the problem of how to produce gaseous ions from surfaces in a way that is characteristic of that material.

There are a variety of experimental techniques for performing mass characterization after ion production, such as time-of-flight mass separation (to be discussed somewhat further here), the quadrupole mass filter, the magnetic sector mass filter, the Wien filter, and so on. Although it is important to choose the mass analysis so that it is well suited to the form of ionization, the hardware for making the mass analysis is rather secondary to the fundamental question of how to make these gaseous ions from the material.

There is a significant history of mass spectrometry applied to surface analysis. For example, some 30 years ago Honig investigated the emission of ions from a surface as a result of ion beam sputtering.[1] This indeed was an early example of the now popular technique known as secondary ion mass spectrometry (SIMS).[2] "Secondary ions" are those ions that are derived from irradiation of a surface whether by an ion, electron, or laser beam, though historically SIMS is associated only with ion sputtering. The term "secondary" is used in distinction to the "primary" beam, that is, the beam directly irradiating the surface.

Another method of mass spectral surface analysis has been to use an intense pulsed laser irradiating the surface to create ions.[3] This method dates back to the advent of powerful lasers in the early 1960s. Though secondary ions are produced, the technique has several names. We will refer to it here as laser ionization mass spectrometry (LIMS), although it is sometimes called the laser microprobe or known by commercial instrument names. For elemental analysis of materials LIMS frequently causes removal of material to depths of a micron or so, thus not being strictly a surface analytical technique. For the analyses of large organic molecules lower laser powers are often used, resulting in instances of negligible surface damage.

For both the SIMS and the LIMS techniques, the material removal is itself responsible for the ionization. This means that these two processes are intrinsically coupled. A strength of these approaches therefore is a simple experimental implementation, that is, no laser is needed. The drawbacks are that there is little control over the ionization, the probabilities of ionization (secondary ionization coefficients) are difficult to predict and can range over orders of magnitude for given situations, and these probabilities can vary greatly depending on the local chemical environment or matrix. The latter problem is sometimes referred to as "matrix effects." An inherent reason for the difficulty in obtaining predictable and/or uniform ionization probabilities of secondary ions is inherently linked to a fundamental physical observation that usually the great majority of emitted particles are neutral in charge. Relative variations in the great majority of particles are small and thus the yield of neutral particles ("secondary neutrals") is a relatively stable phenomenon in moving from one chemical matrix to another.

Therefore by monitoring the secondary neutral atoms and molecules removed from a surface by stimulating radiation such as an ion or laser beam, a more uniform measure of the surface can be expected and therefore a more readily quantifiable and stable measure of that material surface. This concept is not new. The basic physical observation of the dominance of secondary ion emission was first reported upon by Thompson in 1910.[4] Again, about 30 years ago, Honig was the first to clearly comprehend the advantageous nature of observing secondary neutrals when he performed his first mass spectrometry experiments[1] on both sputtered neutral and ionized atoms and molecules. The ionization of sputtered neutral atoms and molecules or laser desorbed atoms and molecules is often called "postionization" because it is ionization after the removal step.

Although the motivation for postionization of secondary neutrals has been appreciated for some time, this concept has taken a practical form only recently. One of the important characteristics of a postionization technique that bears on practicality is that of efficiency and the resultant potential for high sensitivity. If high efficiency could be obtained for ionization and subsequent mass analysis, then the separation of the material removal and ionization steps would undoubtedly also lead to great flexibility in the choice of characteristics of the irradiating beam for desorption (also called the probe beam) while using a relatively low fluence or gentle surface irradiation to minimize surface damage. Although a method that inherently removes material to perform the analysis can be considered rather destructive, it is worth noting that if sensitivity is high enough, successful analyses can be performed while consuming a miniscule fraction of one atomic layer thereby resulting in negligible surface damage. If material information is needed in depth, then the material removal does of course result in damage over the analysis area.

The first approach to postionization used by Honig[1] and one that is still in use[5] is based on electron bombardment ionization. Although very useful results have been obtained, sensitivity is a distinct limitation because of the low ionization probabilities, typically on the order of 10^{-4}. Another postionization approach is based on a radio frequency (rf) plasma held in low-density rare gas above the surface that ionizes sputtered neutral atoms and molecules.[6] This technique is associated with a modest increase in efficiency for the ionization but rather counterbalanced by losses in instrument transmission. For both of these general ionization techniques, overall efficiencies are often lower by a factor 10^{-5} from the approach to be discussed in detail here. One thing worth noting in particular is that both the electron bombardment and the RF plasma postionization approaches can be considered general ionization techniques in that essentially any atom or molecule can be ionized with roughly equal sensitivity (i.e., generally within a factor of 10, except for a few species with ionization potentials above the plasma energy). Fragmentation of molecules must be considered though.

Another approach to sensitive postionization of secondary neutrals, particularly for atoms or small molecules, is the use of resonantly enhanced multiphoton ionization.[7] In this case sensitivities can be much higher than for the electron bombardment or RF plasma postionization approaches, even to the point of saturation (nearly complete ionization within the laser beam). The resonantly enhanced multiphoton ionization is selective, to varying degrees, which can be advantageous in many circumstances.

The approach taken in this laboratory resembles the general nonselective ionization techniques referred to earlier in that a uniformity of ionization is sought, with the mass spectrometer serving as the means of distinguishing various species. Yet the high sensitivity often found for resonantly enhanced multiphoton ionization is indeed desirable. We have found that it is possible to obtain high sensitivity and general ionization that are appropriate for many analytical situations.[8] As a rule, for inorganic materials in which elemental analysis and a modest amount of

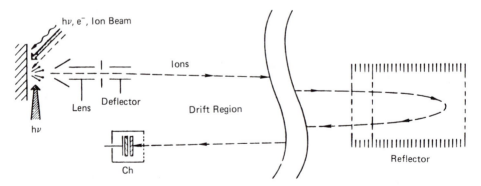

Fig. 4.1. Schematic of the SALI (surface analysis by laser ionization) instrument. Shown are three alternative probe beams (upper left), the ionizing laser beam ($h\nu$, lower left), and the path of the photoions, focused to compensate for transverse velocities, and deflected slightly into the reflecting TOF analyzer leading to a chevron microchannel plate detector (Ch). Republished with permission from Ref. 20.

molecular information are required, we use an intense focused untuned UV pulsed laser for *nonresonant* multiphoton ionization of secondary neutral particles; here rare gas ion sputtering for the desorption of neutrals is usually used under ultrahigh vacuum to maintain cleanliness of the sample surface. For organic compounds and polymeric materials, we use an untuned coherent VUV light source. Thus, the emphasis in our work is based on using a properly chosen *untuned* laser source as a general and straightforward analytical tool, striving for uniform ionization with the mass spectrometer serving the function of chemical distinction. This approach of ionization of secondary neutral particles by untuned laser radiation followed by time-of-flight mass spectrometry is called surface analysis by laser ionization or SALI.[8] A schematic of the SALI experimental arrangement is shown in Fig. 4.1. Various aspects of the experimental procedure will now be discussed in modest detail, followed by a few examples of applications for both inorganic and organic systems.

EXPERIMENTAL TECHNIQUE

Desorption Probes

For inorganic material analysis, our work has concentrated on the use of ion beam sputtering by kilovolt energy rare gas ions, typically argon ions at 5 kV or so. The ion beam may be pulsed, which can result in submonolayer sampling during a measurement. Such submonolayer sampling is often referred to as a "static" analysis. In contrast, with the ion beam on continuously many atomic layers can be removed and the analysis is referred to as a "dynamic" analysis. The sampling depth for ion beam sputtering of a single given event is 1 to 2 atomic layers, which is an isolated event typically lasting $<10^{-12}$ sec. However, with continuous sputtering by the ion beam it is possible to mill into a material and perform analyses as a function of time, or equivalently depth, to obtain what is commonly referred to as a depth profile through a material. Ion beam sputtering of inorganic materials can result in relatively high depth resolution. For example, we have obtained a depth resolution of approximately 25 Å after sputtering to 1000 Å.[9] Such resolution is not uncommon. In general, depth resolution depends on the amount of previously sputtered material, the nature of the material, and the kinetic energy and mass of the bombarding ions. With regard to lateral resolution, which is limited by the spatial extent of the probe beam, ion beams can be focused to less than 1000 Å and lasers can be focused readily to on the order of 1–10 μm. Work on submicron analysis with a liquid metal ion gun[10] is in progress in our laboratory.

Ionization by Laser Radiation

For inorganic analysis we generally have used UV radiation from an excimer laser for ionization though an Nd:YAG-based laser system offers a viable alternative. This is particularly true when the frequency is converted into the ultraviolet by nonlinear crystals. The pulsed powerful laser is focused to pass above the surface so as not to interact directly with the surface, but passes close, typically about 1 mm above the surface. The most commonly employed wavelengths in our laboratory for inorganic analysis have been 248 and 193 nm (equivalent to 5.0 and 6.4 eV/photon, respectively), with a laser pulse width of ~ 10 nsec and a focused power density of $\sim 10^{10}$ W/cm^2. With this power density in the ultraviolet, for species that are ionized nonresonantly by the absorption of two photons, typically the ionization within the focal volume is complete. However, those atoms requiring three or four photons (such as N, O, F, He, Ne, and Ar at 193 nm) typically are not efficiently ionized at such power densities, though a viable alternative commonly used is the larger signals for molecular ions containing N, O, or F atoms. Nevertheless, laser systems with much higher power densities than we have been using are becoming more common. We can expect, for example, that with the chirped-pulse amplification techniques[11] for picosecond lasers, power densities of 10^{13}–10^{14} W/cm^2 will be routine. This will mean that for elemental analysis essentially everything, even helium, can be ionized to completion. One of the outcomes of efficient ionization is uniformity in detection efficiency and a high degree of quantitation for a given species from within one material form (chemical matrix) to another.

For organic molecular detection, excessive photofragmentation that is common with the multiphoton ionization[12] approach under efficient conditions must be avoided. An example of this difficulty will be presented later. We have found that a single-photon ionization using a wavelength in the vacuum ultraviolet is a very practical way to solve this problem and maintain a high degree of uniformity in detection efficiency. Indeed, the use of single-photon ionization for mass spectrometry is a relatively old technique.[13] However, it has not been performed until now with the type of sensitivity as demonstrated below, which is required for the analysis of nonvolatile and thermally labile compounds. A very simple way to generate coherent VUV radiation is based on the ninth harmonic of an Nd:YAG laser. The Nd:YAG laser fundamental first is frequency tripled with standard optical crystals to reach 355 nm. That output at 355 nm is then in turn tripled to yield a third harmonic at 118 nm or 10.5 eV per photon. This is a nonlinear process accomplished by focusing in a gas cell of Xe phase matched with Ar.[14] We found optimization with a rather long focal length lens that avoided gas breakdown observed with short focal length lenses at low laser powers. A 50-cm focal length input lens was used in a cell filled with ~ 15 torr Xe and 150 torr Ar for an input energy of 20 mJ at 355 nm. A conversion efficiency was determined with an absolute acetone photoionization detector[15] to be about 10^{-4} or equivalent to about 10^{12} photons per pulse. This conversion gives a collinear coherent beam of the 355- plus the 118-nm light. To avoid the much more intense 355-nm light from causing excessive fragmentation the two light beams are separated by a lithium fluoride or magnesium fluoride optic.

Reflecting Time-of-Flight Mass Spectrometry (TOF-MS)

Because pulsed lasers are used for ionization, time-of-flight mass spectrometry is a very natural tool. The short laser pulse serves as the start pulse and TOF-MS has an inherent multiplex advantage, that is, for every laser pulse all masses can be recorded. The method is also characterized by high transmission and high mass resolution. We typically operate at a mass resolution of ~ 1000 using our reflecting time-of-flight mass spectrometer; improvements are anticipated. The transmission of such mass spectrometers frequently range above 50%.

The reflecting version of time-of-flight mass spectrometry was originally developed by Mamyrin et al.[16] who used electron impact ionization of gaseous compounds. The reflector was designed as a means of improving mass resolution by compensation for kinetic energy spreads and hence spreads in the arrival time of the ions at each mass at the detector. Basically the way the device works is that higher energy, faster ions penetrate deeper into the electrostatic reflecting field and take longer to turn around than the lower energy slower ions. The electrostatic potentials of the device are adjusted so that the fast ions lag behind the slow ions after reflection but catch up again to the slow ions at the point of detection after a second drift section. Hence a "bunching" of the ions of each mass-to-charge ratio takes place at the detector.

In our initial work we found a second important use for the reflector. By adjusting the back-reflecting potential to an intermediate value between the surface potential in which any secondary ions are formed and the region of photoionization above the sample, then photoions are reflected whereas secondary ions pass through the back of the reflector and thus an electrostatic separation between secondary ions and photoions takes place. Alternatively, if desired the laser can be turned off, the mass spectrometer's potentials simply adjusted, and secondary ions can be detected in the reflecting mode. The separation of any secondary ions from photoions is valuable as a way of reducing background and avoiding any confusion.

Another variation with the reflector is the use of a second detector placed behind the reflector. Although our laboratory has not pursued this in detail, other groups have used such an arrangement to perform studies of the metastable decomposition of certain molecular ions. Interested readers can refer to Ref. 17 for more information on this topic.

Ions are accelerated into a particle multiplier detector at the end of the flight tube. We use microchannel plates in our experiments because they have a planar surface that provides for a good length definition for the drift region. Also microchannel plates can accept a large flux of particles before reaching saturation. After the particle multiplication from a pair of plates in a chevron configuration[18] with a typical gain of 10^6, the electrons exiting the particle multiplier impinge on a metal anode that leads to the signal line from the mass spectrometer. This signal is amplified across a 50-Ω input impedance and then fed into either a transient digitizer or a time-to-digital converter.

The transient digitizer is an analog device. It digitizes the analog waveform of the output from the multiplier and amplifier. For large signals, as is commonly encountered, there are simply too many ions in a very short time to be detected in a pulse counting (strictly digital) mode. Analog detection therefore is often the only method of choice. For situations in which the count rate is sufficiently low, pulse counting is advantageous because it will give the best signal-to-noise ratios and usually with the highest time resolution. Low count rates can occur, for example, with the detection of impurities within a specified time region in the mass spectrum or when the primary ion beam or even for that matter the laser beam is weak.

Modern time-to-digital converters typically have a time resolution of 1 nsec or less. In contrast, most transient digitizers have time resolutions of 5–10 nsec, although there are newer models out with 1-nsec sampling periods.

ANALYSIS OF INORGANIC MATERIALS

Using nonresonant multiphoton ionization with reflecting time-of-flight mass spectrometry, the SALI technique has been able to achieve an overall efficiency ("useful yield") on the order of 10^{-3}. That is, for every 1000 atoms on the surface, one ion is detected. Although at first this may not sound particularly efficient, when compared with other nonselective postionization techniques mentioned in the introduction, in which many orders of magnitude lower efficiencies are found, it can be seen that such a useful yield is actually very high. One loss in efficiency is caused by the probability of the sputtered atom or molecule falling within the laser focal volume at the time of the laser pulse. This can be considered mathematically as the fraction of the velocity distribution and limited subtended solid angle. Although the laser is focused, the close proximity of the laser beam to the sample still results in a fairly good solid angle of about $\frac{1}{2}$ sr. The additional losses are typically small: the ionization step (which is often saturated), the transmission through the time-of-flight mass spectrometer, which, as mentioned before, is often on the order of about 50 percent, and the microchannel plate response that is about 50% efficient.

Because SALI generally can achieve a uniform ionization efficiency by driving the ionization to completion regardless of the cross section, it is found that often little or no calibration is required to obtain data of at least semiquantitative value. Nevertheless it is important to realize that a high degree of quantitation, say 10% accuracy, generally requires the use of a calibrant. The calibrant need not be closely matched to the sample, however. The factors that can cause difficulty for quantitation without calibrants even under conditions of saturation of the ionization are the neutral particles' angular and velocity distributions and the size and shape of the laser beam volume. With regard to the laser beam volume, even if saturation occurs within the center focal region, different species will be more or less efficiently ionized in the wings of the laser beam. This effect can be minimized by using a laser beam with a sharp falloff in its spatial profile. For example, a Gaussian beam has a fairly rapid falloff in intensity. In general, these limitations do not add up to relative sensitivities of more than a factor of 2 or 3.

Two review articles on SALI are available summarizing various issues for SALI of inorganic materials.[19,20] They present some examples as well as cite other earlier analyses from the literature. We now consider two recent examples of depth profiling analyses performed on high critical temperature (T_c) superconducting thin films and the Si/SiO_2 interface.

Depth profiles were obtained with a raster-scanned Ar^+ beam incident at 60° from normal from a duoplasmatron ion gun. Typical ion beam parameters for depth profiling are 0.05–2 μA, 5 keV, 50–200 μm diameter, and raster areas of 2×10^{-3} to 4×10^{-2} cm^2. The laser is gated in time relative to the ion beam position to sample atoms and molecules originating from the center of the raster pattern, which is the flattest location and thus associated with the best depth resolution. In these studies an excimer laser operating on KrF (248 nm, 5.0 eV) was used, focused to a power density of $\sim 10^{10}$ W/cm^2.

High-T_c Superconducting Thin Films

The recent discovery of high-T_c superconductors has major technological implications. Although rapid advances are being made, there is much that must be learned with regard to both the fundamental physics and proper materials processing.

Examples of SALI are shown in Fig. 4.2 of thin (1-μm-thick) superconducting YBa$_2$Cu$_3$O$_{7-\delta}$ films grown by electron beam codeposition[21] on SrTiO$_3$ (100). Electrically, these films are of some of the highest quality produced to date, with critical current densities of $\sim 10^6$ A/cm^3 at 4.2 K and $\sim 10^6$ A/cm^2 at 77 K.

The SALI depth profiles[22] in Fig. 4.2 show considerable structure in the composition as a function of depth, indicating the presence of additional (nonsuperconducting) phases and also the presence of impurities and substrate diffusion in the films, with particularly substantial amounts of impurities within the first few 100 Å. Indeed, contact resistance problems and atmospheric degradation (notable by H$_2$O) are an issue. Examination of several locations on these films, separated by millimeter dimensions, generally revealed little compositional variation across the sample. More recently prepared samples no longer show these substantial oscillations.

As is typically found with SALI, the uncorrected measured relative intensities for nearly all elemental components particularly those ionized by two photons at these laser intensities, are within about a factor or two of true value. The data of Fig. 4.2 are plotted as the measured relative intensities with the following exceptions: in Fig. 4.2a the Cu and Ba signals have been scaled by multiplying by factors of 2.1 and 1.9, respectively, to give the approximate 1:2:3 ratio and for visual clarity, and in Fig. 4.2b the BaO$^+$ and BaOH$^+$ signals have been multiplied by 6.3 (they probably have been depleted from photodissociation) to give the 1:2 ratio of Y:Ba.

An example[23] of the actual mass spectrum recorded from within a 0.2-μm-thick high-T_c superconducting YBa$_2$Cu$_3$O$_{7-\delta}$ thin film deposited on SrTiO$_3$ (100), in this case by magnetron sputtering,[24] is shown in Fig. 4.3. The spectrum shows many cluster ions [e.g., BaCuO, BaCu, SrCuO, Y$_2$O$_2$, Y$_2$O$_3$, Y$_3$O$_4$ (present but off the mass scale in Fig. 4.3), YBaO$_2$, YSrO$_2$, Ba$_2$O, Ba$_2$O$_2$]. Some of these are undoubtedly associated with phases other than the high-T_c superconducting phase. For example, the relative intensities of the mass peaks at m/e 210 (Y$_2$O$_2$), 226 (Y$_2$O$_3$), and 331 amu (Y$_3$O$_4$) are very close to those taken from a Y$_2$O$_3$ reference powder. The observation of the peak at 331 amu in particular is difficult to rationalize as coming from the accepted structure of the superconducting lattice, although sputter damage might conceivably cause this. By monitoring the relative intensities of these peaks it should be possible to measure relative amounts of the various phases as a function of material depth. Experiments are underway to make these correlations between known compositional phases serving as calibrants [such as BaO, Ba(OH)$_2$, BaCO$_3$, Y$_2$O$_3$, BaCuO$_2$ and more complex compounds with Y, Ba, Cu, and O including single crystal YBa$_2$Cu$_3$O$_{7-\delta}$] and the measured relative cluster intensities.

The depth profiles in Fig. 4.4 are selected clusters. Clear evidence of several phases can be found by the differences in the depth profiles. Particularly notable

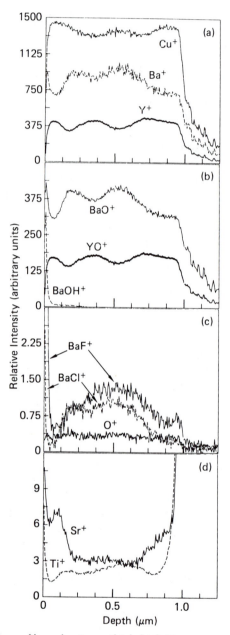

Fig. 4.2. SALI depth profiles of a 1-μm-thick high-T_c superconductor with nominal composition of $YBa_2Cu_3O_7$, deposited on $SrTiO_3$. In (c), the BaF^+ signal peaks off scale at 12. Republished with permission from Ref. 22.

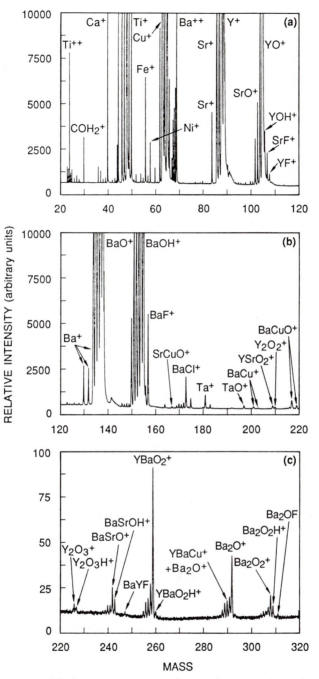

Fig. 4.3. SALI time-of-flight mass spectrum taken with 5000 pulses of 248-nm radiation from the central depth of an 1800-Å-thick high-T_c superconducting thin film with nominal composition $YBa_2Cu_3O_7$ under static conditions (pulsing the Ar^+ beam), showing lower intensity components. Republished with permission from Ref. 23.

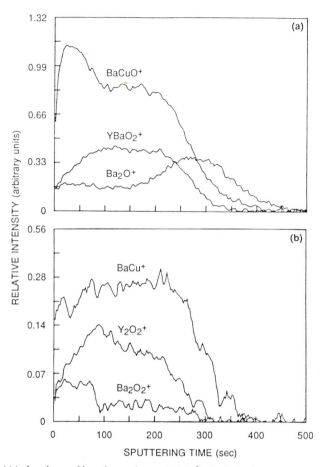

Fig. 4.4. SALI depth profiles through an 1800-Å-thick high-T_c superconducting film with nominal composition $YBa_2Cu_3O_7$ deposited on $SrTiO_3(100)$ taken with Ar^+ sputtering and photoionization at 248 nm with the following multiplicative scaling for visual clarity: BaCu 1.5; Y_2O_2, Ba_2O_2, Ba_2O, $BaYO_2$, and BaCuO all 1.0. Republished with permission from Ref. 23.

variations are seen in the profiles of Ba_2O vs Ba_2O_2, BaCu vs BaCuO, and Y_2O_2 vs $YBaO_2$, although of course in a more general sense all six of the profiles in Fig. 4.4 are distinct. As another point of comparison, the Ba_2O_2 and BaCuO profiles are relatively close in shape to that for BaO (not shown), all having a maximum within the first ~600 Å (the first 100 sec of sputtering) of the film, whereas Y is very much depleted there. It is suggested that this is a region relatively high in the unwanted phase BaCuO.

Auger electron spectroscopy (AES) depth profiles have been measured[25] for these films and also for the films made earlier that have had SALI depth profiles taken (Fig. 4.2).[22] The AES results, consisting of atomic Y, Ba, Cu, and O profiles, corroborate the SALI results quantitatively for the relative variations of the Y, Ba, and Cu with depth. The atomic O profiles show fair agreement also, but, as mentioned before, greater laser intensity than is presently available to us is desirable. It should be noted that SALI, in contrast to AES, can detect hydrogen, make simultaneous measurements on low concentrations, and provide chemical information by the observation of molecular emission.

The Si/SiO₂ Interface

Because of the demonstrated high useful yields (10^{-3} for all masses simultaneously), the SALI technique has significant advantages for examining sharp interfaces in materials in which concentration gradients might occur over a small distance and the number of atoms available for examination is small. This is particularly true of contaminants or minor constituents segregating to an interface. By comparison, SIMS measurements could suffer from rapid changes in sensitivity/quantitation at the interface (a matrix effect very difficult because of the problem of producing an appropriate comparison standard for the situation), and electron bombardment postionization methods would lack sufficient sensitivity.

Although higher power density laser conditions will allow saturation of the ionization for every element, as mentioned there are currently a few elements of high ionization potential having low atomic ionization sensitivities with our laser (a Lambda-Physik EMG103-MSC), specifically N, O, F, He, Ne, and Ar. However, nonrare gas elements of this group can still be studied by looking at an easily ionized molecule containing the element.

An example is shown in Fig. 4.5, in which depth profiles of F and O concentrations are monitored by the two sputtered molecules ^{28}SiF and ^{29}SiO. These are plotted for a three-layer structure implanted with F at 93 keV and 10^{15}/cm². The upper layer is a P-doped polycrystalline Si (~2500 Å thick), the middle layer is SiO₂ (410 Å thick), and the substrate is single crystal Si. This sample is technologically relevant because selective chemical vapor deposition of W by WF₆ is being used for interconnections in VLSI circuitry. The effect of F from this process needs study. Thus, these F implanted gate structures were made to better understand F migration in correlation with electrical measurements.

The atomic F signal (four-photon ionization at 248 nm) is very small; the SiF signal is much larger and much more convenient for monitoring the F intensities. Although atomic oxygen is detected with substantially higher intensities than atomic F, SiO is quite convenient for recording the oxide layer (which is at a slightly different depth in the two samples). Note that the sputter yields for Si and SiO₂ are approximately equal. The lower plot shows an unannealed specimen with the F intensity peaking in the center of the upper layer and dropping relatively smoothly through the interface region (SIMS measurements on this sample show a spurious oxygen-induced matrix enhancement of F^+ intensity at the interface). The interface region for the polycrystalline Si and SiO₂ is where the SiO signal rises at about 370 sec sputtering for Fig. 4.5b). The upper sample (annealed) shows significant F migration to the interfaces and also to the bulk of the SiO₂, accompanied by some loss of fluorine from the sample. More details on this measurement are presented elsewhere.[26]

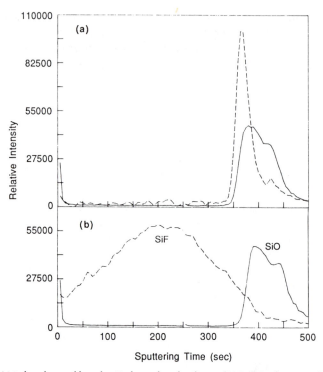

Fig. 4.5. SALI depth profile of a P-doped polysilicon/SiO_2/single crystal silicon electronic test device that has been implanted with F at 93 keV and 10^{15}/cm^2. (a) unannealed. (b) Annealed at 800°C for 30 min in Ar; A KrF laser (248 nm) operating at 50 Hz and 60 mJ/pulse was used for ionization, with Ar^+ sputtering at 7 keV. The depth profiles represent the photoions SiF^+ (dashed) and $^{29}SiO^+$ (solid).

ANALYSIS OF NONVOLATILE ORGANIC MATERIAL

Mass spectroscopic analysis of nonvolatile organic materials can be associated with fragmentation that occurs during both the desorption and ionization processes. This fragmentation may be excessive and not give characteristic mass peaks when using multiphoton ionization for the detection of desorbed neutrals (whether or not there is an intermediate resonance at the laser frequency) particularly under high sensitivity conditions in which fairly high intensity laser fields must be used. Because the molecular ion that is formed typically absorbs single photons (to fragmenting states by bound–bound electronic transitions) more strongly than the neutral molecule absorbing into the ionization continuum (a bound–free electronic transition), it is often difficult to limit fragmentation while efficiency is being sought. For low primary beam doses (low degree of damage) and/or small amounts of sample, efficiency is an important issue, and sampling may be limited to $\sim 10^{-12}$ mol. Exemplary of this difficulty, the branching ratio between parent molecular ion (if even observable) and fragment molecules generally changes dramatically with laser intensity making quantitation and even spectral assignment difficult.

By using VUV single-photon ionization these fragmentation processes are greatly reduced as demonstrated in the following examples. Single-photon ionization using a VUV source is generally considered a "soft" ionization method. In this process only modest amounts of excess internal energy are available to be deposited in the resulting photoion and efficient ionization can occur with relatively low light intensities so that multiple photon absorption is unlikely. (The larger the molecule, the longer internal energy can remain in the molecule without causing fragmentation,[27] thus making this type of ionization increasingly soft with increasing mass.) Furthermore, single-photon ionization cross sections are much more uniform from molecule to molecule relative to multiphoton ionization (MPI),[28] resulting in more quantitative detection probabilities.

Spectra of organic compounds have been taken with rare gas ion sputtering, being recorded under low-dose irradiation conditions (less than 10^{13} incident ions/cm^2) and subsequent single-photon ionization. However, laser desorption is expected perhaps to yield superior results for biochemicals and environmental toxins, and possibly for bulk polymers too.

Biochemical Analysis

As an example,[29] the purine 7-methylguanine was placed on a clean silicon wafer allowing the solvent H_2O to dry before introduction into the ultrahigh vacuum chamber; small crystallites were observed.

The spectrum of 7-methylguanine shown in Fig. 4.6 exhibits a strong parent ion signal (165 amu) as well as the loss of hydrogen (164 amu) and highly characteristic fragment peaks at $m/e = 149$ and 134 amu due to the loss of the NH_2 and CH_3 groups. The fairly strong feature at $m/e = 124$ amu is likely due to the decomposition of the imidazole ring leaving a pyrimidine ring with an NH group attached; the peak at 93 amu then corresponds to the further loss of NH_2 and NH groups. The maximum sensitivity can be estimated from a signal-to-noise ratio of about 10 for a single pulse and a desorption yield of 1 to 10 molecules per incident ion, which leads to a limit of 3×10^{-17} to 3×10^{-16} mol.

The extent of fragmentation from ion bombardment vs photoionization is not presently known, although future comparative studies with laser desorption will help to clarify this issue. Additional structural information for unknown molecules can be obtained by photofragmentation with the addition of variable amounts of a more intense second laser beam.[30]

Bulk Polymer Analysis

For polymers of complex molecular structure, there is a great need to obtain structural information using mass spectrometry. At the same time, this information has been particularly difficult to obtain for bulk material (in distinction to dispersed or monolayer thick samples). The SALI technique together with single-photon ionization promises to make a significant contribution in this area. As an example,[31] Fig. 4.7 shows the SALI spectrum of the random polyamide copolymer nylon 66 taken with 1000 pulses of Xe^+ bombardment at 7 keV and 118-nm radiation for ionization.

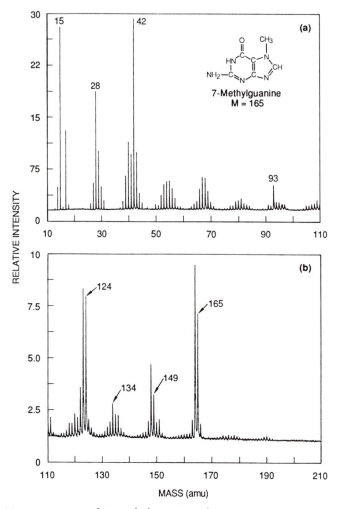

Fig. 4.6. Mass spectrum of 7-methylguanine. The spectrum was accumulated over 1000 pulses with 118-nm radiation for single-photon ionization following pulsed Ar⁺ sputtering. Republished with permission from Ref. 29.

The low mass range of the spectrum ($m/e \lesssim 100$) is dominated by small fragments, containing C, H, N, and O. Larger fragments of the polymer chain give definite structural information. Distinct higher mass peaks appear at $m/e = 368$ and 353, which are greater in mass than the polymer repeat unit ($m/e = 226$). These features can easily be related to the structure of this particular polymer formed from the diamine (monomer subunit $M_1 = 114$ amu, see Fig. 4.7) and the

Fig. 4.7. SALI mass spectrum of untreated nylon 66 using 1000 pulses of Xe⁺ bombardment and 118-nm radiation for one-photon ionization at approximately 3×10^3 W/cm². Particularly noteworthy mass peaks are discussed in the text. The slope of the baseline is due to rf pickup by the particular amplifier used here. Republished with permission from Ref. 31.

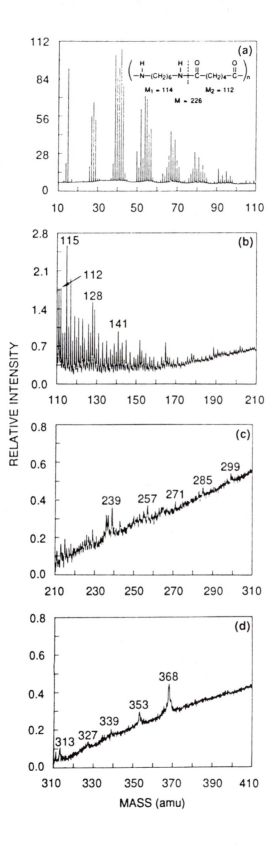

decarboxylic acid (corresponding monomer subunit $M_2 = 112$ amu). Although the repeat unit ($M = M_1 + M_2 = 226$ amu) is not seen in the spectrum, for example, the peak at 239 amu can be assigned to $M + CO-NH$, and in particular the three highest mass peaks at 339, 353, and 368 amu show evidence for preferential cleavage of the $N-C$ bonds leading to fragments of MM_2H, MM_2NH, and $HNMM_2NH$, respectively. These last three assignments give definite structural information and also provide some insight into the sputtering/cleavage process for this material. In fact, it is worth noting that the nonselective single-photon ionization will be an excellent general tool for providing insight into the desorption process itself. Some other noteworthy mass peaks include m/e 115 (M_1H) and m/e 128 (M_2NH_2). These assignments reveal the structure of nylon 66.

MPI spectra obtained using 248-nm laser light for a range of intensities were unsuccessful, giving only small fragment peaks at about 80 amu and below. No satisfactory spectrum could be obtained with MPI in the sense that no significant structural information could be discerned.

After examining numerous other bulk polymer systems,[31,32] such as polyethylene, poly(methylmethacrylate), polystyrene, and poly(tetrafluoroethylene), comparisons of SALI spectra with spectra obtained by SIMS reveal that for the systems studied to date, the single-photon ionization of sputtered neutral molecules gives at least comparable quality spectra and more often the more readily interpretable characteristic spectra.

CONCLUSIONS

The combination of stimulated desorption, properly chosen *untuned* laser radiation, and time-of-flight mass spectrometry is opening new opportunities in surface and material chemical analysis. This approach rests on the fact that the desorption yields of neutral atoms and molecules are relatively invariant quantities characteristic of the surface under investigation. For inorganic chemical systems and elemental analysis, the laser beam for nonresonant multiphoton ionization above the surface is focused to very high power densities, aiming to completely ionize the intersected desorbed species, independent of their photoionization cross sections. For organic chemical systems, a coherent light beam of a sufficiently short wavelength is used to produce single-photon ionization with relatively high efficiency and minimal fragmentation; unlike at longer wavelengths, molecular photoionization cross sections have been found in this case not to exhibit large variations at a fixed wavelength. Thus, under both types of untuned radiation for the two types of analyses, a general, high-fidelity, uniform, and sensitive surface analysis can be obtained while taking advantage of reflecting time-of-flight mass spectrometry.

ACKNOWLEDGMENTS

The author is indebted to Drs. Joan Pallix, Udo Schühle, and Keith Gillen for their many valuable contributions. Financial support from the National Science Foundation, Division of Materials Research and Perkin-Elmer Corporation, Physical Electronics Division is gratefully acknowledged.

REFERENCES

1. R. E. Honig, *J. Apply. Phys.* **29,** 549 (1958).

2. See, e.g., the series *SIMS II* through *SIMS V,* edited by A. Benninghoven et al. Springer-Verlag, Berlin (1979–1986).

3. R. J. Conzemius and J. M. Capellen, *Int. J. Mass Spectrom. Ion Phys.* **34,** 197 (1980); I. D. Kovalev, G. A. Maksimov, A. I. Suchkov, and N. V. Larin, *Int. J. Mass Spectrom. Ion Phys.* **27,** 101 (1978).

4. J. J. Thomson, *Phil. Mag.* **20,** 752 (1910).

5. H. Gnaser, J. Fleischhauer, and W. O. Hofer, *Appl. Phys.* **A37,** 211 (1985).

6. J. R. Woodyard and C. B. Cooper, *J. Appl. Phys.* **35,** 1107 (1964); H. Oechsner and W. Gerhard, *Surf. Sci.* **44,** 480 (1974).

7. N. Winograd, J. P. Baxter, and F. M. Kimock, *Chem. Phys. Lett.* **88,** 581 (1982).

8. C. H. Becker and K. T. Gillen, *Anal. Chem.* **56,** 1671 (1984).

9. J. B. Pallix, C. H. Becker, and N. Newman, *J. Vac. Sci. Technol. A* **6,** 1049 (1988).

10. J. Melngailis, *J. Vac. Sci. Technol. B* **5,** 469 (1987).

11. D. Strickland, P. Maine, M. Bouvier, S. Williamson, and G. Mourou, in *Ultrafast Phenomena V,* edited by G. R. Fleming and A. E. Siegman. Springer-Verlag, Berlin, 1986, p. 38.

12. J. P. Reilly and K. L. Kompa, *J. Chem. Phys.* **73,** 5468 (1980).

13. F. P. Lossing and I. Tanaka, *J. Chem. Phys.* **25,** 1031 (1956); H. Hurzeler, M. G. Inghram, and J. D. Morrison, *J. Chem. Phys.* **28,** 76 (1958); N. W. Reid, *Int. J. Mass Spectrom. Ion Phys.* **6,** 1 (1971).

14. A. H. Kung, J. F. Young, and S. E. Harris, *Appl. Phys. Lett.* **22,** 301 (1973); **28,** 239 (erratum); L. J. Zych and J. F. Young, *IEEE J. Quant. Electron* **QE-14,** 147 (1978); A. H. Kung, *Opt. Lett.* **8,** 24 (1983).

15. J.A.R. Samson, *Techniques of Vacuum Ultraviolet Spectroscopy,* Chap. 8. John Wiley, New York, 1967.

16. B. A. Mamyrin, V. I. Karataev, D. V. Shmikk, and V. A. Zagulin, *Sov. Phys.–JETP* (Engl. trans.) **37,** 45 (1973).

17. X. Tang, W. Ens, K. G. Standing, and J. B. Westmore, *Anal. Chem.* **60,** 1791 (1988).

18. J. L. Wiza, *Nucl. Instrum. Method.* **162,** 587, (1979).

19. C. H. Becker, *Scanning Electron Microsc.* **IV,** 1267 (1986).

20. J. B. Pallix, C. H. Becker, and N. Newman, *Mat. Res. Soc. Bull.* **XII**(6), 52 (1987).

21. M. Naito, R. H. Hammond, B. Oh, M. P. Hahn, J.W.P. Su, P. Rosenthal, A. F. Marshall, M. R. Beasley, T. H. Geballe, and A. Kapitulnik, *J. Mat. Res.* **2,** 713 (1987).

22. J. B. Pallix, C. H. Becker, N. Missert, M. Naito, R. H. Hammond, and P. Wright, *Secondary Ion Mass Spectrometry SIMS VI,* edited by A. Benninghoven, A. M. Huber, and H. W. Werner. John Wiley, New York, 1988, p. 817.

23. J. B. Pallix, C. H. Becker, N. Missert, K. Char, and R. H. Hammond, in *Thin Film Processing and Characterization of High-Temperature Superconductors,* edited by J.M.E. Harper, R. J. Colton, and L. C. Feldman. Am. Instit. Phys. Conf. Proc. No. 165, New York, 1988, p. 413.

24. K. Char, A. D. Kent, A. Kapitulnik, M. R. Beasley, and T. H. Geballe, *Appl. Phys. Lett.* **51,** 1370 (1987).

25. N. Missert, unpublished results.

26. J. B. Pallix, C. H. Becker, and K. T. Gillen, *Appl. Surf. Sci.* **32,** 1 (1988).

27. P. J. Robinson and K. A. Holbrook, *Unimolecular Reactions.* Wiley-Interscience, New York, 1972.

28. J. Berkowitz, *Photoabsorption, Photoionization, and Photoelectron Spectroscopy.* Academic Press, New York, 1979.

29. U. Schühle, J. B. Pallix, and C. H. Becker, *J. Am. Chem. Soc.* **110,** 2323 (1988).

30. D. F. Hunt, J. Shabanowitz, and J. R. Yates, III, *J. Chem. Soc., Chem. Commun.* 548 (1987).

31. U. Schühle, J. B. Pallix, and C. H. Becker, *J. Vac. Sci. Technol. A* **6** 936 (1988).

32. J. B. Pallix, U. Schühle, C. H. Becker, and D. L. Huestis, *Anal. Chem.,* **61**, 805 (1989).

5

Laser Microprobe Mass Spectrometry: Ion and Neutral Analysis

ROBERT W. ODOM and BRUNO SCHUELER

Laser microprobe mass spectrometry is a microanalytical materials analysis technique employed in the characterization of a material's elemental and, in some cases, molecular composition.[1-3] The laser microprobe technique is based on performing a mass and intensity analysis of the ionic species formed in the high-power density laser irradiation of a solid sample. The microanalytical feature of the laser microprobe technique is achieved by utilizing a finely focused (~ 1 μm in diameter) laser pulse to initiate the vaporization and ionization of a materials constituents. There are a number of different types of laser and mass spectrometer combinations[4] (some of these are described elsewhere in this book) that provide chemical characterization of a solid sample. The laser microprobe technique discussed in this chapter employs a pulsed, Q-switched, frequency quadrupled Nd:YAG laser (λ = 266 nm) and a time-of-flight mass spectrometer (TOF-MS). The combination of this finely focused, high-irradiance (variable between $\sim 10^8$ and 10^{12} W/cm^2) laser pulse with the "simultaneous" mass detection capabilities of TOF-MS enables microanalytical survey analysis of a wide range of sample materials to be performed. The laser microprobe technique is referred to by several acronyms including LAMMA, LIMA, LAMMS, and one that has found favor in our laboratory, LIMS, which stands for laser ionization mass spectrometry.

In terms of analytical utility, the microanalysis of sophisticated materials is playing an ever important role in modern materials research and development. Analysis techniques that have sensitive elemental and/or molecular detection capabilities within analytical areas on the order of a few micrometers (or less) in diameter are being increasingly utilized for the characterization of materials such as semiconductor devices, integrated optical components, novel metallurgical alloys, ceramic composites, as well as biological materials. The LIMS technique is finding increasing utilization in these types of materials analyses because it can provide high sensitivity survey characterizations. This laser-based technique augments and complements the microanalytical capabilities of the more established materials analysis techniques such as secondary ion mass spectrometry (SIMS),[5] Auger electron spectrometry (AES),[6] and electron probe microanalysis (EPMA).[7] The unique microanalytical capabilities that the LIMS technique adds to these existing techniques include the following:

1. Rapid, sensitive, elemental survey microanalysis,
2. Ability to analyze electrically insulating materials, and
3. Potential of providing molecular or chemical bonding information from the analytical microvolume.

As a result of these microanalytical capabilities, LIMS is employed extensively in various survey microanalysis applications and the range of these applications is very large. The only practical limitations to performing a qualitative laser microprobe analysis are the requirements that the sample material be reasonably vacuum compatible and that the Nd:YAG laser beam be focused on the analytical area of interest. In our laboratory, we have performed informative laser microprobe analyses on sample materials ranging from dehydrated sewage solids to liver tissues to a variety of sophisticated semiconductor materials and devices. There is a small but dedicated community of scientists performing both fundamental and applied research using the laser microprobe technique, and their analytical applications are quite diverse.

The first commercial laser microprobe instrument was enthusiastically introduced (by Leybold Heraeus, Cologne, West Germany) approximately 10 years ago. To the interested observers (those in the materials analysis community), it might have been construed from the initial interest and applications that a short development period would be required before the technique became a *quantitative* microanalysis method. Although today there are a number of unique quantitative or semiquantitative methods[8] for this technique, the development of a general LIMS quantitative methodology is still being pursued. As will be discussed in more detail, some of the difficulties in quantitative methods development relate to the incomplete understanding of the vaporization and ionization mechanisms occurring in the pulsed laser irradiation of a materials surface, and some of these difficulties result from instrumental limitations. It is correct to say, however, that progress in this area of quantitative LIMS methods development is being made and, as was the case for many other materials analysis techniques, a rather comprehensive understanding of the pertinent fundamental processes is required before quantitative methods can be developed.

This chapter is divided into two rather distinct parts. The first part describes the conventional laser microprobe instrumentation and discusses examples of various applications of this technique. The majority of these applications are taken from work performed at Charles Evans & Associates; however, references to the work performed by other laser microprobe researchers are included in this section. The second part of this chapter discusses the experimental configuration and selected analyses performed with a postablation ionization (PAI) laser microprobe system. This system employs a second laser pulse to ionize the neutral plume vaporized by the conventional laser microprobe laser beam. These neutrals are ionized via nonresonant multiphoton ionization (NRMPI) processes. The decoupling of the laser ablation and ionization steps in a laser microprobe analysis generally improves the ion yield uniformity and increases the surface sensitivity of this microanalytical technique.[9]

EXPERIMENTAL

The Laser Microprobe Mass Spectrometer

Figure 5.1 illustrates a schematic diagram of a reflection mode laser microprobe mass spectrometer. This particular instrument is the model LIMA 2A originally manufactured by Cambridge Mass Spectrometry, Ltd (Cambridge, England), which is now a part of Kratos Analytical. A similar instrument is manufactured by Leybold Heraeus and is referred to as the LAMMA 1000. Leybold also manufactured a transmission mode laser microprobe system (LAMMA 500). The reflection and transmission mode instruments will be described in more detail. Vacuum Generators (East Grinstead, England) has recently introduced a TOF-MS/SIMS system that can be equipped with a focused laser beam for laser microprobe analyses. The basic components of the laser microprobe illustrated in Fig. 5.1 include

1. A Q-switched, frequency quadruped Nd:YAG laser system, laser focusing optics, and various laser beam monitors.
2. An He:Ne pilot laser and visible light optics. The He:Ne beam is coaxial with the Nd:YAG laser output and the pilot laser/optical microscope system enables the analytical area of interest to be located.
3. A sample mounting stage having precision vacuum manipulators that provide three-dimensional positioning of the sample within the ionizing laser beam.
4. A time-of-flight mass spectrometer having a flight patch ~2m in length and

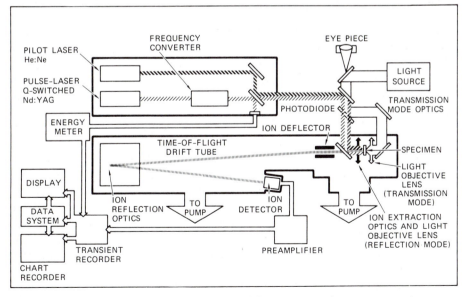

Fig. 5.1. Schematic diagram of the model LIMA 2A laser microprobe mass spectrometer.

equipped with a single- or dual-stage ion reflection optics[10] that enhance the mass resolving power of the spectrometer.

5. An ion detection system that includes an electron multiplier (EM) detector, various signal amplifier circuits, a multichannel transient recorder, and a computer acquisition and data processing system.

This basic instrumental configuration permits the identification of the mass and measurement of the intensity of the ionic species produced in the high-power density laser irradiation of a materials surface. The microanalytical feature of this technique arises from the ability to focus the laser beam to a small spot size (theoretically set by the diffraction limit of $\sim 2\lambda$), whereas the survey analysis capability results from the coupling of the pulsed ionization produced by the pulsed laser with TOF mass analysis using the transient recorder.

In a typical laser microprobe, the analytical area of interest is positioned in the focal spot of the He:Ne laser beam, the transient recorder is armed to record, and the Nd:YAG laser is fired. The short duration of this laser pulse ($5 \leq \tau_p \leq 15$ nsec) produces a packet of ions that is accelerated from the sample surface and injected into the drift region of the mass spectrometer. After acceleration, all ions have approximately the same kinetic energy and, hence, ions of different mass will have different velocities. These ions will disperse in time and arrive at the EM detector in discrete packets. The amplified currents produced by the ion detector [which correspond to different mass to charge ratio (m/z) ions] are converted to amplified voltage signals that are then input to the transient recorder. This device rapidly digitizes and stores these analog voltage signals over a selected time range after the firing of the ionizing laser pulse. The time of arrival of the various ions at the detector is measured by the transient recorder and the intensity of these digitized analog voltages is proportional to the number of ions detected. The transient recorder acquires a "complete" mass spectrum over a large mass range (e.g., from 0 to ~ 300 amu) for each laser pulse fired onto the sample surface. The intensity–time spectrum stored on the transient recorder is read into the computer system in which the time is converted to a mass scale and various data processing routines are performed. This basic sequence of events is then repeated at other sample locations to build up statistical data and/or survey other sample regions.

The process of events described occurs very rapidly as the following example illustrates. Assuming $\tau_p = 5$ nsec, a 2-m flight path, and a 3-kV ion acceleration potential, an H^+ ion would arrive at the electron multiplier in ~ 2.6 μsec whereas a $^{238}U^+$ ion would arrive in ~ 41 μsec. The width of these ion signals can be as short as several tens of nanoseconds if the spectrometer performance (particularly the ion reflection optics) is optimized and these fast signals require high-speed components in the detection system to accurately record the mass spectrum. A critical component in this detection system is and has been the frequency bandwidth of the transient recorder.[11] The early model microprobe instruments were equipped with transient recorders having effective bandwidths on the order of 50 MHz, whereas more recent instruments have recorders with 100- to 200-MHz bandwidths.

The full-width half-maximum (FWHM) mass resolution of the TOF-MS system on the laser microprobe system is given by

$$\frac{M}{\Delta M} \approx \frac{t_d}{2\Delta t} \tag{5.1}$$

where t_d is the total drift time of the ions from the sample surface to the detector and Δt is the width of the detected ion signal. This width of the detected ion pulse depends on the ion formation time ($\geq \tau_p$), the kinetic energy spread of the ions, as well as any time dispersion caused by the ion optics in the mass spectrometer and/ or the EM detector. The ion reflection optics illustrated in Fig. 5.1 minimize the time spread of the detected ion signal with respect to the ion kinetic energy distribution[10] (which can be hundreds of eV under high laser irradiance conditions[12]) so that the detected width of the ion signals reflects (primarily) their formation time. Assuming the instrumental conditions previously discussed and a 5-nsec ion formation time, the m/z 100 ion would ideally exhibit an FWHM mass resolution of ~2600. Although mass high resolutions of several thousand have been observed with TOF-MS systems equipped with ion reflection optics,[13] typical values observed on our LIMA 2A laser microprobe instrument range between 250 and 750 at $m/z = 100$. The difference between the ideal value calculated and the values typically observed are due (primarily) to the 16-nsec time resolution of the transient recorder used with our system (Sony-Tektronix Model 390 AD), possible time broadening due to space charge effects, and the fact that the ion formation times can be longer than the 5-nsec laser pulse width. The practical mass resolution values of 250 to 750 are sufficient to readily distinguish the isotope signals for all elements as well as many protonated $(M + H)^+$ ions from molecular ions. This mass resolution is not sufficient to separate signals of isobaric ions such as $^{56}Fe^+$ and $^{28}Si_2^+$ that would require a mass resolution of ~3000.

The signal voltages recorded on the transient recorder are proportional to the number of ions detected in each peak of the mass spectrum. Thus, in principle at least, the measurement of the peak height voltage or, better yet, the area under each peak, will provide a quantitative measure of the number of detected ions. There are, however, two significant factors that limit the accurate conversion of the ion signals into detected voltages. The first factor is the possibility of nonlinear outputs from the electron multiplier at high intensity input ion signal levels and the second factor relates to the imprecision in the voltage digitization by the transient recorder. The EM detectors employed on laser microprobe instruments are typically discrete dynode types having relatively large (~2.5-cm) diameter input apertures. Since the pulse of ions produced by the laser irradiation process can reach relatively high instantaneous currents at the detector, these ion signals can drive the EM output into a nonlinear output regime. For example, if the multiplier gain is G, the instantaneous output current from the EM is given by

$$i_p = \frac{NGq}{\Delta t} \tag{5.2}$$

where N is the number of ions contained in the pulse of width Δt and q is the charge on the ion. At a multiplier gain of 1×10^5 and a peak width of 25 nsec, each ion

($N = 1$) in the peak would produce an output current of ~0.6 μA. Thus, a peak containing 10,000 ions would produce an output current of ~6 mA. Many electron multiplier designs cannot maintain constant gain characteristics at mA output current levels and thus a nonlinear response of the EM for higher intensity signals is observed. There are a number of factors that limit the output current of electron multipliers including space charge limits on the amount of charge that can be extracted from any given dynode.[14] The nonlinear response of typical discrete dynode multipliers is well recognized by laser microprobe users and several research groups are actively attempting to increase the linear response characteristics of selected detectors. We at Charles Evans & Associates have recently begun to evaluate the performance characteristics of mesh-type discrete dynode multipliers for LIMS. It is too early to tell whether this type of detector offers significantly better pulsed output current linearity than the more conventional venetian blind or linear focus EM designs. The mesh EM design has, however, exhibited linear, mA level pulsed outputs currents when used as photomultiplier tubes (PMT).[15]

An alternate method for improving the accuracy of the ion detection process is to attempt to "calibrate" the EM response over a fairly large input signal range.[11] Figure 5.2 illustrates representative raw data that could be used for such a detector calibration. This plot is the ratio of the measured $^{113}In^+/^{115}In^+$ peak area intensities as a function of the $^{115}In^+$ peak area intensity produced in the analysis of an In foil target. The detected ion intensities were varied in this analysis by varying the multiplier gain using a nearly constant laser irradiance. Thus, the ion intensity variation represents a variation in the EM output current for essentially constant input

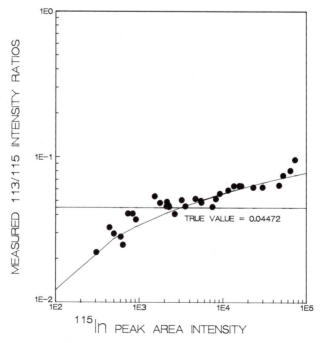

Fig. 5.2. Semilogarithmic plot of $^{113}In^+/^{115}In^+$ intensity ratio vs $^{115}In^+$ peak area.

signals. The true ^{113}In/^{115}In isotope ratio is 0.04472, which requires a detector dynamic range of better than 20:1 for accurate measurements. At low (<1000) ^{115}In$^+$ peak area intensities, the measured isotope ratio is low primarily because of the low ^{113}In$^+$ signal levels. The measured isotope ratios are within $\sim$20% of the true value over an ^{115}In$^+$ intensity range from 1000 to 10,000, whereas at the higher ^{115}In$^+$ intensities ($>10,000$), the measured ratio is higher than the true value because of the nonlinear response of the EM at these higher ^{115}In$^+$ output current levels. The line drawn through some of the data points is a least-squares fit of the data and this type of fitting procedure could be used to correct for the nonlinearity of the EM output as a function of output current. This calibration approach for improving the dynamic range of the EM detector may require separate calibration curves for different mass ions (which appears to be the case for the few examples that we have studied), and thus does not represent as viable an approach as one that attempts to extend the dynamic range through hardware performance improvements.

A second factor that limits the accuracy and precision of the ion intensity measurements in LIMS analyses is the errors associated with the digitization of the output voltage by the transient recorder.[11] The transient recorders employed with the laser microprobe generally have either 8- or 10-bit digitizers. Thus, under ideal conditions, the 8-bit digitizer can provide a maximum of 256 discrete voltage levels whereas the 10-bit digitizer has 1024 levels. The number of effective bits for high-frequency, nonsinusoidal inputs such as those generated in the laser microprobe is always less than the optimum specifications quoted by the manufacturers. One speaks of the levels of "dynamic" digitization for the transient recorder and, for example, the 10-bit Sony-Tektronix 390 AD has a dynamic digitization level of 6.5 bits for 10-MHz signals. This device therefore provides approximately 90 discrete levels at the input frequency of typical ion signal pulses. The newer, faster transient recorders have similar reductions in the effective number of bits for "real" signals. This reduction in the number of effective bit levels reduces the precision of any voltage measurement.

There are several methods by which the precision of these measurements can be improved including the use of two transient recorders operating at different sensitivities and employing a logarithmic preamplifier in the detection circuit.[9] The laser microprobe detection system in our laboratory is equipped with a logarithmic amplifier (MVP Electronics, Tustin, CA, Model DCL-80-30) that provides a full four decades of voltage measurement over the 2-V input range of the 390 AD transient recorder. This logarithmic amplifier has a 30-MHz bandwidth that does not appear to severely distort the input signal waveforms. In addition to providing four decades of dynamic range, this logarithmic amplifier significantly improves the precision of the voltage digitization compared to a linear preamplifier/single transient recorder configuration. The relative standard deviation (RSD_{LOGTR}) of the voltage measurements using the logarithmic amplifier is given by

$$RSD_{LOGTR} = \pm(10^{1/n} - 1) \times 100\% \tag{5.3}$$

where n is the number of digitizer levels per decade. Assuming that the 390 AD has a total of 90 effective levels ($n = 22.5$), the precision within each decade is $\pm11\%$.

This precision is to be compared to the values associated with a linear preamplifier (RSD_{LTR}) given by

$$RSD_{LTR} = \pm \frac{Y}{nX} \times 100\% \qquad (5.4)$$

where Y is the maximum input voltage to the recorder and X is the measured voltage. Thus, the precision of the linear system scales with the detected voltage. For the 390 AD transient recorder having an effective dynamic range of 90 levels, the RSD_{LTR} can reach values on the order of $\pm 100\%$ for X values that are one-twentieth the maximum input value.

These two factors are the most critical impediments to the development of general quantitative laser microprobe applications; however, a number of interesting and useful quantitative or semiquantitative applications have been developed within these limiting factors and several of these applications are discussed in the next section.

Mention was made at the beginning of this section of the reflection and transmission modes of the laser microprobe. These modes refer to the direction of the incident laser beam with respect to the sample surface. On the reflection instruments (illustrated in Fig. 5.1), the laser beam is focused onto the *front* surface of the sample and this configuration can analyze essentially any type of sample material that is vacuum compatible. Typical laser spot diameters are on the order of 1–2 μm with this configuration. For the transmission mode instrument, the Nd:YAG laser beam is focused onto the *backside* of the sample surface (i.e., the surface facing away from the ion extraction optics), and this configuration can analyze only thin samples (≤ 1 μm thick) or the edges of thicker samples. The laser focusing optics and optical microscope on this instrument are, however, better than those on the reflection mode instruments and the laser spot diameter can be focused to a spot diameter of ~ 0.5 μm on the transmission instruments. The transmission instrument was the first to be introduced commercially and has been utilized primarily for biomedical and particle analysis applications.[1]

The laser microprobe can be operated under laser irradiance conditions that provide either elemental or molecular microanalysis capabilities. Elemental analyses are generally performed at relatively high irradiance values ($\geq 10^9$ W/cm^2) and the ionization produced at these irradiances is primarily composed of atomic and or atomic cluster ions. This mode of operation is referred to as laser ionization (LI)[16] and part-per-million atomic (ppma) elemental detection sensitivities have been observed under these irradiance conditions.[8] "Molecular" LIMS analyses are typically performed at irradiances on the order of 10^8 W/cm^2 and this mode of operation is referred to as laser desorption (LD).[17] The ionization produced at these lower irradiances often includes molecular or pseudomolecular (protonated, deprotonated, cationized) ions and structurally significant fragment peaks.[18] Both the LI and LD ionization conditions can be employed to analyze for positive or negative ions emitted from the sample. The choice of analysis polarity depends on the sample composition and the efficiency of ion formation from the various analyte ions. For example, the elements in groups 1 through 13 form almost exclusively positive ions (with the notable exception of H), whereas the group 14 through 17 elements readily form negative ions. The detection polarity employed for molecular LIMS analyses depends critically on the sample composition, although many molecular species appear to produce more structurally significant ions in the positive ion analysis mode.[19,20]

LASER MICROPROBE MASS SPECTROMETRY APPLICATIONS

The introduction briefly discussed the range of applications of the laser microprobe analysis technique. These applications can be generally grouped into the categories of either elemental or molecular analyses. For elemental analyses, the applications can be further divided into qualitative or quantitative (semiquantitative), surface (survey) or in-depth analyses, whereas the molecular analyses are generally qualitative, surface studies. The examples presented in this section are taken from analytical studies performed at Charles Evans & Associates on the Cambridge Mass Spectrometry LIMA 2A laser microprobe.

Elemental Survey Analysis

A large proportion of practical materials analyses are concerned with determining the materials' surface or in-depth elemental composition. Examples of these survey analyses include measurements of the major, minor, and trace element composition, chemical stoichiometry, and presence or absence of chemical contaminants. Examples in which these types of survey analyses are important include chemical certification of various starting materials used in semiconductor, optical, metallurgical, and polymeric materials processing, evaluation of the chemistry of various processing steps, failure analysis studies of products in which "chemical" problems are suggested as the causative factors of the failure, and the study of various environmental and biological processes. Microanalytical survey techniques address these same analysis areas while performing the analysis in analytical volumes on the order of 1 μm^3 or less.

By far the greatest application of the laser microprobe in our laboratory has been in this area of elemental survey microanalysis. Two of these types of analyses are discussed below.

Contamination Microanalysis

Figures 5.3 and 5.4 through 5.6 illustrate two examples of contaminant microanalyses that are readily performed with the laser microprobe. Figure 5.3 is a negative ion LIMS mass spectrum produced from the 4-μm square window etched through a photoresist layer deposited onto an HgCdTe substrate. An aluminum layer was deposited onto this window to make electrical contact with the substrate; however, a number of the Al contacts on certain devices exhibited poor adhesion. The spectrum illustrated in Fig. 5.3 was produced from a window region on one of these devices that exhibited poor Al adhesion. The HgCdTe substrate in this spectrum contains relatively high Cl$^-$ and Br$^-$ signals that are not observed in spectra produced from similar regions on devices exhibiting good Al adhesion. Because this

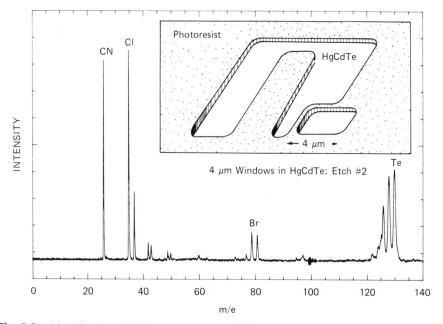

Fig. 5.3. Negative ion LIMS mass spectrum of the surface of a contaminated HgCdTe surface.

photoresist was etched with solutions containing Cl and Br, the analysis indicates that these contaminated surfaces were incompletely cleaned after the etching. This example illustrates several unique features of the laser microprobe technique including its ability to perform a near-surface microanalysis within an electrically insulating material (photoresist).

Figures 5.4 through 5.6 illustrate a "depth profile" microanalysis of an Al run on a semiconductor device. These metallizations exhibited high leakage currents and it was suspected that there was a buildup of mobile cations, most probably Na, at the metal/passivation (SiO_2) interface. Laser microprobe depth profiles are performed by firing multiple laser pulses onto the same area of the sample. Each laser shot excavates deeper into the sample and thus samples the composition as a function of depth. Since the craters formed by the laser are not uniform in depth, LIMS depth profiling has rather coarse depth resolutions by conventional standards. The depth resolution for a high-irradiance (sensitive) LIMS analysis is on the order of 0.1 μm. The positive ion spectrum in Fig. 5.4 was produced by the first laser shot onto an Al region and the crater formed by this shot is illustrated in the scanning electron microscope (SEM) image in the figure. The crater is ~5 μm in diameter and 0.5 μm deep. The mass spectrum in this figure is displayed over the mass range from 0 to 100 and contains intense cation signals of Na, Al, Al·HF$^+$ along with lower intensity signals of Si, K, and SiOH. The Al·HF$^+$ signal is most probably produced from a residue of the HF solution employed to remove the top glass from the device. The Na$^+$/Al$^+$ peak area intensity ratios at the estimated sampling depth are displayed in the lower left of this figure and these ratios represent a semiquantitative measure of the relative concentration of these two species. Figure 5.5 is a positive ion mass spectrum produced after firing two laser shots onto the Al run at a location near the one displayed in Fig. 5.4. The crater formed in this analysis is illustrated in the figure along with the abbreviated mass spectrum. The sampling depth for this analysis is ~1 μm, which is the approximate thickness of the Al layer. Thus, this analysis has sampled through the metallization into the underlying SiO_2

SEM Micrograph of Laser Microprobe Crater at Depth #1

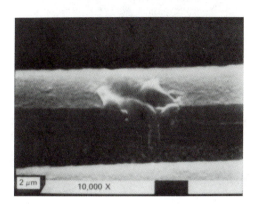

Na/Al Intensity Ratio at Different Depths

Positive Ion LIMS Spectrum at Depth #1

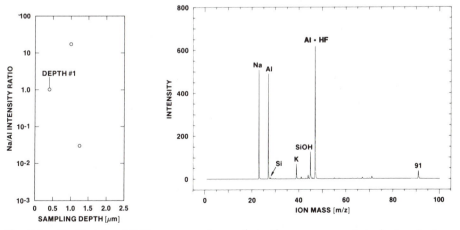

Fig. 5.4. Positive ion LIMS mass spectrum of an Al run on a semiconductor device: first laser shot onto surface.

**SEM Micrograph of Laser
Microprobe Crater at Depth #2**

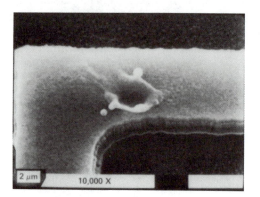

**Na/Al Intensity Ratio
at Different Depths**

Positive Ion LIMS Spectrum at Depth #2

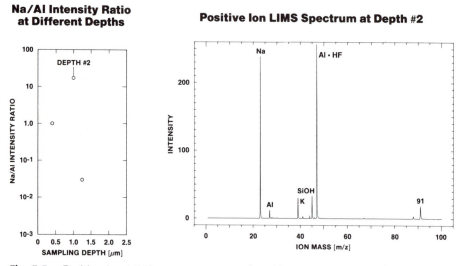

Fig. 5.5. Positive ion LIMS mass spectrum of an Al run on a semiconductor device: second laser shot onto surface.

region. The Na^+/Al^+ ratio for this analysis is approximately a factor of 10 greater than observed in the near-surface analysis (Fig. 5.4) suggesting that Na concentration is high at the Al/glass interface. Figure 5.6 illustrates the results of firing the laser three times onto another region of this sample. This analysis has sampled through the Al region into the underlying SiO_2 as evidenced by the onset of intense Si signals. The Na^+/Al^+ ratio at this sampling depth is relatively low suggesting low Na concentrations at this ~1.5 μm depth.

Quantitative Elemental Analysis

Laser microprobe researchers and analysts have investigated the quantitative or semiquantitative elemental analysis of the technique since its introduction. These investigations have been primarily in the areas of biological materials analysis,[21-23] the characterization of various fibers,[24] and the analysis of various particulates.[25] Recent interest has also centered on the ability of LIMS to determine

**SEM Micrograph of Laser
Microprobe Crater at Depth #3**

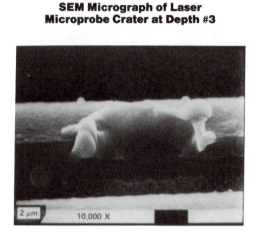

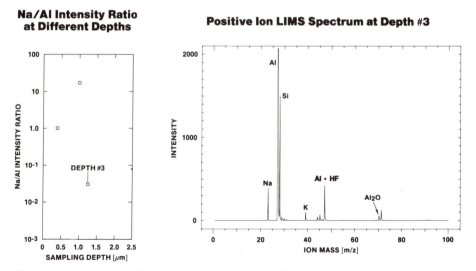

Fig. 5.6. Positive ion LIMS mass spectrum of an Al run on a semiconductor device: third laser shot onto surface.

the stoichiometry or elemental formula of various homologous compounds.[26] One factor that has impeded the general development of quantitative methods for this technique has been the observed variation in the ion yields for a given element contained in various solids (matrices).[27] Although it is been demonstrated that nearly uniform ion yields can be achieved using very high-irradiance ($\geq 10^{11}$ W/cm^2) laser pulses,[28] the signal intensities produced at these irradiances for the major constituents in the sample usually exceed the dynamic range of the TOF-MS detection system. Recent work performed at Cambridge Mass Spectrometry has demonstrated relatively good quantitative microanalysis of various glass samples by operating the laser microprobe at these high irradiance levels and limiting the energy bandwidth of the detected ion signals to a few eV.[29] However, these analysis conditions produce larger analytical craters (≥ 10 μm in diameter) and may reduce the overall detection sensitivity of the technique. The most common approach in the development of quantitative LIMS methods has been the determination of relative sensitivity factors (RSFs) for a particular element in a specific matrix.[8] The RSF of element X with respect to a (matrix) element Y is given by

$$\text{RSF}(X/Y) = \frac{^iI_Y/N_Y{}^if_Y}{^iI_X/N_X{}^if_X} \tag{5.5}$$

where ^iI_Y and ^iI_X are the ion intensities ($+$ or $-$) of the ith isotopes of elements Y and X having isotopic abundances ^if_Y and ^if_X and atomic number densities N_Y and N_X, respectively. This equation is the inverse of the relative ion yields for these two elemental species and the evaluation of the RSFs generally requires preparing standards of the different species. One problem associated with the determination of RSF values in laser microprobe applications is the requirement that these standard samples have a known composition that is chemically homogeneous down to dimensions on the order of 1 μm^3. The preparation of these microanalytical standards is no simple task, and the most widely used methods include doping epoxy resins with known concentrations of organometallic compounds,[30] the preparation of single-phase glass formulations,[31] and the preparation of ion implant standards.[32] This latter method does not produce a uniform depth distribution of the dopant; however, this depth distribution can be ascertained using a sensitive depth profiling technique such as SIMS.[33] The most common method of standard sample preparation for the LIMS technique has been the doping of epoxy resins since many early applications were in that area of biomaterials analyses.[23] Charles Evans & Associates has recently embarked upon a program to develop thin film standards of dielectric materials using ion implantation techniques for controlled doping of these films.[9]

    Figures 5.7 and 5.8 illustrate data from which an Fe RSF in doped epoxy resin can be determined. As will be discussed, the RSF derived from these data was used to quantitate the Fe levels in human liver tissue. This work was performed in our laboratory in collaboration with physicians at the University of California, San Francisco (UCSF) medical center. The epoxy standards were prepared by doping a tissue-embedding resin (Spurrs Resin) with known concentrations of iron acetyl-acetonate, curing the resin, and cutting 1.0-μm-thick sections. Standards having Fe concentrations over a range from 1000 to 0.1 ppm by weight (ppmw) were prepared

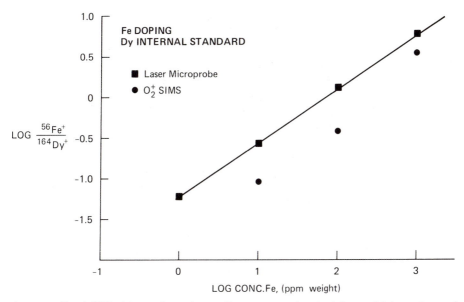

Fig. 5.7. $^{56}Fe^+/^{164}Dy^+$ intensity ratios vs Fe concentration in 1.0-μm-thick sections of Spurrs resin.

in this manner. These elemental standards were also doped with an internal standard (Dy) added as a tetramethyl heptane dionato complex at a concentration of 1000 ppmw. The precision of the data obtained with the Dy internal standard was substantially better than the data produced using C as the matrix species. Figure 5.7 presents the relative $^{56}Fe^+/^{164}Dy^+$ peak area intensities produced from these standards as a function of the Fe loading in the sample. Secondary ion mass spectrometry was also employed in the analysis of these standards, both to ascertain the expected uniformity of the dopants in the various specimens and as a comparison of technique. The response curve of a SIMS analysis using an O_2^+ primary ion beam is also presented in this figure. The Fe/Dy RSF derived from these data is constant over this Fe concentration range and the Fe detection limit for this type of epoxy sample is less than 1 ppmw.

This RSF data were used to analyze thin sections of both healthy and diseased human liver tissue samples. The tissue samples were prepared by embedding the fixed, dehydrated specimens in a Dy doped Spurrs resin and microtoming 1.0-μm-thick sections. This "chemical" sample preparation procedure can introduce chemical contaminants and may disturb the spatial distribution of the tissue's elemental constituents.[34] There was no indication that the preparation of these samples introduced either Fe contamination or m/z 56 spectral interferences. The Ca ion did not form a detectable CaO^+ signal under the analysis conditions employed. Although it is difficult to ascertain with complete confidence that the Fe did not transmigrate during the sample preparation, the results presented indicate that this element did not significantly redistribute during sample preparation.

The results of a LIMS analysis of a section prepared from a specimen obtained from a patient exhibiting the symptoms of hemochromatosis (iron overloading) is

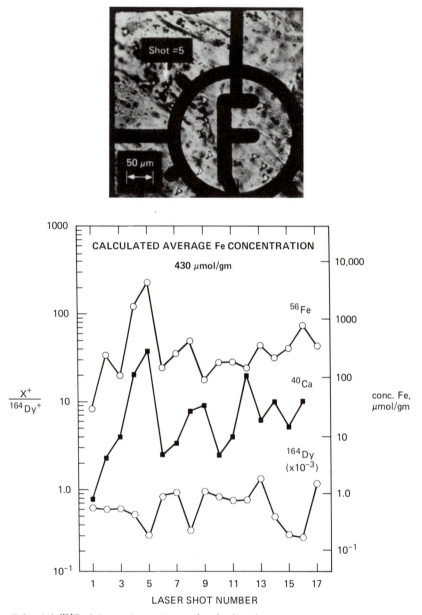

Fig. 5.8. $X^+/^{164}Dy^+$ intensity ratios and calculated Fe concentrations observed in a LIMS analysis of human liver tissue from a patient exhibiting hemochromatosis.

illustrated in Fig. 5.8. The relative X^+/Dy^+ peak area intensities are plotted in this figure as a function of the position on the specimen. Each position (laser shot number) corresponds to a single laser shot into this region in which <25 pg of sample was vaporized. The optical micrograph displayed at the top of this figure was obtained after the LIMS analysis by staining the section with toluidine blue dye. This micrograph shows the various laser shot positions (the number 5 shot is indicated by the arrow), along with the letter "F," which is part of the TEM finder grid on which the thin section was mounted. The vertical axis on the right-hand side of the plot contains an Fe concentration scale derived from the analysis of the Fe/Dy standard. The highest Fe^+/Dy^+ intensity ratio is observed at shot number 5, which appears to coincide with a hemosiderin granule in the tissue specimen. These granules are one of the iron storage sites in liver cells. The average Fe concentration (~430 μmol/g) in this sample was calculated by averaging the measured values at the 17 analysis sites and correlates quite well with the bulk (atomic absorption) Fe concentration levels for this type of tissue. The analysis of a normal human liver specimen yielded average Fe concentrations on the order 2 μmol/g, which also agrees quite well with the bulk analysis values for this type of tissue. It is important to emphasize that the total LIMS analysis of this type of material consumed ~0.5 ng of sample, whereas bulk analysis techniques generally consume milligrams of sample.

This is one example of a semiquantitative laser microprobe analysis. Many others are discussed in the literature cited in this section.

Organic Microanalysis

The laser microprobe technique has the potential for performing a sensitive molecular microanalysis of a vast number of organic materials including polymers[35] and biological specimens.[35] The characterization of a materials organic composition could also find extensive application in the analysis of a whole host of suspected organic thin film contaminants introduced during various materials processing steps. Organic LIMS analysis could easily become the most common application for the technique because of its unique combination of the microanalytical feature with the high mass detection capability of TOF-MS. Before this type of application becomes more routine, however, a better understanding of the types of mass spectra produced from various organic materials at a variety of laser irradiance conditions will be required. This understanding implies to a certain extent the development of a laser microprobe organic mass spectral data base similar to the ones developed for electron impact (EI) and, to a lesser extent, chemical ionization (CI) mass spectrometry. A number of research groups are pursuing the development of such a LIMS spectral data base,[37] however, this work will certainly require many man-years of effort by a number of laser microprobe users. We have recently initiated a research program (funded by the National Science Foundation) to develop such a data base for polymer analysis. A key feature of this research is the evaluation of various pattern recognition processes to classify the mass spectra produced from various polymer formulations analyzed under high-laser irradiance (pyrolytic) conditions. Since organic LIMS analyses are discussed in more detail elsewhere in this monograph, this section will present only one example of an organic LIMS analysis performed in our laboratory.

Figure 5.9 shows positive and negative ion spectra produced in the analysis of a commercial photoresist (PR) layer on a semiconductor device. The spectra produced from this type of polymeric sample is of particular interest to many of our clients because potential causes of device failure could originate with problems in the photoresist. Factors such as incomplete curing, nonuniform coverage, chemi-

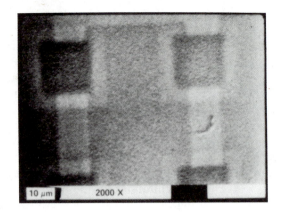

Positive Ion LIMS Spectrum

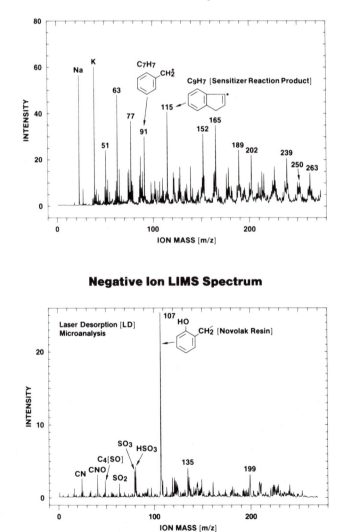

Negative Ion LIMS Spectrum

Fig. 5.9. Positive and negative ion LIMS mass spectra of a photoresist surface.

cally contaminated materials, or incomplete photoresist removal could all lead to device performance problems. The ability to detect the presence or absence of various photoresists has very relevant materials analysis applications. The spectra illustrated in Fig. 5.9 were obtained by focusing a relatively high-irradiance ($\sim 10^9$ W/cm^2) pulse onto the underlying device that produced the distorted crater shape at the photoresist surface. The positive ion mass spectrum exhibits a number of C-based fragment ion peaks, several of which may be structurally significant signals of the photoresist. In particular, the peaks at masses 91 and 115 could be various reaction products of this particular PR formulation. The negative ion mass spectrum in this example exhibits an intense signal at mass 107, which is undoubtedly formed from the Novolak resin in the photoresist. This resin represents about 10% by weight of the PR composition. Although this is only one example of the qualitative organic microanalysis capability of LIMS, it represents a very pertinent one. These types of organic characterizations are being actively pursued in our laboratory and elsewhere.

LASER MICROPROBE POSTABLATION IONIZATION

The technique of postablation ionization (PAI) in laser microprobe mass spectrometry refers to the ionization of the neutral species vaporized by the conventional laser on this instrument. The ionization of these ablated neutrals is most efficiently performed by irradiating the neutral plume with a second laser pulse. To detect only one neutral species in this vapor plume for each ablating laser pulse, the most efficient laser ionization process would be resonant multiphoton ionization (RMPI) in which the ionizing laser is tuned to an absorption line of a particular neutral species.[38] However, the RMPI process is inefficient for survey analysis because the ionizing laser wavelength must be changed for each neutral species of interest. Nonresonant multiphoton ionization (NRMPI), by contrast, provides essentially uniform ionization efficiencies for a large number of elements (see following discussion), but requires high photon fluxes.[39] The observed variability in relative ion yields of various elements analyzed by the LIMS technique was previously mentioned, and a primary motivation for developing a PAI configuration for the laser microprobe was to improve the ionization uniformity. The system described employs NRMPI processes to achieve this improvement.

To achieve the maximum ionization efficiency within the postionizing region for an NRMPI process, the photon flux density of the ionizing laser beam must be as high as possible. NRMPI processes that require the simultaneous, nonresonant absorption of n photons to effect ionization have an ionization probability given very approximately by

$$P(n) = \sigma(n)F^n\tau_p \qquad (5.6)$$

where $P(n)$ is the probability for the n-photon absorption, $\sigma(n)$ is the n-photon absorption cross section, F is the photon flux density, and τ_p is the laser pulse duration. For a two-photon process, a $\sigma(2)$ is on the order of 10^{-49} cm^4 sec, and the photon flux density required for 100% ionization [$P(2) = 1$] is on the order of $F = 5 \times 10^{28}$ cm^{-2} sec^{-1}. This flux density corresponds to a laser irradiance of about 3

$\times$ 10^{10} W cm^{-2} at a wavelength of 266 nm. These photon flux densities are easily achieved using a focused, frequency-quadrupled Nd:YAG laser.

The experimental configuration used in the laser PAI experiments is based on the LIMA 2A laser microprobe previously described. A second high-irradiance, frequency-quadruped Nd:YAG laser ($\lambda = 266$ nm; pulse width $\tau_p = 5$ nsec), focusing optics, and pulse time delay circuity have been added to the basic laser microprobe system to perform PAI analyses. The principle of the experiment is schematically shown in Fig. 5.10. The ablating laser is directed through the Cassegranian optics and focused to an ≈ 2-μm-diameter spot on the target surface, producing the emission of charged and (mostly) neutral particles. At time interval Δt later ($0 \leq \Delta t \leq 2.5$ μsec), the ionizing laser is fired parallel to the sample surface. This ionizing laser beam is focused to a spot size of approximately 100 μm in diameter the center of which passes ~ 0.6 mm above the sample surface. The ionizing laser typically operates at photon flux densities in excess of 7×10^{28} cm^{-2} sec^{-1}, which are sufficient to ionize all the neutrals within the photon beam that have an ionization potential ≤ 9.32 eV. It should be noted that because of both the velocity distribution(s) of the vaporized neutrals and the small interaction volume of the ionizing laser, only a fraction of the total number of ablated neutrals is within the ionization region during any 5-nsec pulse of this ionizing laser.

Ion mass and intensity analysis are performed in a manner similar to conventional laser microprobe analyses, except that it is usually necessary to separate the

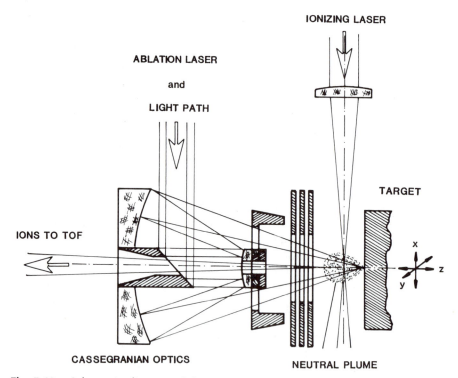

Fig. 5.10. Schematic diagram of the experimental setup used for laser microprobe postablation ionization.

ions formed by the ablating laser pulse from those produced by the ionizing laser. The failure to separate these essentially distinct mass spectra would result in overlapping peaks having different mass scales. These two ion signals are effectively separated in the following manner. Because the ablated ions are formed at or near the sample surface and experience the full acceleration potential whereas the postionized signals are formed above the sample surface, the ablated ions have a higher kinetic energy than the PAI ions. By increasing the sample accelerating potential appropriately, the mass spectrometer can be adjusted such that the ablated ions have sufficient kinetic energy to pass through the ion reflection optics and strike the spectrometer walls, whereas the postionized species are reflected in the reflection optics toward the detector. All other characteristics of the detection process are identical to those previously discussed. The remainder of this chapter presents various applications of the PAI technique.

Laser Microprobe and Laser Microprobe Postionization Analysis of GaAs

One of the first materials evaluated in the PAI analysis mode was GaAs.[40] This compound was chosen because of its relevance to the semiconductor industry, its simple stoichiometry, and the fact that the ionization potentials (IP) of Ga and As have quite different values. The IP of Ga is 5.97 eV and NRMPI of this element requires the simultaneous absorption of two photons at 266 nm (4.66 eV photon energy). Arsenic has an IP of 9.81 eV and the ionization of this element requires a three-photon absorption process at this wavelength. Since GaAs is a binary compound, a number of typical features of the conventional LIMS and PAI analyses are easily demonstrated with this target. This statement is not meant to imply that the analysis of a chemically simple material necessarily produces simple results!

Figures 5.11 and 5.12 are typical positive ion mass spectra obtained from a GaAs target surface in the conventional LIMS and PAI analysis modes, respectively. The ablating laser irradiance was of the order of 5×10^8 W cm^{-2} for both analyses and the $\lambda = 266$ nm for both lasers. The photon flux density of the ionizing laser was approximately 10^{29} cm^{-2} sec^{-1} corresponding to an irradiance of 7×10^{10} W/cm^2. The ionizing laser was fired with a delay time $\Delta t = 800$ nsec after the ablating laser pulse.

The most obvious difference between these two spectra is the complete absence of any signal related to As in the ablation only mass spectrum (Fig. 5.11). This mass spectrum by itself would suggest that the sample was composed of essentially pure Ga. Positive ion signals for As are detected in conventional LIMS analyses only at relatively high irradiance ($\geq 10^{10}$ W cm^{-2}). These high-irradiance conditions often produce large craters (≥ 10 μm in diameter) that are not acceptable in certain microanalysis applications. Arsenic does, however, readily form a negative ion under ablation only conditions.

The mass spectrum of postionized neutrals clearly suggests that a GaAs target has been analyzed. This spectrum illustrates a general observation that the PAI technique provides a more accurate compositional survey analysis of the elemental constituents in a material than does the conventional LIMS technique. The total Ga$^+$ signal intensity in this spectrum is about 65% larger than the As$^+$ intensity, which undoubtedly is a reflection of the lower ionization probability of the three-

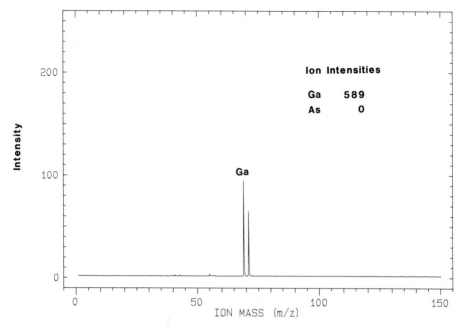

Fig. 5.11. Typical laser microprobe positive ion mass spectrum obtained from GaAs using an ablating laser irradiance $E_a = 5 \times 10^8$ W/cm² and $\lambda = 266$ nm (Ref. 40).

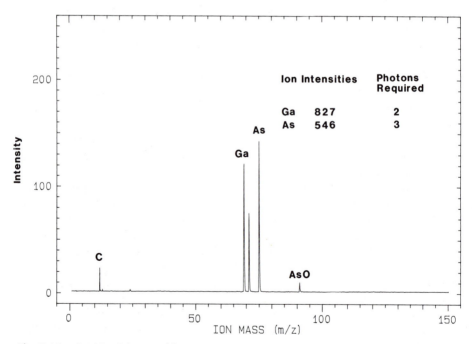

Fig. 5.12. Positive ion postablation ionization mass spectrum from GaAs. The ionizing laser was fired $\Delta t = 800$ nsec after the ablating laser pulse. The laser irradiance within the ionization region was $E_1 = 7 \times 10^{10}$ W/cm² and $\lambda = 266$ nm ($E_a = 5 \times 10^8$ W/cm², $\lambda_a = 266$ nm) (Ref. 40).

photon NRMPI process for As. This type discrimination of two- vs three-photon ionization processes is expected from simple geometrical considerations of the photon intensity distribution in the ionization region. The experimental results do show, however, that the As$^+$ intensity can exceed the Ga$^+$ intensity by more than a factor of two at delay times $\Delta t > 1.2$ μsec.

Because the ablated neutrals can have a range of emission velocities from the target surface, a complete PAI analysis requires measuring the ion signals as a function of delay time, Δt, between the firing of the two laser pulses. Figures 5.13 and 5.14 illustrate the results of such a delay time analysis of GaAs performed at three different ablating laser irradiances. The data in Fig. 5.13 are the postionized Ga$^+$

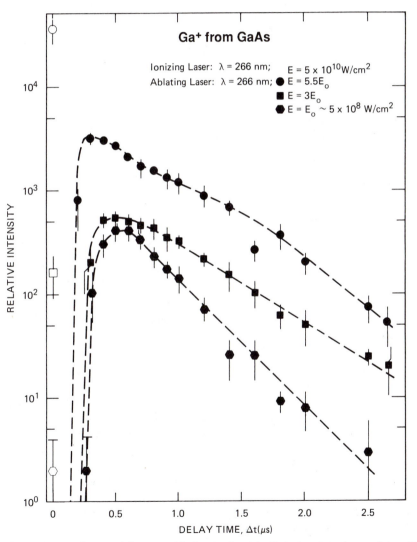

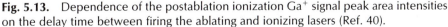

Fig. 5.13. Dependence of the postablation ionization Ga$^+$ signal peak area intensities on the delay time between firing the ablating and ionizing lasers (Ref. 40).

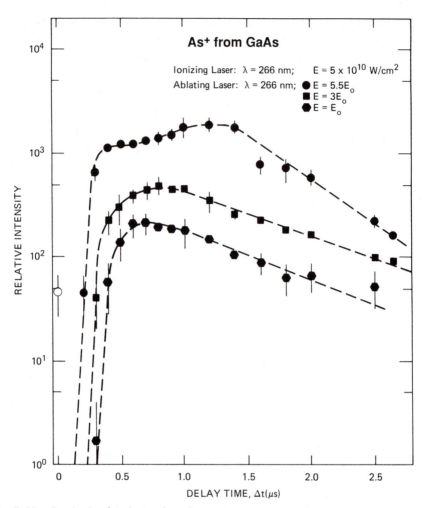

Fig. 5.14. Postionized As⁺ signal peak area intensities as a function of the delay time between firing the ablating and ionizing lasers (Ref. 40).

intensities as a function of Δt and Fig. 5.14 presents the As⁺ delay time data. Each data point in these plots is the average of at least seven PAI analyses performed at seven different locations on the GaAs sample surface. The ionizing laser irradiance was held constant in these analyses and the ionizing laser photon flux density was $\sim 7 \times 10^{28}$ cm⁻² sec⁻¹. The lower intensity set of data points was obtained at an ablating laser irradiance $E_0 = 5 \times 10^8$ W cm⁻², the intermediate intensity values were produced at $E = 3E_0$, and the highest intensity signals were produced at $E = 5.5E_0$. The ion intensities produced in the ablating laser only analysis are plotted at $\Delta t = 0$.

These figures illustrate several typical features of the postionization analysis that have been observed with many sample materials. For example, the onset of a PAI signal for Ga at the ablating laser irradiance $E = E_0$ occurs at $\Delta t \approx 250$ nsec. This Δt represents the time it takes the "fastest" neutrals desorbed at this irradiance to reach the ionizing zone. The Ga^+ PAI signal is observed to increase steeply above this onset delay time, reach a maximum value, and then decrease slowly at longer delay times. The conventional LIMS (ablating laser only) analysis produced detectable Ga^+ signals in only 10% of the mass spectra obtained at this ablating irradiance. At lower ablating irradiances, the PAI technique exhibits substantially higher absolute detection sensitivities than conventional LIMS for all elements except the alkalis. Since lower ablating irradiances vaporize a smaller volume of material, the PAI technique generally provides a more sensitive near-surface microanalysis than the conventional LIMS technique. We have been able to detect relatively intense postionized signals from all the materials analyzed to date at ablation irradiances below the level at which optical damage is observed (laser crater diameter ≤ 0.5 μm). The magnitude of the increase in the PAI ion yield depends on the materials' composition, and ion yield enhancement factors between 5 and 10^4 have been observed.[40]

The data produced at higher ablating laser irradiances start and peak at shorter delay times, indicating that faster neutral species are being emitted from the target at this higher laser irradiance. The ablated ion intensities also increase strongly with increasing ablating irradiance although the ablating only As^+ intensity remains severely underrepresented at all irradiances (Fig. 5.14).

The postionized As^+ intensities exhibit qualitative behavior similar to Ga^+ as a function of Δt. However, three important differences are observed for these two elements. Both the onset and the peak of the PAI As intensity are shifted toward longer Δt values and the shapes of the As^+ and Ga^+ intensity distributions with respect to Δt are different. The fact that the postionized Ga^+ signal is detected ~ 150 nsec earlier than As^+ (at $E = E_0$) cannot be accounted for by the mass difference of these two species assuming they have the same emission energy from the target.

The simplest interpretation of these observations is to assume that the velocity distribution of the neutrals evaporated by the ablating laser can be described by a Maxwell–Boltzmann (MB) velocity distribution of an ideal gas having some (constant) "temperature" T. This distribution is given by

$$dN/N \, dv = (2/\pi)^{1/2}(m/kT)^{3/2}v^2 \exp[-(mv^2/2)/kT] \qquad (5.7)$$

where N is the number of particles of mass m having a velocity v, T is the temperature of the ideal gas, and k is the Boltzmann constant. If the particles follow an MB distribution, then a plot of $\ln[v^{-2}(dN/N \, dv]$ vs v^2 should be a straight line with a slope determined by mass and temperature. The PAI experimental values can be easily transformed into the required quantities since the velocity is simply given by the ratio of the distance of the ionization zone from the target and the delay time Δt between ablating and ionizing laser. The measured postionized neutral signal at a given Δt is the fraction of particles within a given velocity interval ($dN/N \, dv$), assuming they are generated at the same time.

Figure 5.15 illustrates the data displayed in Figs. 5.13 and 5.14 plotted as $\ln[v^{-2}(dN/N \, dv)]$ vs v^2. The data plotted in Fig. 5.15 were obtained at ablating laser irradiance $E = E_0$. The postionized Ga and As signals do indeed follow the expected straight line dependence; however, these two elements exhibit different slopes. If kT is assumed to be constant, then the difference in the slopes would have

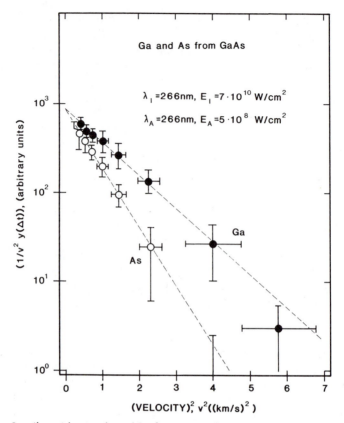

Fig. 5.15. Semilogarithmic plot of $[v^{-2}y(\Delta t)]$ vs v^2 for the postionized As^+ and Ga^+ signals according to Eq. (5.7) for the lower sets of data points of Figs. 5.13 and 5.14.

to be caused by a difference in the mass of the emitted neutrals. The slope of the fitting line for As is approximately twice the slope of the Ga fitting line. The difference in the slopes strongly suggests that if the Ga^+ PAI signals are produced from atomic Ga neutrals, then As^+ PAI signals have to originate from a cluster As_2 or GaAs. The best fit for the "temperature" in these analyses was $T = 4000$ K ($kT = 0.33$ eV) for both species.

It should also be noted that both fitting lines have an intersection point as the lines are extrapolated toward $v = 0$. This intersection point indicates that equal numbers of Ga and As are emitted from the target. Although this is obviously true, the experimental observation of this is somewhat surprising considering the differences in IP for Ga and As. One explanation of this observation is that various dissociation/ionization processes occur in the As cluster species that produce this result.

The apparent formation of As^+ from As cluster species in this PAI analysis is also supported by the results of a study of the ion yields as a function of ionizing laser flux density [F in Eq. (5.6)]. The NRMPI of atomic As requires a three-photon process whereas the ionization of atomic Ga requires a two-photon process. Therefore a quadratic dependence [$n = 2$ in Eq. (5.6)] of the Ga^+ signal and a cubic dependence ($n = 3$) for As$^+$ formation as a function of F for PAI analysis performed at any fixed Δt and ablating laser irradiance should be observed. The results of these experiments, however, show that both species exhibit nearly the same power dependence and $n \approx 1.5$ for both signals. These results strongly suggest that the ionization process(es) is saturated for both species.

Similar fitting lines can be prepared from the data obtained at the higher ablating irradiances. Assuming that the same species are emitted at higher irradiance, then kT must increase with increasing ablating laser irradiance. It should be noted that the Ga signals can again be fitted to an MB distribution assuming that only neutral Ga atoms are emitted from the target. In the case of As, the signals obtained at the intermediate ablating laser irradiance can be fitted with the Ga kT values by assuming that mainly As$_2$ neutrals leave the target. At $E = 5.5E_0$, an additional contribution from even heavier neutral species has to be assumed to fit the observed dependence of the As signals on the delay time.

Although we do not claim that this model rigorously explains the process of neutral particle emission from the laser-irradiated surfaces, it does provide a reasonable phenomenological approximation to the observed emission characteristics and the data presented will be discussed keeping this simple model in mind.

Postionization Analysis of a Steel Sample

A second example is the PAI analysis of a National Bureau of Standards (NBS) microprobe steel standard. SRM 479a is a Cr, Fe, Ni alloy that has been certified by the NBS for both its elemental composition and homogeneity down to ~1 μm lateral dimensions. The atomic percentage ratios for this standard are ^{52}Cr/^{56}Fe = 0.225 and ^{58}Ni/^{56}Fe = 0.0972. Figure 5.16 shows the dependence of various PAI signal intensities as a function of Δt obtained at fixed ablating and ionizing laser irradiances. Each data point in the figure is the average of 15 analyses at 15 different locations on the sample surface. The error bars are the 1 σ standard deviation (SD) of these average values. The ion intensities obtained using the ablating laser only are also shown in Fig. 5.16 as data points at $\Delta t = 0$.

An analysis of the ablated ion intensities ratios yields concentration ratios of ^{52}Cr/^{56}Fe = 0.58 and a ^{56}Ni/^{56}Fe = 0.063. The Cr/Fe ratio obtained from these ablation only experiments is a factor of ~3 too high whereas the Ni/Fe ratio is a factor of ~1.5 too low. An analysis of the PAI results performed by taking the ratios of the total peak intensities for the different elemental time distributions yields a Cr/Fe ratio that is about a factor 1.6 lower than the certified value, whereas the Ni/Fe ratio is within about 10% of the quoted value. The dotted lines in Fig. 5.16 are best fit MB velocity distributions calculated for atomic neutrals with $kT = 0.75$ eV. The approximation shows good agreement with the experimental data except for the low-level signals of Ni at $\Delta t > 1$ μsec.

Figure 5.17 illustrates this same PAI data plotted as a semilogarithmic plot of the ratios of the PAI signal intensities in a given velocity range vs v^2. Assuming that two different neutral species have the same MB energy, this intensity ratio is proportional to the relative number of neutrals vaporized [Eq. (5.7)]. The Cr/Fe ratio has a straight line dependence in this representation whereas fairly large deviations are found in the Ni/Fe ratios at low velocities as would be expected from the MB

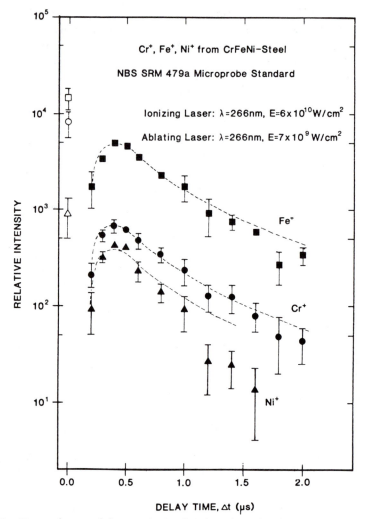

Fig. 5.16. Dependence of the postionized Cr^+, Fe^+, and Ni^+ signal peak area intensities on the delay time between firing the ablating and ionizing lasers obtained from the NBS SRM 479a microprobe standard. Data points at $\Delta t = 0$ were obtained using the ablating laser only.

fits in Fig. 5.16. The extrapolation of the fitting lines to $v = 0$ gives ratios of $^{52}Cr/^{56}Fe = 0.14$ and $^{58}Ni/^{56}Fe = 0.10$, the latter being in excellent agreement with the quoted value.

Postionization Analysis of $Hg_{0.78}Cd_{0.22}Te$

$Hg_{1-x}Cd_xTe$ is II–VI ternary material utilized as an infrared light sensor. The microanalysis of this type of material is difficult mainly because of the relatively high volatility of Hg. The surface-sensitive, PAI survey analysis of this type of ternary material is a very useful analytical application of the PAI technique. The sample analyzed in this study had a quoted value of $x = 0.22 \pm 0.01$. Figure 5.18 illustrates a typical positive ion mass spectrum produced from this material using the conventional (single laser) mode of analysis. The laser irradiance was $\sim 3.5 \times 10^9$ W/cm^2 ($\lambda = 266$ nm) and the laser damage craters were about 4 μm in diameter under these conditions. Ion species related to Te and Cd only are observed in the ablated ion mass spectrum. The most intense signals in the mass spectrum are peaks corresponding to CdTe$^+$ and Te$_2^+$, and it would be very difficult to identify the atomic composition of this sample from this mass spectrum. The corresponding postionization mass spectrum in Fig. 5.19 was obtained with the same ablating laser irradiance and an ionizing laser irradiance of 6.4×10^{10} W/cm^2 at a delay time $\Delta t = 1.8$ μsec. This PAI mass spectrum exhibits significantly more signal intensity ($\sim$factor of 30) than the ablating laser only spectrum, and the PAI spectrum contains readily identifiable signals for the three elements in this sample.

Figure 5.20 plots the velocity normalized intensities of the Te$^+$, Cd$^+$, and Hg$^+$ signals observed in the PAI delay time analysis of this sample in the form of a $\ln[v^{-2}(dN/N\ dv)]$ vs v^2 curve. These data were obtained using the laser irradiance

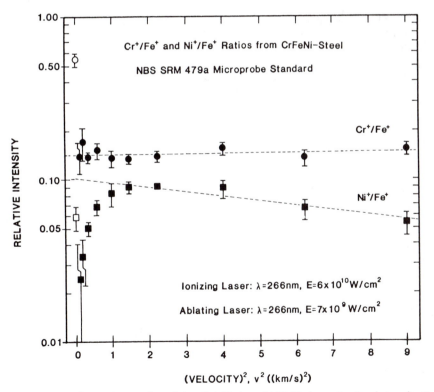

Fig. 5.17. Semilogarithmic plot of the Cr$^+$/Fe$^+$ and Ni$^+$/Fe$^+$ postionized signal ratios obtained from the NBS SRM 479a microprobe standard on v^2.

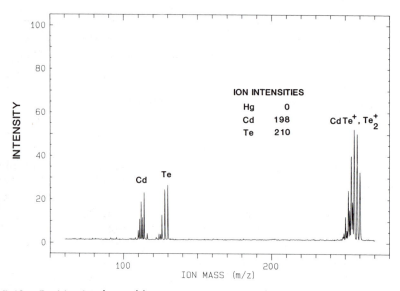

Fig. 5.18. Positive ion laser ablation mass spectrum of $Hg_{0.78}Cd_{0.22}Te$ obtained with E_a = 3.5 × 10⁹ W/cm² and λ = 266 nm (Ref. 40).

conditions employed to generate the mass spectrum in Fig. 5.19. It is obvious that none of the sets of data points can be represented by a single straight line dependence but appear to be the superposition of two fitting lines of different slope. One possible interpretation is that the data are comprised of MB distributions of different mass neutrals such that first lighter and then later heavier neutral species are

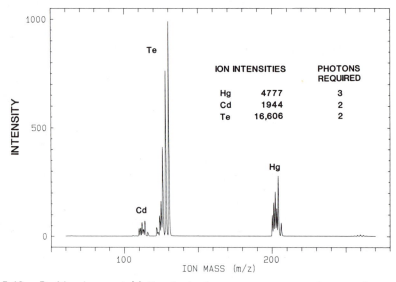

Fig. 5.19. Positive ion postablation ionization mass spectrum of $Hg_{0.78}Cd_{0.22}Te$ taken with E_a = 3.5 × 10⁹ W/cm², E_i = 6.4 × 10¹⁰ W/cm², and $λ_i$ = $λ_a$ = 266 nm. The ionizing laser was fired 1.8 μsec after the ablating laser.

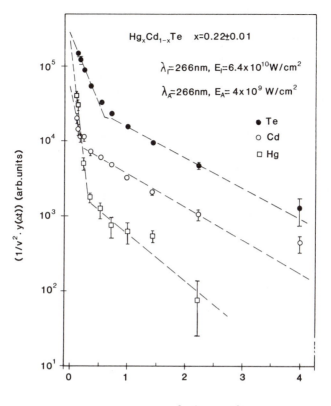

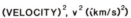

Fig. 5.20. Dependence of $[v^{-2}y(\Delta t)]$ for postionized Te^+, Cd^+, and Hg^+ from $Hg_{0.78}Cd_{0.22}Te$ on v^2.

emitted and ionized. The ratios of the fitting line slopes suggest that the faster neutrals ($v > 2.5$ km sec^{-1}) are mainly atomic Te, Cd, and Hg, whereas the slower ones are correlated to heavier cluster species.

The rather complicated behavior of the signal intensities suggests three ways to determine the materials' approximate stoichiometry. Since the Cd/Te mass ratio is close to unity, representative Cd/Te ratios can be readily calculated in the region in which mainly atomic species should be found. For delay times between $\Delta t =$ 300 and 700 nsec, the measured ratio is Cd/Te = 0.23 ± 17%, which is in relatively good agreement with the expected value. The Hg/Te ratio cannot be determined in this way because of the large slope difference.

An alternative procedure involves extrapolating the fitting lines for the data below $\Delta t = 800$ nsec ($v \geq 0.7$ km sec^{-1}) to $v = 0$, which yields the ratios Cd/Te = 0.22 ± 16% and Hg/Te = 0.06 ± 20%. Only the Cd/Te value is in good agreement with the expected value using this procedure. Of course, an incorrect Hg/Te ratio is expected if only atomic Te and Hg are ionized by the laser beam since NRMPI for Te is two-photon process whereas Hg requires a three-photon absorption process. If the ion intensities observed at longer delay times ($\Delta t > 800$ nsec) are extrapolated to $v = 0$, the calculated ratios are Cd/Te = 0.20 ± 25% and Hg/

Te = 0.77 ± 20%, both of which are in reasonable agreement with the expected stoichiometry. The PAI determination of the stoichiometry for this relatively complex material is not trivial. This analysis clearly illustrates that the phenomenological description based on an MB formalism works best for those species that are emitted as elemental particles.

The data discussed were obtained at relatively high ablating laser irradiances that produced substantial damage to the target. Reducing the ablating laser irradiance by a factor of ~7 reduces the crater diameters to ~0.5 μm. At this lower ablating irradiance, no ions are observed in the ablation only mode of analysis; however, the PAI intensities decrease by only a factor of two. Reducing the ablating laser irradiance further eliminates all visible surface damage, but the PAI signals are still easily detected. Thus, as was the case for the GaAs material, the PAI technique provides significantly enhanced elemental *surface* sensitivity compared to the conventional laser microprobe technique.

Organic Background Reduction with Postionization

The analysis of organic compounds adsorbed onto solid surfaces represents an important application of solids mass spectrometry.[41] In many elemental analyses, however, signals produced from organic species severely complicate the mass spectra and often prevent a useful elemental analysis of a solid material. These organic "contaminants" on solid surfaces are typically introduced by various chemical

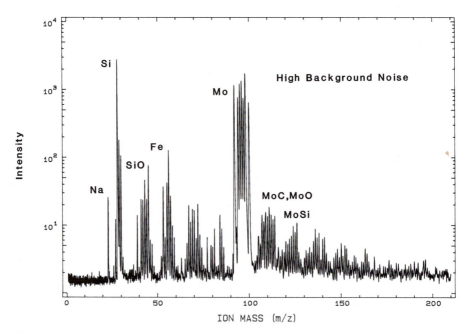

Fig. 5.21. Positive ion laser ablation mass spectrum obtained from organically contaminated surface of $MoSi_{2.25}$.

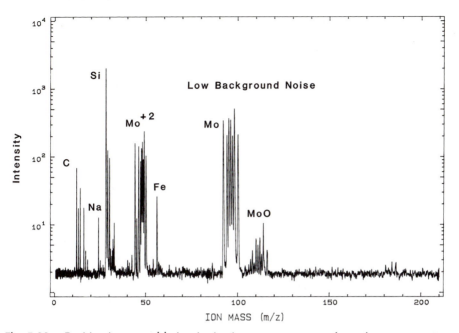

Fig. 5.22. Positive ion postablation ionization mass spectrum from the same $MoSi_{2.25}$ surface as in Fig. 5.21 using the same ablating laser irradiance. The ionizing laser was fired 400 nsec after the ablating laser and had an irradiance of $E_1 = 5 \times 10^{10}$ W/cm² at $\lambda = 266$ nm.

processing steps and/or inadvertent handling. Some of our early investigations of the PAI technique demonstrated that the ion signals produced from chemical contaminants on surfaces typically consisted of C or low-mass carbon cluster ions. Thus, the NRMPI process appears to effectively eliminate higher mass interferences arising from these contaminants, which further suggests the utility of the PAI technique in survey elemental analyses.

Figure 5.21 shows a positive ion mass spectrum produced by a conventional LIMS analysis of an $MoSi_{2.25}$ surface. This mass spectrum is complex and any low-intensity signals in the mass range between m/z 60 to 90 and 110 to ~150 would be very difficult to identify. The detection of isotopic patterns for low-level signals in these mass ranges would, in most cases, be virtually impossible. The postionization mass spectrum illustrated in Fig. 5.22 was obtained at the same ablating irradiance and the PAI analysis area was close to the one that produced the spectrum in Fig. 5.21. The ionizing laser irradiance was 5×10^{10} W/cm². The PAI mass spectrum shows a dramatic reduction in the "organic" background intensity along with strongly enhanced C, C_2, and C_2H cation signals. The most intense signals in the PAI mass spectrum are those of Mo and Si. Postionization did, however, generate "new" background signals of the Mo^{2+} ions. The PAI mass spectrum contains significantly fewer background mass interferences than the conventional LIMS spectrum and this background reduction is observed in all PAI analyses of "contaminated" sample surfaces.

ACKNOWLEDGMENTS

The authors wish to thank Mr. Charles J. Hitzman, Ms. Ilsabe C. Niemeyer, and Dr. Filippo Radicati di Brozolo for the contributions they have made to the work described herein. Support for portions of the LIMS research performed at Charles Evans & Associates was provided by the National Science Foundation (Grants ISI 8560203 and ISI 87000019) and the National Institutes of Health (Grant GM33123-03). This support is gratefully acknowledged.

REFERENCES

1. Excellent papers on early laser microprobe applications are contained in *Fres. Zt. Anal. Chem.* **308**, 1981.

2. *Proceedings of Second International Laser Microprobe Mass Spectrometry Workshop,* edited by U. Seydel and B. Lindner. Borstel, West Germany, 1983.

3. *Proceedings of Third International Laser Microprobe Mass Spectrometry Workshop,* edited by F. Adams and L. Van Vaeck. Antwerp, Belgium, 1986.

4. R. J. Conzemius and J. M. Capellen, *Int. J. Mass Spectrom. Ion Phys.* **34**, 197 (1980).

5. *Secondary Ion Mass Spectrometry,* edited by A. Benninghoven, F. G. Rudenauer, and H. W. Werner. John Wiley, New York, 1987.

6. R. J. Blattner, in *Microstructural Science,* edited by D. Stevens et al. Elsevier North Holland, New York, 1980, p. 63.

7. *Scanning Electron Microscopy and X-Ray Microanalysis,* edited by J. I. Goldstein, D. E. Newbury, P. Echlin, D. C. Joy, C. Fiori, and E. Lifshin. Plenum Press, New York, 1981.

8. R. Kaufmann, in *Microbeam Analysis–1982,* edited by K.F.T. Heinrich. San Francisco Press, San Francisco, 1982, p. 341.

9. R. W. Odom and B. Schueler, *Thin Solid Films* **154**, 1 (1987).

10. B. A. Mamyrin, V. I. Karatev, D. V. Shmikk, and V. A. Zagulin, *Sov. Phys.–JETP* **37**, 45 (1973).

11. D. S. Simons, *Int. J. Mass Spectrom. Ion Process* **55**, 15 (1983).

12. J. F. Ready, *Effects of High-Power Laser Irradiation.* Academic Press, New York, 1971.

13. E. Niehuis, T. Heller, H. Feld, and A. Benninghoven, in *Secondary Ion Mass Spectrometry SIMS V,* edited by A. Benninghoven, R. J. Colton, D. S. Simons, and H. W. Werner, Springer-Verlag, New York, 1985, p. 188.

14. R. W. Engstrom, *Photomultiplier Handbook.* RCA Corp., Lancaster PA, 1980.

15. Hamamatsu Corp. Technical Data Sheet.

16. I. D. Kovalev, G. A. Maksimuv, A. I. Suchov, and N. V. Lamin, *Int. J. Mass Spectrom. Ion Phys.* **27**, 101 (1978).

17. R. J. Cotter and A. L. Yergey, *Anal. Chem.* **53**, 1306 (1981).

18. D. H. Hardin and M. Vestal. *Anal. Chem.* **56**, 81 (1984).

19. L. Van Vaeck, J. Claereboudt, J. De Waele, E. Esmans, and R. Gijbels, *Anal. Chem.* **57**, 2944 (1985).

20. Z. A. Wilk and D. M. Hercules, *Anal. Chem.* **59**, 1819 (1987).

21. G. L. Fain and W. H. Schroeder, *J. Physiol.* **368**, 641 (1985).

22. U. Seydel and B. Lindner, in *Microbeam Analysis–1987,* edited by R. H. Geiss. San Francisco Press, San Francisco, 1987, p. 353.

23. P. F. Schmidt, in *Analysis of Organic and Biological Surfaces,* Chap. 3, edited by P. Echlin. John Wiley, New York, 1984.

24. J. De Waele, P. Van Espen, E. Vansant, and F. Adams, in *Microbeam Analysis–1982,* edited by K.F.T. Heinrich. San Francisco Press, San Francisco, 1982, p. 371.

25. R. Kaufmann and P. Wieser, in *Particle Characterization in Technology, Vol. I,* Chap. 2, edited by J. K. Beddow. CRC Press, Baton Rouge, FL, 1984.

26. E. Michels and R. Gijbels, *Anal. Chem.* **56,** 1115 (1984).

27. P. Surkyn and F. Adams, *J. Trace Micro. Tech.* **1,** 79 (1982).

28. J.A.J. Jansen and A. W. Witmer, *Spectrochim. Acta* **37B,** 483 (1982).

29. T. Dingle and B. W. Griffiths, in *Microbeam Analysis–1985,* edited by J. T. Armstrong. San Francisco Press, San Francisco, 1985, p. 315.

30. P. Wieser, R. Wurster, and H. Seiler, in *Scanning Electron Microscopy, IV.* AMF O'Hare, Chicago, IL, 1982, p. 1435.

31. D. S. Simons, in *Secondary Ion Mass Spectrometry SIMS IV,* edited by A. Benninghoven, J. Okano, R. Shimizu, and H. W. Werner. Springer-Verlag, New York, 1983, p. 101.

32. R. G. Wilson and G. R. Brewer, *Ion Beams with Applications to Ion Implantation,* Chap. 3. John Wiley, New York, 1973.

33. P. Williams, in *Scanning Electron Microscopy, II.* AMF O'Hare, Chicago, IL, 1985, p. 553.

34. F. D. Ingram and M. J. Ingram, in *Scanning Electron Microscopy IV.* AMF O'Hare, Chicago, IL, 1980, p. 147.

35. J. A. Gardella and D. M. Hercules, *Spectrosc. Lett.* **13,** 13 (1980).

36. U. Seydel et al., *Eur. J. Biochem.* **145,** 505 (1984).

37. R. A. Fletcher and L. A. Currie, in *Microbeam Analysis–1987,* edited by R. H. Geiss. San Francisco Press, San Francisco, 1987, p. 369.

38. G. S. Hurst, M. G. Payne, S. D. Kramer, and J. P. Young, *Rev. Mod. Phys.* **51,** 767 (1979).

39. J. Morellec, D. Normand, and G. Petite, in *Advances in Atomic and Molecular Physics.* Academic Press, New York, 1982, p. 97.

40. B. Schueler and R. W. Odom, *J. Appl. Phys.* **61,** 4652 (1987).

41. W. Lange, M. Jirikowsky, and A. Benninghoven, *Surf. Sci.* **136,** 419 (1984).

6

Laser Micro Mass Spectrometry for Direct Analysis of Coal Macerals

JOHN J. MORELLI

The physical nature of coal has long been examined on a microscale; however, examination of the microchemistry of coal is a more recent development. The microchemical nature of inorganic (minerals) and organic (macerals) components of coal is fundamental to understanding coalification processes. Several techniques such as electron spin resonance,[1] gas chromatography/mass spectrometry,[2] nuclear magnetic resonance,[3,4] Curie-point pyrolysis mass spectrometry,[5] and Fourier transform infrared spectroscopy[6] are used to obtain structural information on bulk coal. Microchemical information on coal can be obtained with techniques such as proton-induced X-ray emission,[7] secondary ion mass spectrometry (SIMS),[8] X-ray microanalysis, and laser micro mass spectrometry (LAMMS).[9] Proton-induced X-ray emission provides elemental information on coals, whereas the techniques based on mass spectrometry allow chemical and molecular information to be obtained. The high-incident ion densities resulting from microprobe secondary ion mass spectrometric investigations preclude structural determinations of large organic functionalities; more typically, low m/z fragments such as C, CH, CO, and OH are observed. Alternatively, a laser focused to less than 10 μm can be successfully employed to study a large variety of organic compounds[10] yielding structural and molecular weight information.

The first use of a laser as an ion source in mass spectrometry was demonstrated by Honig in 1963.[11] Laser ionization was demonstrated to be applicable to organic compounds in 1966 by Vastola and Pirone.[12] In spite of the 20-year history of the use of the laser ionization source for mass spectrometry, practical applications have not developed rapidly. The slow development of laser mass spectrometry (LMS) is due, in part, to the complexity of the laser–solid interaction. Since the laser radiation employed in the LMS technique is often focused to the micron level, spatial inhomogeneities of the analyte also increase the difficulty of many LMS investigations. Irrespective of the problems associated with the LMS technique, there are many advantages to acquiring mass spectra by laser ionization; these include, but are not limited to, the ability to conduct microprobe investigations of organic substituents.[13-16]

Microprobe investigations using LAMMS have a distinct disadvantage to comparable SIMS studies. Unlike incident ion beams, incident laser radiation is not amenable to rapid positional manipulation using electron optics. Instead, rastering of a sample during a LAMMS analysis requires repetitive sample repositioning. For example, analyzing a 1 mm^2 area by use of LAMMS with a spatial resolution of 10 μm generates 1000 spectra requiring 100 sample manipulations. To make this operation practical, automated sample positioning and data acquisition are required. Commercial LAMMS systems are not equipped with automated mapping capability; however, reports on modification of commercially available hardware to allow automated ion mapping of coals exist in the literature.[15]

Laser micro mass spectrometry (LAMMS) is the terminology recommended by the International Union of Pure and Applied Chemistry (IUPAC) for "any technique in which a specimen is bombarded with a finely focused laser beam (diameter less than 10 μm) in the UV or visible range . . . , and the ions generated are recorded with a time-of-flight spectrometer"[19]; however, we need to differentiate between investigations that attempt to distinguish coal components spatially, and those characterizing neat compounds and materials. To maintain this distinction, this discussion uses the term LAMMS in referring only to those investigations requiring spatial resolution. The term laser mass spectrometry (LMS) is similarly restricted to descriptions of studies employing the LAMMA-1000 instrument (described in the following text) without any spatial monitoring of chemical variations. LMS typically involves accumulating and averaging more than 10 spectra from an undefined area of the sample.

It is important to understand the term chemical homogeneity to fully appreciate the complications introduced when quantitative analysis is attempted with any microprobe technique. Chemical homogeneity is a relative property that depends on the spatial resolution and precision of the analytical procedure applied to the investigation.[20] An analytically homogeneous solid is defined as a material whose chemical composition throughout the sample volume does not significantly fluctuate by more than the error of the analytical procedure.[21] The lack of a standard for coal of well-characterized homogeneity on a microscale complicates LAMMS investigations of coals; these complications involve the inability to distinguish between the variations in the chemical/physical nature of coal constituents and the variation introduced by the precision of the LAMMS technique. Such complications must be circumvented prior to using LAMMS to obtain quantitative or semiquantitative information.

Quantification with the LAMMA-1000 instrument has been difficult because of the complex nature and inadequate modeling of the ion formation processes occurring during laser ionization. Factors that affect shot-to-shot ion yield include real physical and chemical variations of coal components and surrounding matrix. Isolation and evaluation of all factors affecting shot-to-shot ion yield are difficult or impossible. There has been some success with the use of internal standards.[22] Furthermore, much can be learned from previous reports regarding quantitative and semiquantitative analysis by the use of SIMS.[23-25] In both LAMMS and SIMS, there have been two basic approaches to convert qualitative data into semiquantitative information: (1) the use of sensitivity factors and (2) the use of theoretical models combined with raw data to determine fundamental parameter values. For coal characterization, relative elemental sensitivity factors for LAMMS cannot currently be acquired because of the inability to monitor elemental concentrations on a microscale with the accuracy necessary to certify coal maceral "standards" in the sub-ppm range. The inability to obtain reliable sensitivity factors limits our approach to a semiquantitative one, in which empirical normalizations or semi-theoretical correction factors are used to convert the qualitative data into semi-quantitative information.

A study comparing several theoretical and empirical methods for quantitative SIMS analysis was made by Rudat and Morrison[25] who evaluated simple normal-ization techniques with and without correction for instrumental discrimination effects. The comparative study emphasized the dubious nature of the theoretical models addressed and showed that normalization of uncorrected signals resulted in accuracy approaching the theoretical methods. A "simulated" bulk analysis of elements dispersed throughout the organic portion of coal has been made with LAMMS by Morelli et al.[9] The LAMMS results were compared with results of DCAS and INAA. DCAS and INAA are routinely used in quantitative analysis of the bulk elemental content of coals. The inability to obtain reliable quantitative LAMMS data forced Morelli et al.[9] to normalize each elemental concentration of one coal to that of a chosen coal; thus, relative concentration variations were mon-itored. By comparing the results of "bulk" LAMMS to DCAS and INAA, the expected accuracy of LAMMS on a microscale was estimated.

This review focuses on laser mass spectrometric investigations employing the LAMMA-1000 Laser Microprobe Mass Analyzer (Leybold-Heraeuss[17]) conducted by the Chemistry Department of the University of Pittsburgh in collaboration with the U.S. Geological Survey under the supervision of Dr. Paul C. Lyons of the Geo-logical Survey and Dr. David M. Hercules of the University of Pittsburgh. The LAMMA-1000 is a commercially available system for ionizing microvolumes of matter by a laser beam and analyzing the ions in a time-of-flight spectrometer.[18] The results obtained to date regarding qualitative and semiquantitative investiga-tion of coal constituents, including vitrinite, resinites, and pyrite are presented. Also discussed are preliminary results regarding the automation of the LAMMA-1000 instrument for spatially mapping coal. The difficulties encountered during microprobe investigations of coal, a heterogeneous material, and the approaches for circumventing these difficulties are indicated. Many of these LAMMS meth-odologies presented may be found useful to future investigators of coal and will provide some general insight for studies of heterogeneous solids by use of a micro-probe. The ultimate goal of laser micro mass spectrometric investigations of coal is to obtain information on the microscopic chemical nature of the mineral and maceral components. Before such studies are attempted, coal must be demon-strated to be amenable to LAMMS investigation. Furthermore, procedures yielding useful information must be established.

A recent review outlines the LMS techniques, instruments, and applications, while also comparing the microprobe and bulk instruments[26]; therefore, this review will not detail the variety of methodology available for these studies. The evalua-tion of several coals by use of LAMMS has been reported.[9,13] In most cases, dc arc optical emission spectroscopy (DCAS), instrumental neutron activation analysis (INAA), and petrographic microscopy were employed for comparative purposes.

EXPERIMENTAL

Laser mass spectra and laser micro mass spectra were obtained by the use of a commercially available instrument, the LAMMA-1000.[18] The ionization source consists of an Nd:YAG laser. The fundamental frequency (1060 nm) is quadrupled to 265 nm. The beam is Q-switched, producing a full-width half-maximum (FWHM) pulse of 20 nsec. The intensity of the laser can be varied continuously by a pair of twisted polarizers. The diameter of the laser beam at the focal point is approximately 2 μm. Samples are mounted on an x, y, z micromanipulator having a maximum scanning range of 70 $\times$ 50 $\times$ 50 mm, respectively. The sample can be viewed by a microscope (standard equipment with the LAMMA-1000) at a magnification of 250$\times$, and the area of interest for analysis can be positioned using the sample manipulators. The laser impinges upon the sample at an angle of 45° to the sample surface. Ions formed from the laser pulse interaction with the sample are extracted normal to the sample surface. These ions are then accelerated (4 kV) into a time-of-flight (1.8-m flight length) mass spectrometer equipped with an ion reflector to compensate for the spread of initial ion kinetic energies. Figure 6.1 shows the relationship of the laser and ion optics relative to the sample. The ions are ultimately postaccelerated to 7 kV and detected by a Cu–Be secondary electron multiplier. The 17-stage electron multiplier has a gain of 10^6 with a good signal-to-noise ratio.[27] The signal is preamplified (10$\times$) and fed into a Biomation 8100 transient record digitizing at 100 MHz. The digitized data are temporarily stored in the transient recorder having 2 kbytes of memory. The data are ultimately transferred and processed in a Hewlett-Packard 1000 E-series computer. The resolution of the instrument under ideal conditions is approximately 800 (FWHM) at mass 350.

Figure 6.2 illustrates the event sequence used during automated LAMMS ion mapping. Also shown is a block diagram of the hardware configuration for ion mapping using the LAMMA-1000. The HP-1000 E-series computer controls all operations without operator intervention. The sample is scanned in a sequential fashion along the x and y axes. Initially, the sample is stepped along the x axis beginning at the origin ($x = 0$, $y = 0$). When movement along the x axis is complete, the sample is returned to $x = 0$, and the y axis is incremented. The movement is controlled by the computer via a Daedel Inc. Model PC410 stepper motor controller, which is interfaced to three Superior Electric Corp. stepper motors for incremental movement in the x, y, and z directions. Movement in the z direction changes the sample distance to the laser focal point and is not currently automated. Focus is adjusted during a preliminary LAMMS evaluation of the sample. The stepper motor controller is currently configured for a minimum step size of 2.50 μm; any multiple of 2.50 μm can be entered (i.e., 2.50, 5.00, 7.50, etc.). Complete details regarding the hardware, software, and capabilities of the automated LAMMS system have been published.[15]

The samples selected for LAMMS analysis are presented in Table 6.1. Sample CLB-1 was cut with a band saw, polished, and rinsed with deionized water. A thin section of the coal ball and associated banded coal was prepared for examination in reflected and transmitted light. Acetate peels were made of the coal ball coun-

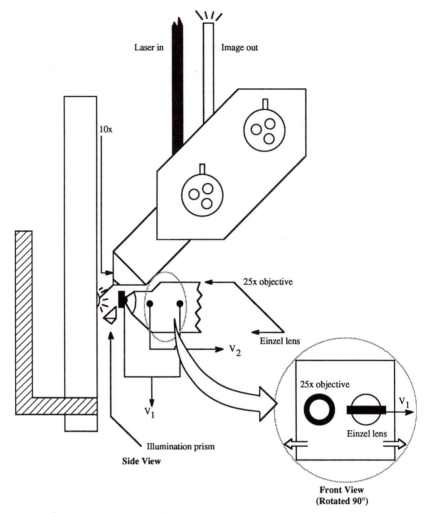

Fig. 6.1. Schematic representation of the laser and ion optics within the sample chamber of the LAMMA-1000. Inset illustrates front view of the movable carriage allowing the remote exchange of the observational and analytical optics.

terpart for histological analysis. A polished block containing the pyrite coal ball was prepared and characterized for pyrite, macerals, and microlithotypes by the use of an automated petrographic microscope. Reflectances of the vitrinite were determined with an automated image analysis system (AIAS) with polarized light (546 nm, 1.518 RI oil[28]). The polished block containing the coal ball (Fig. 6.3) was photographed, and a Mylar overlay was marked with sites for LAMMS analysis. A preliminary qualitative evaluation using LAMMS was performed. Three sites showing optically homogeneous banded vitrinite from two bands above and below the coal ball were also selected. Data on C, H, N, O, and ash content were obtained from the resins (LP-25A through LP-44A of Table 6.1 at the U.S. Geological Survey using elemental analytical techniques described previously[29]). All resins were analyzed in duplicate except for LP-25A, which was analyzed in quadruplicate. The "modern resin" (LP-30) was obtained from a mature living tree. A vitrinite concentrate from LP-18 was analyzed for trace and minor elements by INAA. The small amounts and size of the submaceral colloresinite (Fig. 6.4) make conventional separation and chemical analysis impossible. Alternatively, the study of LP-18 is very amenable to chemical analysis by microprobe techniques such as LAMMS. Samples designated as LP-2 through LP-8 were analyzed by using LAMMS, INAA, DCAS, and petrographic microscopy. A 250- or 500-μm-diameter circle was inscribed around the sites of interest with a diamond marker microscope accessory. The circular marking allowed easy location of the specific sites of interest in subsequent LAMMS analysis. Three spots on each vitrinite were identified as V1, V2, and V3. Details on DCAS and INAA analytical procedures have been published[30,31] and will not be presented here.

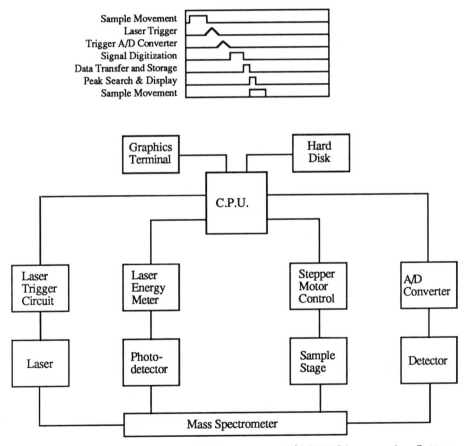

Fig. 6.2. Top: event sequence used during automated LAMMS ion mapping. Bottom: block diagram of the hardware configuration for ion mapping using the LAMMA-1000. Republished with permission from Ref. 15.

Table 6.1. Samples selected for analysis

Sample I.D.	Rank	Country/state	Coal basin or field	Maceral/ mineral type	Coal bed	Age
CLB-1	Mv bituminous	USA/Maryland	Castleman	Coal ball	Lower Bakerstown	Upper Pennsylvanian
LP-1	Meta- anthracite	?[a]	?	Pyrite grain	?	?
LP-2	Anthracite	USA/ Pennsylvania	Western Middle	Vitrinite	Mammoth	Middle Pennsylvanian
LP-3	Anthracite	USA/ West Virginia	Meadow Branch	Vitrinite	Unnamed	Early Mississippian
LP-4	Lv bituminous	USA/Maryland	Georges Creek	Vitrinite	Pittsburgh	Late Pennsylvanian
LP-5	Mv bituminous	USA/Maryland	Castleman	Vitrinite	Lower Bakerstown	Late Pennsylvanian
LP-6	hvA bituminous	USA/Alabama	Warrior	Vitrinite	Cobb	Early Pennsylvanian
LP-7	hvA bituminous	USA/ West Virginia	—	Vitrinite	Raymond (Pittsburgh)	Late Pennsylvanian
LP-8	hvC bituminous	USA/Kentucky	Illinois Basin	Vitrinite	No. 9	Middle Pennsylvanian
LP-18	LvA bituminous	England	Westphalia	—	Swallow Wood	Upper Carboniferous
LP-25A	A lignite[b]	Australia	Latrobe Valley	Fossil resin	Yallourn	Miocene
LP-26A	Lignite	New Zealand	Waikato	Fossil resin	Renown	Eocene
LP-28	Subbituminous	New Zealand	Waikato	Fossil resin	Kupakupa	Eocene
LP-30	—	New Zealand	—			Modern
LP-43A	Lignite	Australia	Latrobe Valley	Fossil resin	Yallourn	Miocene
LP-44A	Lignite	Australia	Latrobe Valley	Fossil resin	Yallourn	Miocene

[a]Information not available.
[b]Brown coal.

RESULTS AND DISCUSSION

Histological and optical analysis of the acetate peels of the pyrite coal ball (Fig. 6.3) indicated several kinds of structurally preserved tissue and round bodies. The details of the histological study have been published previously.[13] For LAMMS analysis, three morphologically distinct areas were chosen for initial qualitative evaluation. These sites were the banded vitrinite of Fig. 6.3 (V), the pyritic areas [for example, (P) in Fig. 6.5], and the nonbanded vitrinite known as corpocollinite [(C) in Fig. 6.5]. Each site was analyzed 12 times for both positive and negative ions and two averages were obtained for each site and each ion polarity.

Average mass spectra of the pyrite and vitrinite sites indicated in Figs. 6.3 and 6.5 are shown in Figs. 6.6, 6.7, and 6.8. For simplicity in subsequent discussions, elemental signals yielding characteristic isotopic distributions are described by listing only the most abundant isotope. The pyrite (Fig. 6.6) yielded characteristic signals at m/z 32, 56, 88, 112, 120, 144, and 176, corresponding to S^+, Fe^+, FeS^+, Fe_2^+, Fe_2S^+, and $Fe_2S_2^+$, respectively. The LAMMS spectra from the banded vitrinites (Fig. 6.7) yielded Li^+, Na^+, Mg^+, Al^+, Si^+, K^+, Ca^+, Ti^+, Fe^+, Ga^+, Sr^+, and Ba^+ at m/z 7, 23, 24, 27, 28, 39, 40, 48, 56, 69, 88, and 138, respectively. The nonbanded vitrinite (corpocollinite) yielded similar elemental peaks, except for Li^+, Ti^+, Sr^+, and Ba^+ (Fig. 6.8).

The negative-ion LAMMS of the pyritic areas yielded signals at m/z 32, 64, and m/z 120, corresponding to S^-, S_2^-, and FeS_2^-. The positive-ion spectra of both banded and nonbanded vitrinite yielded similar carbon cluster maxima and minima. The maxima fall in the hydrocarbon cluster series C_nH when n is odd (m/z 37, 61, 85, 109, 133) and the minima fall in the hydrocarbon series C_nH_2 when n is even (m/z 50, 74, 98, 122, 146). This series appears to be typical of the vitrinites and is not found in resinites.

The negative-ion spectra of the banded and nonbanded vitrinites were indistinguishable and shall not be discussed in detail here. The results from the elemental analysis of resins (LP-25 through LP-44A of Table 6.1) are given in Table 6.2. Notably, all resins analyzed had similar H:C ratios (1.53–1.57), but fossil resins from the subbituminous coals had significantly lower O:C ratios of 0.05 and 0.06 compared to 0.10 and 0.12 for the modern resin, and the resins from brown coals (lig-

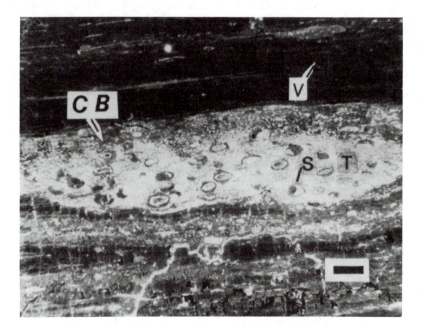

Fig. 6.3. Pyrite coal ball (CB) containing the seed fern *Myeloxylon*. S, secretory structure; T, tissue; V, banded vitrinite. Scale = 1.25 mm. Republished with permission from Ref. 13.

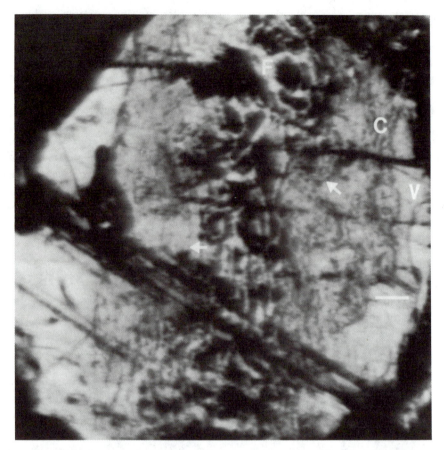

Fig. 6.4. Photomicrograph of colloresinite (c), vitrinite (V), and fusinite (F) in the British Swallow Wood coal bed, Yorkshire, England. Arrows point to cells in colloresinite and vitrinite (telinite). Reflected light, air. Scale = 20 μm.

nites) of Victoria. All fossil resins show similar trends in their positive-ion spectra. One series shows peaks at m/z 65, 77, 91 (most intense within this series), 105, 119, and 133–134. A second series occurs at m/z 27, 39 (most intense within this series), 51, and 63. A third series, lower in abundance than the previous two, occurs at m/z 128 and 143. Alternatively, the modern resin from the conifer *Agathis australis* (LP-30) yielded a different major positive-ion series of peaks at m/z 86 (most intense within this series), 98, 110, 122, and 133–134. A second series at m/z 65, 77, 91 (most intense within this series), 105, and 119 was observed with less intensity.

The ion observed at m/z 39 was not assigned to K^+. The ratio of m/z 41 to m/z 39 (approximately 1:1) is much greater than that predicted by the natural abundances of each isotope. Furthermore, the ash content of the samples (Table 6.2) is very low, indicating that the amount of potassium present within the sample should be minimal. The lack of elemental information within the spectra obtained from each resinite suggests that ionization of the organic portions of the coal is

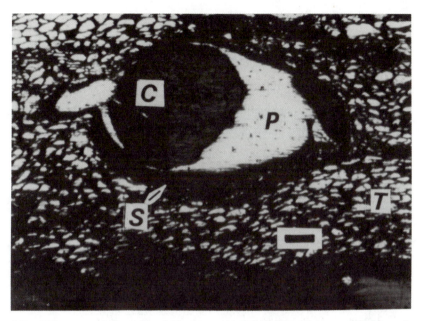

Fig. 6.5. Enlargement (from Fig. 6.3) of secretory structure (S) surrounded by scleren-chyma cells (T), pyrite (P), nonbanded vitrinite or corpocollinite (C). Reflected light. Scale = 80 μm. Republished with permission from Ref. 13.

favored. The presence of m/z 39 from the organic portion of the coal is consistent with the results obtained from other pyrolysis mass spectrometric investigations of coals.[5,32,33]

The lack of LAMMS spectra of coals and coal macerals in the literature neces-sitates the examination of mass spectra obtained with other techniques. Meuzelaar

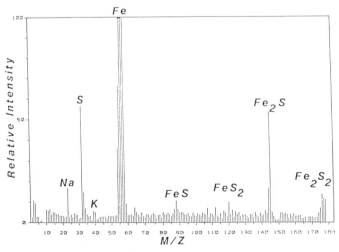

Fig. 6.6. LAMMA-1000 positive ion mass spectrum of iron sulfide (pyrite) in coal ball of Fig. 6.3. Na and K are possible contaminants. Republished with permission from Ref. 13.

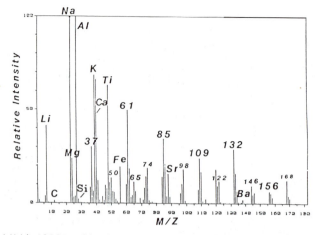

Fig. 6.7. LAMMA-1000 positive ion mass spectrum of banded vitrinite from the Lower Bakerstown coal bed. The larger numbers indicate the C_nH^+ peaks where n is odd and the smaller numbers indicate C_nH^+ peaks where n is even. Republished with permission from Ref. 13.

et al.[33] used Curie-point pyrolysis mass spectrometry to show that major alkyl benzene and naphthalene series also occur in bituminous and anthracitic vitrinites and inertinites; both of these macerals are more aromatic than resinite. Fossil resins are composed primarily of resinite. Meuzelaar et al. concluded that dihydroxybenzenes decrease and naphthalene increase with increasing rank of bituminous and anthracitic coals. The occurrence of a major series of positive-ion peaks belonging to two series, a C_xH_3 ($x = 2$–5) and $C_6H_5(CH_2)_x$ ($x = 1$–4), with the most intense peaks in each series at m/z 39 and 91, respectively, is typical of the mass spectra from n-

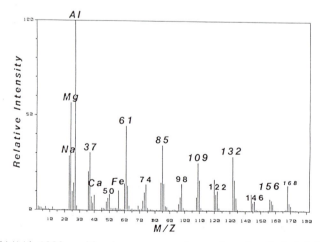

Fig. 6.8. LAMMA-1000 positive-ion mass spectrum of nonbanded vitrinite (corpocollinite) from the Lower Bakerstown coal bed. Note lack of peaks at m/z 44 and m/z 65, and the similarity of the hydrocarbon ion intensities to those of Fig. 7. Republished with permission from Ref. 13.

alkyl benzenes.[32] The m/z 65 to 133–134 positive-ion series is characteristic of alkyl benzenes having alkyl side chains with up to four carbons. The m/z series in the range m/z 27 to 63 is probably due to aliphatic components of the resin and/or fragmentation and rearrangement of the alkyl side chain from alkyl benzenes.[34] Meuzelaar et al. attribute the peaks at m/z 128 and 142–143 to naphthalenes ($C_{10}H_8$) and methylnaphthalenes ($C_{11}H_{10}$).[33]

The major positive-ion series of the modern resin occurring in the range m/z 86 to 134 is interpreted as oxygenated aliphatic hydrocarbons belonging to the series $C_xO_2H_6$ ($x = 4$–8). The lowest mass member of this series would have the formula $C_4H_6O_2$, a possible pyrolysis produce of the resin. The presence of this major series in the modern resin is consistent with the presence of oxygen-containing functional groups such as carboxyl (COOH), found in terpenes. Terpenes are abundant in resins.

Py-MS of fossil resins[5,35–37] show similarities to the LAMMS spectra reported here. Py-MS and LAMMS spectra show pronounced peaks at m/z 128 and 142, attributable to naphthalene and methylnaphthalene class compounds. The LAMMS spectra imply that the early stages of resin coalification, up to and including a rank of subbituminous B coal, involve the production and concentration of alkyl benzenes with up to four-carbon alkyl side chains. Also, the production of naphthalenes and methylnaphthalenes is indicated. The presence of an alkyl benzene and naphthalene series has been noted previously by Curie-point Py-MS of maceral concentrates from vitrinite and inertinite maceral groups of bituminous coal.[5] Elemental data (Table 6.2) indicate that between the brown coal (lignite) stage and the subbituminous B stage of resin coalification there is a measurable loss of oxygen-rich functionalities with no attendant change in H/C. Cyclic aliphatic compounds such as terpenes are found in biological fresh ("modern") resins; therefore, early resin coalification involves the conversion of polycyclic aliphatic alcohols and acids to alkyl benzenes with four or more carbons in the alkyl side chains. The maintenance of a H/C in the range of 1.53 and 1.57 and a major drop of O/C from 0.11 to 0.6 during coalification from the modern resin to subbituminous B rank indicate that the loss of H and C content is approximately 1:1, while oxygen loss is greater. The elemental information supports COOH loss as a major process in early coalification.

Figure 6.4 is a photomicrograph of a low volatile A bituminous coal from the British Swallow Wood coal bed, Yorkshire, England. This sample (LP-18 of Table 6.1) consists of three intimately associated submacerals, colloresinite (C), vitrinite

Table 6.2.   Results of elemental and ash analysis of resins

Sample number	C	H	N	O[a]	Ash	H:C	O:C
LP-30	78.4	10.2	0.63	10.8	0.0	1.55	0.10
LP-43A	77.2	9.9	0.28	12.6	0.0	1.53	0.12
LP-44A	78.0	10.1	0.60	11.3	0.0	1.54	0.11
LP-25A	77.6	10.0	0.08	12.4	0.0	1.53	0.12
LP-26A	82.2	10.7	0.49	6.61	0.0	1.55	0.06
LP-28	83.0	11.0	0.27	5.73	0.0	1.57	0.05

[a]By difference.

(V), and fusinite (F). The following elements were detected in all three submacerals: Li, Na, Mg, Al, Si, S, K, Ca, V, Cr, Fe, Ga, Ba, La, F, Cl, corresponding to major isotopic ions at m/z 7, 23, 24, 27, 28, 32, 39, 40, 51, 52, 56, 69, 138, and 139 in positive-ion spectra and 19 and 35 in negative-ion spectra. Titanium was detected as intense signals at m/z 46, 47, 48, 49, and 50 in vitrinite and colloresinite but was not consistently observed in the fusinite. The colloresinite was further distinguished by having a ^{39}K signal greater in intensity than the ^{40}Ca signal. The ^{39}K/^{40}Ca was reversed in the vitrinite (i.e., the ^{40}Ca signal was greater).

The sensitivity of LAMMS to chemical and physical variation of the matrix analyzed has been previously noted.[9] To best compensate for these variations, a preliminary qualitative LAMMS investigation of a variety of coal macerals, clays, and minerals was undertaken. Six elements were found to have minimum variation among the vitrinites analyzed: Ba, Cr, Ga, Sr, Ti, and V. The elements Cr, Ga, Ti, and V are known to have probable organic association in some coals.[38,39] A bulk semiquantitative LAMMS analysis on coals LP-2 through LP-8 was attempted. The relative elemental variations from coal to coal, as indicated by either DCAS or INAA data obtained from vitrinite concentrate (in practice, a weighted average of the INAA and DCAS results was used), should agree with those variations indicated by the bulk LAMMS data obtained from similar material, provided the LAMMS signals are acquired from several vitrinitic microareas. Acquiring LAMMS signals from several microareas minimizes sampling errors.

The LAMMS data obtained for seven vitrinites (LP2 to LP8) of varying rank (anthracite to high-volatile C bituminous, respectively) were treated in the following manner. First, the overall signal for each specific element, Ba, Cr, Ga, Sr, Ti, and V, relative to the signal for m/z 85, was obtained. The six elements are assumed to have high organic affinity based on the work of Horten and Aubery[38] and Zubovic et al.[39]; therefore, a reference peak, m/z 85, which consistently appeared and originated from the organic matrix of the vitrinite was chosen. The nature of this organic affinity has not been established, but the presence of organometallic species and/or highly dispersed mineral matter is possible. The signal from the chosen monoionic ions of each element in each spectrum for every vitrinite site was added to obtain the overall signal for each element. The signal for m/z 85 was also added in each spectrum for every sample. The total elemental signal obtained for an individual coal maceral was then normalized to the total m/z 85 signal for the same coal. For example, LP5 had eight sites analyzed with 12 shots averaged at each site, and nine sites analyzed with one shot. Ideally, the data could be related to a relative elemental sensitivity factor, and the concentration of the element in question could be determined and compared to that obtained using DCAS and INAA; however, as previously specified, no such sensitivity factors exist for a coal matrix. A second step allowing the determination of relative variations in elemental concentration from coal to coal was necessary. Whatever the unknown sensitivity for each element, it was assumed that the sensitivity would remain constant over a given range of similar coal rank (e.g., high-volatile C bituminous coal to medium volatile bituminous). Because variations in coal matrices do exist between coal ranks, this assumption is not completely valid; however, it was demonstrated that with proper care the approach is viable.[9] As is true for SIMS,[24] the accuracy of quantitative analysis utilizing LAMMS improves when the material used as a reference has a matrix similar to that being studied. To minimize possible matrix effects, medium volatile bituminous coal LP5, a middle rank, was chosen for a reference coal when comparing elemental variations among the bituminous class LP4 to LP8 and the anthracite coal LP2 for a reference when comparing the elemental variation in the anthracitic class coals (LP2 and LP3). LP2 was of slightly higher rank with greater aromaticity than LP3. The resulting LAMMS data obtained were therefore calculated as

$$\frac{(Z)_{LPx}}{(Z)_{LP5}} = \frac{\sum^{n}[I_{(m/z)A_Z}]/\sum^{n}[I_{(m/z\ 85)}]_{LPx}}{\sum^{n}[I_{(m/z)A_Z}]/\sum^{n}[I_{(m/z\ 85)}]_{LP5}} \tag{6.1}$$

for the bituminous coals. The values of $x = 4$, or 6 through 8; $Z =$ Ba, Cr, Ga, Sr, Ti, or V. A_Z is the chosen monoionic mass of element Z (138[Ba], 52[Cr], 69[Ga], 88[Sr], 47[Ti], or 51[V]). I_{A_Z} is the intensity (peak height) of mass A_Z. n is the number of spectra taken for each maceral (LP2, $n = 38$; LP3, $n = 34$; LP4, $n = 40$; LP5, $n = 105$; LP7, $n = 41$; LP8, $n = 36$; LP9, $n = 36$). The anthracites were treated similarly; however, the reference coal was LP2 rather than LP5 in Eq. (6.1).

Each value from Eq. (6.1) was compared to its corresponding value obtained from DCAS and INAA. The DCAS and INAA data were calculated by normalizing the reported absolute elemental concentration from each coal to the value reported for LP2 or LP5 (based on the coal's rank as anthracitic or bituminous, respectively). Table 6.3 presents the results of these calculations. It is reassuring to observe the

Table 6.3. Relative elemental concentration variation within the LPx series

Technique	Element	LP4/LP5	LP6/LP5	LP7/LP5	LP8/LP5	LP3/LP2
INAA	Ba	3.2 ± 0.87	5.3 ± 1.40	1.5 ± 0.51	0.2 ± 0.2	0.3 ± 0.11
DCAS		3.8 ± 0.95	5.7 ± 0.80	1.7 ± 0.25	0.05 ± 0.03	0.2 ± 0.03
LAMMS		1.1 ± 0.54	3.2 ± 1.70	0.5 ± 0.27	0.4 ± 0.22	0.3 ± 0.45
INAA	V	1.5 ± 0.13	0.6 ± 0.07	1.4 ± 0.14	1.60 ± 0.13	0.3 ± 0.02
DCAS		2.7 ± 1.35	<1.0	1.4 ± 0.70	1.90 ± 0.85	0.3 ± 0.08
LAMMS		0.9 ± 0.20	0.5 ± 0.09	1.4 ± 0.32	1.2 ± 0.29	0.3 ± 0.09
INAA	Cr	1.2 ± 0.06	0.4 ± 0.02	0.8 ± 0.17	1.00 ± 0.03	0.3 ± 0.03
DCAS		1.4 ± 0.38	0.3 ± 0.17	0.7 ± 0.30	1.00 ± 0.35	0.3 ± 0.08
LAMMS		1.4 ± 0.32	0.6 ± 0.12	0.7 ± 0.17	1.1 ± 0.26	1.0 ± 0.32
INAA	Sr	1.5 ± 0.33	1.8 ± 0.39	0.8 ± 0.23	0.4 ± 0.12	>0.4
DCAS		—[a]	—	—	—	—
LAMMS		0.7 ± 0.25	2.5 ± 0.89	1.5 ± 0.58	1.4 ± 0.49	50. ± 25.
INAA	Ga	—	—	—	—	—
DCAS		2.5 ± 1.90	<1.0	<1.0	1.60 ± 1.10	—
LAMMS		1.6 ± 0.58	0.7 ± 0.27	0.7 ± 0.27	2.0 ± 0.72	2.5 ± 1.12
INAA	Ti	0.4 ± 0.11	<1.2	0.3 ± 0.30	0.20 ± 0.05	0.1 ± 0.10
DCAS		0.5 ± 0.07	0.1 ± 0.04	0.2 ± 0.03	0.10 ± 0.07	0.2 ± 0.03
LAMMS		0.03 ± 0.11	0.3 ± 0.45	1.2 ± 0.30	0.8 ± 0.30	0.6 ± 0.30

[a]—, data not available.

general correlation that exists between the INAA/DCAS results and those from LAMMS. The supposition of greatest concern in our method is that matrix variation will not significantly affect the accuracy. This assumption appears to be validated in light of the general agreement between the INAA/DCAS and LAMMS data.

The "error factor"[24,25,40−43] is a useful measure of precision and accuracy in any quantitative SIMS analysis employing relative sensitivity factors. Error factors are determined by dividing the experimentally measured concentration by the true concentration. Here the weighted average (wa) of the INAA and DCAS measurements is used as an estimate for the true value. By definition, if the error factor is less than one, the negative reciprocal is used.

Table 6.4 presents the error factors derived from the LAMMS data. LAMMS determinations of Ba and Ti have considerable error. We have observed that Ti is a component in some clays, and it is also present in coal as the mineral anatase (TiO_2). It is quite possible that the errors involved in LAMMS determination of both Ba and Ti arise from the selection of sites for analysis based on visual appearance as purely vitrinitic in nature and devoid of mineral or clay matter. For further comparison, semiquantitative elemental determinations using SIMS generally produces error factors in the range of $1.2–5.0$[24,25] for well-characterized standards. Although relative error factors in coal studies have not been determined by SIMS, LAMMS is likely comparable to SIMS in ability to provide semiquantitative data on coal constituents.

The investigations to date show that the largest contrast in LAMMS occurs when spectra obtained from mineral inclusions are compared to spectra of organic components of coal. For example, the spectra obtained from the pyrite of CLB-1 (Fig. 6.6) are much different[15] from spectra of the banded vitrinite (Fig. 6.7). Because of its prominent pyrite inclusion (Fig. 6.9), sample LP1 was chosen to demonstrate the capability for automated ion mapping using our modified LAMMA-1000. The region containing the inclusion was mapped and is outlined in Fig. 6.9. Positive-ion maps were obtained for Fe^+ (m/z 56) and Fe_2S^+ (m/z 144) (Fig. 6.10a and b, respectively). Both the Fe^+ (Fig. 6.10a) and Fe_2S^+ (Fig. 6.10b) maps clearly define the position of the inclusion. Although the intensity of Fe_2S^+

Table 6.4. Error factors for LAMMS semiquantitative analysis of LPx series[a]

Element	LP4	LP6	LP7	LP8	LP3
Ba	−3.2	−1.75	−3.4	8.0	1.5
V	−1.7	−1.20	1.0	−1.3	1.0
Cr	1.2	1.50	−1.1	1.1	3.3
Ti	−1.7	3.00	6.0	4.0	3.0
Sr	−2.1	1.40	1.9	3.5	—[b]

[a]Error factors are based on weighted average (wa) of results from both DCAS and INAA, where

wa = [sum(x/variance)]/[sum(1/variance)].

x = data for each element in each LPx/LP5 (or LP3/LP2)
determination for INAA and DCAS. For example, wa Ba
(LP4/LP5) = (3.2/0.76 + 3.8/0.90)/(1/0.76 + 1/0.90).

Data herein were derived from data presented in Table 6.2. See text for details.

[b]Data not available due to imprecision of measurements.

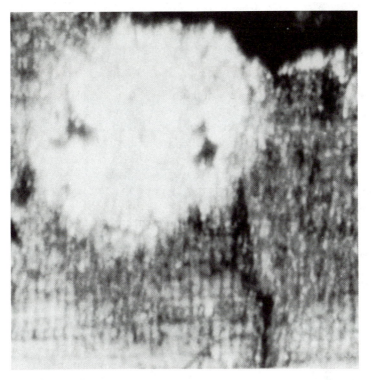

Fig. 6.9. Optical photograph of coal sample. Maceral contains bright circular area 150 mm. The square area was analyzed and visible within are track marks from laser ablations. Republished with permission from Ref. 15.

ions is much lower than that of the Fe^+ ions (see Fig. 6.6, for example), the Fe_2S^+ ion can still be correlated with the position of the inclusion. The exact nature of the iron- and sulfur-containing species comprising the inclusion could not be determined unequivocally using LMS alone; however, the ability to obtain chemical information was clearly demonstrated.

CONCLUSIONS

The variation observed in the organic structural information from resinites of different coal ranks can be related to processes important to coalification. More work is needed to unequivocally determine the structural components of the resins studied. LAMMS can be used to distinguish between intimately associated macerals on the scale of ~ 20 μm, and possibly less. Comparisons made between the results of LAMMS and those of DCAS and INAA indicated that the accuracy of the elemental information obtained is comparable to the SIMS technique. The fundamental problem in quantitative LAMMS analysis is the difficulty in resolving and evaluating all sources of variation in shot-to-shot ion yield. This inherent difficulty can

(a)

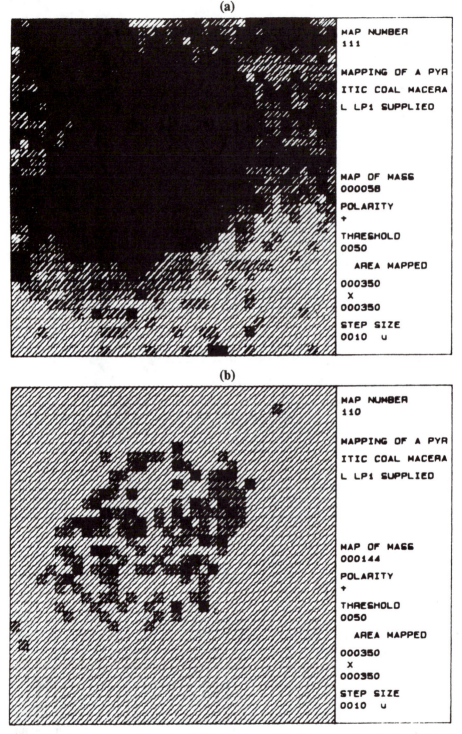

(b)

Fig. 6.10. (a) Ion map of Fe$^+$ at m/z 56. (b) Map of Fe$_2$S$^+$ ion at m/z 144. Republished with permission from Ref. 15.

be circumvented by monitoring relative changes in elemental concentrations of similar matrices. In this study, coals of related rank were analyzed. The ability to monitor spatial variations in ion signals using LAMMS and the modifications made to the LAMMA-1000 allow automated ion mapping of specific areas in coals of interest.

REFERENCES

1. H. L. Retcofsky, G. P. Thompson, M. Hough, and R. A. Friedel, *Organic Chemistry of Coal* (ACS Symposia Series #71). American Chemical Society, Washington, DC, 1978, p. 142.

2. J. Bimer, P. H. Given, and R. Swadesh, *Organic Chemistry of Coal* (ACS Symposia Series #71). American Chemical Society, Washington, DC, 1978, p. 86.

3. P. G. Hatcher, I. A. Grefer, and W. L. Earl, *Org. Geochem.* **3**, 49 (1982).

4. P. G. Hatcher, I. A. Breger, N. Szeverenyi, and G. E. Maciel, *Org. Geochem.* **4**, 9 (1983).

5. H.L.C. Meuzelaar, A. M. Harper, R. J. Pugmire, and J. Karas, *Int. J. Coal Geol.* **4**, 143 (1984).

6. J. O. Lephardt, *Analytical Pyrolysis.* Butterworths, London, 1984, p. 95.

7. J. A. Minkin, E.C.T. Chao, C. L. Thompson, R. Nobling, and H. Blank, *Scan. Electron. Microsc.* **I**, 175 (1982).

8. R. R. Martin, B. Wiens, N. S. McIntyre, B. I. Kronberg, and J. A. MacPhee, *Fuel* **65**, 1024 (1986).

9. J. J. Morelli, D. M. Hercules, P. C. Lyons, C. A. Palmer, and J. D. Fletcher, *Mikrochim. Acta* **III**, 105 (1988).

10. D. M. Hercules, K. Balasanmugam, T. A. Dang, and C. P. Li, *Anal. Chem.* **38**, 280A (1982).

11. R. E. Honig, *Appl. Phys. Lett.* **2**, 138 (1963).

12. F. J. Vastola and A. J. Pirone, Symposium on Pyrolysis Reactions of Fossil Fuels, in *151st Meeting of the Division of Fuel Chemistry.* American Chemical Society, Pittsburgh, 1966.

13. P. C. Lyons, D. M. Hercules, J. J. Morelli, G. A. Sellers, D. Mattern, C. L. Thompson-Rizer, F. W. Brown, and A. Millay, *Int. J. Coal Geol.* **7**, 185 (1987).

14. F. P. Novak, Z. A. Wilk, and D. M. Hercules, *J. Trace Micro. Tech.* **3**(3), 149 (1985).

15. Z. A. Wilk and D. M. Hercules, *Anal. Chem.* **59**(14), 1819 (1987).

16. F. P. Novak and D. M. Hercules, *Anal. Lett.* **18**(A4), 503 (1985).

17. The use of trade or brand names is for descriptive purposes only and does not constitute endorsement of products by McDonnell Douglas Research Laboratories, the U.S. Geological Survey, or the University of Pittsburgh.

18. H. J. Heinen, S. Meier, and H. Vogt, *Int. J. Mass. Spectrom. Ion Phys.* **47**, 19 (1983).

19. M. Grasserbauer, K.F.J. Heinrich, and G. H. Morrison, *Pure Appl. Chem.* **55**(12), 2023 (1983).

20. K. Danzer, K. Doerffel, H. Earhardt, M. Geisler, G. Erhlich, and P. Gadow, *Anal. Chim. Acta* **105**, 1 (1979).

21. L. Moenke-Blankenburg, *Prog. Anal. Spectrosc.* **9**, 335 (1986).

22. P. Wieser, R. Wursher, and R. Wechsung, *LAMMA Workshop.* Leybold-Heraeus, GmbH, 1983, p. 29.

23. S. Tamaki, *Mikrochim. Acta* **3**, 1 (1985).

24. G. O. Ramseyer and G. H. Morrison, *Anal. Chem.* **55**, 1963 (1983).

25. M. A. Rudat and G. H. Morrison, *Anal. Chem.* **51,** 1179 (1979).

26. R. J. Cotter, *Anal. Chim. Acta* **195,** 45 (1987).

27. H. Vogt, H. J. Heinen, S. Meier, and R. Wechsung, *Z. Anal. Chem.* **308,** 195 (1981).

28. E.C.T. Chao, J. A. Minkin, and C. L. Thompson, *Int. J. Coal Geol.* **2,** 113 (1982).

29. P. C. Lyons, P. G. Hatcher, J. A. Minkin, C. L. Thompson, R. R. Larson, Z. A. Brown, and R. N. Pheifer, *Int. J. Coal Geol.* **3,** 257 (1984).

30. J. D. Fletcher and D. W. Golightly, *The Determination of 28 Elements in Whole Coal by Direct-Current Arc Spectrography.* Tech. Report, Branch of Analytical Chemistry, Office of Mineral Resources, U.S. Geological Survey, 1985. Open-File Report 85-204, pp. 1–10.

31. C. A. Palmer and P. A. Baedecker, *The Determination of 41 Elements in Whole Coal by Instrumental Neutron Activation Analysis.* Tech. Report, Branch of Analytical Chemistry, Office of Mineral Resources, U.S. Geological Survey, 1988. To be published as a U.S. Geological Bulletin.

32. W. Brackman, K. Spaargaren, J.P.C.M. Van Dongen, P. A. Couperus, and F. Bakker, *Geochim. Cosmochim. Acta* **48,** 2483 (1984).

33. H.L.C. Meuzelaar, J. Haverkamp, and F. D. Hileman, *Pyrolysis Mass Spectrometry of Recent and Fossil Biomaterials.* Elsevier, New York, 1982, p. 293.

34. N. S. McIntyre, R. R. Martin, W. J. Chauvin, C. G. Winder, J. R. Brown, and J. A. MacPhee, *Fuel* **64,** 1705 (1985).

35. N. E. Vanderborgh and C.E.R. Jones, *Anal. Chem.* **55,** 527 (1983).

36. P. K. Dutta and Y. Talmi, *Fuel* **61,** 1241 (1982).

37. P. A. Schenck, J. W. de Leeuw, T. C. Viets, and J. Haverkamp, *Petroleum Geochemistry and Exploration of Europe.* Blackwell Scientific Pubs., New York, 1983, p. 267.

38. L. Horton and K. V. Aubery, *J. Soc. Chem. Ind.* **69,** S41 (1950).

39. P. Zubovic, T. Stadinchenko, and N. B. Shefey, Chemical basis of minor element associations in coal and other carbonaceous sediments, in *Short Papers in the Geologic and Hydrologic Sciences.* United States Printing Office, Washington, DC, 1961, p. D-345–348.

40. D. E. Newbury, *Scanning* **3,** 110 (1980).

41. J. D. Ganjie and G. H. Morrison, *Anal. Chem.* **50,** 2034 (1978).

42. D. H. Smith and W. H. Christie, *Int. J. Mass Spectrom. Ion Phys.* **26,** 61 (1978).

43. J. D. Ganjie, D. P. Leta, and G. H. Morrison, *Anal. Chem.* **50,** 285 (1978).

7

Postionization of Laser-Desorbed Neutrals for the Analysis of Molecular Adsorbates on Surfaces

DONALD P. LAND, DIANA T.S. WANG, TSONG-LIN TAI, MICHAEL G. SHERMAN, JOHN C. HEMMINGER, and ROBERT T. McIVER, Jr.

In the last 20 years there has been a tremendous amount of activity in the field of surface science. Much of this activity has been aimed at the development of new and more precise probes of the solid–gas interface. The degree of success of these efforts is revealed by the fact that techniques now exist that can provide the elemental composition of a surface layer for virtually any solid sample.[1] Among these techniques are Auger electron spectroscopy, X-ray photoelectron spectroscopy, ion scattering spectroscopy, and Rutherford backscattering. In addition, the static structure of a surface can be determined by diffraction techniques (electron or He atom), or modern methods of electron microscopy (scanning tunneling microscopy or high-voltage electron microscopes).

In contrast, determining the identity of *molecular* species on a surface is far more difficult. This is true for species even of only moderate complexity, but is particularly a problem for surfaces with more than one molecular component, as might be encountered in problems in catalysis, lubrication, adhesion, corrosion, or microelectronics fabrication. Yet, for a surface chemist, the identity of molecular species is the starting point for developing an understanding of a wide variety of surface phenomena.

In this chapter we describe our efforts to develop the techniques of laser-induced thermal desorption (LITD) and Fourier transform mass spectrometry (FTMS) to characterize molecular species adsorbed at solid–gas interfaces. Our approach has been to use low laser power densities to desorb neutral molecular species and postionization with an electron beam to produce mass spectra. One of the advantages of this approach is that elucidation of the molecular identity of the surface species is aided by the fact that in most cases the laser desorption mass spectra are very similar to the conventional electron ionization mass spectra of the species. Another feature is that the sensitivity of the method is enhanced by using a Fourier transform mass spectrometer that can capture a complete mass spectrum for each shot of the laser.

THERMAL DESORPTION OF MOLECULES FROM A SURFACE

Conceptually, the combination of desorption of a molecular adsorbate with sensitive, high-resolution mass spectrometry detection should provide a method for identifying molecular adsorbates. Indeed, thermal desorption spectroscopy (TDS), in which a sample is heated to drive off molecular species that are then ionized and mass analyzed, is a highly developed method that has provided a significant amount of information on the chemistry of adsorbate systems.[2] One of the limitations of TDS experiments is that surface reactions can alter the identity of molecular adsorbates before they desorb. For example, bimolecular reactions of the adsorbed species can cause problems because heating the surface results in increased mobility and reaction probability of the surface species. Another problem is that the rates of unimolecular decomposition reactions increase as the surface temperature is increased and can become competitive with desorption. The first of these problems is minimized by working at low coverages, but this is not always of practical interest. To minimize the second problem, the surface–adsorbate bond must be broken before the molecule undergoes unimolecular decomposition. This is an important concept because it suggests that the time scale for desorption must be very short, since even weak van der Waals bond vibrational frequencies are on the order of 10^{12} Hz. Thus, the principal advantage of pulsed laser desorption experiments is that the desorption is completed before competing unimolecular and bimolecular reactions alter the identity of the surface species.

A generalized laser-induced thermal desorption (LITD) experiment can be divided into three steps: desorption of the intact molecular species, postionization with minimal fragmentation, and high-resolution, sensitive mass analysis. In this section we describe the important factors that affect the desorption process in a laser-induced thermal desorption experiment.

The surface temperature as a function of time for the situation of a laser pulse in the nanosecond time scale absorbed at the surface of a solid is found by solving the differential equation for heat flow[3]:

$$\nabla^2 T(x, y, z, t) - (1/k)\, \partial T(x, y, z, t)/\partial t = -A(x, y, z, t)/K \qquad (7.1)$$

where $T(x, y, z, t)$ is the temperature as a function of time and position, k is the thermal diffusivity, K is the thermal conductivity, and $A(x, y, z, t)$ represents the heat source. Solution of this equation, assuming a square temporal profile for the laser pulse, shows that the peak temperature jump at the surface is proportional to the absorbed power per unit area. Figure 7.1a shows the calculated temperature jump, $\Delta T(t)$, at the center of the laser beam for 10 MW/cm^2 absorbed power on Si (a good thermal conductor), whereas Fig. 7.1b shows the temperature jump on SiO$_2$ (a poor thermal conductor) for the same absorbed laser power. The plots show that the temperature jump at the surface of Si is an order of magnitude lower than that at the surface of SiO$_2$, and with Si the surface cools quickly once the laser is off, whereas the surface of SiO$_2$ remains hot for an order of magnitude longer time. Also

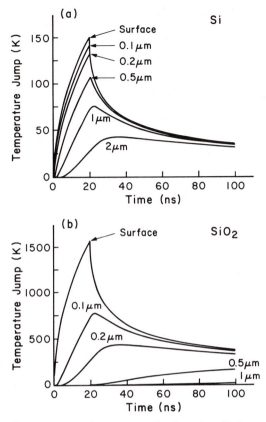

Fig. 7.1. Calculated temperature increase under laser irradiation assuming 20 MW/cm² absorbed power (a) using the thermal parameters for Si and (b) using the thermal parameters for SiO₂.

notice that the thermal penetration depth is much greater for the conductor than for the insulator. Thus, for materials with high thermal conductivity the incident energy moves rapidly from the surface into the bulk of the material, whereas with thermal insulators the energy is confined at the surface near the point of irradiance. The thermal diffusion lengths during the laser pulse are on the order of a few micrometers, which is much smaller than the 500-μm diameter of the laser beam. This implies that the heated area is localized to the region directly irradiated by the laser. As a result, in LITD the surface can be probed at specific positions without modifying the remainder of the sample.

Another important factor in LITD experiments is the rate of desorption as a function of temperature. For the simple case of first-order desorption kinetics, this can be modeled by writing an Arrhenius equation of the form

$$\text{Rate}(t) = dC/dt = vC \exp[-E/RT(t)] \qquad (7.2)$$

where C is the adsorbate coverage, v is the frequency factor, and E is the desorption activation energy. Figure 7.2 shows the calculated desorption rate and integrated number of molecules desorbed from a Ni substrate as a function of time for the case in which the adsorbate either desorbs or remains intact on the surface.[4] The assumed parameters for this calculation are for a beam diameter of 1 mm, an absorbed laser power density of 10 MW/cm², and an initial coverage of one mono-layer (10^{15} cm^{-2}). Under these conditions, about one-half of the original molecules desorb. Figure 7.2 also shows that the molecules desorb from the surface in a burst that is even shorter than the width of the laser pulse. This results because of the exponential dependence of the desorption rate on temperature. This aspect of the desorption provides excellent time resolution.

The yield of desorption products in an LITD experiment is also strongly influenced by the competition between desorption and reactions of the molecules on the surface. The branching ratio for these two processes can be modeled using Arrhenius equations, such as Eq. (7.2) with appropriate activation energies and frequency factors. Figure 7.3 shows how the branching ratio varies with heating rate for a system such as methanol adsorbed on Ni. When the surface is heated slowly, the low-energy, decomposition pathway dominates the branching ratio, but under laser heating conditions, the desorption channel, which is the high activation energy process, becomes important. This effect can be understood qualitatively by realizing that as the temperature, $T(t)$, rapidly becomes large, the exponential terms in the rate equations [Eq. (7.2)] approach unity before the species can react. Thus, at high heating rates the branching ratio is strongly influenced by the ratio of the frequency factors, and under these conditions the desorption channel generally dominates.[5]

It is widely recognized from TDS experiments that faster heating rates cause surface species to desorb at higher temperatures.[2] This is an important considera-

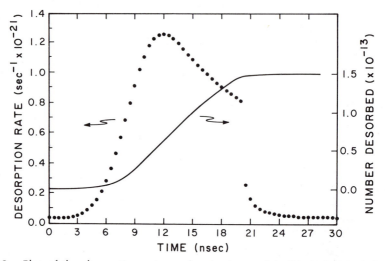

Fig. 7.2. Plot of the desorption rate, molecules/second (solid circles) and the integrated number of molecules desorbed (solid line) for an adsorbate with a desorption activation energy of 20 kcal/mol and a preexponential of 10^{-13} sec^{-1}. The temperature jump used is that calculated for a Ni sample absorbing 10 MW/cm² over 20 nsec, achieving a peak temperature increase of 1000 K.

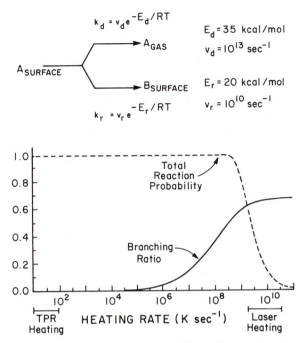

Fig. 7.3. (Top) Model reaction scheme showing direct desorption of adsorbate A competing with a surface reaction to form the surface species B. The kinetic parameters shown are those used to generate the curves. (Bottom) Plot of the calculated branching ratio for desorption, A(gas)/[A(gas) + B(surface)], and the total reaction probability, [A(gas) + B(surface)] divided by the initial amount of adsorbate as a function of heating rate. A temperature jump of 1000 K and a starting temperature of 300 K are used for all heating rates.

tion for LITD experiments because it implies that for substrates with low melting points, the substrate may melt before the desorption temperature for a certain species is reached. Thus, the use of low-melting substrate materials such as Cu, Ag, and Au may not be as effective for laser desorption experiments as high melting materials such as Pt and graphite.

EXPERIMENTAL

The pulsed laser used in our experiments produces a burst of neutrals from the surface that lasts for less than a microsecond. Ionizing and detecting such a short burst cannot be accomplished efficiently with a scanning-type mass spectrometer, such as a magnetic sector or quadrupole instrument, because only a single m/z value is detected at a time. A complete mass spectrum could be obtained only by pulsing the laser many thousands of times while the spectrometer is scanned. An additional complication is that if the energy of the laser pulses varies from shot to shot, the relative abundances of ions in the mass spectrum would fluctuate.

A more efficient approach for LITD experiments is to employ a mass spectrometer that operates in a pulsed mode and produces a full mass spectrum for each laser pulse. Two types of mass spectrometers that have this capability are the time-of-flight (TOF) and the Fourier transform (FT) mass spectrometers. TOF mass spectrometers are inexpensive and simple to construct, but have much lower mass resolution than FT mass spectrometers. Another difference is that with FT mass spectrometers it is possible to store the ions and perform additional experiments such as collision-induced dissociation and laser photodissociation of the ions to determine their structures.

The Fourier transform mass spectrometer used in our studies was built in our laboratory using a 1.2 T Varian electromagnet, an IonSpec model 2000 data system, and a vacuum chamber that is pumped to a base pressure of 2×10^{-10} torr by an ion pump and a turbomolecular pump.[6] A Pt(s) [7(111) $\times$ (100)] single crystal, which is 6 mm in diameter, or a similarly sized piece of an unknown material, is positioned in front of a hole in one of the electrodes of the FTMS analyzer cell by a Varian sample manipulator equipped with heating and cooling capabilities. Molecular adsorbates of interest can be exposed to the sample by backfilling the vacuum chamber. Different coverages can be prepared by varying the pressure of the gas or the time of the exposure or both. A typical exposure time is on the order of 3 min. After an exposure, the chamber is pumped down to a pressure of approximately 5×10^{-10} torr before the laser desorption experiments are conducted.

Figure 7.4 shows the sequence of events for a typical LITD–FTMS experiment. In Fig. 7.4a, a pulsed laser beam rapidly heats the near-surface region of a sample that is supported adjacent to an FT mass spectrometer analyzer cell. Low-power densities are used, as previously described, so that the temperature jump results in desorption of intact molecular adsorbates as neutrals, which then expand into the FTMS cell. As Fig. 7.4b illustrates, the desorbed neutrals pass through an electron beam (or a second laser beam) that is several centimeters away from the sample surface. In Fig. 7.4c, ions formed by electron ionization are trapped in the analyzer cell by a strong magnetic field and small dc voltages applied to the plates of the cell. After the burst of neutrals has passed through the analyzer cell, the electron beam is turned off and the ions undergo cyclotron motion at a frequency given by the cyclotron equation

$$\omega = qB/m \qquad (7.3)$$

where m/q is the mass-to-charge ratio and B is the magnetic field strength. In Fig. 7.4d the ions are accelerated by a pulsed radiofrequency electric field and their cyclotron frequencies are measured. The mass-to-charge ratio of an ion is calcu-

---

Fig. 7.4. The sequence of events in a laser-induced thermal desorption FTMS experiment. (a) The laser beam enters the cell and strikes the crystal. (b) Some of the desorbed molecules are ionized by an electron beam. (c) Ions are trapped in the analyzer cell by the magnetic and electric fields. (d) Ions are accelerated by an rf pulse and the resulting coherent image current is detected.

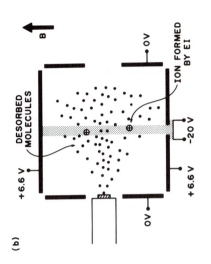

(a) TRAPPING PLATE
+6.6 VOLTS

ELECTRON
BEAM

LASER BEAM

CRYSTAL

SUPPORT

O VOLTS
TRANSMITTER PLATE

+6.6 VOLTS
TRAPPING PLATE

O VOLTS
TRANSMITTER
PLATE

−20 VOLTS
FILAMENT

B

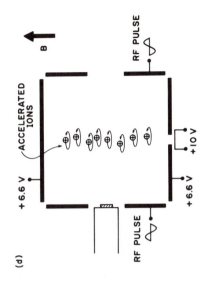

(b)

DESORBED
MOLECULES

ION FORMED
BY EI

+6.6 V

0 V

0 V

+6.6 V

−20 V

B

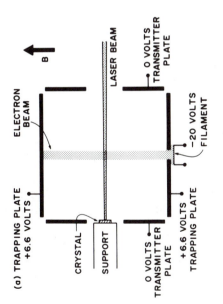

(c)

TRAPPED
IONS

+6.6 V

0 V

0 V

+6.6 V

+10 V

B

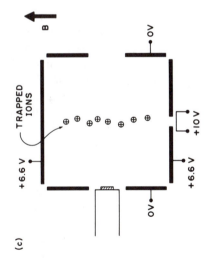

(d)

ACCELERATED
IONS

RF PULSE

+6.6 V

RF PULSE

+6.6 V

+10 V

B

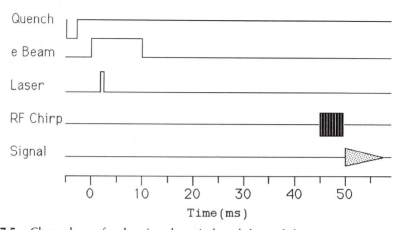

Fig. 7.5. Chronology of pulses in a laser-induced thermal desorption FT mass spectrometry experiment with electron postionization. At the beginning of each experiment the quench pulse removes any ions still stored in the cell. The electron beam fires 5 msec later, followed by the laser pulse after the electron beam has reached its full 70-eV potential. The electron beam remains on for several milliseconds after the laser has fired. After a short delay to allow the burst of neutrals to be pumped away, the rf chirp scans through the frequency range corresponding to the mass range of interest and shortly thereafter the transient image current signal is collected and digitized.

lated from the measured cyclotron frequency (ω) and the strength of the magnetic field (B).[7]

The sequence of pulses for a typical LITD–FTMS experiment is shown in Fig. 7.5. During the quench pulse period, a large negative voltage pulse is applied to the trapping plates of the cell to remove any positive ions that are still trapped. Next, a 10-μA electron beam, which traverses the cell, is turned on for 2 msec before the laser pulse and is left on for 8 msec afterward. The third pulse from the computer triggers the laser. Both the 248-nm (KrF*) wavelength laser radiation and the 350-nm (XeF*) wavelength laser radiation from an excimer laser (Lambda Physik EMG 103 MSC), as well as the 1.06-μm line from an Nd:YAG (Quanta-Ray DCR-2) have been used for our laser desorption experiments. For all wavelengths the laser power density at the crystal was adjusted to be below that necessary to ablate the substrate and produce ions. The power density is adjusted by varying the degree of focusing at the surface or by attenuating the output of the laser.

The detect pulse shown in Fig. 7.5 causes all ions in the analyzer cell to be accelerated to larger radii of gyration. This is accomplished by rapidly programming a frequency synthesizer to scan a range of frequencies corresponding to the cyclotron frequencies of the ions of interest. After the acceleration period, ions in the cell cyclotron coherently and induce small image current signals in a pair of receiver plates. This transient signal is amplified, digitized, and stored in a computer where it is subjected to a fast Fourier transform (FFT) analysis to recover the individual cyclotron frequencies. At pressures below 10^{-8} torr, the transient image current signal lasts for several tenths of a second, and under these conditions mass resolution of greater then 10,000 is possible with FTMS.[8]

RESULTS

LITD of Organic Molecules from a Platinum Surface

Laser desorption of a number of adsorbates from platinum has been studied in our laboratory using both 248- and 350-nm radiation with electron beam postionization. Table 7.1 lists many of the species that were successfully desorbed. The column on the far right shows the four or five most abundant ions produced by electron ionization of the laser-desorbed neutrals. In each case, the mass spectra obtained for laser-desorbed species closely resemble the reference electron ionization mass spectra of the compounds, indicating that desorption of intact molecular species was accomplished.[9] For many of these species, TDS experiments from Pt produces extensive decomposition and desorption of fragments of the parent molecule. For example, when naphthalene is adsorbed onto a clean Pt crystal it completely decomposes under TDS heating conditions to give only H_2 desorption with a layer of carbon remaining on the surface. Using LITD, however, desorption of the intact molecular species dominates over surface decomposition, and the mass spectrum for laser-desorbed naphthalene is essentially the same as its gas-phase EI fragmentation pattern.

Table 7.1. Adsorbates studied by laser desorption with FTMS detection

Adsorbate	Formula	Ionization potential[a] (eV)	Molar absorptivity[b] (248 nm)	Laser desorption masses	
				MPI (248 nm) (m/z)	EI (m/z)
Carbon monoxide	CO	14.0		None	28
Carbon dioxide	CO_2	13.8		None	28,44
Acetone	C_3H_6O	9.7	8	None	41,42,43,58
Acetic anhydride	$C_4H_6O_3$	10.2[c]	30[d]	None	42,43
Ethylene	C_2H_4	10.5	<1[e]	None	26,27,28
Cyanogen	C_2N_2	13.4		None	26,52
Trimethylamine	C_3H_9N	7.8	100[f]	30,42,58,59	30,42,58,59
Isobutane	C_4H_{10}	10.6		None	28,41,42,43
Methanol	CH_3OH	10.8	<1[g]	None	28,29,31,32
Tetrahydrofuran	C_4H_8O	9.4	<1[h]	None	41,42,43,71,72
Benzene	C_6H_6	9.1	58	27,36,37,39,78	50,51,52,77,78
Naphthalene	$C_{10}H_8$	8.5	2100	40,63,102,128	50,51,102,128
Toluene	C_7H_8	8.8	110	36,37,38,92	39,65,91,92
Phenol	C_6H_5OH	8.5	3719	27,37,39,94	39,65,66,94
Methyl benzoate	$C_8H_8O_2$	9.4	878	39,50,51,77,105	51,77,105,136

[a]H. M. Rosenstock, K. Draxl, B. W. Steiner, and J. T. Herron, Energetics of gaseous ions. *J. Phys. Chem. Ref. Data,* Suppl. 1, **6** (1977).

[b]Sadtler Standard Ultraviolet Spectra. Sadtler Research Laboratories, Philadelphia, 1984.

[c]Appearance potential for mass 43.

[d]H. Ley and B. Arends, *Z. Phys. Chem.* **B17** 194 (1932).

[e]D. F. Evens, *J. Chem. Soc. (London)* 1739 (1960).

[f]J. A. Moede and C. Curran, *J. Am. Chem. Soc.* **71** 855 (1949).

[g]B. Hampel, *UV Atlas of Organic Compounds,* Vol. 5, p. M/1.

[h]B. Hampel, *UV Atlas of Organic Compounds,* Vol. 5, p. M/9.

While performing the above experiments, we observed that with certain adsorbates 248-nm laser radiation produced ions directly even when the electron beam was turned off. The pulse sequence for an experiment of this type is the same as that shown in Fig. 7.5 except that the electron beam is turned off. Listed in Table 7.1 are the ionization potentials, molar absorptivities at 248 nm, and fragments seen by LITD with MPI. Examination of these data show that species with an ionization potential below the energy of two photons (9.98 eV) and which have a sufficient molar absorptivity at this wavelength are ionized directly by the laser pulse. Presumably the ionization mechanism is due to resonance-enhanced multiphoton ionization (REMPI).[10,11] The fragmentation patterns resulting from REMPI ionization were observed to be quite different from the gas-phase electron impact fragmentations. In addition, we observed that the REMPI fragmentation patterns and yield of ions changed greatly (>100%) from shot to shot, presumably due to variations in the excimer laser output. This is consistent with what others have observed because the fragmentation caused by multiphoton ionization is very sensitive to the laser power densities applied.[12] In contrast, for species that were not ionized directly by the laser, the mass spectra produced by electron ionization varied very little from shot to shot, and even substantial changes in the laser power density did not change the fragmentation patterns significantly.

For many of the compounds in Table 7.1, LITD with REMPI ionization by a single laser pulse can be over an order of magnitude more efficient than our standard method of ionization with an electron beam. In addition, REMPI is more selective than electron ionization because in a mixture only those species that absorb the incident laser radiation are ionized. The major drawback of REMPI is that the variable fragmentation patterns make the mass spectra more difficult to interpret.

The mechanism of the laser-desorption REMPI process is not completely understood at this time. During the 20-nsec period of the laser pulse, thermal energy molecules can move only a few microns away from the surface of the sample, and once ionized the magnetic field should cause them to move around in a circular cyclotron orbit and collide with the surface. Therefore, there must be a mechanism that allows the ions to move perpendicular to the magnetic field and into the center of the analyzer cell where they are detected. The most likely explanation is that multiple ion-neutral and ion–ion collisions occur in the high-density plume of desorbed material, and these collisions move the ions to the center of the cell.

Analysis of Computer Magnetic Hard Disk Surfaces

We have recently investigated the surfaces of computer magnetic hard disks to identify the molecular species adsorbed on the surface.[13] The magnetic hard disks are multilayer metal structures (Fig. 7.6) with a final layer of carbon sputter deposited on top of the magnetic layer to act as a protective film and lubricant. Adsorption of molecular species in and on this carbon layer has been found to be of great importance to the performance characteristics of the magnetic disk.[14] Figures 7.7 and 7.8 show typical mass spectra obtained for these samples using LITD with 248-nm laser radiation followed by 70-eV electron ionization. The effect of different

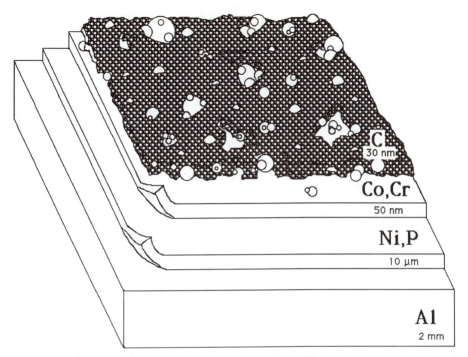

Fig. 7.6. Schematic of the multilayer structure of thin film computer magnetic hard disks. The disks have an aluminum base 2 mm thick onto which a 1-μm film of phosphorus-doped nickel is evaporated. The magnetic layer, a 50-nm-thick evaporated film of a Co alloy, is then covered with 30 nm of sputter deposited carbon.

laser power densities was investigated systematically. Figure 7.7a shows that at the lowest laser power densities utilized (1 MW/cm^2), only CO_2 and water are seen in the mass spectrum. Comparison of Fig. 7.7a with a background mass spectrum obtained with the laser beam blocked showed that most of the water signal at m/z 18 (approximately 90%) is from the chamber background. The CO_2, however, is only a minor background constituent, and the background contribution to the m/z 44 peak in Fig. 7.7a is only about 20% of the total peak height.

At slightly higher incident laser power densities (between 2 and 20 MW/cm^2), a multitude of peaks are observed in the mass spectrum up to 341 u. Figure 7.7b is a typical mass spectrum that was taken at an incident power density of about 5 MW/cm^2. At these low laser powers, ions are not formed by the laser alone and postionization with an electron beam is required. Peaks labeled by a star have been identified using accurate mass measurement as poly(dimethylsiloxane) compounds, which are common to many silicone oils. A possible structure for the ion at mass 221, for example, would be

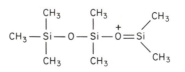

Also, there is evidence that the clusters of peaks between 30 and 110 amu are the fragments of an aliphatic alcohol. At this power level, essentially the same pattern of peaks is observed after as many as 600 laser shots at the same spot. This indicates that the laser beam is not significantly decomposing these surface species and is not ablating the carbon film. The large amount of sample present indicates that the silicone oil was probably placed on the disk intentionally during or after manufacturing. Inadvertent contamination of the disks in our vacuum chamber is unlikely because most pump oils are phenyl-substituted silicone oils, which give very different mass spectra. In addition, samples of hard disks from different sources showed no evidence of silicone compounds under the same experimental conditions, but some did show evidence of various hydrocarbons on the surface. These data demonstrate the utility of LITD–FTMS to monitor the surface composition of sputtered carbon films.

Figure 7.8 shows typical mass spectra obtained with a laser power density of 25

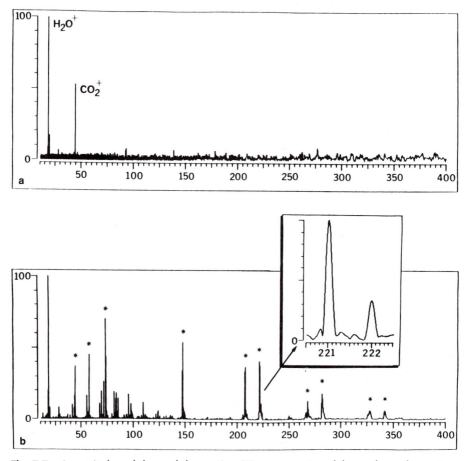

Fig. 7.7. Laser-induced thermal desorption FT mass spectra of the surface of a section of a computer hard disk platter showing the effect of different incident laser powers (P_i). (a) $P_i < 1$ MW/cm^2; (b) $1 < P_i < 20$ MW/cm^2.

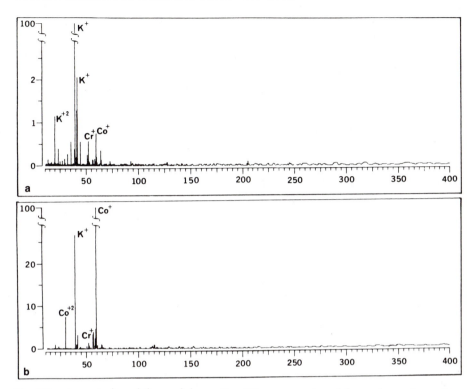

Fig. 7.8. Laser-induced thermal desorption FT mass spectra of the surface of a section of a computer hard disk platter showing the effect of multiple shots at high laser power density (25 MW/cm^2). (a) The second shot at high power for a particular spot; (b) the fifth shot at the same spot at high power.

MW/cm^2. Under these conditions, the carbon overcoat is penetrated and the underlying metal layers of the hard disks are ablated. Figure 7.8a is the mass spectrum obtained after the second shot at a spot with a power level of 25 MW/cm^2. The major peak in the spectrum is K$^+$, and there are traces of cobalt and chromium from the magnetic layer. Potassium is most likely present in the carbon layer as a trace impurity, but because of its low ionization potential, it dominates the spectrum of ions produced in the desorbed plasma, similar to effects seen in laser microprobe experiments using high laser power densities.[15] The FT mass spectrum obtained with the fifth high-power laser pulse at this same spot is shown in Fig. 7.8b. Cobalt is the dominant peak in the mass spectrum and the absolute intensity of the potassium has decreased.

Attempts at further depth profiling were hampered by the widely varying ablation thresholds for the different layers. We found that laser power densities useful for slowly etching the metallic layers caused catastrophic ablation of the carbon layer. On the other hand, lower power densities did not appear to etch the carbon layer at a significant rate. Apparently the laser ablation thresholds for different materials differ significantly and tend to be very sharp. A better method for depth profiling might be to use an ion beam to etch the surface followed by laser desorption for the composition analysis. We are building a new surface analysis instrument that will have the capability to perform such an experiment. The instrument consists of a bell jar with an ion bombardment gun, Auger electron spectroscopy, low-energy electron diffraction, and sample dosing capabilities, connected to an LITD–FTMS chamber. The sample is shuttled between the two chambers by a long z-motion manipulator.

Catalytic Dehydrogenation of Ethylene on Platinum

The LITD–FTMS technique has applications in the study of catalytic reactions as well as the analytical applications already mentioned. Some important catalytic reactions occur when unsaturated hydrocarbons are in contact with platinum. The decomposition of ethylene on platinum, for instance, has been studied in some detail in our laboratory.[16] It is known from previous studies utilizing HREELS[17] and LEED[18] that ethylene (C_2H_4) exists intact on platinum at low temperatures. As the crystal temperature is raised, however, the molecule loses hydrogen and rearranges to a $C-CH_3$ intermediate (ethylidyne) and surface hydrogen. Ethylidyne has been shown to be stable on the surface to temperatures above 450 K, whereas the surface hydrogen generated in the formation of ethylidyne desorbs at about 350 K. On further heating, ethylidyne loses more hydrogen to produce surface carbon and desorbed H_2.

Our experiments were performed by focusing the beam of an excimer laser (248 nm) to 0.5 mm diameter onto a Pt(lll) surface that was dosed at low temperature with ethylene. For all of these experiments, incident laser power densities less than 50 MW/cm^2 were used so that only neutral species, not ions, were desorbed from the surface. The desorbed neutrals were ionized by a pulsed electron beam, as previously described.

Typical data for the laser desorption of ethylene adsorbed on Pt are shown in Fig. 7.9. Initially, a clean Pt crystal held at 190 K was dosed with 2 L of ethylene. Figure 7.9a is the mass spectrum obtained for species desorbed from the surface by a single shot of the laser immediately after the adsorption. Over the mass range 1.5–230 u, the only peaks observed are molecular ethylene (m/z 28) and its 30-eV electron ionization fragments. Ablation of the Pt surface is not observed, as evidenced by the absence of Pt$^+$ peaks in the range 190–198 u. After heating the dosed platinum crystal to 400 K for 30 sec and then recooling to 190 K, the laser was directed at a different spot on the surface and the mass spectrum in Fig. 7.9b was obtained. Comparison of these two spectra shows several important differences:

1. The total ion current in Fig. 7.9b is much smaller than in Fig. 7.9a. This results because some of the ethylene desorbs as the platinum crystal is heated above 250 K and only a small fraction (lower limit of 1%) reacts to form ethylidyne. The laser may drive some of the ethylidyne to surface carbon or some other species that does not desorb.
2. Figure 7.9b shows that after heating there is a peak at mass 2 due to H_2 generated by laser-driven decomposition of ethylidyne, as discussed below.
3. In Fig. 7.9b the relative abundances of 26, 27, and 28 are quite different from those obtained by the 30-eV electron ionization of ethylene. The large 28

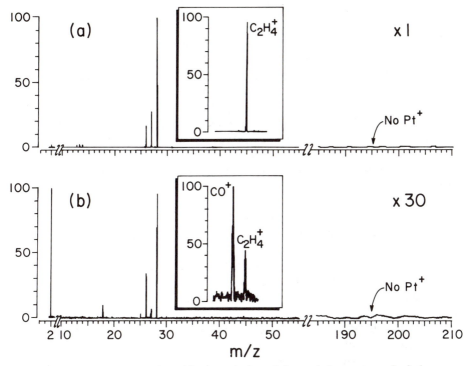

Fig. 7.9. Mass spectra produced by laser-induced thermal desorption of ethylene on Pt before and after annealing to 400 K. (a) Broadband and narrowband Fourier transform mass spectra obtained after adsorption of ethylene at 190 K. (b) Broadband and narrowband spectra obtained after annealing the sample to 400 K and recooling to 190 K.

peak is misleading because, as the expanded view shows, there are two peaks at mass 28. The expanded view shows that the CO^+ signal at 27.994 u is about 2.5 times larger than the $C_2H_4^+$ signal at 28.031 u. Apparently, CO in the vacuum chamber at a partial pressure about 1×10^{-10} torr adsorbed onto the crystal while the temperature was being cycled. The expanded view spectrum was obtained using a narrowband, high-resolution acquisition mode. In Fig. 7.9a the high-resolution mass spectrum shows no CO^+.

4. Mass 26 is a prominent peak in Fig. 7.9b, and we believe it results from the laser-driven decomposition of ethylidyne.

Although laser desorption of ethylidyne might be expected to yield a large peak at mass 27, we observe C_2H_2 and H_2 instead. It appears that the fast heating rate produced by the laser is driving a surface reaction not accessed using slower heating rates. Instead of direct desorption or decomposition, ethylidyne is converted on the surface into hydrogen atoms, which recombine and desorb, and either $C=CH_2$ or $HC\equiv CH$, which also desorbs. The ratios of masses 26:25:24 are consistent with the electron ionization pattern for $HC\equiv CH$. However, determination of which of the two carbon species desorbs would be difficult since $C=CH_2$ most likely undergoes a facile, gas-phase rearrangement to HC≡CH before detection.[19] This model accounts for the predominance of mass 26 and the peak at mass 2 in Fig. 7.9b.

Figure 7.10 combines our observations for the laser-driven reactions with the present model for the dehydrogenation of ethylene on Pt. At low temperatures (<200 K) ethylene adsorbs onto Pt intact, and LITD produces mass spectra that are nearly identical to gas-phase ethylene. Under slow heating conditions, some of the ethylene desorbs at about 250 K. Between approximately 260 and 450 K, the remaining ethylene on the surface is converted irreversibly to ethylidyne ($C-CH_3$) and the liberated hydrogen atoms recombine and desorb as H_2. LITD at this point gives mass spectra resembling H_2 and acetylene as previously discussed. Further heating results in decomposition to surface carbon and liberation of additional hydrogen.

Additional experiments have shown that the H_2^+ peak in Fig. 7.9b is not due to residual surface hydrogen. When 40 L of H_2 was adsorbed onto the clean platinum surface at a temperature below 200 K, laser desorption produced a large peak at m/z 2 due to recombination and desorption of H_2. After annealing the sample to 400 K for 30 sec, laser desorption resulted in no detectable H_2 signal. This shows that the H_2^+ signal in Fig. 7.9b must have come from recombination of H atoms and desorption of H_2 formed by a surface reaction during the laser-induced temperature jump.

Other possible mechanisms for formation of mass 26 following laser desorption of ethylidyne are ruled out by the following experiments. Photochemical decomposition of gas-phase ethylidyne is unlikely because laser desorption at different wavelengths (248 nm, 350 nm from the excimer laser, and 1.06 μm from a Quanta-Ray DCR-2 Nd:YAG laser) give the same results. Bimolecular gas-phase reactions of ethylidyne to give C_2H_2 are unlikely because, with the low coverages used in these experiments, there are few molecular collisions before detection.[20] Finally, experiments in which the ionizing electron beam energy was varied from 15 to 75 eV showed no ethylidyne parent ion (mass 27) even at the lowest energies, so it is unlikely that m/z 26 is the major ion produced by electron ionization of ethylidyne. All of these alternative mechanisms for the formation of m/z 26 are inconsistent with our observation of molecular hydrogen because they would require gas-phase recombination of H atoms. This also rules out direct desorption of internally excited ethylidyne followed by unimolecular decomposition to give C_2H_2.

We propose that the C_2H_2 observed in our laser desorption experiments after warming the ethylene-covered platinum surface to 400 K is the result of a laser-driven surface reaction of ethylidyne that is not observed under slow heating conditions. With the new apparatus being constructed in our laboratory, we plan to use Auger electron spectroscopy to quantify the amount of carbon remaining on the surface after the laser pulse.

Propylene Dehydrogenation on Platinum

Propylene can be thought of as a methyl-substituted ethylene, and both might be expected to react similarly on Pt. Somorjai and co-workers have proposed the formation of propylidyne ($C-CH_2CH_3$) analogous to the formation of ethylidyne after comparison of LEED and HREELS data for the two systems.[21] Propylidyne is

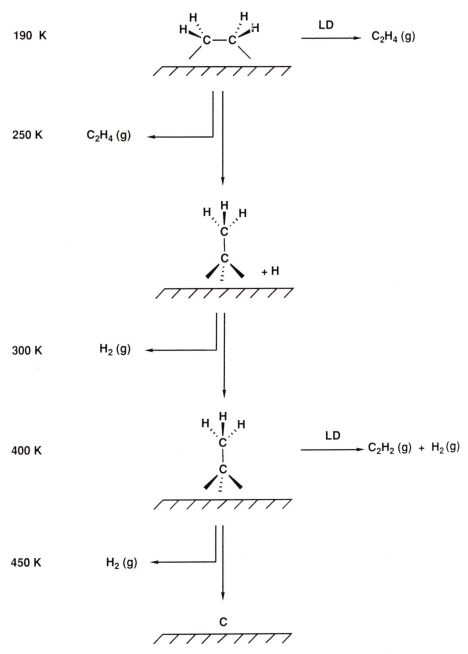

Fig. 7.10. Sequence of reactions in the dehydrogenation of ethylene on Pt. On the left side are the results under slow heating conditions. On the right are the reactions induced by fast laser heating.

believed to form above 270 K on Pt(lll) and is stable up to about 400 K. In our studies, propylene was dosed onto the crystal at 170 K. LITD at this temperature yielded molecular desorption of propylene. The Pt sample was then warmed to 300 K and held at that temperature. The laser was redirected at a different point on the surface and 2 min after heating had begun another LITD–FTMS spectrum was obtained. The relative abundances of the parent ion, m/z 42, and the M-1 peak at m/z 41 both had decreased slightly, whereas m/z 26 and 40 both had increased slightly over their abundances from the cold crystal. These trends continued, leveling off after about 14 min at which time the relative abundance of m/z 42 and 41 had decreased by a factor of about two, whereas m/z 26 and 40 had increased by a factor of about five. The ions at m/z 42 and 41 are derived only from the parent propylene molecule and their decrease indicates a decrease in the amount of propylene on the surface. By analogy to ethylidyne, there are two possible channels for the decomposition of the propylidyne during the laser-induced temperature jump: (1) m/z 40 corresponds to the loss of H from propylidyne, and (2) m/z 26 corresponds to the loss of CH_3 from propylidyne. In addition, another possible channel for the formation of m/z 26 involves loss of CH_3 from adsorbed propylene to directly give ethylidyne, which would desorb as m/z 26 during the laser-induced temperature jump.

Our analysis by LITD–FTMS of the propylene/Pt system confirms the formation of a propylidyne intermediate analogous to the ethylidyne intermediate seen for ethylene/Pt. It also shows, along with the data presented above, the generality of the LITD–FTMS technique for the study of surface composition at the *molecular* level.

DISCUSSION

The previous examples have demonstrated the utility of the LITD–FTMS technique for the analysis of a wide variety of adsorbates on two very different types of surfaces. LITD–FTMS is distinguished because of its capability to obtain molecular information from complicated systems. In this section we compare LITD–FTMS with other surface analysis techniques that are available and summarize the features of LITD–FTMS.

Vibrational spectroscopy (Raman, HREELS, and infrared) methods provide molecular information and have been used extensively for the study of surface reactions and reaction intermediates.[22] However, these methods have mostly been applied to small molecules on surfaces because the spectra produced by complex polyatomic molecules are very difficult to interpret. For example, the poly(dimethylsiloxane) compounds we found on the computer hard disk could not have been unambiguously identified by vibrational spectroscopy.

Several surface analysis techniques utilize mass spectrometry to detect the *ions* that are desorbed when a surface is bombarded by energetic particles or a high-power laser pulse. Methods such as secondary ion mass spectrometry (SIMS), ^{252}Cf plasma desorption mass spectrometry (PDMS), and fast atom bombardment (FAB) mass spectrometry utilize energetic particles, and high-power laser desorption experiments are usually referred to as laser desorption mass spectrometry

(LDMS).[10,15,23,24,26] In contrast, LITD is distinguished from these methods because it utilizes low laser power densities to desorb only *neutral* species from the surface. The distinction between the desorption of ions and neutrals is important because ions are much more reactive than neutrals and can undergo gas-phase reactions that complicate interpretation of the mass spectra. For example, techniques that desorb ions directly often produce adduct ions that form from the adsorbate and substrate ions or alkali impurities, according to the reaction

$$Y^+ + R \rightarrow [Y-R]^+ \tag{7.4}$$

where R is the adsorbate. This is a gentle ionization method, and some researchers even add alkali salts to their samples to increase the yield of cation-attached ions for species that do not form molecular ions from the desorption beam alone.[24]

Thermal desorption spectroscopy (TDS) is another surface analysis method based on the desorption of neutral species. In TDS the sample is placed on a resistively heated probe, and a mass spectrometer with electron ionization is used to monitor the species that desorb as a function of surface temperature. The temperature that produces the maximum desorption flux for a particular species can be related to its adsorption energy on the surface.[25] Since relatively slow heating rates are used in TDS (10°C/sec), the surface species are generally in equilibrium with the temperature of the surface and surface reactions can alter the composition of the species before they desorb. This complication makes it difficult to identify the original surface species. Much faster temperature jumps are available using LITD, and desorption of intact molecular species is often possible.

Another characteristic of TDS is that the entire sample is heated, and the surface species are desorbed over a period of several seconds. In contrast, with LITD only the region under the laser beam is heated, and the surface species desorb in a microsecond or less. By changing the size of the laser beam incident on the surface, one can effectively obtain an average of the surface composition using a large spot size, or interrogate specific defects by using a small beam diameter. This capability is particularly important in the analysis of microelectronics or similar contamination oriented experiments.

Since the instantaneous flux of desorbed material is much higher in LITD than in TDS, a variety of methods can be utilized to ionize the desorbed neutrals. Electron ionization, the method used most often in our laboratory, is useful for identifying unknown adsorbates because at 70 eV all desorbed species are ionized and the EI fragmentation mechanisms are well understood. In addition, we have found it very useful to do a computer search of the EI mass spectral data bases to identify peaks that are prominent in the LITD–FTMS data. One of the disadvantages of EI is that excess energy imparted to the molecules causes extensive fragmentation and makes it difficult to identify the molecular ion. As an alternative, laser multiphoton ionization can be used to ionize the desorbed neutrals.[10,26,27] Laser postionization is more specific than EI and, under the right conditions, can produce minimal fragmentation of large molecules.[27]

Another method of postionization that produces little fragmentation is chemical ionization.[28] Cotter[29] has used chemical ionization of laser-desorbed species in a high-pressure source with a scanning mass spectrometer. The authors and Dr. I. J. Amster have recently developed this method to ionize laser-desorbed neutrals using the FT mass spectrometer.[30] A pulsed valve is used to generate a 100-msec pulse of reagent gas, such as methane, in the FTMS cell prior to the laser experiment. The electron beam is fired during this time and produces large numbers of ions that, at these pressures, undergo numerous collisions and self-protonate to produce reagent ions such as CH_5^+ that are stored in the analyzer cell. After waiting for the neutrals to be pumped away (about 1 sec) the laser is fired, and the burst of desorbed neutrals expands into the cell where some undergo collisions with the stored reagent ions and abstract a proton producing $[M + H]^+$ ions. This method has been successful using acetophenone, gramicidin-s, and heptadecylacridine adsorbed on Pt. LITD with chemical ionization offers promise as a technique for the analysis of large compounds and, in particular, biomolecules.

CONCLUSIONS

We have shown that the rapid temperature jumps produced by laser-induced thermal desorption can achieve the molecular desorption of a wide variety of organic adsorbates. The use of electron beam postionization and a Fourier transform mass spectrometer to collect a complete mass spectrum for every laser pulse is a versatile detection method that is compatible with the pulsed nature of the laser desorption experiment. LITD–FTMS appears to be most useful for obtaining molecular information for submonolayer coverages of complex species, with greater temporal and spatial resolution than has been previously possible.

Future applications of LITD–FTMS may be in the areas of electrode contamination, corrosion studies, and lubrication and adhesion at metal and ceramic surfaces. In addition, it may be possible to apply the technique to the analysis of polymeric surfaces such as prosthetic implants or contact lenses to understand how the human body reacts to the presence of foreign materials. We are very encouraged at the scope of our results so far and with the potential applications of this technique.

ACKNOWLEDGMENTS

Support for this research was provided by the National Science Foundation under Grant CHE8511999 and the donors of the Petroleum Research Fund, administered by the American Chemical Society. D.P.L. wishes to acknowledge support in the form of an IBM Graduate Research Fellowship.

REFERENCES

1. *Low Energy Electrons and Surface Chemistry,* edited by G. Ertl and J. Kueppers. Verlag Chemie, Weinheim, 1974.

2. (a) L. D. Schmidt, *Cat. Rev.-Sci. Eng.* **9**(1), 115 (1974). (b) P. A. Redhead, *Vacuum* **12**, 203 (1962).

3. J. F. Ready, *Effects of High Power Laser Radiation.* Academic Press, New York, 1971, pp. 70–88.

4. M. G. Sherman, J. R. Kingsley, R. T. McIver, Jr., and J. C. Hemminger, in *Catalyst Characterization Science* (ACS Symposium Series No. 288), edited by M. L. Deviney and J. L. Gland. American Chemical Society, Washington, DC, 1985, pp. 238–251.

5. R. B. Hall, *J. Phys. Chem.* **91,** 1007 (1987).

6. M. G. Sherman, J. R. Kingsley, J. C. Hemminger, and R. T. McIver, Jr., *Anal. Chim. Acta* **178,** 79 (1985).

7. M. L. Gross and D. L. Remple *Science* **226,** 261 (1984).

8. A. Marshall, M. B. Comisarow, and G. Parisod, *J. Chem. Phys.* **71,** 4434 (1979).

9. R. T. McIver, Jr., M. G. Sherman, D. P. Land, J. R. Kingsley, and J. C. Hemminger, in *Secondary Ion Mass Spectrometry, SIMS V,* edited by R. J. Colton and D. S. Simmons, Springer-Verlag, Berlin, 1986.

10. S. E. Egorov, V. S. Letokhov, and A. N. Shibanov, in *Surface Studies with Lasers* (Springer Series in Chem. Phys.: 33), edited by F. R. Aussenegg, A. Leitner, and M. E. Lippitsch. Springer-Verlag, New York, 1983, pp. 156–170.

11. (a) D. M. Lubman and R. Naaman, *Chem. Phys. Lett* **95,** 325 (1983). (b) R. B. Opsal and J. P. Reilly, *Chem. Phys. Lett.* **99,** 461 (1983). (c) J. T. Meek, S. R. Long, R. B. Opsal, and J. P. Reilly, *Laser Chem.* **3,** 19 (1983).

12. (a) D. H. Parker, R. B. Bernstein, and D. A. Lichtin, *J. Chem. Phys.* **75**(6), 2577 (1981). (b) D. A. Lichtin, R. B. Bernstein, and K. R. Newton, *J. Chem. Phys.* **75,** 5728 (1981).

13. D. P. Land, T.-S. Tai, J. M. Lindquist, J. C. Hemminger, and R. T. McIver, Jr., *Anal. Chem.* **59,** 2924 (1987).

14. I. Sato, *IEEE Transl. J. Magn. Jpn.* **2,** 4 (1987). Translated from I. Sato, *J. Magn. Soc. Jpn.* **10,** 6 (1986).

15. K. Balasanmugam, S. K. Viswanadham, and D. M. Hercules, *Anal. Chem.* **58**(6), 1102 (1986).

16. M. G. Sherman, D. P. Land, J. C. Hemminger, and R. T. McIver, Jr., *Chem. Phys. Lett.* **137,** 298 (1987).

17. H. Ibach and S. Lehwald, *J. Vac. Sci. Technol.* **15,** 407 (1978).

18. L. L. Kesmodel, L. H. Dubois, and G. A. Somorjai, *Chem. Phys. Lett.* **56,** 267 (1978).

19. (a) O. P. Strausz, R. J. Norstrom, A. C. Hopkinson, M. Schoenborn, and I. G. Csizmadia, *Theoret. Chim. Acta (Berl.)* **29,** 183 (1973). (b) C. E. Dykstra and H. F. Schaefer III, *J. Amer. Chem. Soc.* **100**(5), 1378 (1978).

20. J. P. Cowin, D. J. Auerbach, C. Becker, and L. Wharton, *Surf. Sci.* **78,** 545 (1978).

21. R. J. Koestner, M. A. Van Hove, and G. A. Somorjai, *Surf. Sci.* **116,** 85 (1982).

22. See, for example: *Vibrational Spectroscopies for Adsorbed Species,* edited by A. T. Bell and M. L. Hair. American Chemical Society, Washington, DC, 1980, or *Vibrational Spectroscopy of Molecules on Surfaces,* edited by J. T. Yates, Jr. and T. E. Madey. Plenum Press, New York, 1987.

23. (a) H. J. Heinen, S. Meier, H. Vogt, and R. Wechsing, *Adv. Mass Spec.* **8A,** 942 (1979). (b) P. G. Kistemaker, M.M.J. Lens, G.J.Q. van der Peyl, and A.J.H. Boerboom, *Adv. Mass Spec.* **8A,** 928 (1979). (c) B. Spengler, M. Karas, U. Bahr, and F. Hillenkamp, *J. Phys. Chem.* **91**(26), 6502 (1987). (d) R. J. Cotter and J. C. Tabet, *Int. J. Mass Spectrom. Ion Phys.* **53,** 151 (1983). (e) E. C. Apel, N. S. Nogar, C. M. Miller, and R. C. Estler, *Inst. Phys. Conf. Ser.* **84** (Reson. Ioniz. Spectrosc.) 179 (1986). (f) R. S. Brown and C. L. Wilkins, *Anal. Chem.* **58**(14), 3196 (1986). (g) C. E. Brown, P. Kovacic, R. B. Cody, R. E. Hein, and J. A. Kinsinger, *J. Polym. Sci., Part C: Polym. Lett.* **24**(10), 519 (1986). (h) J. A. Gardella, S. W. Graham, and D. M. Hercules, *Adv. Chem. Ser.* **203** (Polym. Charact.), 635 (1983).

24. (a) R. Stoll and F. W. Roellgen, *Org. Mass Spectrom.* **14,** 642 (1979). (b) E. D. Handin, T. P. Fan, C. R. Blakley, and M. L. Vestal, *Anal. Chem.* **56,** 2 (1984). (c) L. G. Wright, R. G. Cooks, and K. V. Wood, *Biomed. Mass Spectrom.* **12**(4), 159 (1984).

25. J. R. Kingsley, D. Dahlgren, and J. C. Hemminger, *Surf. Sci.* **139,** 417 (1984).

26. S. E. Egorov, V. S. Letokhov, and A. N. Shibanov, *J. Chem. Phys.* **85,** 349 (1984).

27. (a) R. Tembreull and D. M. Lubman, *Anal. Chem.* **59**(8), 1082 (1987). (b) U. Boesl, J. Grotemeyer, K. Walter, and E. W. Schlag, *Inst. Phys. Conf. Ser.* **84** (Reson. Ioniz. Spectrosc., 223 (1986). (c) J. H. Hahn, R. Zenobi, and R. N. Zare, *J. Am. Chem. Soc.* **109**(9), 2842 (1987).

28. F. W. McLafferty, *Interpretation of Mass Spectra,* 3rd ed. University Science Books, Mill Valley, CA, 1980, p. 6.

29. R. J. Cotter, *Anal. Chem.* **52,** 1767 (1980).

30. I. J. Amster, D. P. Land, J. C. Hemminger, and R. T. McIver, Jr., *Anal. Chem.,* **61,** 184 (1989).

8

Applications of Laser Ablation: Elemental Analysis of Solids by Secondary Plasma Source Mass Spectrometry

PETER ARROWSMITH

The techniques and instrumentation available for elemental analysis may be divided into two broad categories: those suitable for analysis of samples in solution form and those developed for direct analysis of solid materials. The number of techniques routinely used for trace elemental analysis of solutions is relatively small, although thousands of instruments around the world perform this kind of work. In part, this is the result of the high sensitivity of established methods and the uniform physical nature of liquid samples. Direct introduction of solid samples into the atomization and excitation sources of spectrometers designed for solution analysis is of interest because it broadens the capability of existing instrumentation and improves sample throughput, since the often laborious sample dissolution process is avoided. In addition to the advantage of speed, sample introduction by laser ablation gives compositional analysis at medium (~ 10 μm) spatial resolution and is applicable to all materials without restriction on physical properties, such as electrical conductivity. Although the laser ablation sample introduction system described here was developed for plasma source mass spectrometry, it is intended for use with other types of excitation sources, and also for optical spectrometers, without modification. The system is capable of analyzing all materials with few restrictions on sample size and shape, and with little or (in most cases) no sample preparation.

The emphasis of this chapter will be on areas that have been of particular concern and interest in the development of a new laser ablation system, but have received relatively little attention in the literature. These topics include the design and evaluation of the ablation cell, gas flow entrainment and transport of ablated particles to the secondary excitation source, and the importance of the particle size distribution of the ablated material. The final part of the chapter discusses the analytical performance and applications of the technique.

ELEMENTAL ANALYSIS OF SOLUTIONS

Solution analysis involves introduction of the sample, generally in the form of a spray of aerosol droplets transported by gas flow, to a suitable source such as a furnace, flame, or plasma for formation of free atoms and excited species. Analysis is conventionally performed with single or multichannel optical spectrometers. Ground state atoms are detected by atomic absorption or fluorescence spectrometry (AAS or AFS) and excited atoms by atomic emission spectroscopy (AES). The standard source for AES is the inductively coupled plasma (ICP) because it gives high excitation temperatures, is relatively immune to chemical interferences, and has a large linear dynamic range.[1]

A recent development in solution analysis has been to extract the ions formed in the ICP into a vacuum for mass spectrometric detection. The ICP is a good source of positive ions since the electron temperature generally exceeds the gas kinetic temperature of ~ 5000 K, giving ionization efficiencies $>90\%$ for many elements. The atmospheric pressure plasma (usually running on Ar) is sampled into vacuum by supersonic expansion through a differentially pumped nozzle and skimmer mounted on a water-cooled flange. Because of rapid expansion, ion recombination is small and ions transmitted through the mass analyzer (a relatively low-resolution quadrupole filter in commercial instruments) to the detector are representative of the ICP. Since the introduction of the first commercial instruments 7 years ago, ICP-mass spectrometry (ICP-MS) has proven to be extremely sensitive, with solution detection limits in the range 10^{-9}–10^{-12} g/mL (1–1000 ppt), typically 10–100 times lower than those for ICP-AES. An important advantage of ICP-MS is the simplicity of the mass spectra. The stable isotopes of all the elements give rise to only ~ 200 peaks (at unit mass resolution) compared to upward of 1000 lines often found for a single heavy element in optical emission spectra. ICP-MS is becoming widely accepted as the most sensitive and versatile method for trace elemental analysis of solutions. Descriptions of the technique and its applications can be found in recent review articles.[2,3]

ANALYSIS OF SOLIDS

In comparison to solution analysis, there are a large number of different techniques for direct compositional and chemical analysis of solids. Many of these tend to be complementary in nature, having advantages and disadvantages in terms of surface and bulk sensitivity, spatial and depth resolution, and ease of quantification. Examples of techniques that offer surface sensitivity are X-ray photoelectron spectroscopy (XPS, also known as electron spectroscopy for chemical analysis or ESCA) and Auger electron spectroscopy. Techniques suitable for bulk analysis of solids include high-energy ion scattering (Rutherford backscattering spectroscopy or RBS), electron microprobe with X-ray detection, and X-ray fluorescence (XRF). Mass spectrometric techniques tend to have the highest sensitivity and widest dynamic range because of the relatively high efficiency of ion collection, transmission, and detection (often compensating for poor ion formation efficiency). They may be divided into two types, those in which the ions are formed directly, as in secondary ion mass spectrometry (SIMS), and those in which neutral species are sputtered from the sample and subsequently ionized in a different process, as in glow discharge mass spectrometry (GDMS) and secondary neutrals mass spectrometry (SNMS). As these techniques have been widely reviewed they will not be discussed further, except for purposes of comparison. Spark source atomic emission and spark source mass spectrometry have also been used for routine analysis of solids, particularly for quality assurance and comparative work. In these techniques, sample material is vaporized and excited by an electric discharge between the sample and an electrode. As with GDMS, spark sources are restricted to samples that are, to some extent, electrically conducting, or that can be made conducting by grinding to a powder and mixing with a high-purity conducting binder. Spark sources tend to have poor precision. In addition, multiply charged atomic ions and combination (molecular) ion species are produced, requiring a reasonably high-resolution mass spectrometer to remove mass interferences.

ANALYSIS OF SOLIDS BY DIRECT SPECTROSCOPY OF LASER PLUMES

Lasers have played an important role in elemental analysis of solids because the focused beam of a pulsed laser is capable of vaporizing or ablating all materials. The ability of a laser to precisely target a small feature of interest is an important advantage compared to techniques with little or no spatial resolution such as spark source and GDMS. At moderate laser irradiances ($>10^7$ W/cm^2) interaction between the laser beam and the sample causes rapid heating and ejection of a plume of material from the surface of the sample. The plume contains vaporized atomic and molecular species as well as liquid and solid particles and it is usually associated with visible emission, particularly at higher laser irradiances, because optically induced breakdown generates a microplasma. Ground and excited state atomic species and ions formed in the microplasma can be detected by AAS, laser-induced fluorescence, AES, or MS performed directly on the plume. In the original (25-year-old) laser microprobe analyzer, emitted light is dispersed in a polychromator and detected by a photographic plate. Improved sensitivity may be obtained by secondary excitation of ablated species with an electric spark; however, the technique suffers from poor precision.[4] In mass spectrometric versions of the laser microprobe, ions formed in the microplasma have been mass analyzed by means of sector, time-of-flight (TOF), or Fourier transform (FTMS) instruments. Commercially available instruments include the laser microprobe mass analyzer (LAMMA) and the laser ionization mass analyzer (LIMA), both utilizing TOF, and laser ablation/ionization sources for FTMS. By adjusting the laser power, desorption and ionization can, to some extent, be selected over ablation and dissociation in the microplasma, enabling these techniques to be used to detect molecular species in addition to elemental analysis. Several excellent reviews covering the subjects of optical and mass spectrometry performed directly on the laser plume have been published.[5-9]

ANALYSIS BY SEPARATE ABLATION AND EXCITATION

If suitable instrumentation is available, the chosen approach to elemental analysis of solids is generally to apply one or more of the direct techniques previously mentioned. However, many analytical laboratories are equipped only for solution analysis and cannot justify the expense of purchasing additional instruments. In this case the analyst has to digest the sample with suitable reagents to get it into solution. This approach has several problems. Dissolution is prone to contamination and is often time consuming (or impossible) for ceramics, semiconductors, plastics, and other, similar materials. Information concerning the spatial distribution of elements and sample homogeneity is lost. Perhaps more importantly, solution analysis may require ~1 g of material, often causing sufficient damage to the sample so that analysis cannot be performed on pieces that will be used in a manufacturing process. Finally, in the case of ICP-MS, molecular ions arising from the mineral acids often present in aqueous solutions may give rise to mass spectral interferences. For these reasons there is much interest in introducing solid samples (in suitable form) into the secondary sources (furnaces, flames, and plasmas) of dual purpose optical and mass spectrometers that may also be used for solution analysis. This approach separates "sample preparation" performed in a primary source (i.e., conversion of the solid sample into a form suitable for transport) from the vaporization, atomization, and excitation processes that take place in the secondary source. The two-stage technique has several advantages compared to direct spectroscopy of the primary source. Since sample preparation, transport, and secondary excitation take place sequentially and in different locations, each process is reasonably independent and the resulting "separation of variables" allows separate optimization and simplifies methods of quantification (see below). Also, atom and ion formation in secondary sources such as the ICP is generally insensitive to the chemical and physical state of the sample material. By comparison, the single-stage techniques of SIMS, spark source, and laser microprobe mass spectrometry have ion yields that often depend strongly on the nature of the sample. Finally, many of the well-characterized advantages of the excitation sources commonly used for solution analysis are also realized when the transported sample is in solid form.

Several types of primary source have been used to generate vapor and particles from solids (so called "solids nebulization") for transport by gas flow via transfer tube to the secondary source. Arc discharges have been used with both ICP-AES and ICP-MS and spark sample introduction is commercially available for ICP-AES.[10,11] Arcs are restricted to conducting materials and sample preparation may be required to form an electrode from the sample. Laser ablation does not have these limitations and it has been widely used in the laboratory for sample introduction to furnaces and flames for AAS and plasmas for AES.[12-27] Because of the sensitivity of ICP-MS, the combination of laser ablation and ICP-MS appears to be particularly useful and concentration detection limits of 0.1–1 μg/g, or better, have been reported.[28-30] Since the amount of material ablated is typically 1–10 ng/laser pulse, this represents absolute detection limits of <10 fg. Because the sample is at atmospheric pressure, laser ablation ICP-MS may be used to analyze volatile materials such as greases and oils. At the present time, two instrument manufacturers (Sciex-Perkin-Elmer and VG Elemental) have developed add-on laser ablation systems for their ICP-MS instruments.[31,32]

LASER ABLATION AND SAMPLE INTRODUCTION

Although both focused continuous-wave (CW) and pulsed lasers may be used for ablation, most analytical work has been done with solid-state pulsed lasers, such as the ruby and the Nd:YAG. Pulsed CO_2 and N_2 gas lasers have not been widely used in analytical applications because they suffer from low repetition rate and/or poor pulse stability; however, there is some recent interest in the use of excimer lasers.[26] The current trend is toward cheaper, compact Nd:YAG lasers having good pulse reproducibility at repetition rates up to 20 Hz and output energies of ~ 100 mJ, sufficient for ablating most materials when focused to spot diameters < 100 μm. The work described here was performed with a 180 mJ Nd:YAG laser operating at the fundamental wavelength of 1.06 μm. The following description of the laser output and, to a lesser extent, the ablation process, refers to flashlamp-pumped solid-state lasers.

The Ablation Process

Laser ablation has been described in detail previously, although the mechanism is complex and not fully understood.[5-8] Briefly, there are two distinct processes corresponding to two modes of pulsed laser operation, yielding the lowest and highest output powers required for ablation.

Free-Running Ablation

Low irradiance ($\leq 10^8$ W/cm^2), relatively long duration pulses (180 μsec for the Nd:YAG laser) are obtained from lasers operating in the free-running mode, also known as non-Q-switched, fixed-Q, or normal operation. The free-running laser output comprises many short duration "spikes" occurring at random intervals and the arrival rate of these pulses at the focused area is sufficient to produce steady-state ablation over the duration of the pulse. Thermal conductivity into the bulk of the sample is significant and a temperature gradient is established across the target area, melting occurs around the edges, and material is vaporized at the center. Rapid expansion due to vaporization causes molten material to be pushed outward and upward within the walls of the crater and material is ejected as a plume of vapor and liquid particles. This is the melting–flushing mechanism first proposed by Ready.[33] The free-running Nd:YAG laser tends to produce craters $\sim 50\%$ larger than the focused laser spot diameter; however, a relatively large amount of material is removed (100–1000 ng/pulse) and deep cavities can be drilled with successive pulses. The nature of the ablated material depends on the properties of the sample. Metals generally produce molten droplets having sufficient size and velocity to remain liquid and spatter on the cell window, located a few centimeters above the sample surface. Ceramic materials tend to give substantial numbers of solid particles having diameters of < 10 μm (suitable for gas flow transport, see below) in addition to larger particles that are not entrained by the gas stream and deposit on the surface of the sample. Free-running ablation of refractory materials with the 10-Hz repetition rate Nd:YAG laser can produce enough particles to partly block the nozzle inlet of the ICP-mass spectrometer after a period of ~20 min.

Q-Switched Ablation

Much higher irradiances ($>10^{10}$ W/cm^2) are obtained by electrooptically Q-switching the laser to compress the output into a single short pulse (~8 nsec duration for the Nd:YAG laser). Interaction of the focused Q-switched pulse with a solid produces multiphoton ionization and the resulting free electrons, coupled to the intense electric field of the laser radiation, cause further ionization leading to fast avalanche breakdown and formation of a microplasma at the surface. Compared to free-running ablation, the amount of material ejected is less, typically 1–10 ng/ pulse for the Nd:YAG, partly because of absorption of the incident light by the plasma. The craters produced by a Q-switched laser pulse are relatively shallow (typically <1 μm deep), but up to three times the diameter of the focused laser spot, because of secondary heating and vaporization by the microplasma. Examination of ablated material by light scattering performed directly on plumes and particle counting of transported material has shown that Q-switched Nd:YAG laser ablation of metals at near atmospheric pressure produces large numbers of solid particles.[34,35] Particles may be formed by direct ablation, rapid cooling of small liquid droplets, and coalescence of atomic and molecular species. It is important to note that particle size distribution is likely to be influenced by the physical properties of the sample material, the nature and topology of the sample surface, the ambient pressure, and the wavelength and irradiance (determined by the output energy, focused spot area, and pulse duration) of the laser. Q-switched ablation of a variety of metals, glasses, and ceramics appears to generate particles over a size range convenient for entrainment and transport by gas flow and transport efficiencies of ~50% have been measured (see below). In contrast to the free-running laser, metals do not spatter on the cell window and there is no observable deposition of particles around the base of the cell after ablation of ceramics. Figure 8.1 shows the temporal evolution of particles along the axis of a plume (normal to the surface of the sample) produced by Q-switched Nd:YAG laser ablation of a metal. The influence of ablation conditions upon the spatial and angular distribution of particles in the plume, and particularly the particle size distribution, is not understood because of the complexity of the ablation process and the experimental difficulty of determining spatial and size distributions on a short time scale.

There is some evidence of fractional volatilization during the ablation process, leading to enrichment of volatile elements and depletion of refractory elements in ablated material compared to the bulk.[8] This problem is substantially reduced with a Q-switched laser because (in the absence of secondary heating by the microplasma) thermal conductivity into the bulk is negligible for nanosecond duration pulses and the width of the melted zone around the target area is small compared to that produced by free-running laser ablation. Because of the various problems associated with free-running laser ablation, the Q-switched Nd:YAG laser was used for all the analytical work described here, although the amount of ablated material, and hence the sensitivity, can be one to two orders of magnitude higher with the free-running laser.

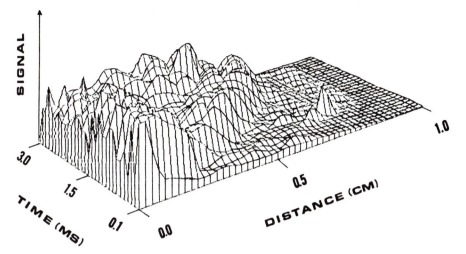

Fig. 8.1. The temporal and spatial evolution of a plume of ablated particles observed by light scattering. The plume was generated by single pulse ablation of a Cu surface with a Q-switched laser at an irradiance of 2.5×10^8 W/cm^2. Republished with permission from Ref. 34.

The Ablation System

The ablation system consists of two parts, an outer box and a small ablation cell that encloses the plume but does not form a seal to the sample surface. The non-critical gap between the base of the cell and the surface of the sample enables non-uniform surfaces to be ablated with few restrictions on sample size or shape.

The Outer Box

To allow sampling at any point over a relatively large (10×10 cm) area, the XYZ sample translation stages and associated stepper motors are contained in the box (side 31 cm) and the sample is translated relative to the fixed laser beam, viewing optics, and ablation cell. The cell is centered on the optical axis and supported above the sample surface by means of an adjustable sliding flange to accommodate focus optics with focal lengths in the range of 4–10 cm. A schematic of the system is shown in Fig. 8.2. Here the ablation cell is located within the dashed area. The outer box is gas tight and filled with plasma support gas (normally Ar) at ~4 torr above atmospheric pressure to provide a pressure gradient along the transfer tube for transport of ablated material into the ICP. Depending on the ablation cell in use, gas can flow into the box via two inlets with extension tubes to direct the flow toward the open ends of the cell (see Fig. 8.3a and b) or through a tube, mounted on a second adjustable flange, leading directly into the cell (Figs. 8.2 and 8.3c). The total inlet flow rate is fixed at 5–10 L/min by mass flow controller and both the pressure in the box and the flow through the transfer tube (1–2 L/min) are regulated by means of two flow-control valves vented to the atmosphere. The purposes of the excess inlet flow and the vent are to flow clean gas in and around the cell and to provide an alternate outlet for any ablated material that escapes the cell (see

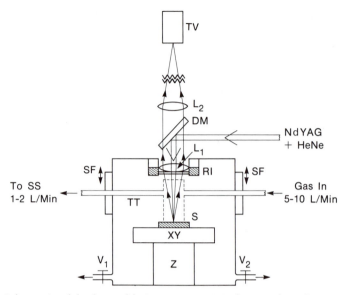

Fig. 8.2. Schematic of the laser ablation system. TT is the transfer tube connecting the ablation cell (located within the dashed area) to the secondary source SS, SF are sliding flanges, DM is a dichroic mirror, RI is a ring illuminator, V_1 and V_2 are flow control valves, L_1 (behind a window) and L_2 are lenses. The sample S sits on the XYZ translation stage. Republished with permission from Ref. 35.

below). The transfer tube is a straight, horizontal length of smooth-walled polyflow or Teflon tubing (i.d. 4.5 mm, overall length 80 cm) connected to the sample injector inlet of the ICP torch without intervening taps. This tube length allows the outer box to be located close to the ICP torch without modification of the mass spectrometer, enabling rapid changeover from laser ablation to conventional solution sample introduction. The front of the box is removable and sealed by an O-ring to allow access to the sample table. All three axes of sample translation are separately controlled by PC with rates from single step to 5000 sec^{-1} and step resolution of 3 μm. Control software allows the sample to be translated to any point for single area sampling, or continuously rastered in the horizontal (XY) plane for ablation by

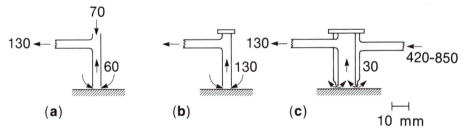

Fig. 8.3. The three cell designs evaluated for entrainment of ablated material; cells are to scale and approximate average flow velocities (in cm/sec) are indicated. Republished with permission from Ref. 35.

separate or overlapping laser spots (depending on the laser repetition and step rates).

The Ablation Cell

Three basic cell designs, shown in Fig. 8.3, were evaluated. All have cylindrical geometry and were constructed in glass to provide smooth walls. Several cells of each design were made, internal volumes ranged over $1-3$ cm^3, internal diameters were $5-10$ mm, and heights were $20-40$ mm. Although dimensions did not appear to be critical within the limits tested, they may not be optimal.

Laser Beam Focusing

In the absence of spherical aberration, a focused laser spot is diffraction limited and the diameter is given by either the expression $4\lambda f/\pi D$ for a Gaussian (pure TEM_{00}) beam or by the product θf for a mixed mode beam with significant angular divergence θ. At focal lengths $f > 40$ mm a simple singlet lens may be used to focus the unexpanded Nd:YAG laser beam (diameter $D \sim 4$ mm) since spherical aberration is negligible for $f/D \geq 10$. In this case the spot size is limited by the divergence of the beam (~ 1 mrad) and is estimated to be 40 μm or more, depending on the focal length. The spot diameter can be decreased by expanding the beam to reduce the divergence; however, the spot will be larger than expected due to aberration (significant for $f/D < 10$). (Schemes to reduce the divergence of the laser by spatial filtering of the focused beam to remove high order modes are impractical because of the inevitable plasma breakdown at the focus.) Use of a $7\times$ beam expander without an aperture, in conjunction with an achromat-meniscus lens combination to reduce aberration, gives an estimated focus spot diameter of ~ 20 μm for the Nd:YAG laser. Smaller diameters are obtained, at the cost of reduced output energy, by placing an aperture in the expanded beam. Since the amount of material ablated is proportional to the spot area, the unexpanded laser beam was focused with a single lens in the present work for increased sensitivity. An HeNe laser is aligned coaxially with the Nd:YAG laser beam to act as a beam spot locator for optical viewing of the sample surface via a TV camera and monitor, with overall magnification of $300\times$. It should be noted that transmissive optics also have chromatic aberration and therefore it is impossible to focus two or more widely different wavelengths at the same plane without some form of compensation.

PARTICLE ENTRAINMENT AND TRANSPORT

General Considerations

Particles are likely to account for a significant fraction of the ablated mass and the sensitivity of any laser ablation–secondary excitation technique will depend on the efficiency with which particles are entrained, transported, and subsequently vaporized. The overall efficiency of the plume entrainment device (the ablation cell), the transfer tube, and the secondary source may be defined as the fraction of ablated material that is atomized or ionized in the secondary source. In principle, the efficiency is determined by convolution of three particle size functions: the size distribution of particles in the primary source, the transfer function of the cell and tube (which together determine the transport efficiency), and the response function of the secondary source. Since large particles will be incompletely vaporized in the secondary source, it has a response function that falls off for particles larger than about 10 μm, depending on the physical properties of the particles and the temperature distribution and residence time in the source (for a discussion of particle vaporization in plasmas, see Ref. 36). Hence it would appear that the cell and tube need transport particles only up to some size limit, corresponding to the fall-off in the response function of the secondary source. However, a practical system that can be used for a variety of sample types, ablation conditions, and secondary sources should have a wide transfer function to transfer particles of different size (possibly having different composition) with equal efficiency and to give maximum overall sensitivity. If problems of incomplete vaporization or excessive loading do occur in the secondary source, it is preferable to modify the primary source distribution to favor smaller particles by, for example, using a short duration laser pulse, rather than depositing out large particles with consequent reduction in transport efficiency and possible poor temporal response and memory effects.

Some combinations of primary source particle size distribution, tube transfer function, and secondary source response function are shown in Fig. 8.4. If the upper end of the primary distribution, denoted by the hatched area of Fig. 8.4a,

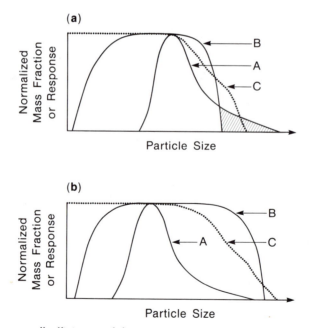

Fig. 8.4. The overall efficiency of the system represented by convolution of particle size functions for A, the primary source, B, the transfer function of the cell and tube, and C, the response of the secondary source. Cases of partial and complete overlap of the primary source distribution are shown in (a) and (b), respectively.

extends beyond the transfer function or the response function, then the transport efficiency and the overall sensitivity are reduced. More importantly, any pulse-to-pulse variation in the particle size distribution, due to fluctuations in the laser output power or the ablation process, will cause a corresponding change in the signal, although the mass ablated by each laser pulse is constant. This source of noise is removed, and transport efficiency and sensitivity are maximized, when the transfer and response functions encompass the primary source particle distribution. In this ideal situation, shown in Fig. 8.4b, only pulse-to-pulse fluctuations in the ablated mass will cause variation in the observed signal. Although cells and tubes have been widely used to transfer ablated material to flames and furnaces for AAS and to microwave, direct current, and inductively coupled plasmas for AES, no published design has addressed the subject of entrainment and transport and the dependence upon particle size. An additional objective of the present work was to develop a cell for fast transfer of ablated material with short signal decay time to allow rapid sampling at different locations in samples of arbitrary size and shape.

Ablation Cell Design

It is instructive to describe the evolution of the cell in terms of the experimental approach, which involved testing three designs.

The First Cell Design (Fig. 8.3a)

This cell is windowless and eliminates possible problems such as attenuation of the laser beam due to deposition of material. More importantly, it eliminates the back reflection that can damage the focus lens if the window is located midway between the lens and the beam focus. The operating principle of this cell is that ablated particles, ejected upward from the sample surface, will be sufficiently slowed by the relatively slow upward flow of gas to be entrained and transported into the side tube. The temporal response of cell 8.3a to multiple 10-Hz laser shots, obtained by peak-hopping the mass spectrometer (see below), is shown in Fig. 8.5. There is an initial rise in signal of 10^6 as the laser is turned on, a fast decline of one order of magnitude when it is turned off, followed by a slow decline over a period of minutes. The fast decline is the result of clearing of ablated material from the small volume cell, and the slow decline arises from reentry of particles (at low concentration) that escaped from the cell and mixed with the gas in the outer box. The half-life of the slow signal decay is 1.9 min, in reasonable agreement with the estimated half-life for purging the outer box at a flow of 10 L/min. The temporal response of cell 8.3a is described exactly by a pair of rate equations:

$$\frac{dC_c}{dt} = \frac{F_eR}{V_c} - \frac{C_cQ_c}{V_c} + \frac{C_bQ_c}{V_c} \tag{8.1}$$

$$\frac{dC_b}{dt} = \frac{(1 - F_e)R}{V_b} - \frac{C_b(Q_c + Q_b)}{V_b} \tag{8.2}$$

where t is the time from turning on the laser, V_c, Q_c, and C_c are, respectively, the effective volume, outlet gas flow, and concentration of ablated material for the cell, V_b, Q_b, and C_b are equivalent quantities for the outer box, R is the ablation rate,

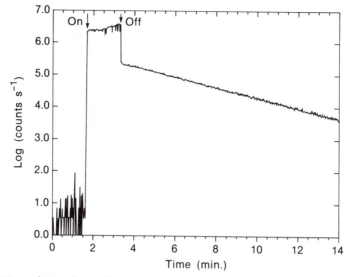

Fig. 8.5. Typical time dependent response of cell 8.3a showing the Mn signal obtained by ablating aluminum alloy (NBS SRM 1241a, 0.38% Mn) with an inlet flow rate of 10 L/min. The points at which the 10-Hz repetition rate laser was turned on and off are indicated by arrows. Republished with permission from Ref. 35.

and F_e is the fraction of ablated material entrained and removed by the cell. The last term on the right-hand side of Eq. (8.1) represents the return to the cell of ablated particles from the outer box. The quantities V_c and V_b were determined in separate experiments from the exponential signal decays of the cell and outer box upon turning off the laser. These effective volumes were different from the corresponding physical volumes, particularly for the cell (see cell 8.3c, below), implying there is incomplete mixing of ablated particles. The only unknown parameter in Eqs. (8.1) and (8.2) is the cell entrainment efficiency F_e and this was determined by fitting expressions for C_c and C_b to the observed temporal response of a cell (see below). The slow signal decay of Fig. 8.5 is undesirable since it prevents rapid independent analysis of different samples (or different areas on the same sample) and could give rise to memory effects.

The Second Cell Design (Fig. 8.3b)

To prevent loss of ablated material, the top of the cell was sealed with a window. The cell was held ~1 mm above the sample surface. In the absence of the inlet extension tubes, a slow signal decay similar to that shown in Fig. 8.5 was obtained, showing that material is ejected sideways, close to the surface, with sufficient velocity to escape through the inward gas flow at the base of the cell. The relative amounts of material ejected sideways and upward are likely to depend upon the ablation conditions, the aspect ratio of the ablation crater, and the sample topography. With two inlet flow tubes directed from opposite sides at the base of the cell, the transient response was greatly improved and the signal rapidly decreased to the background level after turning off the laser. Hence, material that escapes sideways is unable to return to the cell because of the flow of purge gas around the base. Only clean gas enters the cell and the final term of Eq. (8.1) is zero, effectively decoupling the cell and the outer box.

The Final Cell Design (Fig. 8.3c)

The inlet extension tubes used with cell 8.3b were eliminated by flowing the inlet gas through an annulus to form a symmetric gas sheath around the base of the cell. The width of the annulus at the base of the cell is ~1 mm and the i.d. of the inner tube was increased to 10 mm to reduce deposition on the walls. The temporal response to a burst of 10-Hz laser pulses is shown in Fig. 8.6; the signal drops five orders of magnitude to the background within 1–2 sec of switching off the laser. The signal and the transient response were observed to be insensitive to changes in the inlet gas flow over the range 5–10 L/min (at fixed transfer flow rate) and to gap distances of 0.5–3 mm between the base of the cell and the sample surface. Hence this cell design prevents reentry of ablated material without the need for inlet extension tubes and is ideal for use with nonuniform and nonflat sample surfaces. The optimum transfer gas flow rate, corresponding to the maximum observed signal, is ~1.5 L/min for this cell. However, the optimum flow is likely to depend on the ablated particle size distribution (see below) and therefore may vary with sample material and ablation conditions. Figure 8.7 shows a typical response to a single laser pulse, obtained by peak-hopping the mass spectrometer (see below). The FWHM of the transient is ~270 msec and the half-life of the signal decay is ~180 msec, depending on the volume of the cell and the transfer gas flow. The observed half-life is about three times that calculated for the 2.4-cm^3 cell at this flow rate (1.7

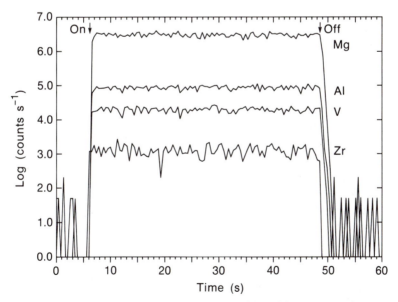

Fig. 8.6. Temporal response of cell 8.3c obtained by ablating NBS aluminum alloy (~95% Al, 4.54% Mg, 0.016% V, and 0.002% Zr) with the 10-Hz laser. The Al signal was dynamically attenuated by ~10^3 in the mass spectrometer to bring it onto scale. Republished with permission from Ref. 35.

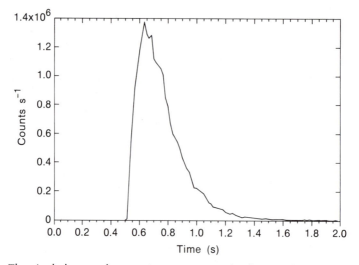

Fig. 8.7. The single laser pulse transient response of cell 8.3c, showing the Mg signal from aluminum alloy. Republished with permission from Ref. 35.

L/min). An effective volume larger than the physical volume indicates incomplete mixing of particles within the cell and possible trapping of particles in the region between the window and the entrance to the transfer tube. The transient response of cell 8.3c is sufficiently long to give a steady signal at laser repetition rates >10 Hz, convenient for acquiring data at many m/z values in peak-hopping mode or for mass scans with relatively long measurement time for improved precision (see below). Cell 8.3c gave signals comparable to those obtained with an enclosed cell,[28] and was therefore used for all subsequent work. This cell should be of general use for all laser ablation–secondary excitation techniques since it is similar to two earlier designs for capturing ablated material for transport to a flame source for AAS and to an ICP for AES.[14,15,27]

Entrainment Efficiency

Particles are entrained only if the upward gas flow velocity v_g exceeds the terminal sedimentation velocity v_s (the velocity at which a particle falls in static gas) and the steady-state velocity reached by an entrained particle in the vertical direction is $v_{ss} = v_g - v_s$. Sedimentation velocities range from 8.4×10^{-3} to 9.8 cm/sec for heavy Mo metal particles (density 10.2 g/cm^3) of diameter 0.5–20 μm. Entrainment was modeled by solving the equations of motion for the vertical and horizontal velocity components of spherical Mo particles, slowed by viscous drag in a stream of Ar gas moving upward at 25 cm/sec under atmospheric pressure. Particles were given an initial upward velocity of $\sim 10^4$ cm/sec (similar to that expected for ablated particles) at an angle of 30° to the vertical. (Details of the model and the inherent assumptions are given in Ref. 35.) The model showed that small particles <1 μm diameter are entrained over distances of 1–2 mm or less, and particles up to $\sim$5 μm diameter are entrained within the 30 mm height of the cell, but particles >3 μm impact the sides of a 10-mm-i.d. cell before entrainment. Although large particles impact either the cell walls or the cell window, it is likely they will be entrained if they rebound with reduced velocity, since v_g of 25 cm/sec is sufficient to transport upward even relatively large particles. Calculated times for entrainment of heavy particles are of the order of 10 msec, much shorter than the duration of the single laser pulse transient signal (see Fig. 8.7), suggesting the latter is governed by the flow pattern of entrained particles. The predictions of the model are supported by the observations that large liquid particles (produced by free-running laser ablation of metals) are not entrained within a 30-mm cell and spatter on the window, and deposition may occur on the inside of the cell after ablation of some materials for long periods (>1 hr).

Experimentally, it is possible to estimate F_e, the fraction of ablated material entrained by the cell, by solving Eqs. (8.1) and (8.2). The ratio C_c/C_b is a sensitive function of F_e and is independent of the rate of ablation. C_c/C_b is given by the ratio of the signal intensities from the cell and the outer box after ablation for a known period. This was obtained for cell 8.3c by rapidly switching the 10 L/min flow from the cell to the inlet of the outer box immediately after ablating a piece of Mo metal with the Q-switched laser. F_e was determined to be 92% with good agreement between different ablation periods. Hence, only 8% of ablated particles escape to the outer box and are not detected.

Theory of Particle Transport

Particles transported by gas flow through a horizontal tube are lost by gravitational settling and diffusion to the walls. Although exact solutions have been obtained for laminar parabolic gas flow, neither the equations nor the dependence of transport efficiency on particle size have apparently been discussed in the analytical literature, even though it is stated that particles may be transferred over long distances.[8,25] The assumption of fully formed parabolic velocity distribution without end effects is reasonable since the transition length to parabolic flow, given by 0.1RRe, where R is the tube radius and Re is the flow Reynolds number, is short compared to the overall tube length.[37] Particles are assumed to be spherical and at low concentration so coalescence is negligible. Gravitational deposition is important for relatively large particles with significant sedimentation velocity v_s. In addition to v_s acting vertically, a transported particle has a horizontal velocity component determined by the gas velocity, which varies with the radial position of the particle. A particle starting at the top of the tube will travel the farthest distance before settling out. The critical distance at which all particles of given size settle out is

$$L_{cr} = \frac{8\bar{v}R}{3v_s} \tag{8.3}$$

where $\bar{v}$ is the average bulk flow velocity. The fraction of particles transmitted through a tube of length L is given by Thomas[38]:

$$F = \frac{2}{\pi}(\alpha\beta + \sin^{-1}\beta - 2\alpha^3\beta) \tag{8.4}$$

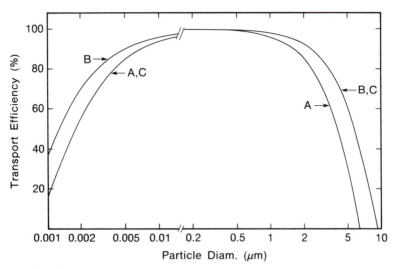

Fig. 8.8. Calculated transport efficiency for Mo particles (density = 10.2 g/cm³) for various flow rates, tube inner diameter, and lengths (A, 1.5 L/min, 4.5 mm, 80 cm; B, 1.5 L/min, 4.5 mm, 40 cm and 3.0 L/min, 4.5 mm, 80 cm; C, 1.5 L/min, 2.25 mm, 80 cm). Republished with permission from Ref. 35.

where $\alpha = (L/L_{cr})^{1/3} = (3\pi v_s LR/8Q)^{1/3}$ for a given flow rate Q, and $\beta = (1 - \alpha^2)^{1/2}$.

Loss by diffusion is significant for small particles. This problem was solved by Gormley and Kennedy for parabolic flow assuming the concentration of particles at the walls of the tube is zero.[39] The ratio of the number of particles at the exit of a tube to the number at the entrance is

$$n/n_0 = 0.8191e^{-7.314h} + 0.0975e^{-44.6h} + 0.0325e^{-114h} \qquad (h > 0.0156) \quad (8.5)$$

$$n/n_0 = 1 - 4.07h^{2/3} + 2.4h + 0.446h^{4/3} \qquad (h < 0.0156) \quad (8.6)$$

where $h = LD_P/2\bar{v}R^2$ and D_P is the particle diffusivity, a function of particle diameter.[40] The product of F and n/n_0 is shown in Fig. 8.8 as a function of particle diameter for various flow conditions in Ar. Even for dense Mo particles, sizes over the range 0.005–2 µm are transported with >80% efficiency, but losses become serious for significantly smaller and larger particles. As expected, transport efficiency is improved for larger particles with short, small-diameter transfer tubes and high flows. However, the theory predicts there is no optimum tube diameter and length for a fixed flow rate. The better the overlap of the primary source particle size distribution and the transfer function of the tube, the higher the transport efficiency. Usually the particle size is unknown and it is best to use the shortest, narrowest tube possible, giving a wide transfer function. For example, an efficiency of 80% [a value of $\alpha = 0.504$ in Eq. (8.4)] for a 10-µm particle of density 10.2 g/cm³ at the optimum flow of 1.5 L/min, corresponds to an LR of 1.1 cm², or a tube i.d. of 1.1 mm and length of 20 cm.

Experimental Particle Size and Transport Efficiency

As discussed previously, overall sensitivity could depend critically on the presence of a few relatively large particles in the primary source, since particles that exceed either the upper limit of the transfer function of the cell and tube or the fall-off in the secondary source response function are likely to account for a high proportion of the ablated mass. A particle size distribution was obtained by ablating Mo metal with the Q-switched Nd:YAG laser and transporting ablated material to a particle counter for sizing by light scattering. The distribution, shown in Fig. 8.9, appears to peak at $\leq$0.2–0.3 μm (the lowest channel of the particle counter) and declines rapidly to 0.5 μm. Since entrainment and transport efficiency are not expected to vary with particle size in this region, Fig. 8.9 should resemble the primary source distribution. The most probable size is in agreement with a preliminary size estimate of >0.1 μm for particles observed by light scattering in plumes of ablated Cu.[41] Although a relatively small number of particles (<0.05% of the total) was detected at 5–12 μm, they account for ~50% of the ablated mass. These particles are larger than the predicted 5 μm size limit for transport (Fig. 8.8) and they may arise from coalescence of primary particles within the transfer tube. Since similar flow conditions were used for particle sizing and transport these particles will also reach the secondary source.

Mass transport efficiency was determined to be ~40% for ablated Mo with an average of ~3 ng removed per laser shot. If all this material is ablated as a single particle, the diameter would be ~8 μm, and therefore this is expected to be the

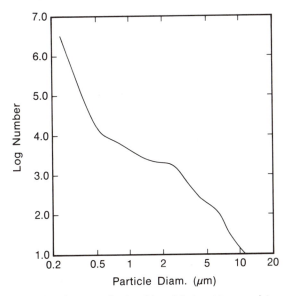

Fig. 8.9. Particle size distribution obtained by ablating Mo metal (mean of two determinations, each normalized to 10^4 laser pulses). Ablated material was entrained and transported to the particle counter with cell 8.3c, under flow conditions similar to those used for transfer to the secondary source. Republished with permission from Ref. 35.

upper size limit for particles in the primary source. A small number of particles larger than 8 μm were observed by impacting ablated material on a surface placed close to the plume. Hence, there is evidence of infrequent fluctuations in ablation conditions, possibly caused by pulse-to-pulse fluctuation in the laser output or variation in the nature of the sample surface. Allowing for entrainment loss of 10% (see above), the remaining 50% of ablated material is lost by either gravitational or diffusive deposition. A wash of the walls of the entrainment cell and transfer tube confirmed most of the lost material to be in the transfer tube. Without knowledge of the primary source size distribution for a sample size of $\sim 10^5$ particles it is not possible to determine the relative contributions to transport loss from gravitational deposition of heavy particles (formed in the primary source and by coalescence within the transfer tube) and diffusive removal of small (<5 nm) particles. However, gravitational deposition appears to be most important since a small number of relatively large (>5 μm) particles was observed in impacted plume material and at the exit of the transfer tube.

The importance of the concept of transfer function is demonstrated by the dependence of signal intensity and precision on transport tube length and diameter, observed for laser ablation ICP-AES.[22,23] The previously unexplained trends of increased signal and improved precision with reduced tube length and diameter (at fixed gas flow rate) can be understood in terms of increased transport efficiency due to a broadened tube transfer function giving greater overlap with the primary source particle distribution. When the transfer function completely overlaps the upper end of the primary distribution no further improvement in intensity and precision will be obtained. The observed dependence of signal on tube length shows the predicted convergence at short lengths to a common intensity level, close to that for the smallest (2-mm-i.d.) tube.[22] However, the observation of a minimum (best) precision corresponding to an optimum tube length, in the plots of signal relative standard deviation, is unexpected. The increased noise observed at short tube lengths may arise from fluctuations in the ICP caused by transport and possible incomplete vaporization of relatively large particles.

ANALYSIS MODES AND APPLICATIONS OF LASER ABLATION ICP-MS

Data Acquisition

The current commercial ICP-MS instruments (manufactured by Sciex-Perkin-Elmer, VG Elemental, and, most recently, Nermag, introduced in early 1988) all use a quadrupole mass filter to transmit an ion species of selected m/z ratio to the detector. Quadrupoles can be scanned very rapidly under electronic control and their high transmission and versatility compensate, to some extent, for the single channel limitation of the device. Although quadrupoles have relatively low resolution (for elemental analysis $m/\Delta m$ is varied from 1 to 250, giving unit peak width across the mass range), the abundance sensitivity, defined as the contribution from an intense peak at mass m to adjacent peaks at $m \pm 1$, is typically better than 1 part in 10^6. The various manufacturers have used different techniques to control the quadrupole, allowing the user to perform rapid sequential analysis on a sample that may contain a large number of elements of interest. The present work was performed on a Sciex model 250 ELAN with digital computer control of the quadrupole scan. The instrument can be operated in two modes. In the mass scan mode, signal intensity is obtained as a function of m/z over one or more mass regions with 1–20 data points per mass unit. Since the instrument has no facility for rapidly adding scans, a single scan is usually made with sufficient measurement time (typically 5–100 msec) at each data point to give reasonable precision. The alternative mode of operation is multiple ion monitoring, whereby the quadrupole is rapidly peak-hopped between a number of previously selected masses. Signal may be recorded at 1–20 points over each mass unit and a cycle is completed when each mass has been visited once. This process is usually repeated for a number of cycles to give the signal intensity as a function of time for each of the selected masses (or isotopes). Since the dead time spent between data points is fairly short ($\sim$5 msec) and independent of their mass separation, peak-hopping can give a reasonably high duty cycle and sample rate per mass, depending on the number of requested masses and the measurement time. The Sciex instrument also has the ability to attenuate the signal at a given mass "on the fly" by applying, under computer control, a retarding potential to the quadrupole for the period of data acquisition at that mass. This is a powerful feature since it allows a major element that would normally cause detector saturation to be brought within the dynamic range for use as an internal standard (see below).

In the VG PlasmaQuad instrument, the quadrupole is driven by a multichannel analyzer. This is ideal for making rapid scans over one or more mass ranges with very fast accumulation of successive scans. A single mass scan can be divided into a maximum of 4096 channels and the complete mass range of 1–250 Da can be scanned in a minimum time of $\sim$100 msec. At high scan rates the time spent at each mass is <1 msec and it is necessary to increase the time spent at each channel and/or sum several hundred scans to improve the dynamic range of detection. The duty cycle of the multichannel analyzer can be poor, particularly when the peaks of interest are widely separated in mass, since a large proportion of the time is spent scanning over the intermediate region. The situation is improved with multichannel analyzers that can slew scan through regions of no interest at a high rate. The dynamic range of the VG instrument can also be extended to overcome saturation on intense peaks by switching from pulse-counting to analog detection, thereby increasing the dynamic range by a factor of 100.

The challenge for these instruments is to acquire data with reasonable precision from a transient signal <500 msec in duration, such as that produced by single pulse laser ablation. The approach is to sample the signal at the highest possible rate to cancel systematic errors due to the changing signal. For the ELAN this is best approached by peak-hopping, with the disadvantage that only a limited number of masses (<10) can be detected, without severely reducing the data sample rate. Furthermore, these masses have to be selected before the analysis is performed, although in practice the nature of the sample is often known or specific elements are of interest. Similarly, for the PlasmaQuad, it is preferable to sum a number of fast scans rather than take a single scan. The advantage of the multichannel analyzer is that the complete mass scan is obtained without prior knowledge of the sample composition. However, even at the maximum scan rate, fewer than five complete scans can be obtained over a short duration transient. For these reasons, most analytical work has been performed with either steady-state signals produced with the 10- to 20-Hz repetition rate laser (the author) or with extended single laser pulse transients of ~ 10 sec duration (see below).

Modes of Analysis

The various modes of analysis arise from the different ways of operating the laser and the mass spectrometer in conjunction with sample translation. Schemes for triggering the laser, the spectrometer, and the translation stage have been discussed.[28]

Single Laser Pulse Transient Signal

Possible applications include single area analysis of adsorbed contaminants or particulates (at reduced laser irradiance to minimize ablation of the substrate) and investigation of sample homogeneity by single pulse sampling at low repetition rate (≤ 1 Hz) in conjunction with sample translation. As discussed in the previous section, the data acquisition rate should be as high as possible for single pulse ablation and peak-hopping is normally used. Figure 8.7 shows a typical single laser pulse transient signal obtained by peak-hopping with a measurement time of 5 msec. Since the decay time of the transient is proportional to V_c/Q_c [see Eq. (8.1)], its duration may be increased (at fixed flow rate) by increasing the volume of the ablation cell. This appears to be the approach used to extend the duration of the transients for data acquisition by multichannel analyzer, since the reported decay time of ~ 10 sec is in good agreement with the expected half-life (ablation cell diameter 5 cm, height 6.5 cm, at a flow of 0.5 L/min).[30,42] Extending the transient will reduce the signal-to-background ratio and the sensitivity. However, detection limits of $\sim$ 0.01 μg/g were reported, presumably because a large amount of material was ablated by the relatively high power (0.5–1 J) free-running laser.

Multiple Laser Pulses

In the absence of sample translation, ablation tends to produce "hole-drilling." The rate of drilling depends on the physical properties of the sample material and the irradiance, pulse repetition rate, and depth of focus of the laser beam. At repetition rates ≥ 10 Hz a continuous signal is obtained since the signal decay time of the ablation cell is sufficiently long to give good overlap of the individual transients. The signal gradually decreases for most materials over a period of several minutes due to defocusing of the laser beam as the cavity deepens. At repetition rates of ≤ 1 Hz the transients are separate and independent and it is possible to perform course depth profiling. Figure 8.10 shows the variation in the intensity of the maximum of the transient signal for several elements during ablation of a multilayer thin film material. The Q-switched laser was slightly defocused to reduce the irradiance and improve the depth resolution. The film has a Co–Pt layer close to the surface and this is clearly shown by the variation of the Co and Pt signals and the appearance of the Al substrate after 30–40 laser pulses. From the known thickness of the Co–Pt layer, the depth ablated per shot is estimated to be ~ 50 Å. However, there is an indication that the crater becomes wider with successive laser pulses,

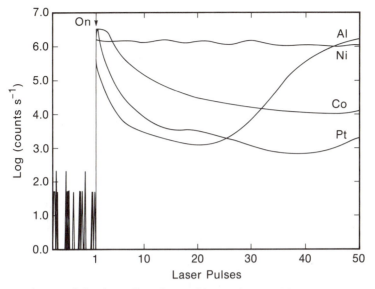

Fig. 8.10. Elemental depth profiles obtained by single area ablation of a multilayer thin film material at a pulse repetition rate of 1 Hz.

probably because of the Gaussian intensity profile of the laser beam, since the (laser-off) baseline is not reached for Co and Pt. It should be noted that this type of analysis, involving repeated single pulse sampling, will be tedious with extended duration transients because a 20–30 sec delay between pulses is required to obtain independent analysis for each laser pulse.

Steady Signal

As discussed in the previous section, use of the annular cell (Fig. 8.3c) with the $\geq$10-Hz laser produces a continuous signal. The signal may be maintained constant and steady over any desired period by simultaneous translation of the sample to overcome defocusing of the laser beam by hole-drilling. There are several applications for this mode of operation. Peak-hopping the mass spectrometer gives a constant signal (for homogeneous materials) at each of the selected masses, convenient for optimizing the operating conditions, such as transport gas flow rate, ion optical lens voltages, focus position of the sample, and ICP power. This is also the preferred mode for quantitative analysis (see Fig. 8.6 and the following section) because relatively slow systematic changes in signal, affecting all elements, can be corrected by internal standardization. For example, change in laser focus normally causes variation in the ablated mass over a time longer than the period between repeat sampling of a mass peak. More rapid changes in signal, arising from pulse-to-pulse variation in ablation conditions, may not be corrected since only one or two of the elemental intensities will be affected. Although laser pulse rates <10 Hz can be used, the observed signal contains a periodic oscillation due to the rise and decay of the time dependent signal of the ablation cell. If the data sample rate per mass is similar to the repetition rate, each mass may be sampled on the same part of the signal cycle causing further systematic error. As previously discussed, this is overcome by use of high data sample rates or a larger cell to smooth the signal (with the disadvantage of memory effects). All these problems are more serious for single-channel instruments, such as quadrupole-based plasma source mass spectrometers and improved precision is to be expected with multichannel optical emission instruments. Peak-hopping mode may also be used to investigate spatial variation in composition and inhomogeneous materials. The number of laser pulses placed within the area of the focused laser spot is determined by the laser pulse rate and the step speed of sample translation. It might be expected that ablation by overlapping laser pulses will be particularly susceptible to variation in elemental composition caused by fractional vaporization of the bulk or redeposition of material close to the ablation zone. No obvious problems of this nature have been observed with the Q-switched laser.

Sequentially scanning the spectrometer with a constant ablation signal yields a detailed mass spectrum of the sample. Since a single mass scan is influenced by any variation in the ablated mass, it is unsuitable for quantitative analysis; however, it enables an unknown sample to be rapidly surveyed and is useful for estimating concentrations (accuracy to within a factor of 10). Figures 8.11 and 8.12 show the low- and high-mass sections of a spectrum of an Al_2O_3-TiC-ZrO_2 ceramic. The total data acquisition time was excessive ($\sim$8 min) to achieve high signal-to-background ratio. Although the signals at Al and Ti are both saturated, the sensitivity can be estimated from the Zr signal (concentration $\sim$1% in the ceramic), giving a background equivalent concentration of $\sim$0.1 ppm. The ratio MO^+/M^+ for refractory oxides is generally $<$0.1% in mass spectra of ablated samples, compared to 1–5% typically observed for ICP-MS of aqueous solutions, because of the reduced

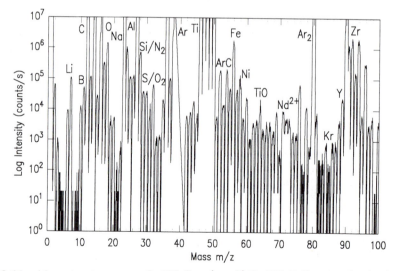

Fig. 8.11. Mass spectrum over 0–100 Da of an Al_2O_3-TiC-ZrO_2 ceramic obtained by 10-Hz Q-switched laser ablation with simultaneous sample translation. Approximately 15 laser pulses were overlapped within the $\sim$100 μm diameter of the focused beam; the measurement time was 100 msec per data point.

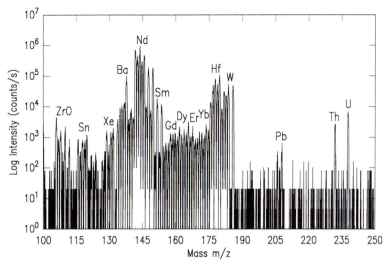

Fig. 8.12. Mass spectrum over 100–250 Da of the Al_2O_3-TiC-ZrO_2 ceramic obtained under the same conditions as Fig. 8.11.

amount of oxygen in the dry ICP. Other potential mass interferences due to atomic and molecular ions (N^+, O^+, OH^+, H_2O^+, O_2^+, NO^+, ArN^+, ArO^+, etc.) arising from impurities in the Ar plasma gas and entrainment of air in the ICP are also substantially reduced. For comparison, a spectrum of the same ceramic material was obtained by ablation with the free-running laser. The intensity of the refractory oxides was greatly increased ($MO^+/M^+ \sim 10\%$ for Nd and Th), probably due to the larger amount of material ablated, causing significant cooling and loading of the ICP with consequent reduction in the extent of oxide dissociation.

Quantitative Analysis

The separation of ablation, sample transport, and secondary ionization in the ICP allows the signal intensity $I(E,S)$ for a given element E in a sample S to be related to the concentration $C(E,S)$ by the product of separate variables. Under conditions in which ablated material is completely vaporized by the ICP:

$$I(E,S) = K \cdot C(E,S) \cdot \alpha(E) \cdot R(E) \cdot A(E,S) \qquad (8.7)$$

where K is a constant that relates signal count rate to concentration and includes the transport efficiency (assumed to be element independent), $\alpha(E)$ is the ionization efficiency in the ICP, a function of the ionization energy of the element and the temperature T_{ICP}, $R(E)$ is the response of the mass spectrometer, and $A(E,S)$ is the ablation yield. Assuming $\alpha(E)$ is 100%, an $A(E,S)$ of 1 ng/laser pulse and a typical signal of 10^6 counts/sec for an element at 1% concentration, the efficiency of laser ablation ICP-MS is approximately one count for every 10^6 ablated atoms at 10-Hz pulse rate. The form of Eq. (8.7) suggests a scheme for quantitative analysis in the absence of matrix-matched standard samples. Since the absolute ablation yield is

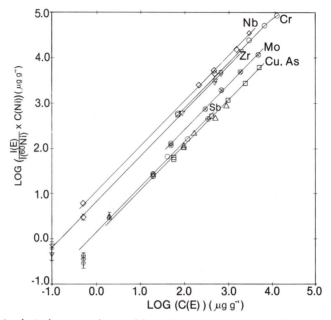

Fig. 8.13. Analytical curves obtained for NBS microprobe steels (SRMs 661–665) with internal standardization on ^{60}Ni. Republished with permission from Ref. 28.

unknown and dependent on the ablation conditions, it is necessary to ratio the observed intensity to that of an internal standard element IS present in the same sample. Cancellation of the sample dependent terms gives

$$\frac{I(E)}{I(IS)} = \frac{C(E)}{C(IS)} \cdot \frac{\alpha(E)}{\alpha(IS)} \cdot \frac{R(E)}{R(IS)} \cdot \frac{A(E)}{A(IS)} \tag{8.8}$$

The ratio $A(E)/A(IS)$ allows for fractional vaporization during ablation and is likely to be close to unity for the Q-switched laser. Both the relative response of the instrument $R(E)/R(IS)$ and the parameter T_{ICP}, that determines the ratio $\alpha(E)/\alpha(IS)$, may be determined by fitting observed intensities obtained by ablation of a standard material. Since the $R(E)/R(IS)$ varies smoothly over the mass range, it is not necessary for the standard to contain the same elements or matrix as the unknown. The expression for $\alpha(E)$ is given by the Saha equation for ionization in a plasma with appropriate electronic partition functions for the neutral and the ion.[43] Apart from the concentrations to be determined $C(E)$, the only remaining unknown in Eq. (8.8) is the concentration of the internal standard $C(IS)$. It is convenient to make the internal standard a matrix element (for example, an isotope of Fe for analysis of steel) since the concentration is usually known to within a few percent. The analysis is performed by peak-hopping the mass spectrometer in constant signal mode (see previous section) with dynamic attenuation of the internal standard signal to overcome detector saturation. This mode of operation is illustrated in Fig. 8.6 for analysis of aluminum alloy by internal standardization on Al.

The validity of the internal standardization technique was tested by ablating a set of steel standard reference materials (SRMs) and performing analysis on elements covering a wide concentration range. The resulting logarithmic analytical curves, derived from Eq. (8.8) with internal standardization on a minor isotope of Ni without dynamic attenuation, are shown in Fig. 8.13. The plots are linear over a dynamic range of $\sim 10^5$ and for the favorable situation of matrix-matched samples and standards, precision and accuracy appear to be approximately $\pm 5\%$. Detection limits ranged from 2 ppm for As to 0.2 ppm for Nb.[28] The intercept of the analytical curves is determined by the final three terms on the right-hand side of Eq. (8.8). In practice, both the intercept and the detection limits were found to closely follow the ionization energy for each element, showing the variation in the relative instrument response and ablation yield between elements is small. The technique of internal standardization by dynamic attenuation of the matrix element signal is currently being evaluated for analysis of materials in the absence of matrix-matched standards.

CONCLUSIONS

The combination of laser ablation and plasma source mass spectrometry is an attractive technique for elemental analysis of solid materials. Separation of sample preparation, performed by the laser, from secondary vaporization and ionization in the ICP, gives relative ion signals that are insensitive to the nature of the sample and permits each process to be independently optimized. The technique extends the use of ICP mass spectrometers and is ideal for rapid survey and comparative analysis. It also appears to be useful for semiquantitative analysis in the absence of standard samples, medium-resolution spatial analysis, and low-resolution depth profiling. Several areas require further investigation. The influence of laser parameters and the physical properties of the sample material on the ablation process is poorly understood. Of particular interest is the mechanism of particle formation and the extent to which the plume particle size distribution and fractional vaporization can be controlled by selection of laser energy, wavelength, and pulse duration. Transport efficiency between the ablation cell and the secondary plasma may depend critically on the ablated particle size distribution. Also, noise in the observed signal may arise from pulse-to-pulse variation in the particle size distribution, in addition to fluctuation in the mass of ablated material. Particle vaporization and ionization in the ICP have also received relatively little attention from analytical chemists, although the absence of solvent should simplify the situation, compared to introduction of dissolved samples. Since both laser ablation and plasma vaporization of solids and powders are of current interest for materials processing, it appears that these problems will be investigated in the near future.

REFERENCES

1. V. A. Fassel, *Science* **202**, 183 (1978).
2. D. J. Douglas and R. S. Houk, *Prog. Anal. Atom. Spectrosc.* **8**, 1 (1985).
3. R. S. Houk, *Anal. Chem.* **58**, 97A (1986).
4. W. van Deijck, J. Balke, and F.J.M.J. Maessen, *Spectrochim. Acta* **34B**, 359 (1979).

5. K. Laqua, "Analytical Spectroscopy Using Laser Atomizers," in *Analytical Laser Spectroscopy,* Chap. 2., edited by N. Omenetto. Wiley, New York, 1979.

6. K. Dittrich and R. Wennrich, *Prog. Anal. Atom. Spectrosc.* **7,** 139 (1984).

7. E. H. Piepmeier, "Laser Ablation for Atomic Spectroscopy," in *Analytical Applications of Lasers,* Chap. 19, edited by E. H. Piepmeier. Wiley, New York, 1986.

8. D. A. Cremers and L. J. Radziemski, "Laser Plasmas for Chemical Analysis," in *Laser Spectroscopy and Its Applications,* Chap. 5, edited by L. J. Radziemski, R. W. Solarz, and J. A. Paisner. Marcel Dekker, New York, 1987.

9. R. S. Houk, "Laser Ionization Techniques for Analytical Mass Spectrometry," in *Analytical Applications of Lasers,* Chap. 18, edited by E. H. Piepmeier. Wiley, New York, 1986.

10. J. Y. Marks, D. E. Fornwalt, and R. E. Yungk, *Spectrochim. Acta* **38B,** 107 (1983).

11. S. J. Jiang and R. S. Houk, *Anal. Chem.* **58,** 1739 (1986).

12. R. Wennrich and K. Dittrich, *Spectrochim. Acta* **42B,** 995 (1987).

13. T. Ishizuka,Y. Uwamino, and H. Sunahara, *Anal. Chem.* **49,** 1339 (1977).

14. T. Kántor, L. Polos, P. Fodor, and E. Pungor, *Talanta* **23,** 585 (1976).

15. T. Kántor, L. Bezur, E. Pungor, P. Fodor, J. Nagy-Balogh, and Gy. Heincz, *Spectrochim. Acta* **34B,** 341 (1979).

16. F. Leis and K. Laqua, *Spectrochim. Acta* **33B,** 727 (1978).

17. T. Ishizuka and Y. Uwamino, *Anal. Chem.* **52,** 125 (1980).

18. P. G. Mitchell, J. Sneddon, and L. J. Radziemski, *Appl. Spectrosc.* **40,** 274 (1986).

19. P. G. Mitchell, J. Sneddon, and L. J. Radziemski, *Appl. Spectrosc.* **41,** 141 (1987).

20. M. Thompson, J. E. Goulter, and F. Sieper, *Analyst* **106,** 32 (1981).

21. J. W. Carr and G. Horlick, *Spectrochim. Acta* **37B,** 1 (1982).

22. H. Kawaguchi, J. Xu, T. Tanaka, and A. Mizuike, *Bunseki Kagaku* **31,** E185 (1982).

23. T. Ishizuka and Y. Uwamino, *Spectrochim. Acta* **38B,** 519 (1983).

24. D. A. Cremers, F. L. Archuleta, and H. C. Dilworth, *Proc. SPIE Int. Soc. Opt. Eng.* **540,** 542 (1985).

25. R. Jowitt and I. D. Abell, U.S. Patent 4,598,577 (1986).

26. M. E. Tremblay, B. W. Smith, M. B. Leong, and J. D. Winefordner, *Spectrosc. Lett.* **20,** 311 (1987).

27. A. R. Barringer, U.S. Patent 4,220,414 (1980).

28. P. Arrowsmith, *Anal. Chem.* **59,** 1437 (1987).

29. P. Arrowsmith, in *Advanced Characterization Techniques for Ceramics,* Proc. Am. Ceramic Soc., San Francisco, 1988.

30. A. L. Gray, *Analyst* **110,** 551 (1985).

31. R. Ediger and J. Hager, *Steel Times* 238 (May 1988).

32. C. T. Tye and P. Barrett, *Steel Times* 240 (May 1988).

33. J. F. Ready, *Effects of High-Power Laser Radiation.* Academic Press, New York, 1971.

34. C. W. Huie and E. S. Yeung, *Anal. Chem.* **58,** 1989 (1986).

35. P. Arrowsmith and S. K. Hughes, *Appl. Spectrosc.* **42,** 1231 (1988).

36. M. I. Boulos, *Pure Appl. Chem.* **57,** 1321 (1985).

37. N. A. Fuchs, *The Mechanics of Aerosols,* edited by C. N. Davies. Pergamon, Oxford, 1964, Section 39, p. 204.

38. J. W. Thomas, *J. Air Poll. Control Assoc.* **8,** 32 (1958).

39. P. G. Gormley and M. Kennedy, *Proc. Royal Irish Acad.* **52A,** 163 (1949).

40. H. E. Hesketh, *Fine Particles in Gaseous Media,* 2nd ed., Chaps. 1–3. Lewis, Chelsea, Michigan, 1986.

41. S. M. Kimbrell and E. S. Yeung, *Spectrochim. Acta* **43B,** 529 (1988).

42. VG Elemental, Application Report PQ 705 (1987).

43. L. de Galan, R. Smith, and J. D. Winefordner, *Spectrochim. Acta* **23B,** 521 (1968).

9

Laser Ablation and Ionization Studies in a Glow Discharge

KENNETH R. HESS and WILLARD W. HARRISON

One of the most attractive features of lasers in a scientific laboratory is their ability to combine effectively with other instrumentation and methodology to create a technique of much greater utility than either of the two alone. The interactive nature of lasers has led to the development of many "hyphenated" techniques. So it is that the glow discharge, coupled with a laser, offers unique opportunities for plasma diagnostics and for practical trace element analysis through enhanced atomization and ionization processes. The high power attainable with a laser can be efficiently coupled to the glow discharge in a controlled and selective manner. By controlling wavelength and power, the analytical chemist can optimize the experimental conditions for a particular desired result. In conjunction with the glow discharge ion source, direct elemental analysis of solid samples such as metals, alloys, semiconductors, rocks, and minerals is possible.

The glow discharge is a relatively simple device that has been known in one form or another for over 60 years. More recently, it has found favor as a spectroscopic source for atomic absorption,[1,2] atomic emission,[3,4] and mass spectrometry.[5,6] Figure 9.1 shows the principal components of the glow discharge. It consists basically of a cathode and an anode sealed into a low pressure rare gas environment, normally argon at ~1 torr. The analytical sample comprises the negatively biased cathode; the anode is normally the ion source housing, including the ion exit orifice. Voltage (1–2 kV) is applied across the electrodes, causing breakdown of the gas, and formation of a plasma. Positive argon ions are accelerated across the cathode fall potential (dark space in Fig. 9.1) and strike the sample cathode, causing release of sample atoms by a process known as sputtering.[7,8] As these atoms diffuse into the negative flow (Fig. 9.1), collisions with electrons and energetic metastable atoms cause excitation and ionization, thus providing the basis for atomic emission and mass spectrometric analytical techniques, respectfully.

Lasers offer striking advantages to enhance these basic glow discharge processes. For example, a glow discharge plasma ionizes only about 1% of the sputtered atoms; the laser can ionize the other 99% in a given beam volume. In addition, glow discharge sputtering (atomization) requires a conducting sample; laser ablation atomization can be carried out on nonconductors. There is also the selectivity

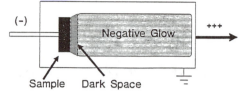

Fig. 9.1. Schematic of glow discharge ion source showing principal components.

intrinsic in laser operation, both in sampling location and photon energy delivered, that allows advantageous application of lasers to glow discharges. This interaction has taken several forms, including atomic fluorescence,[9,10] resonance ionization,[11] optogalvanic spectroscopy,[12] and laser ablation. The last of these will be given particular attention in this report, which will not be directly concerned with analytical measurements, but rather the fundamental processes involved.

INSTRUMENTATION

A diagram of the glow discharge apparatus used in these studies is presented in Fig. 9.2. The glow discharge source consists of a 2.75-in. six-way cross with quartz windows to allow laser interaction with the discharge plasma. The cathodes are machined from the material of interest to form pins 1.5 mm in diameter with 5 mm exposed to the glow discharge. The cathode is mounted in a pin vise that is shielded from the discharge by a nonconducting machinable ceramic (Macor, Corning Glass Works). A stainless-steel plate with a 0.5-mm aperture serves as the anode and ion exit orifice. The discharge is powered by a Kepco OPS-3500 power supply operated in either a dc or pulsed mode. With discharge pressures of 0.2–1.2 torr employed, the current ranged from 1 to 5 mA with voltages of 300–3500 V.

Ions formed in the discharge region are extracted through the ion exit orifice and focused by an einzel-type lens through a Bessel-Box energy analyzer (Extrel,

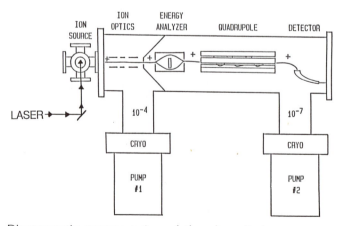

Fig. 9.2. Diagrammatic representation of the glow discharge mass spectrometer system.

Inc.) into a quadrupole for analysis (Extrel, Inc.). The ions are detected with an electron multiplier operated in either an analog mode or in a pulse counting configuration. In the pulse-counting mode, a counter/processor (Princeton Applied Research Model #112) is gated to collect counts from a data gate, a background gate, or a difference between the two gates, allowing for background subtraction. The mass spectrometer and data collection/output is controlled by a DEC-MINC-11 microcomputer system.

The laser system consists of a Lumonics TE-860-4 excimer laser operating with XeCl. The output from the excimer pumped a Lumonics EPD-330 dye laser which generated 10-nsec laser pulses with a 0.003-nm bandwidth and energies generally in the 5–10 mJ range. A frequency doubler (Inrad 5-12) was also available that, with the use of various laser dyes (Exciton, Inc.), allowed access to wavelengths from 265 to 700 nm. The output of the laser system is focused by several different optic systems to produce a variety of beam sizes and directed through the quartz windows in the discharge housing by the use of gimbal mounted mirrors.

LASER-BASED METHODS

Laser resonance ionization takes advantage of the neutral sputtered atomic population in the discharge, creating a secondary method of ionization for enhanced sensitivity and selectivity. Details of the resonance ionization method have been covered in more detail elsewhere.[13,14] Essentially, atoms in the discharge plasma can absorb photons of laser light with a wavelength corresponding to an atomic transition. Absorption of two or more of these photons can result in the photoionization of the atom and these ions may then be extracted and analyzed mass spectrometrically. In addition to the enhanced sensitivity generated by increasing the number of ions created in the discharge, the resonance ionization technique also increases the selectivity of glow discharge mass spectrometry through the wavelength dependence of photon absorption. The method is then both wavelength selective and mass selective. This increased selectivity is useful in reducing the isobaric overlaps caused by background molecular species that are prevalent in glow discharge mass spectra. Further information on the coupling of the laser to glow discharge for resonance ionization is available in previous publications.[11,15]

The laser system has also been employed for investigations into the ionization processes occurring in the discharge. The two dominant mechanisms of ionization for the sputtered atoms are believed to be electron impact and energy transfer from an excited metastable state of the rare gas, termed Penning ionization. By coupling the laser to the discharge, the laser photons may be used effectively to perturb the ionization rates within the discharge. These pertubations affect the electrical resistance of the discharge plasma, altering the current/voltage characteristics of the discharge. The changes may be monitored electrically, generating a technique with electrical detection of an optical effect, optogalvanic effect (OGE) spectroscopy.[16,17] If laser photons corresponding to an atomic transition are absorbed by an atom, this excited atom will undergo enhanced electron impact ionization relative to its ground state and the ionization rate in the discharge will be increased, resulting in decreased discharge resistance and increased discharge current for a constant voltage system. Decreases in discharge current may also occur with laser interaction and were historically the first OGE effects discovered.[18] These signals correspond to decreases in the net ionization rate of the plasma and result from the depopulation of rare gas metastable states. The metastable states may absorb a photon, resulting in an excited state that is no longer metastable, allowing radiative decay of the rare gas atoms back to the ground state, reducing the number of metastables and degree of Penning ionization. If the metastables were to play a large role in the ionization mechanisms of the discharge, then their laser depopulation should result in decreased discharge ionization that may be monitored both electrically and through extraction of the generated ions into a mass spectrometer.

Investigations into the effect of laser metastable depopulation in the glow discharge were performed and the detailed results appear elsewhere.[19] Briefly, laser irradiation of a neon discharge with a brass cathode at a wavelength corresponding to a metastable depopulation transition (594.48 nm) resulted in a net decrease of ion signal intensity from the following discharge species: Ne^+, Ne^{2+}, Ne_2^+, OH^+, N_2^+, N_2H^+, Cu^+, Zn^+, Cu_2^+, and Zn_2^+, as monitored mass spectrometrically. With the exception of the neon compounds, all of these species have ionization potentials below the depopulated neon metastable level of 16.62 eV and would be expected to exhibit ion signal decreases if Penning ionization were an important ionization mechanism. The ion signals of the neon species may be expected to show a metastable dependence since ionization from the neon metastable level is an enhanced process relative to ionization from the ground state[20] and the reduction of the metastable population will therefore reduce the neon ion signals. When the discharge gas was changed to argon with the same brass cathode, the results were markedly different with only argon and those species (Cu and Zn) with ionization potentials below the depopulated argon metastable state at 11.55 eV (696.54 nm transition) showing signal decreases. Species with ionization potentials above 11.55 eV (OH^+, H_2O^+, N_2^+, O^+, N^+) exhibited no change on laser irradiation. This behavior is consistent with the Penning ionization mechanism and indicates the substantial role of metastable gas atoms in the ionization mechanism of glow discharge mass spectrometry sources. These investigations provide an illustration of the advantageous coupling of a laser system to a glow discharge source for fundamental investigations of glow discharge processes.

LASER ABLATION IN THE GLOW DISCHARGE

The glow discharge can take advantage of the laser as a method of secondary atomization. In this case, a high-power laser beam is tightly focused onto a surface, resulting in the ejection of an atomic population. A portion of this atomic population may be ionized in the ablation process and the ions extracted for mass spectrometric analysis, a technique for which there are several reviews.[21-24] In addition to the ions, a substantial number of atoms remain neutral in their ground state or excited states and these atoms may be used for atomic absorption analysis[25,26] or atomic emission spectroscopy.[27,28] This laser atomization process has also been advantageously employed as a solid sample introduction system for a variety of subsequent ionization and excitation sources, including inductively coupled plasma (ICP) emission,[29-31] ICP mass spectrometry,[32,33] microwave discharges,[34] a direct current plasma (DCP),[35] arc/spark emission,[36] resonance ionization sources,[37,38] and ion cyclotron resonance chambers in which ion and atom reactions may be studied.[39] A high-powered laser was coupled to the glow discharge in our laboratory as a method for injecting solid material into the glow discharge for subsequent ionization. This method could be useful in its ability to analyze nonconducting solid materials and in fundamental studies of the discharge and its ionization mechanisms. Initial investigations in the area of direct ionization during the laser ablation process (no discharge present) were carried out, followed by experiments involving ablation of the discharge cathode during discharge operation, ablation of a secondary sample into a discharge, and ablation of material between the pulses of a pulsed discharge.

The mechanism of ablation is very complex and not clearly defined. When the laser beam impacts the target, a portion of the beam is reflected and a portion absorbed. The energy contained in the absorbed portion of the beam is transferred to the target through an exchange of energy from the photons to electrons of the sample atoms, which then may exchange energy with the sample lattice. This thermal heating is believed to be the dominant mechanism of laser–solid interaction. The important parameters that influence this process include material parameters and laser parameters. Depending on the combination of these parameters, with the laser power being the most important, three distinct degrees of laser interaction can occur. These are simple laser heating (power levels $<10^4$ W/cm^2), the melting of the surface with some differential boiling of low vapor pressure elements (10^4–10^6 W/cm^2), and laser vaporization of the sample (10^6–10^8 W/cm^2). Above 10^9 W/cm^2 nearly uniform atomization with 100% ionization of the elements in the sample is observed, resulting in relative sensitivity coefficients very close to unity. The reader is referred to Chapter 8 in this book for further details on the mechanisms of laser ablation.

Our initial investigations into laser ablation effects involved the use of the laser as both the atomization and ionization source. The output of an excimer pumped Rhodamine 590 dye (1–4 mJ, 10-nsec pulse) was focused to 100 μm, resulting in a power density of 10^8–10^9 W/cm^2. This power density is sufficient to cause ionization of the ablated material. Although our system was not optimally designed for the maximum detection of laser ionization signals, the type of species produced and the effects of an inert atmosphere above the ablation surface were of interest.

Sample pins with 45° angled faces were positioned in the cathode holder with a 5-mm distance from the pin tip to the ion exit orifice. The laser beam was directed and focused onto the face of the pin. A laser-generated plasma was formed at the surface of the pin, and the ions created in this plasma were extracted and focused into the mass spectrometer. Figure 9.3 presents the mass spectrum produced for copper samples under a vacuum of 1×10^{-7} torr. This spectrum shows the primary species formed to be Cu$^+$ with some small signals from species such as H$_2$O$^+$ and N$_2^+$ that arise from surface adsorbed gases. Low-level signals, $<0.01\%$, are also observed for the copper oxides, CuO, CuO$_2$, and CuO$_3$. The magnitude of these data is such that ion signals originating from minor species must be accumulated over a large number of scans (several 100), which limits the utility of the technique with our system. Other limits on the usefulness of the method include its uneven

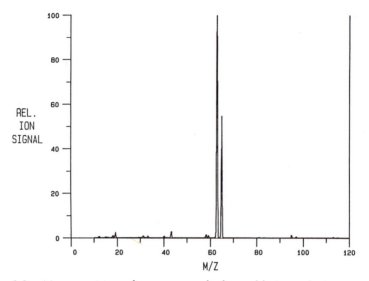

Fig. 9.3. Mass spectrum of copper sample, laser ablation only, in vacuum.

laser shot-to-shot reproducibility and its poor ion signal stability with time. The ion signal produced by the laser steadily decreases with the number of laser pulses absorbed by the sample surface. In most applications of laser ablation mass spectrometry, the sample is rotated to generate a fresh surface between laser pulses or single laser shots are used for analysis. The single-shot technique would require the use of a mass spectrometer that can simultaneously monitor all masses, such as a time-of-flight instrument, rather than a quadrupole scanning instrument. The cause of the signal deterioration with time is believed to be caused by shielding of the laser plasma from the ion exit orifice by the craters formed during the first laser pulse interactions or through defocusing of the beam as the crater forms.

A second area of interest was the influence of an inert atmosphere of argon on the laser-produced ion signal. Other workers have investigated the plasmas produced by laser bombardment of metal targets in the presence of 0.1–10 torr of various gases by atomic emission and found a pressure dependence on the emission signal. The plasma front was observed to spread out further and faster at lower pressures[40–42] with a lower excitation temperature also observed.[42] The optimum pressure for copper emission in air was found to be approximately 1 torr with bombardment of a N_2 laser.[41] The increased pressures also result in lower radial diffusion of the laser-ablated atomic population as measured by atomic absorption. The exact cause of the increased excitation at higher pressures has not been clearly defined in any of this work and any corresponding effects on the ion signal generated by the laser were of interest.

The influence of argon on the laser ion signal would manifest itself in several areas; first, it could spread the ion signal out in time, and second, it could contain and enhance plasma formation and ionization. The mean free paths of the ions will also be affected by the argon, decreasing with increasing pressures. This could inhibit the extraction and collection of the ions formed in the ablation process. The ion signal pulse width produced by the laser is observed to increase on the addition of argon. The higher collision rate with the presence of an argon atmosphere can cause collisional broadening of the ion pulse. As the ions traverse the region from the pin to the ion exit orifice, they undergo collisions that will slow them down and broaden the time profile of the signal.

The mass spectra taken under an argon atmosphere are very similar to those taken under vacuum, with the same type of species observed except for the addition of metal argides, $ArCu^+$, formed in the laser-generated plasma. The effect of argon on the magnitude of the ion signals is plotted for three sample-ion exit orifice distances in Fig. 9.4. As expected, the closer the sample is to the ion exit orifice, the higher the ion signal. For all three distances, a small amount of argon (<0.05 torr) in the sample chamber can have a pronounced effect on the magnitude of the ion signal, increasing it by a factor of three to four. As the pressure continues to be increased, the laser-generated ion signal decreases steadily. This decrease is less dramatic than would be expected from simple mean free path considerations. The mean free path is inversely proportional to pressure, so increasing the pressure would result in fewer ions being able to travel to the ion exit orifice without undergoing a collision, lowering the extraction efficiency of the ions formed by the ablation process. The fraction of ions able to travel a collision-free distance, x, for a mean-free-path value of b, is given by $e^{-x/b}$.[20] Linearly decreasing the mean free path should result in an exponential decline in the number of ions that can reach the ion exit orifice without a collision. The plots of ablated ion signal vs pressure show somewhat less than an exponential decline, suggesting other possible pressure effects that increase the ion signal are important. These effects are not known at the present time, but may involve fundamental laser-generated plasma processes similar to the effects on atomic emission previously discussed.

In addition to ablation of metal samples, the laser may be used to ablate material directly from the discharge cathode during discharge operation as illustrated in

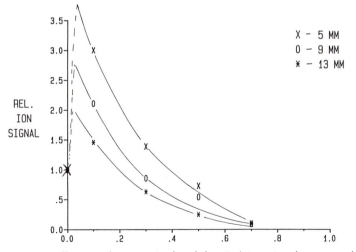

Fig. 9.4. Pressure effect on the magnitude of the Cu^+ ion signal generated by laser ablation only for various sample-ion exit orifice distances, normalized to the signal in a vacuum.

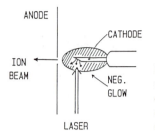

Fig. 9.5. Schematic diagram of the laser/sample configuration for the ablation of a discharge cathode with a discharge present. (Reprinted with permission from *Analytical
Chemistry,* **58,** 341A, 1986. Copyright 1986 American
Chemical Society.)

Fig. 9.5. In this mode, the laser interaction with the cathode causes the rapid release
of a substantial number of electrons and large amounts of sample material into the
discharge. This material release changes the resistance of the discharge during the
time of laser interaction, creating arcs in the operation of the discharge. In our
system, these arcs generate noise spikes that interfere with the triggering of the photon counter/processor, preventing the computer acquisition of mass spectra under
these conditions. Signals were available for analysis from analog time profiles.

To determine if the discharge could ionize material ablated as an atomic population from the cathode, the laser was defocused until a point was reached at
which the ion signal resulting from ablation just disappeared. The power density
was then less than the threshold for ionization but still sufficient for sample atomization. This generates an atomic population for subsequent ionization in the glow
discharge plasma, similar to ablation into an ICP.[33] Such ionization of an ablated
atomic copper population generated from a copper cathode at low laser power was
observed as illustrated in Fig. 9.6, a time profile for the copper ion signal at 2 mA

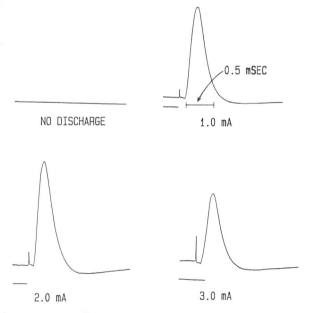

Fig. 9.6. Discharge current effect on ion signal generated from laser ablation of a discharge cathode. Argon discharge, 0.4 torr, Cu cathode.

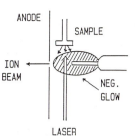

Fig. 9.7. Schematic diagram of the laser/sample config-uration for the ablation of a sample into a secondary glow discharge. (Reprinted with permission from *Analytical Chemistry*, **58,** 341A, 1986. Copyright 1986 American Chemical Society.)

and 0.4 torr argon. The results show that the ablation of an atomic population into an adjacent discharge produces ion signals representative of the ablated material.

The laser ablation of material into a glow discharge from a sample not serving as the discharge cathode was next studied. The sample to be ablated was in the form of a disk positioned adjacent to the glow discharge. The discharge was sustained by a tantalum cathode chosen for its low sputtering characteristics, which serve to limit the amount of cathode material in the discharge. This experimental configu-ration is shown in Fig. 9.7. A copper disk 3 mm in diameter was used as the ablated sample for investigations into the effects of discharge parameters on the observed discharge ionization of the ablated material. The laser beam was defocused to the point at which the ion signal created by the laser just vanished, allowing sufficient power for atomization without ionization. The discharge was then struck and ion signals from the ablated sample observed. The analog signals for the ^{63}Cu species with discharge off and on (Fig. 9.8) demonstrate the fact that the discharge is ion-izing ablated material. A National Bureau of Standards #410A steel sample was ablated into the discharge for subsequent ionization and the iron region of the mass spectrum is presented in Fig. 9.9. The ^{52}Cr peak is present at 2% by mass, showing the relative insensitivity of this method at this time due to the low duty cycle of the laser.

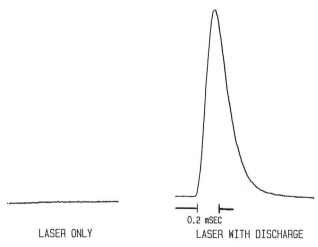

Fig. 9.8. Analog time profile for the ion signal generated by ablation of a copper sam-ple into a secondary discharge. Argon discharge, 0.4 torr, 2 mA, Ta cathode.

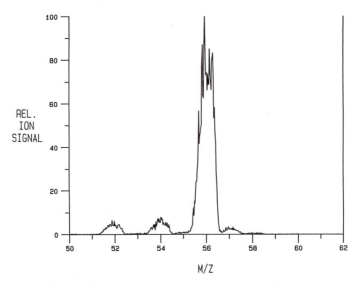

M/Z

Fig. 9.9.   Laser ablation/discharge ionization mass spectrum for NBS #410 steel. Argon discharge, 0.4 torr, 2 mA, Ta cathode.

   In addition to conducting bulk metal samples, the method was applied to non-conducting powdered samples. With laser atomization the analytical sample no longer has to be conducting, which reduces the constraints on the sample type. Nonconducting samples of $CuCO_3$ and $CuSO_4$ were prepared for ablation atomization by pressing 2-cm-diameter disks of the powdered material in a commercial IR pellet press. These disks were then mounted on a holder with silver paint in the same manner as the copper disks. A tantalum cathode was used for the discharge, and the laser power was adjusted to a level just below the threshold at which ions from the ablated material were observed. The mass spectrum is generated by setting a data gate over the laser formed ions. Copper peaks are visible, but the $^{63}Cu/^{65}Cu$ ratio is not that expected from isotopic abundance data. The signal from ^{65}Cu is reduced relative to the ^{63}Cu, due to cratering effects and a time dependence on the ablation process resulting in a lower ion signal from the ^{65}Cu, which is analyzed a number of laser shots after the ^{63}Cu. To solve these problems, either a time-of-flight mass spectrometer that can analyze the entire mass range in one laser shot could be used, or the sample could be rotated to generate a fresh surface for ablation with each laser pulse.

   Sample cratering is a common problem and can affect the ion signals produced through laser ablation.[32,43–45] The deeper the laser produced crater becomes, the greater the plasma produced by the laser is shielded from the extraction optics of the mass spectrometer. As the number of laser shots increases, the depth of this crater will grow and ion extraction will diminish. A second effect of cratering will be the changing focus of the laser. Crater depths can be reached in which the sample surface will no longer be at the optimum focal distance for laser ionization. Both of these effects will lower the ion signal generated by the laser as the number of laser pulses impinging on the sample surface is increased.

We have addressed the cratering problem by studying cathodic ablation between the pulses of a pulsed glow discharge. The laser trigger is output to a delay circuit from which it triggers the pulse generator of the pulsed discharge. The delay allows the discharge pulse to be positioned at any time between the laser pulses. The time profiles of the laser ablation and the pulsed discharge ion signals are shown in Fig. 9.10. The ablation ion signal generated between discharge pulses was observed to remain more stable with increasing numbers of laser shots than the ion signal generated from ablation without a discharge pulse. This is exhibited in Fig. 9.11, which shows the magnitude of the ablation only and ablation with pulsed discharge ion signals vs the number of laser shots. In the case of the ablation only, cratering effects decrease the ion signal as the number of laser shots increase. Ablation between the discharge pulses appears to reduce this effect.

The discharge pulse could serve essentially to "resurface" the sample between each laser pulse, which would lower the effects of the laser produced craters. During each discharge pulse, sample material is being removed through sputtering by the Ar^+. At pressures of 1.0 torr, approximately 99% of the sputtered material is believed to be redeposited onto the sample surface through collisions with the rare gas atoms and ions present above the sample surface.[46,47] Between each laser ablation pulse, the sample surface would have material removed and redeposited by the glow discharge, which could effectively smooth the ablation crater. The next laser shot would then impact a fresh surface of redeposited material, preventing the accumulative build up of cratering effects that could lead to laser plasma shielding and lower ion signals as the number of laser interactions increases. Based on rough approximations of sputter removal rates in the discharge of 200–250 Å/pulse of copper[48] and an approximate crater depth of 500 Å (copper sample at a power density similar to that used in this work, 9×10^8 W/cm^2 [49]), the discharge could sputter and redeposit sufficient material to effectively resurface the laser-produced crater.

A second method of testing the validity of this hypothesis is by monitoring the

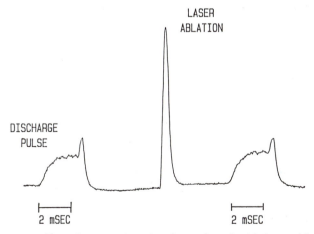

Fig. 9.10. Time profiles of copper ion signals produced with laser ablation between the pulses of a pulsed discharge. Argon discharge, 0.5 torr, 1.6 mA average current, copper cathode.

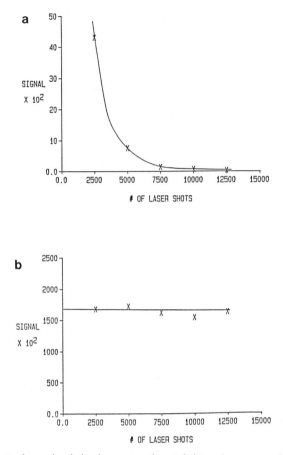

Fig. 9.11. Effect of a pulsed discharge on the stability of a copper laser ablation ion signal. (a) No discharge pulse; (b) with a discharge pulse. Argon discharge, 0.5 torr, 1.6 mA average current, copper cathode.

effect of various discharge parameters on the signal produced by the laser. Pressure variations could be expected to change the redeposition rate and the sputter etch rate of the discharge. This, in turn, would influence the amount of material removed and redeposited during the discharge pulse, which could affect the stability of the laser-ablated ion signal. Studies of the laser-generated ion signal as the pressure was varied between 0.3 and 1.0 torr showed no effect on the signal stability, although it is recognized that these pressure limits reflect a very small pressure difference. Increasing the pulse length of the discharge would be expected to increase the amount of material sputtered, decreasing the cratering effects. When the discharge pulse length was increased, no effect on the stabilities of the ablated ion signal was observed. Changes in ablated ion signal stability with changes in discharge parameters would be expected only if the variations in discharge parameters caused sufficient changes in the sputter removal rate to no longer smooth the laser-generated craters between pulses. Neither the pressure, which has only a small effect on sputtering, nor the pulse length, which would be directly proportional to the amount of material removed, is likely to cause large enough sputter etch rate changes to affect the ablated ion signal stability.

The effect of discharge current on sputter removal is greater than linear for low currents, so relatively small changes in current can have large effects on the sputter removal and redeposition rates. Figure 9.12 shows the effect of changing current on the magnitude and stability of a copper ion signal arising from ablation between discharge pulses. Initially, no discharge is present and the ion signal is seen to decrease with the first 5000 laser shots. During this time period, the cratering effects appear to have a substantial impact on the laser ion signal. When the discharge is turned on at 0.5 mA, the ablated ion signal rises, decreases slightly, and reaches a steady state. When the current is increased to 1.0 mA and then to 1.5 mA, the same process is observed. When the discharge is turned off, the signal decays to the original ion signal level present before the discharge was turned on. On turning the discharge back on at 1.5 mA, the ablated ion signal returns to the level it was before the discharge was turned off. There is little difference in the magnitude of the signal between the different discharge currents. This plot would indicate that the ablation and the discharge sputter removal processes appear to reach an equilibrium at which point the cratering effects no longer decrease the ion signals. At the laser power density used in these experiments (approximately 7×10^8 W/cm^2) the discharge current does not seem to influence the magnitude of the ablated ion signal. Any changes in physical properties of the sample caused by redeposition do not appear to affect the magnitude of the ablation ionization process. At sufficient powers, this would be anticipated since with increasing power densities, the physical properties of the ablated material become less important in determining the magnitude of ionization by ablation.

If the laser power density is decreased an order of magnitude by partially blocking a portion of the dye laser amplifier cell, the physical properties of the material may become important in determining the intensity of the ablated ion signal. A

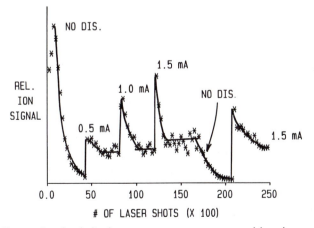

Fig. 9.12. Effects of pulsed discharge average current on ablated copper ion signal stability for high laser power. Argon discharge, 0.5 torr, copper cathode.

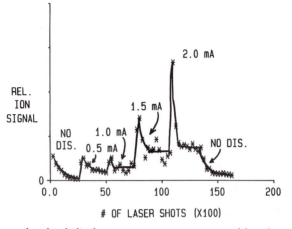

Fig. 9.13. Effects of pulsed discharge average current on ablated copper ion signal stability for low laser power. Argon discharge, 0.5 torr, copper cathode.

second experiment was conducted after decreasing the laser power to 7×10^7 W/ cm^2, generating the results shown in Fig. 9.13. This plot closely resembles Fig. 9.12 with one notable exception. In this case, increasing the current also increases the magnitude of the ablated ion signal. Under these power conditions, the ablation process may be operating in a regime in which the physical properties of the material (heat of vaporization, work function, thermal conductivity, etc.) are important. The material redeposited from sputtering may have different physical properties that allow it to be more efficiently ablated and ionized than the bulk material. The greater the redeposited material present, the higher the ablation ion signal. This could explain the increases in the ablation ion signal observed as the discharge current is increased and more material is removed and redeposited. Again, when the discharge is turned off, the ablated ion signal will fall to the level present before the discharge was initiated.

If the laser is ablating redeposited material, then the composition of the ablated material should be indicative of the redeposited surface composition. Differences in sputter yields among various elements will result in differences in their redeposition concentrations, with elements that have a higher sputter yield also having a higher surface concentration relative to the bulk concentration than elements with a lower sputter rate. Laser ablation with no discharge will give concentration ratios for the bulk material, whereas ablation between discharge pulses might give elemental ratios corresponding to the redeposited material, if it is the redeposited material being ablated. The elemental ratios (relative sensitivity factors, RSFs) arising from the ablation process would then be expected to be similar to the discharge sputter yields and different from the elemental ratios obtained from the ablation of the bulk material with no discharge present. To determine if this were the case, experiments were performed to determine the RSFs from the ablation ionization of a bulk material, the RSFs from ablation ionization of the same material between the pulses of a pulsed discharge, and the RSFs from glow discharge ionization with no laser interaction.

The ratio of ^{62}Ni to ^{63}Cu ion signals in a National Bureau of Standards (NBS)

#410A steel sample for laser ablation only, for laser ablation between a discharge pulse, and the glow discharge only, along with the NBS certified value, are presented in Table 9.1. As can be seen, the $^{62}Ni/^{63}Cu$ ratio for ablation closely matches the certified value, indicating nonselective ionization by the ablation process. When ablation occurs between the discharge pulses, the ratio of $^{62}Ni/^{63}Cu$ appears closer to the ratio present with discharge ionization. The elemental ratios of the redeposited material are evidently different than the ratios in the bulk material, showing the differences in sputter rate among different elements. For the ablation between discharge pulses, copper has been enhanced relative to nickel by a factor of 1.62, indicating preferential sputtering of copper. Sputter weight loss studies in the glow discharge have also shown the preferential sputtering of copper with a relative enhancement of 1.32.[50] Fundamental studies of sputter yields generated by 100- to 300-eV argon ions further confirm the enhanced sputtering of copper vs nickel with observed ratios of approximately 1.67.[51] Since the elemental ratios with glow discharge ionization match the ratios in the redeposited material, this would indicate the differences in ionization observed in the glow discharge are due mostly to differences in elemental sputter yields and not to some selective ionization process. The fact that the glow discharge ionization mechanisms are largely nonselective improves the relative sensitivity coefficients for glow discharge analysis. Similar experiments for an NBS #1263 steel sample further confirmed the results from the NBS #410A standard.

In addition to fundamental investigations of the relative sensitivity coefficients in the discharge, laser ablation may also be used as a technique for investigation into certain ionization processes of the discharge. In argon discharges, ion species of the type ArM^+ where M is the sputtered metal atom are generated at a level as high as 10% of the metal ion signal.[52] These signals can be a source of isobaric interferences and methods for their reduction would be required. Two mechanisms of formation of these argon–metal compounds have been proposed[52] (Ar* represents the argon metastable):

1. Associative Ionization

$$Ar^* + M^0 \rightarrow ArM^+ + e^-$$

2. Three-Body Collision

$$M^+ + 2Ar \rightarrow ArM^+ + e^-$$

The first mechanism is believed to be dominant at low pressures, whereas the second is more important at pressures at 1.0 torr and above. In the glow discharge,

Table 9.1. $^{62}Ni/^{63}Cu$ ion signal ratios obtained from various ion sources for NBS #410A steel sample

Ion source	Ratio $^{62}Ni/^{63}Cu$
Certified NBS value	0.24
Ablation only	0.25
Ablation with pulsed discharge	0.43
Pulsed discharge only	0.39

both of these mechanisms can occur, and it is not clear which is more important. For laser ablation and ionization of a metal target in a low-pressure atmosphere of argon, mechanism #2 can be created in the relative absence of mechanism #1, allowing the respective importance of each mechanism to be estimated provided the ArM^+ ion signal dependence on the metal atom or ion population is normalized in some manner. Mechanism #1 is dependent on both the Ar^* and M^0 populations, which are directly proportional to the M^+ value.[53] Mechanism #2 is directly dependent upon the M^+ concentration, so both mechanisms have an M^+ dependence. If a ratio of M^+/ArM^+ is taken, the dependence upon the M^+ population should be removed, allowing direct comparisons between the amount of ArM^+ formed in the discharge vs that formed with ablation only. The ratio of the M^+/ArM^+ value with ablation only to the M^+/ArM^+ value in the glow discharge will give the ratio of mechanism #2 (from ablation only) to the total formation mechanism (from glow discharge, mechanisms #1 and #2). These ratios would be expected to show a pressure dependence, with mechanism #2 becoming more important as the pressure is increased.

A copper pin was placed in the discharge chamber and the ratio of Cu^+ to $ArCu^+$ was recorded as a function of pressure for the ion signals generated by laser ablation. These same ratios were also recorded as a function of pressure for glow discharge-generated ionization. These values are given in Table 9.2. There was little Ar^+ observed in the mass spectra and other work with emission of laser-created plasmas in argon at 1 torr have shown no emission from the argon gas,[40,42] suggesting that little laser-excited argon exists and that there is little Ar^* with laser ablation only. Under these conditions, mechanism #1 for $ArCu^+$ formation cannot occur and the $ArCu^+$ signal generated is formed solely by mechanism #2. At equivalent pressures in the glow discharge and ablation studies, the same population of Ar^0 exists, so the same relative degree of $ArCu^+$ formation through mechanism #2 is also present. In addition, the discharge creates Ar^*, which will now allow mechanism #1 to occur as well. Dividing the $Cu^+/ArCu^+$ ratio for ablation by the $Cu^+/ArCu^+$ for the discharge will give the relative importance of mechanism #2 in the total $ArCu^+$ formation process. These values, as a percentage of mechanism #2 in the total process, are plotted vs pressure in Fig. 9.14. Mechanism #2 becomes more important as the pressure increases, which increases the number of Ar^0 and therefore the probability of a three-body collision, as is expected. These results may be useful in evaluating the means for reducing such interferences in analytical studies and further illustrate the potential role of the laser in fundamental studies of glow discharge processes.

Table 9.2. $Cu^+/ArCu^+$ ion signal ratios for ablation and discharge ion sources at various argon pressures

Pressure (torr)	Ablation	Discharge	Ratio ablation/discharge
0.25	529.1	371.7	1.42
0.45	483.4	366.8	1.32
0.60	857.6	689.3	1.24
0.80	646.9	582.6	1.11

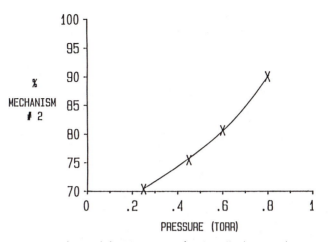

Fig. 9.14. Percentage of ArCu$^+$ formation mechanism #2 (see text) vs pressure. Argon discharge, 2 mA, copper cathode.

REFERENCES

1. B. M. Gatehouse and A. Walsh, *Spectrochim. Acta* **16,** 602 (1960).
2. A. E. Bernhard, *Spectroscopy* **2,** 24 (1987).
3. W. Grimm, *Spectrochim. Acta* **23B,** 443 (1968).
4. K. Wagatsuma and K. Hirokawa, *Spectrochim. Acta* **42B,** 523 (1987).
5. W. W. Harrison, K. R. Hess, R. K. Marcus, and F. L. King, *Anal. Chem.* **58,** 341A (1986).
6. N. Jajubowski, D. Stuewer, and G. Toelg, *Int. J. Mass Spectrom. Ion Phys.* **71,** 183 (1986).
7. W. D. Westwood, *Prog. Surf. Sci.* **7,** 71 (1976).
8. H. Oechsner, *Appl. Phys.* **8,** 185 (1975).
9. N. Omenetto and J. D. Winefordner, *Spectrochim. Acta* **39B,** 1389 (1984).
10. H. Bubert, *Spectrochim. Acta* **39B,** 1377 (1984).
11. K. R. Hess and W. W. Harrison, *Anal. Chem.* **58,** 1696 (1986).
12. R. A. Keller, R. Engleman, Jr., and E. F. Zalewski, *J. Opt. Soc. Am.* **69,** 738 (1979).
13. G. S. Hurst, M. G. Payne, S. D. Kramer, and J. P. Young, *Rev. Mod. Phys.* **51,** 767 (1979).
14. J. P. Young, G. S. Hurst, S. D. Kramer, and M. G. Payne, *Anal. Chem.* **51,** 1050A (1979).
15. P. J. Savickas, K. R. Hess, R. K. Marcus, and W. W. Harrison, *Anal. Chem.* **56,** 817 (1984).
16. J. C. Travis and J. R. DeVoe, in *Lasers in Chemical Analysis,* edited by G. M. Hieftje, J. C. Travis and F. E. Lytle. Humana Press, Clifton, NJ, 1981, pp. 93–124.
17. D. S. King and P. K. Schenck, *Laser Focus* **18,** 50 (1982).
18. F. M. Penning, *Physica* **8,** 137 (1928).
19. K. R. Hess and W. W. Harrison, *Anal. Chem.* **60,** 691 (1988).
20. B. Chapman, *Glow Discharge Processes.* John Wiley, New York, 1980.
21. R. J. Conzemius and J. M. Capellen, *Int. J. Mass Spectrom. Ion Phys.* **34,** 197 (1980).
22. R. J. Cotter, *Anal. Chem.* **56,** 485A (1984).
23. H. vanDoveren, *Spectrochim. Acta* **39B,** 1513 (1984).

24. F. P. Novak, K. Balasanmugam, K. Viswanadham, C. D. Parker, Z. A. Wilk, D. Mattern, and D. M. Hercules, *Int. J. Mass Spectrom. Ion Phys.* **53**, 135 (1983).

25. R. M. Manabe and E. H. Piepmeir, *Anal. Chem.* **51**, 2066 (1979).

26. R. Wennrich and K. Dittrich, *Spectrochim. Acta* **39B**, 657 (1984).

27. G. Dimitrov and T. Zheleva, *Spectrochim. Acta* **39B**, 1209 (1984).

28. K. J. Mason and J. M. Goldberg, *Anal. Chem.* **59**, 1250 (1987).

29. T. Ishizuka and Y. Uwamino, *Spectrochim. Acta* **38B**, 519 (1983).

30. J. W. Carr and G. Horlick, *Spectrochim. Acta* **37B**, 1 (1982).

31. A. Aziz, J. A. C. Broekaer, K. Laqua, and L. Leis, *Spectrochim. Acta* **39B**, 1091 (1984).

32. A. L. Gray, *Analyst* **110**, 551 (1985).

33. P. Arrowsmith, *Anal. Chem.* **59**, 1437 (1987).

34. F. Leis and K. Laqua, *Spectrochim. Acta* **34B**, 307 (1979).

35. P. G. Mitchell, J. Sneddon, and L. J. Radziemski, *Appl. Spectrosc.* **40**, 274 (1986).

36. W. van Deijcke, J. Balhe, and F. J. M. Maesser, *Spectrochim. Acta* **34B**, 359 (1979).

37. D. W. Beekman, T. A. Callcott, S. D. Kramer, E. T. Arakawa, G. S. Hurst, and E. Nussbaum, *Int. J. Mass Spectrom. Ion Phys.* **34**, 89 (1980).

38. S. Mayo, T. B. Lucatorto, and G. G. Luther, *Anal. Chem.* **54**, 553 (1982).

39. B. S. Freiser, *Talanta* **32**, 697 (1985).

40. K. Kagawa and S. Yokoi, *Spectrochim. Acta* **37B**, 789 (1982).

41. K. Kagawa, S. Yokoi, and S. Nakajima, *Optics Commun.* **45**, 261 (1983).

42. K. Kagawa, M. Ohtani, S. Yokoi, and S. Nakajima, *Spectrochim. Acta* **39B**, 525 (1984).

43. R. H. Scott and A. Strasheim, *Spectrochim. Acta* **26B**, 707 (1971).

44. R. A. Bingham and P. L. Salter, *Anal. Chem.* **48**, 1735 (1976).

45. Y. B. Zel'Dovich and Y. P. Raizer, *J. Exp. Theoret. Phys. (U.S.S.R.)* **20**, 772 (1965).

46. E. Nasser, *Fundamentals of Gaseous Ionization and Plasma Electronics.* Wiley-Interscience, New York, 1971.

47. H. Mase, S. Nakaya, and Y. Hatta, *Appl. Phys.* **38**, 2960 (1967).

48. P. J. Savickas, Ph.D. dissertation, University of Virginia, 1984.

49. I. Opauszky, *Pure Appl. Chem.* **54**, 879 (1982).

50. C. M. Barshick, unpublished work, University of Virginia, 1988.

51. N. Laegreid and G. K. Wehner, *J. Appl. Phys.* **32**, 365 (1961).

52. J. W. Coburn and W. W. Harrison, *Appl. Spectrosc. Rev.* **17**, 95 (1981).

53. K. R. Hess, Ph.D. dissertation, University of Virginia, 1986.

10

Lasers for the Generation and Excitation of Ions in Tandem Mass Spectrometry

W. BART EMARY, OWEN W. HAND, and R. GRAHAM COOKS

Generation of gas phase ions for mass spectrometric analysis historically has been performed using particle–particle interactions such as electron impact (EI) of an evaporated organic sample.[1] The high flux of electrons necessary to create nanoamperes of ion current is available by simple resistive heating of a filament. Chemical ionization (CI), for example, by proton transfer from a chemical reagent to the neutral sample molecule (M) to produce $(M + H)^+$, is also widely used.[2] Photoionization, however, was long used much less frequently for analytical mass spectrometry because only low-flux photon sources were available. More significantly, photoionization cross sections are low ($\sim 10^{-19}$ cm^2) relative to those typical of electron-impact or ion/molecule reactions ($\sim 10^{-15}$ cm^2).[3] Lasers produce high fluxes of photons and have found considerable use for creating and exciting ions in mass spectrometry, particularly for more fundamental studies in which the well-defined energy available for processes such as photodissociation is advantageous.

This chapter summarizes experiments in which lasers have been used in conjunction with tandem mass spectrometry (MS/MS).[4–15] The work described is focussed on three areas of research: (1) ionization processes, including laser desorption and photoionization, (2) a comparison of methods for activation and dissociation of gas-phase ions, and (3) MS/MS studies of gas-phase ion–molecule reactions. Several analytical applications of the combination of lasers and tandem mass spectrometers are also presented. Novel MS/MS instrumentation and laser/MS interfaces (e.g., those employing optical fibers) are also described.

LASERS FOR THE PRODUCTION OF GAS-PHASE IONS

The use of lasers for both the desorption of nonvolatile and thermally labile molecules and for photoionization of gas-phase molecules has become well established. In this section the use of photoionization in the quadrupole ion trap mass spectrometer is described, but the production of gas-phase ions via laser desorption (LD) is emphasized. LD mass spectra, in comparison to those obtained by other desorption ionization (DI) methods, as well as mechanistic aspects of LD, are pre-

sented. In the subsequent section, the use of MS/MS for the activation and dissociation of ions produced by LD is emphasized.

Laser Desorption of Solids: A Method for the Analysis of Thermally Labile and Nonvolatile Molecules

In 1969, Beckey introduced a sampling method termed field desorption (FD)[16] that
produces abundant molecular ions from nonvolatile compounds and that set the
stage for the development of a number of other ionization methods. Collectively,
the main group of techniques for sampling thermally labile molecules is known as
desorption ionization (DI), which includes field desorption (FD),[17] ^{252}Cf plasma
desorption,[18] secondary ion mass spectrometry (SIMS),[19] fast atom bombardment
(FAB),[20] and laser desorption (LD).[21] Even though the initial forms in which energy
is transferred to the condensed phase during DI are widely different, they ultimately
produce similar mass spectra. The probe species are as follows: in plasma desorption, MeV energy nuclei (e.g., fission fragments of ^{252}Cf) impact either a solid or
liquid surface; in static SIMS, ($<10^{-9}$ A cm^{-2}) keV ions bombard a solid surface;
in liquid SIMS and FAB, ($>10^{-6}$ A cm^{-2}) keV ions or neutrals, respectively, bombard a vacuum-compatible liquid in which analyte is dissolved or suspended. In
laser desorption,[22] the subject of this section, a solid is irradiated with a laser beam
generating ions and neutrals from the surface. Prior to organic molecular applications, both SIMS and LD had been used in elemental analysis of solids.[23]

Both pulsed and continuous wave (CW) lasers have been used in LD experiments with mass spectrometers. Continuous wave lasers work well with scanning
instruments that separate ions in space because a relatively steady ion current is
generated, allowing time for the mass spectrum to be scanned. However, continuous wave lasers have the disadvantage of increased pyrolysis of the sample as compared to pulsed lasers.[24] Fast heating rates, characteristic of pulsed lasers with their
higher instantaneous power, minimize the decomposition of molecules and reduce
sample consumption.[25] Pulsed lasers frequently produce secondary particles (ions
and neutrals) long after the laser pulse. The production of ions can continue for
microseconds and neutral emission for even longer periods (several hundreds of
microseconds).[26] Since scanning instruments cannot obtain a complete mass spectrum on this time scale (for a single laser pulse), cumulative or time-based separation mass spectrometers are more appropriate. Such devices include the Fourier
transform ion cyclotron resonance (FT-ICR) mass spectrometer,[27] the quadrupole
ion trap,[28] and time-of-flight instruments.[29,30]

Ion Formation Processes in Laser Desorption (LD)

The mechanisms for the formation of gas-phase ions from condensed-phase samples using LD (and the other DI techniques) are not easily deduced and are continuously being refined.[31,32] A better understanding of the desorption ionization mechanism will allow more control in the acquisition of the desired information. Several
important experimental factors that affect laser desorption mass spectra are the
laser power per unit area, the photon wavelength, the electronic absorption spectrum of sample molecules, the orientation of the incoming laser beam to sample,
the sample thickness, and the substrate composition.

A complete description of ion emission[31,32] must account for the following experimental data reported in the literature. (1) Higher kinetic energy (KE) particles (tens of eV) are emitted immediately after a laser pulse in contrast to later times (several μsec). The ions emitted in the early stages are usually atomic or smaller molecular fragments. They appear to represent nonthermal components of desorption because the KEs correspond to unrealistically high temperatures. (2) Pseudomolecular ions [e.g., $(M + H)^+$ and $(M + K)^+$] have relatively low kinetic energies. (3) Many more neutral particles are emitted than ions, and the neutrals are emitted for longer times.

The generation of gas-phase ions from organic salts provides the simplest description for the formation of ions in DI. For laser desorption, photon energy can excite a localized area of the sample. If the sample has an absorption band of an appropriate energy, electronic excitation can occur. For thin layers of nonabsorbing organics, the substrate absorbs most of the energy, transferring it back to the sample as thermal energy. In either case, desorption occurs if the translational energy made available is sufficient to overcome lattice binding energies. Cation–anion (C^+A^-) bond breaking is the key step that, when accomplished, results in C^+ and A^- being liberated as charged particles, making this a very efficient ion formation process. Neutral C^+A^- species are also emitted as are ionic clusters resulting from nonsymmetrical combinations of anions and cations.[33]

The positive ion (LD) mass spectra of KNO_3 and tetrabutylammonium bromide (10^7 W/cm^2; $\lambda = 532$ nm, $t = 10$ nsec), shown in Fig. 10.1a and b, respectively, are typical of organic and inorganic salts. In Fig. 10.1a, K^+ is the most abundant ion and the cluster ion $K[K(NO_3)]^+$ is present. In Fig. 10.1b, the intense ion at m/z 242 corresponds to the intact tetrabutylammonium cation (C^+).[13] Ions of lower relative abundance observed at lower masses correspond to loss of alkene molecules derived from alkyl side chains of the intact cation. The ion currents from both samples are high, which is typical of laser desorption of inorganic and organic salts.[34]

Thermal emission can also be used to describe LD of tetraalkylammonium salts using IR lasers ($\lambda = 10.6$ μm).[26] Substrate thermal conductivity and sample electronic absorption bands also greatly affect the efficiency of the desorption process.[35,36] For nonabsorbing samples on metal substrates, which have high thermal conductivities less laser power is required to form ions than for nonabsorbing samples on substrates which have poor heat transfer properties.[35]

Increasing the total energy available to the system can also greatly affect mass spectra. Table 10.1 lists the relative abundances (RA) of the major ions in the LD mass spectra of tetraphenylphosphonium bromide on silver foil and shows the result of increased energy transfer to the molecules with increasing laser power. The ratio of the unimolecular fragment ions at m/z 262$^+$ and 183$^+$ to the intact cation (m/z 339) increases with increasing laser energy.[13] The increase in the extent of fragmentation can result either from greater internal energy deposited during desorption or "ladder switching" in which the neutral or ionic fragments can absorb further photons and be ionized and/or fragmented.[37] Analogous excitation events occur under multiple collision conditions in MS/MS experiments in which gaseous collisions are used to energize and fragment mass selected ions.[38]

As previously described, ionization of salts is a direct consequence of the

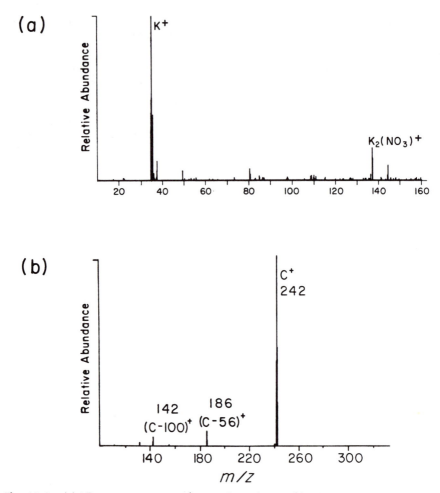

Fig. 10.1. (a) LD mass spectrum of potassium nitrate; (b) LD mass spectrum of tetra-butylammonium bromide (for both, power density = 10^7 W/cm^2, t = 10 nsec, λ = 532 nm).

desorption process during keV bombardment and laser desorption. Ionization processes are greatly affected by the polarity of molecules for nonionic samples. The pseudomolecular ions $(M + H)^+$ and $(M - H)^-$ are often formed if highly polar functional groups such as $-COOH$ and $-NH_2$ are present.[39] Improved ion currents are usually observed during laser desorption when polar samples are admixed with alkali salts to form adduct ions. An example of this is shown in Fig. 10.2 in which an alkali salt (KNO_3) is admixed with the nucleoside guanosine and irradiated with a Nd:YAG (λ = 1060 nm, 10 nsec) laser.[11] An abundant (guanosine + K)$^+$ ion is observed at m/z 322. Very weak signals with little molecular weight information are evident in the LD mass spectrum of a pure sample of guanosine. Cationization and anionization are thought to be the result of ion–molecule reactions in the dense gas phase (selvedge) above the surface.[40] This mechanism is analogous to that suggested[41] to apply in molecular SIMS and FAB.

Table 10.1. Positive ion LD mass spectra of tetraphenylphosphonium bromide at different laser energies[a]

		Laser energy (mJ/pulse)		
m/z	Ion	5	20	30
339	C^+	100	100	100
262	$(C - C_6H_5)^+$	8	10	12
183	$(C - C_{12}H_{12})^+$	14	21	27

[a]Laser wavelength was 532 nm.

The cationization products $(C + M)^+$ formed during LD are closed-shell species, in contrast to the radical molecular cations $(M^{\ddot{+}})$ formed during photoionization. The two types of ions also fragment differently; radical ions frequently lose radical fragments and closed-shell ions lose even-electron fragments, both resulting in the formation of even-electron ions.

This is exemplified in Fig. 10.3, which shows 1,6-naphthalene dicarboxylic acid. Figure 10.3a displays ions resulting from postionization by electron impact of laser-desorbed neutrals to form $(M^{\ddot{+}})$.[42] $M^{\ddot{+}}$ loses the radical fragments ·OH and ·COOH to yield ions at m/z 199 and 171, respectively. Contrast this to the fragmentation

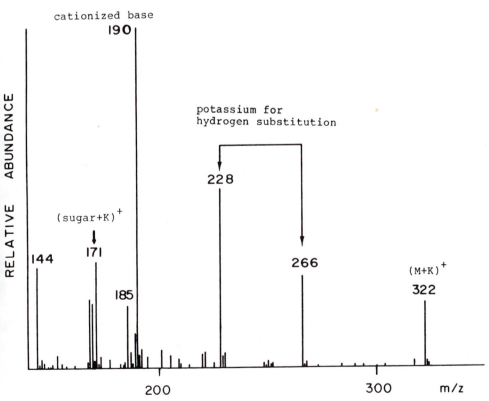

Fig. 10.2. LD mass spectrum of guanosine admixed with potassium nitrate, power density = 10^7 W/cm^2, t = 10 nsec, λ = 1064 nm.

of $(M + H)^+$ formed by post-ionizing (using chemical ionization) laser-desorbed neutrals.[42] H_2O and CO_2 molecules are eliminated to produce ions at m/z 199 and 173 (Fig. 10.3b).

In contrast to polar molecules, nonpolar aromatic compounds such as polynuclear aromatic hydrocarbons (PAH) form less abundant pseudomolecular ions [such as $(M + H)^+$] during LD. The most abundant ions formed from such compounds during LD (as in SIMS) are the radical cations M^+, and they probably result from molecular interactions with sufficiently energetic electrons, photons, and secondary particles produced during irradiation. In laser desorption experiments, photon fluxes are high ($\sim 10^{19}$ photons $cm^{-2} sec^{-1}$), and photoionization becomes sig-

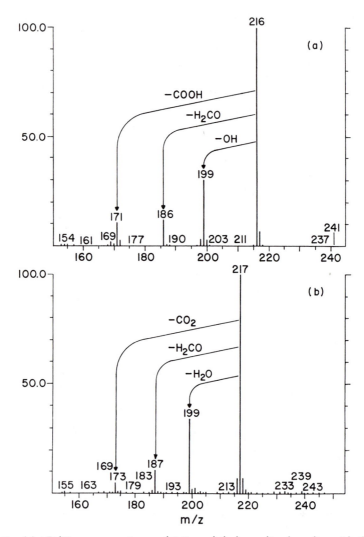

Fig. 10.3. (a) LD/EI mass spectrum of 1,6-naphthalene dicarboxylic acid; (b) LD/CI mass spectrum of 1,6-napthalene dicarboxylic acid (both spectra, power density = 10^7 W/cm², t = 10 nsec, λ = 1064 nm).

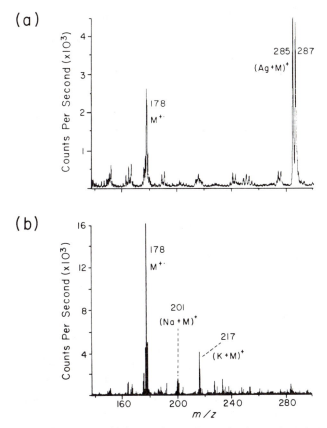

Fig. 10.4. (a) SIMS spectrum of phenanthrene burnished on silver foil, 5 keV Ar⁺, 10^{-10} A/cm²; (b) LD mass spectrum of the same sample using the same instrument, power density = 10^7 W/cm², t = 10 nsec, λ = 532 nm.

nificant. A comparison of the SIMS and LD mass spectra of a PAH (phenanthrene) using the same mass spectrometer and the same PAH sample is shown in Fig. 10.4.[13] Both molecular ions (M⁺) and cationized molecules [(C + M)⁺, where C = cation] are evident in these spectra. The data here indicate that SIMS is apparently the more energetic process, as formation of (Ag⁺ + M) occurs readily. No Ag⁺ is formed during LD at laser power densities used and correspondingly no cationized molecules are formed. If the metal is present in a highly dispersed form, such that photon energy cannot be quickly dissipated to the metal substrate, silver cationization of the organic molecule is seen.[4] The secondary ion currents are low for both DI experiments on this nonpolar sample.

Optimization of the Sample Conditions for Ion Emission during Laser Desorption

The presence of matrix can cause dramatic effects in DI mass spectra.[43] For example, the LD mass spectrum of the crude extract from the cactus plant *Trichocereus pascana* shows greatly improved molecular ion abundance, ion yield, and ion emis-

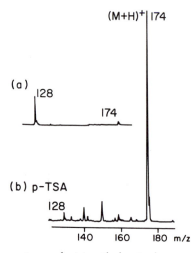

Fig. 10.5. (a) LD mass spectrum of a trimethylaminohexanoate zwitterion; (b) reverse derivatization of the same sample admixed with p-toluenesulfonic acid, power density = 10^7 W/cm², t = 10 nsec, λ = 1064 nm.

sion time when the sample is diluted with excess NH_4Cl (10:1).[7] It is not likely that the matrix contributes directly to the ionization of candicine (an ionic compound present in the mixture). Apparently, the matrix facilitates desorption of the analyte and yields ions that have less internal energy and thus exhibit less fragmentation. Recently, other studies have been reported in which a matrix that resonantly absorbs UV photons (tryptophan) during LD was used to enhance energy transfer to a nonabsorbing analyte.[36] This results in a reduction in the extent of molecular ion fragmentation and in the threshold laser power needed to produce molecular ions. Matrix dilution of an analyte yields similarly favorable results when the sample is bombarded by a keV argon ion beam.[43]

Matrix manipulation and sample preparation are aspects of DI experiments that are receiving growing attention.[44] Indeed, this is the basis for the success of fast atom bombardment (FAB).[19] The vacuum-compatible liquid matrices used during FAB have contributed to the increased success of many types of biomolecular analyses. A common FAB matrix, glycerol, has also been used in LD of sugars in this laboratory (low-power continuous-wave CO_2).[10] The use of the liquid matrix in these experiments resulted in three significant improvements in continuous-wave LD: (1) increased time of ion emission, (2) better precision of measurement of relative abundances of ions, and (3) higher ion yield.

Other matrix manipulation methods use chemically reactive matrices. Significant improvement in the quality of LD mass spectra of zwitterions is obtained by addition of a proton-donating matrix such as p-toluenesulfonic acid (p-TSA). This procedure is sometimes termed reverse derivatization[45] because it reverses the previous trends in sample modification for mass spectrometry, which attempted to facilitate evaporation by making molecules less polar.[46] The reverse derivatization procedure is illustrated for LD in Fig. 10.5a in which the mass spectrum of a neat sample of a zwitterion, trimethylaminohexanoate, displays an $(M + H)^+$ ion at m/z 174 that is weak relative to the fragment ion at m/z 128. Addition of the proton donor *p*-TSA to this sample (Fig. 10.5b) results in a greatly improved mass spectrum.[5] Note the striking increase in the absolute yield and relative abundance of the protonated molecule at m/z 174. Production of charged centers in amines by chemical reactions such as quaternization has been reported[47] and results in lower detection limits.

Photoionization of Gaseous Molecules

Lasers can also be used to produce ions from gaseous organic molecules by photoionization. Because a large percentage of the species desorbed during DI are neutral atoms and molecules, photoionization can greatly increase the yields of gas-phase ions when used in conjunction with desorption experiments. Multiphoton ionization of gas-phase molecules produced by both laser desorption[48–52] and ion bombardment[53,54] has become an active area of research. One great advantage of single and double photon photoionization over other methods, such as EI, is that energy deposition can be controlled more precisely,[51] whereas multiphoton ionization has the advantage of extremely high ionization efficiency (unit efficiency in the volume intercepted by the laser beam).[55]

Since most stable organic neutral molecules have closed-shell electron configurations, removing one electron yields a radical ion ($M^{+\cdot}$, where M represents the molecule). As an example, the photoionization ($\lambda = 532$ nm, power density $<10^8$ W/cm^2) mass spectrum of trans-1,3-pentadiene, shown in Fig. 10.6a, was obtained

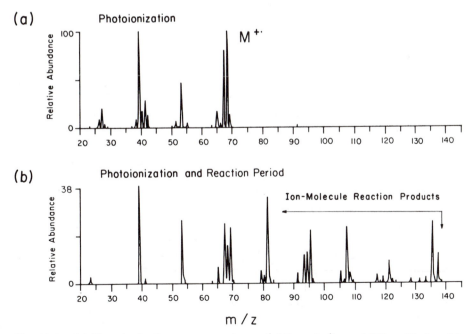

Fig. 10.6. (a) Photoionization mass spectrum of 1,5-pentadiene, MW $= 68$, $t = 10$ nsec, $\lambda = 532$ nm, power density $\leq 10^8$ W/cm^2; (b) ion–molecule reaction product mass spectrum (100 msec) of $M^{+\cdot}$ from 1,5-pentadiene with its neutral precursor.

with a quadrupole ion trap mass spectrometer that was modified with an optical fiber interface (see Fig. 10.7). This type of instrument is particularly well suited to the low cross-section process of photoionization. Since ions can be accumulated in the trap, the molecular ion (M^+) is evident at m/z 68 and fragment ions are displayed with lower m/z values. If the total energy of the photons absorbed is greater than the ionization energy (IE) of the molecule, the excess energy will be partitioned between the kinetic energy of the ejected electron and excess internal energy in the molecular ion. Excess internal energy present in the molecular ions can induce unimolecular fragmentation; typically radical fragments are lost from M^+ resulting in stable even-electron ions and these features are evident in the mass spectrum shown in Fig. 10.6a in which the major fragments are due to the loss of $H^·$, $CH_3^·$ and $C_2H_5^·$. The ion trap allows the ions to be contained for a second or more allowing time for extensive ion–molecule reactions to occur, if an appropriate neutral reagent is present. For example, trapping M^+ from 1,3-pentadiene for 100 msec in the presence of its neutral precursor yields a mass spectrum that shows numerous ion–molecule reaction products (Fig. 10.6b).[56] This particular case is of interest because the self-chemical ionization reactions allow structural differentiation of C_5H_8 isomers that are otherwise difficult to distinguish by mass spectrometry.

TANDEM MASS SPECTROMETRY

In the previous sections, production of gas-phase ions using lasers was described. It is desirable to gain structural information on these ions, and in this section, excitation and dissociation of ions is discussed. In particular, collision-induced dissociation and photodissociation are presented and contrasted.

The concept and benefits of tandem mass spectrometry (MS/MS) have been reviewed[57] and will be only briefly discussed here. Laser ionization capabilities are enhanced by MS/MS as already shown for isomer distinction in the quadrupole ion trap (Fig. 10.6). Figure 10.8 illustrates a general MS/MS ion separation system. The separate operations occur in different regions of sector and quadrupole instruments, whereas they are separated in time in ion trap instruments. The sample is introduced and ions are generated in the source (S). An ion of interest is selected using the first stage of analysis. The selected ion is excited and fragmented by collision-induced dissociation (CID) in the second reaction stage. Alternatively, it can be excited by photoabsorption,[3] by collisions with a surface,[58–61] or transformed to higher mass products by ion–molecule reactions.[62] The product (daughter) ions are analyzed using the second stage of mass analysis.

One of the most valuable analytical applications of tandem mass spectrometry is mixture analysis.[63] Ionization of a multicomponent sample produces molecular and fragment ions from each component in the mixture. Identification of a particular constituent from the ions observed is sometimes difficult when only the mass spectrum is available. This is particularly true when abundant signals also occur due to matrix as typically occur in FAB and liquid SIMS experiments.[64,65] Using MS/MS, the first mass analyzer eliminates or minimizes these chemical interfer-

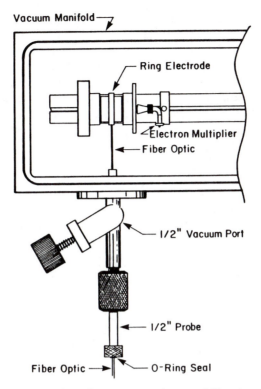

Fig. 10.7. Quadrupole ion trap with optical fiber interface.

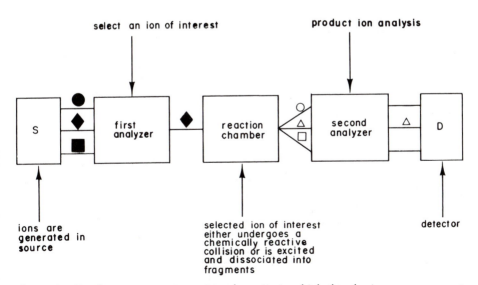

Fig. 10.8. Tandem mass spectrometry schematic in which the abscissa can represent either space or time.

ences. The daughter ions produced from the parent ions are then used to identify the individual component(s) of interest.

Ion structural analysis is a second area in which tandem mass spectrometry is useful. Losses of fragments from an excited parent ion occur and are characteristic of the ion structure. For example, in metabolic studies, O-glucuronide conjugates invariably lose a neutral dehydroxyglucuronic acid moiety, corresponding to loss of 176 Da from the parent ion, whereas sulfate ester conjugates lose 80 Da (SO_3).[45,66-68] These characteristic fragmentations allow for the facile characterization of these classes of conjugates[69,70] in a complex biological sample.

MS/MS Instruments that Separate Ions in Space

A reverse geometry instrument is composed of a magnetic sector followed by an electric sector.[71] For MS/MS experiments, a collision cell can be placed before the magnetic sector or between the two sectors. At keV energies, CID cross sections are high for electronic interactions between ions and target gas. Excited electronic states of the ion cross to the ground vibrationally excited electronic state from which dissociation occurs.[57]

This instrument has been used for several studies in which laser desorption is used for ionization and MS/MS for structural analysis of the gas-phase ions produced.[4,6,7,9] With this combination direct comparison of the ions produced by laser desorption and SIMS has also been performed.[4] The MS/MS spectrum of cationized sucrose ions formed during LD, for example, is very similar to the analogous SIMS spectrum. This suggests that gas-phase unimolecular dissociation occurs in SIMS. MS/MS of laser-desorbed ions has also been used to study Ag^+ affinities of alcohols[6] and for the identification of naturally occurring quaternary compounds.[7]

In contrast to the high-energy collisions described, a tandem quadrupole mass spectrometer that operates at low ion energy (<100 eV) is shown in Fig. 10.9. The triple quadrupole shown is modified for laser photoionization/desorption experiments using a removable fiber optic probe.[11] No modification to the EI/CI ion source is necessary to perform laser desorption experiments. The first quadrupole mass filter serves to select the parent ion. The second quadrupole serves as the collision chamber filled with an inert target gas. Product ions formed in the collision cell are passed through to the third quadrupole in which they are mass analyzed. In our laboratory, optical fibers have proven to be simple and rugged as laser–ion source interfaces. They have the further advantage of providing capabilities of "time sharing" of a single laser between several mass spectrometers.

MS/MS studies of laser-desorbed ions using the triple quadrupole mass spectrometer[8,10,11] have been valuable for the studies of fragmentation patterns of organic cations,[8] cationized sugar,[10] and peptides.[11] The ion chemistry described in these studies will be presented in more detail in a later section of this chapter.

MS/MS Instruments that Separate Ions in Time

The quadrupole ion trap is an instrument (Fig. 10.7) that can be used for mass analysis by operating in a high pass filter or mass instability mode.[28] Recently, MS/MS capabilities have been added[12] by providing the ability to activate particular

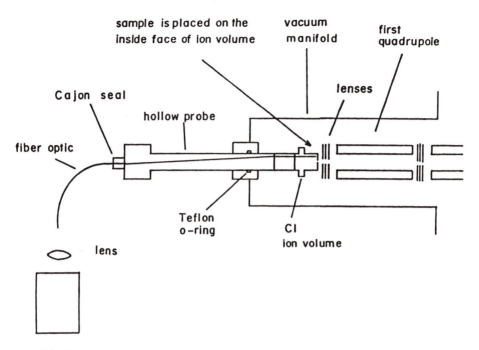

Fig. 10.9. Triple quadrupole mass spectrometer used for LD and photoionization experiments. A removable fiber-optic probe is used as the interface.

selected ions by irradiating them at their resonance frequencies, thereby causing translational excitation and facilitating collision-induced dissociation. The RF voltage applied to the ring electrode determines the minimum mass-to-charge ratios of ions that can be trapped. Daughter ions are produced from a selected parent ion either by photodissociation or by CID. Each parent ion is selected by raising the RF voltage until only $M^{\pm}$ (in some experiments together with its higher mass isotopes) remains in the trap and then lowered to collect daughter ions. For CID, a small (mV) "tickle" voltage is applied to the endcaps at a frequency that matches the fundamental frequency of the parent ion motion in the axial direction. This ion exclusively absorbs energy increasing its translational energy. Higher translational energies promote energy deposition via more energetic collisions with an inert target gas (helium). After excitation and daughter ion formation, the RF ring voltage is raised to eject all daughter ions in the course of performing mass analysis. In photodissociation, no tickle voltage is applied and photons rather than collisions are used to excite the ions. Both experiments can also be performed using a recently developed method of mass selection which combines RF and DC voltages applied to the trap to isolate the parent ion of interest with exclusion of all other ions.[72]

Photodissociation of Gas-Phase Ions

Irradiation of gas-phase ions can result in their excitation and dissociation into smaller ionic fragments in a process known as photodissociation (PD). An experi-

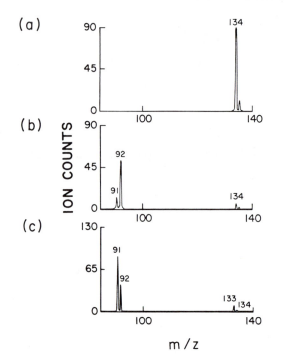

Fig. 10.10. (a) Mass spectrum; (b) CID daughter ion spectrum of M^+ (m/z 134) and (c) photodissociation daughter ion spectrum of n-butylbenzene in a quadrupole ion trap, power density = 10^8 W/cm^2, t = 10 nsec, λ = 532 nm.

mental setup we have used for photodissociation employs the unique optical fiber interface to the quadrupole ion trap (Fig. 10.7).[11] The initial population of ions can be created in a variety of ways (e.g., by EI, CI, or DI). After a parent ion of interest is selected in the ion trap, a pulse of laser irradiation excites these ions and they dissociate.

Photodissociation is a promising ion dissociation technique because the amount of excitation energy available to the ion system can be accurately controlled and varied by control of the photon energy and flux. Several experiments have encouraged the view that large energy depositions may be achieved via PD, particularly with excimer–laser UV photons that may prove useful for dissociating large biomolecules[73] that are otherwise difficult to fragment. The accurate control of energy deposition during photodissociation has already proven useful for isomer distinction.[74]

A photodissociation daughter spectrum of the n-butylbenzene molecular ion (M^+) in the ion trap is shown in Fig. 10.10. The activation energy for the formation of the fragment 91^+ is higher than that for 92^+ and the results follow expectations based on these values.[12] For comparison, excitation of M^+ has also been performed in the same ion trap by CID with an inert target gas (Fig. 10.10c). As is evident in these daughter ion spectra, PD can deposit more energy than CID into an ion in the quadrupole ion trap. It should be noted that the trap is very efficient in collecting daughter ions, but it is limited in the amount of internal energy that can be deposited during CID, thus providing impetus for further development of photodissociation.

ION CHEMISTRY

In addition to activation and dissociation of laser-desorbed gas-phase ions, the chemical reactivity of ions produced by laser desorption has also been studied with the instrumentation previously described. In this section three areas of ion chemical studies will be discussed.

Ion–Molecule Reactions

The different DI methods of volatilization of samples have different energy transfer mechanisms, yet the mass spectra are frequently similar. This suggests that common chemistry underlies the origin of many of the ions observed in the DI mass spectra. Compare the similarities between the SIMS and LD mass spectra of a sample consisting of the disaccharide sucrose and silver (Fig. 10.11a and b).[4] Both spectra display abundant cationized molecules. In the LD mass spectrum, the cationized product also forms an adduct with NH_3 generated by the thermal degradation of NH_4Cl added to the sample to increase the ion emission time. In addition to (Ag + M)$^+$, an abundant ion occurs in the mass spectra due to the cationized monosaccharide at m/z 185. This ion can arise by unimolecular dissociation, supported by CID data, of the cationized molecule. In addition, the daughter ion spectrum of (Ag + sucrose)$^+$ shows both loss of water and the intact sugar (Fig. 10.11c) to yield the ions at m/z 167 and 107, respectively.

Although cationization is evident in LD with samples consisting of polar molecules admixed with alkali halides, other kinds of ion–molecule reactions are often greatly reduced when compared to SIMS. A frequent occurrence in SIMS spectra of tetraalkylammonium salts is the addition of $(CH_2)_n$ units to molecular ions and loss of H_2 from the cation. For example, the SIMS spectrum of tetrabutylammonium bromide displays an intense intact cation [C$^+$] at m/z 242 as expected, as well as products due to $(CH_2)_n$ additions. However, the LD mass spectrum of the same sample examined in the same instrument shows only the intact cation.[13]

An additional example of this interfacial chemistry is evident in the halogen-for-hydrogen substitution reactions in quaternary ammonium salts, which occur to a lesser extent in LD compared to SIMS.[13] Ions resulting from cationization of PAHs on a silver foil are also much less abundant in LD than in SIMS. For most organic analyses, lower power densities during LD preclude formation of Ag$^+$, thus inhibiting formation of metal-cationization products. Note that some formally similar interfacial reactions in LD and SIMS may actually occur by different mechanisms. An example is provided by intermolecular transmethylation of zwitterionic compounds.[75] A review that compares the occurrence of various interfacial reactions in DI has recently appeared.[76]

The most abundant particles desorbed during LD and the other DI methods are neutrals that are typically much more abundant than ions.[77] More efficient use of these neutrals can be made by postionization. One method is to react the neutrals

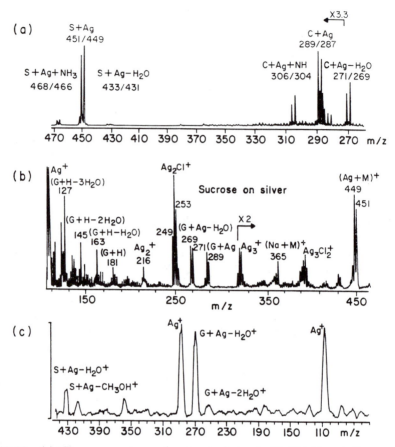

Fig. 10.11. (a) LD mass spectrum of sucrose admixed with NH_4Cl and silver powder, $\lambda = 1064$ nm, power density $= 10^8$ W/cm^2; (b) SIMS spectrum of sucrose and NH_4Cl supported on a silver foil, 5 keV Ar$^+$, 10^{-11} A/cm^2; (c) CID daughter ion spectrum of Ag$^+$ adduct with sucrose generated by laser desorption.

with a chemical reagent ion (LD/CI).[78] Irradiation of a pure sample of the tetrapeptide Gly-Pro-Gly-Gly yields an extremely weak ion signal. However, introduction of a CI plasma above the irradiated surface results in a much stronger signal with proton transfer from the reagent ion to a desorbed neutral yielding an abundant $(M + H)^+$ ion. Another method entrains laser-desorbed neutrals into a supersonic beam with subsequent resonant multiphotonionization of these neutrals.[79]

It is interesting to compare $(M + H)^+$ of the same peptide Gly-Pro-Gly-Gly using LD/CI and liquid SIMS. Collision-induced dissociation of $(M + H)^+$ formed during LD/CI yields a dominant ion at m/z 133 (Fig. 10.12a). However, CID of $(M + H)^+$ formed during liquid SIMS (essentially the same experiment as FAB) shows a much different daughter ion spectrum (Fig. 10.12b).[42] The fragments in both experiments are due to cleavages along the peptide backbone chain. The differences could be due to a different internal energy distribution of the parent ion or a different site of protonation on the peptide.

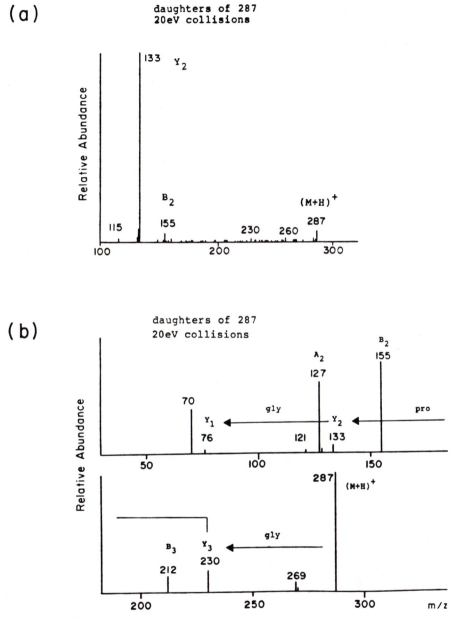

Fig. 10.12. (a) LD/CI daughter ion spectrum of (M + H)⁺ from the peptide Gly-Pro-Gly-Gly; (b) liquid SIMS daughter ion spectrum of (M + H)⁺ form the peptide Gly-Pro-Gly-Gly.

Bonding in Cationized Molecules

Tandem mass spectrometry can be used to probe the relative strength of interaction between a cation and neutral molecule. The charge/ionic radius ratio can be an important factor governing the strength of bond formation. Cationization of a neutral molecule in the ion source to yield $(C + M)^+$ can result in a loosely or tightly bound species. For example, collision-induced dissociation of lithiated dextrose generated using a continuous-wave CO_2 laser shows an extremely strong interaction between the two species.[10] The daughter ion spectrum shows that m/z 187, $(Li + M)^+$, loses masses 18 (water), and then successive losses of 30 (formaldehyde). These neutral losses from the even-electron parent ion are indicative of ring opening of the sugar. This result can be contrasted to the loosely bound sodium adduct of the dextrose molecule, $(Na + M)^+$, m/z 203. Collision-induced dissociation of this ion yields only the alkali cation at m/z 23 and no indication that sugar bonds are disrupted.[10]

The stronger binding of lithium compared with sodium to oxygen containing molecules agrees with other experimentally determined thermochemical data.[80] In FAB, glycerol is a commonly used matrix and has been reported to form covalent bonds with analyte ions.[81] Strong glycerol–adduct interactions are also evidenced by rearrangement of these ions generated during FAB.[82] A number of monoglycerated adducts of organic and inorganic molecules were found to eliminate small neutral fragments during daughter ion scans. The analytical utility of ion–substrate bond strengths is illustrated by studies in which relative alkali cation affinities allow distinction of a number of isomeric sugars to be made.[83]

Metal–Ligand Affinities Determined Using LD MS/MS

Metal ion–ligand interactions are important in the field of organometallic chemistry[84] and are of interest in catalysis. An important parameter of the strength of these interactions is metal–ligand affinities. Affinities can be determined in the gas phase in which interacting species are free from solvent effects and the intrinsic chemistry of the system can be studied.[85] The metal ions can react with gas phase organic ligands L_x introduced into the source to produce many different ions of interest. Information about reaction kinetics and bond strengths can be determined using FTMS and quadrupole ion traps by conventional equilibrium methods.[3,86] MS/MS can also provide information on reaction pathways and structures of metal–organic species.[87]

Metal bound dimers have been studied using LD MS/MS by selecting $(L_1ML_2)^+$ with a magnetic sector of a reversed geometry instrument and subjecting the parent ion to high-energy collisions with helium target gas.[6] Laser desorption is useful because high power densities can generate abundant metal ions from a foil. The major daughter ions produced are $[ML_1]^+$ and $[ML_2]^+$ (Table 10.2). For two structurally similar ligands, the daughter ion currents $[ML_1]^+$ and $[ML_2]^+$ depend only on relative activation energies (kinetic control) and reflect the metal ion affinities. Hence the method allows a rapid ordering of metal ion affinities based on a simple kinetic measurement.

The metals studied include silver cation binding to a series of straight-chain and

Table 10.2. Relative fragment ion abundances from silver-bound dimers[a,b]

Fragment	$(CH_3OH)Ag^+$	$(C_2H_5OH)Ag^+$	$(n\text{-}C_3H_7OH)Ag^+$	$(i\text{-}C_3H_7OH)Ag^+$	$(n\text{-}C_4H_9OH)Ag^+$	$(t\text{-}C_4H_9OH)Ag^+$
$(CH_3OH)Ag^+$	1	2.9	4.9	6.2	9.4	15.8
$(C_2H_5OH)Ag^+$	0.34	1	1.7	2.1	3.6	5.7
$(n\text{-}C_3H_7OH)Ag^+$	0.20	0.59	1	1.2	1.8	3.2
$(i\text{-}C_3H_7OH)Ag^+$	0.16	0.47	0.83	1	1.6	2.6
$(n\text{-}C_4H_9OH)Ag^+$	0.11	0.29	0.55	0.62	1	1.7
$(t\text{-}C_4H_9OH)Ag^+$	0.063	0.18	0.31	0.37	0.59	1

[a]Numerator in abundance ratio shown in row at top, denominator in column at left.
[b]In some cases ratios are an average of both indirect and direct measurements.

branched alcohols ranging from one to four carbons. Larger and more highly branched alcohols generally have the highest Ag^+ affinities compared to smaller alcohols or their straight-chain isomers (Table 10.2). The nature of the heteroatom of a ligand is also very important in metal ion affinity measurements as evidenced by the fact that loss of H_2O is five times more favorable than loss of isoelectronic NH_3 from $H_2OAgNH_3)^+$. Cu^+ and Na^+ centered dimers were also used to study metal cation affinities.[88] Comparisons of ion abundance ratios indicate that the basicities of alcohols for Ag^+ and Na^+ are very similar and are lower than those for Cu^+. This is expected since Cu^+ exhibits greater reactivity in the ion source.[88]

ANALYTICAL APPLICATIONS

Chemical Noise Reduction during Laser Desorption

One of the uses of tandem mass spectrometry is in complex mixture analysis. Both MS/MS and LD have been used together to identify quaternary alkaloids such as candicine in crude extracts from the cactus plant *Trichocereus pasacana*.[7] The LD mass spectrum of a plant extract (inset in Fig. 10.13) shows the mass range in the vicinity of the molecular ion. Information from the mass spectrum regarding the presence or absence of candicine (C^+ = 180) in this plant extract cannot be obtained because of equally abundant ions derived from other sources.

However, the CID daughter (MS/MS) spectrum of the parent ion at m/z 180 (Fig. 10.13) can clearly be used to identify the presence of candicine in the *T. pasacana* extract.[7] The determination of the presence of candicine in *T. spachianus* has also confirmed the reproducibility and applicability of LD/MS/MS. In addition, structurally characteristic fragment ions are produced. This method of combining laser desorption with MS/MS has also allowed new natural products, methylated analogs of candicine, to be discovered in other cactus species. Although these natural products are of limited intrinsic importance, these experiments are important in establishing the applicability of LD/MS/MS to the characterization of ionic organic compounds in biological samples.

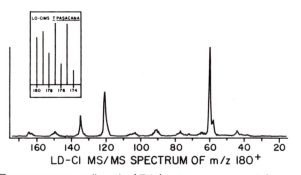

Fig. 10.13. LD mass spectrum (inset) of *Trichocereus pasacana* taken on a reverse sector instrument; The LD/CID daughter ion spectrum of m/z 180, power density = 10^8 W/cm², λ = 1064 nm, t = 10 nsec.

Depth Profiling

By interfacing a laser to a SIMS instrument via an optical fiber it is possible to perform depth profiling of multilayered samples. The laser is used to ablate surface material and static SIMS (10^{-10} A cm^{-2}, Ar$^+$ ion beam) to provide surface molecular information.[13] An advantage of the combination is that widely differing rates of profiling can be selected using the laser. High-flux ($>10^{-6}$ A cm^{-2}) ion beams have previously been used in atomic depth profiling. However, they appear to have limited use for molecular depth profiling because high primary ion currents rapidly degrade molecular species and induce diffusion of bulk molecules. Laser desorption in combination with FTMS has recently been used to gain molecular depth information on multilayered samples.[89]

Multilayered samples consisting of a substrate (Ag or graphite), approximately 100–200 monolayers of an electrosprayed organic salt (e.g., tetraethylammonium perchlorate, TEA), and 100–300 monolayers of a sputtered copper overlayer can be used to illustrate depth profiling capabilities. Even after 1 hr of continuous ion bombardment at the maximum primary ion flux with the argon gun (10^{-8} A cm^{-2}), little change in the mass spectrum was evident.

However, with only a few laser pulses ($\sim 10^6$ W cm^{-2}, $\lambda = 532$ nm) the metal overlayer is partially removed, revealing the underlying organic layer. It is interesting to note that the substrate (Ag) does not yield any ions. This is true even after a few thousand laser pulses. The effect of the laser dose is shown in Fig. 10.14. The SIMS mass spectra after 0, 7, and 400 laser pulses show the intact organic cation ($C^+ = m/z$ 130) with increasing abundance whereas copper ion signal diminishes. Much higher rates of profiling can easily be obtained by increasing the laser power or repetition rate.

FUTURE DIRECTIONS

Novel developments in instrumentation will probably be at the forefront of the future work in the areas previously discussed. For example, the relatively new ion excitation method of surface-induced dissociation[56–59] has proven fruitful, and may be combined with laser ionization in the future. Another rapidly developing area is that of quadrupole ion traps. These mass spectrometers are well suited to pulsed laser experiments because of their trapping capabilities and access to time-dependent phenomena.

Normally, ions have to be generated within the confines of the ion trap. However, recent data have been shown in which ions generated externally to the trap can be injected and the products formed can be isolated and dissociated by PD (Fig. 10.15).[14] Ion injection allows a wider variety of ionization experiments (such as laser desorption) to be performed with the trap. For example, ion injection of metal ions produced by laser desorption has been achieved and the result of ion injection of Au$^+$ produced by LD of a foil[14] is shown in Fig. 10.16. The metal ions were reacted with organic ligands (benzene) to produce (benzene$_2$·· Au)$^+$, (benzene·· Au)$^+$, and (benzene·· Au·· H$_2$O)$^+$. In addition, the ion trap allows MS^n experiments to be performed. Currently, this instrument has performed (MS)^7 experiments.[72]

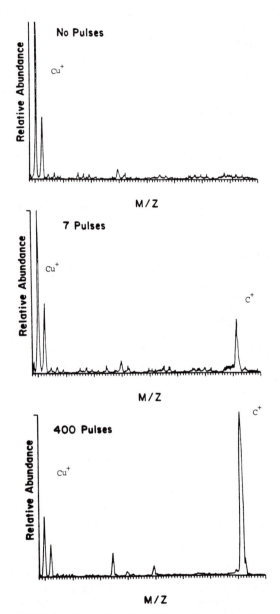

Fig. 10.14. Laser/SIMS depth profile of multilayered sample after 0, 7, and 400 laser shots; laser power density = 10^6 W/cm^2, λ = 532 nm, t = 10 nsec; SIMS, 5 keV Ar$^+$, 10^{-11} A/cm^2.

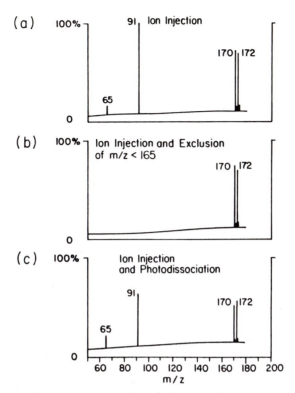

Fig. 10.15. (A) Ion injection into a quadrupole ion trap after electron ionization (EI) of 3-bromotoluene; (B) exclusion of all ions but M$^+$; (C) photodissociation of M$^+$.

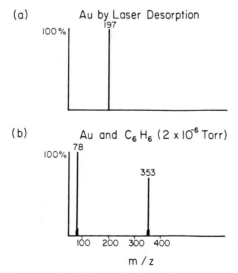

Fig. 10.16. (A) Ion injection of gold ions after laser desorption; (B) 100-msec reaction time period of gold metal ions with benzene.

These instrumental developments will allow a wider range of experiments to be performed in the future, including other laser/tandem mass spectrometry combinations.

ACKNOWLEDGMENTS

This work was supported by the National Science Foundation (CHE 87-21768). Thanks go to Jae Schwarz and Brian Winger for helpful comments in preparation of this manuscript.

REFERENCES

1. F. W. Aston, *Mass Spectra and Isotopes.* Green & Co., New York, 1933.
2. M.S.B. Munson and F. H. Field, *J. Am. Chem. Soc.* **85,** 2621 (1966).
3. M. Bowers (ed.), *Gas Phase Ion Chemistry.* Academic Press, London, 1979.
4. D. Zakett, A. E. Schoen, R. G. Cooks, and P. H. Hemberger, *J. Am. Chem. Soc.* **103,** 1295 (1981).
5. K. L. Busch, S. E. Unger, A. Vinze, R. G. Cooks, and T. Keough, *J. Am. Chem. Soc.* **104,** 1507 (1982).
6. S. A. McLuckey, A. E. Schoen, and R. G. Cooks, *J. Am. Chem. Soc.* **104,** 848 (1982).
7. D. V. Davis, R. G. Cooks, B. N. Meyer, and J. L. McLaughlin, *Anal. Chem.* **55,** 1302 (1983).
8. K. Busch, B.-H. Hsu, K. V. Wood, R. G. Cooks, C. G. Schwarz, and A. R. Katritzky, *J. Org. Chem.* **49,** 764 (1984).
9. J. L. Pierce, K. L. Busch, R. G. Cooks, and R. A. Walton, *Inorg. Chem.* **21,** 2597 (1982).
10. L. G. Wright, R. G. Cooks, and K. V. Wood, *Biomed. Mass Spectrom.* **12,** 159 (1985).
11. W. B. Emary, K. V. Wood, and R. G. Cooks, *Anal. Chem.* **59,** 1069 (1987).
12. J. N. Louris, J. Brodbelt, and R. G. Cooks, *Int. J. Mass Spectrom. Ion Process.* **75,** 345 (1987).
13. O. W. Hand, W. B. Emary, B. E. Winger, and R. G. Cooks, *Int. J. Mass Spectrom. Ion Process.* in press.
14. J. N. Louris, J. W. Amy, T. Y. Ridley, and R. G. Cooks, *Int. J. Mass Spectrom. Ion Process,* **88,** 97 (1988).
15. J. S. Brodbelt and R. G. Cooks, *Spectra (Finnegan MAT)* **11,** 30 (1988).
16. H. D. Beckey, *Int. J. Mass Spec. Ion Phys.* **2,** 500 (1969).
17. H. Beckey, *Principles of Field Ionization and Field Desorption Mass Spectrometry.* Permagon Press, New York, 1977.
18. R. MacFarlane, *Anal. Chem.* **55,** 1247A (1983).
19. A. Benninghoven, F. Rudenauer, and H. Werner, *Secondary Ion Mass Spectrometry.* John Wiley, New York, 1977.
20. R. Bordoli, G. Elliot, R. Sedgwick, and A. Tyler, *Anal. Chem.* **54,** 645a (1982).
21. M. A. Posthumus, P. G. Kistemaker, H.L.C. Meuzelaar, and M. C. Ten Noever de Brauw, *Anal. Chem.* **50,** 985.
22. D. M. Hercules, R. Day, K. Balasanmugam, and C. P. Li, *Anal. Chem.* **54,** 280A (1982).
23. R. J. Conzemius and J. M. Capellen, *Int. J. Mass Spectrom. Ion Phys.* 34, 197 (1980).
24. R. Stoll and F. W. Röllgen, *Org. Mass Spectrom.* **14,** 642 (1979).

25. R. J. Cotter, *Anal. Chem.* **56**, 485A (1984).

26. J.-C. Tabet and R. J. Cotter, *Int. J. Mass Spectrom. Ion Phys.* **54**, 151; (a) R. B. Van Breeman, M. Snow, and R. J. Cotter, *Int. J. Mass Spectrom. Ion Phys.* **49**, 35 (1983).

27. M. Comisarow and A. G. Marshall, *Chem. Phys. Lett.* **25**, 282 (1974).

28. G. Stafford, P. Kelley, J. Reynolds, and J.F.J. Todd, *Int. J. Mass Spectrom. Ion Process.* **60**, 85 (1984).

29. M. Yang and J. Reilly, *Anal. Instrum.* **16**, 133 (1987).

30. U. Boesl, J. Grotemeyer, K. Walter, and E. Schlag, *Anal. Instrum.* **16**, 151 (1987).

31. F. P. Novak, K. Balasanmugan, K. Viswanadham, C. D. Parker, Z. A. Wilk, D. Mattern, and D. M. Hercules, *Int. J. Mass Spectrom. Ion Phys.* **53**, 135 (1983).

32. F. Hillenkamp, M. Karas, and J. Rosmarinowsky, in *Desorption Mass Spectrometry,* Chap. 4, edited by P. A. Lyon. American Chemical Society: Washington, DC, 1985.

33. J. E. Campana, *Mass Spec. Rev.* **6**, 395 (1987).

34. D. A. McCrery, D. A. Peake, and M. L. Gross, *Anal. Chem.* **57**, 1181 (1985).

35. P. G. Kistemaker, G. Q. Vander Peyl, and J. Haverkamp, in *Soft Ionization Biological Mass Spectrometry,* edited by H. Morris. Heyden & Sons, London, 1980.

36. M. Karas, D. Bachmann, and F. Hillenkamp, *Anal. Chem.* **57**, 2935 (1985).

37. H. Neusser, *Int. J. Mass Spectrom. Ion Process.* **79**, 141 (1987).

38. V. H. Wysocki, H. I. Kenttämaa, and R. G. Cooks, *Int. J. Mass Spectrom. Ion Process.* **75**, 181 (1987).

39. C. D. Parker and D. M. Hercules, *Anal. Chem.* **58**, 25 (1986).

40. G.J.Q. Van Der Peyl, K. Isa, J. Haverkamp, and P. G. Kistemaker, *Org. Mass Spectrom.* **16**, 416 (1981).

41. S. J. Pachuta and R. G. Cooks, *Chem. Rev.* **87**, 647 (1987).

42. W. B. Emary, Ph.D. Thesis, Purdue University, 1988.

43. B.-H. Hsu, Y. Xie, K. Busch, and R. G. Cooks, *Int. J. Mass Spectrom. Ion Phys.* **51**, 225 (1983).

44. K. Busch and R. G. Cooks, *Science,* **218**, 247 (1982).

45. P. L. Jacobs, L. P. Delbressine, F. M. Kaspersen, and G. J. Schmeits, *Biomed. Environ. Mass Spect.* **14**, 689 (1987).

46. D. R. Knapp, *Handbook of Analytical Derivitization Reactions.* Wiley-Interscience, New York, 1979.

47. R. J. Colton, M. M. Ross, and D. A. Kidwell, *Nucl. Instr. Meth. Phys. Res.* **B13**, 259 (1986).

48. R. Tembreull and D. M. Lubman, *Anal. Chem.* **59**, 1003 (1987).

49. R. Tembreull and D. M. Lubman, *Anal. Chem.* **59**, 1082 (1987).

50. R. Tembreull and D. M. Lubman, *Anal. Chem.* **58**, 1299 (1986).

51. J. Grotemeyer, U. Boesl, K. Walter, and E. W. Schlag, *Org. Mass Spectrom.* **21**, 645 (1986).

52. F. Engelke, J. H. Hahn, W. Henke, and R. N. Fare, *Anal. Chem.* **59**, 909 (1987).

53. F. M. Kimock, J. P. Baxter, D. I. Poppas, P. H. Kobrin, and N. Winograd, *Anal. Chem.* **56**, 2782 (1984).

54. D. L. Donohue, W. H. Christie, D. E. Goeringer, and H. S. McKown, *Anal. Chem.* **57**, 1193 (1985).

55. S. H. Lin, Y. Fujimura, H. J. Neusser, and E. W. Schlag, *Multiphoton Spectroscopy of Molecules.* Academic Press, Orlando, 1984.

56. C. Kascheres and R. G. Cooks, *Anal. Chim. Acta,* **215**, 223 (1988).

57. F. W. McLafferty (ed.), *Tandem Mass Spectrometry.* Wiley-Interscience, New York, 1983.

58. M. E. Bier, J. Amy, R. G. Cooks, J. Syka, P. Ceja, and G. Stafford, *Int. J. Mass Spectrom. Ion Process.* **77**, 31 (1987).

59. K. L. Schey, R. G. Cooks, R. Grix, and H. Wollnik, *Int. J. Spectrom. Ion Process.* **77**, 49 (1987).

60. Md.A. Mabud, M. DeKrey, and R. G. Cooks, *Int. J. Mass Spectrom. Ion Process.* **67**, 285 (1985).

61. M. DeKrey, H. I. Kenttämaa, V. H. Wysocki, and R. G. Cooks, *Org. Mass Spectrom.* **21**, 193 (1986).

62. R. Pachuta, H. I. Kenttämaa, R. G. Cooks, T. Zennie, C. Ping, C.-J. Chang, and J. Cassady, *Org. Mass Spectrom.* **23**, 10 (1988).

63. R. Kondrat and R. G. Cooks, *Anal. Chem.* **50**, 81A (1978).

64. M. Barber, R. S. Bordoli, and R. D. Sedgewick, in *Soft Ionization Biological Mass Spectrometry,* edited by H. Morris. Heyden & Sons, London, 1980.

65. C. Fenselau and R. J. Cotter, *Chem. Rev.* **87**, 501 (1987).

66. R. H. Bieri and J. Greaves, *Biomed. Environ. Mass Spectrom.* **14**, 555 (1987).

67. I. Jardine, G. Scanlan, V. Mattox, and R. Kuma, *Biomed. Mass Spectrom.* **11**, 4 (1984).

68. S. Gaskell, B. Brownsey, P. Brooks, and B. Green, *Biomed. Mass Spectrom.* **10**, 215 (1983).

69. C. Fensalau, L. Yelle, M. Stogniew, D. Liberato, J. Lehman, P. Feng, and M. Colvin, *Int. J. Mass Spectrom. Ion Phys.* **46**, 411 (1983).

70. K. Straub, in *Mass Spectrometry in Biomedical Research,* edited by S. J. Gaskell. John Wiley, New York, 1986, p. 115.

71. J. Beynon, R. G. Cooks, J. Amy, W. E. Baitinger, and T. Ridley, *Anal. Chem.* **45**, 1023A (1973).

72. R. J. Strife, P. E. Kelley and M. Weber-Graubau, *Rapid Comm. Mass Spectrom.* **2**, 105 (1988).

73. W. D. Bowers, S. Delbert, R. Hunter, and R. McIver, Jr., *J. Am. Chem. Soc.* **106**, 1288 (1984).

74. T. Morgan, E. Kingston, F. Harris, and J. Beynon, *Org. Mass Spectrom.* **17**, 594 (1982).

75. O. W. Hand, B.-H. Hsu, and R. G. Cooks, *Org. Mass Spectrom.* **23**, 16 (1988).

76. L. D. Detter, O. W. Hand, R. G. Cooks, and R. A. Walton, *Mass Spec. Rev.,* **7**, 465 (1988).

77. J. Campana and R. Freas, *J. Chem. Soc., Chem. Commun.* 1414 (1984).

78. R. J. Cotter, *Anal. Chem.* **52**, 1767 (1980).

79. D. C. Lubman, *Anal. Chem.* **59**, 31A (1987).

80. P. Kebarle, *Rev. Phys. Chem.* **28**, 445 (1972).

81. G. Glish, P. Todd, K. L. Busch, and R. G. Cooks, *Int. J. Mass Spectrom. Ion Phys.* **56**, 177 (1984).

82. W. B. Emary, T. Toren, and R. G. Cooks, *Anal. Chem.* **58**, 1218 (1986).

83. G. Puzo, J. J. Fournie, and J. C. Prome, *Anal. Chem.* **57**, 892 (1985).

84. D. B. Jacobson and B. S. Freiser, *Organometallics* **3**, 513 (1984).

85. J. L. Beauchamp, *Annu. Rev. Phys. Chem.* **22**, 527 (1971).

86. J. S. Brodbelt-Lustig and R. G. Cooks, *Talanta,* **36**, 255 (1989).

87. B. S. Freiser, *Talanta* **32**, 697 (1985).

88. S. A. McLuckey, Ph.D. Thesis, Purdue University, 1982.

89. D. Land, T. Tai, J. Lindquist, J. Hemminger, and R. McIver, *Anal. Chem.* **59**, 2927 (1987).

11

Combining Lasers with Fourier Transform Ion Cyclotron Resonance Mass Spectrometry: Applications to the Study of Gas-Phase Metal Ion Chemistry

ROBERT R. WELLER, TIMOTHY J. MacMAHON,
and BEN S. FREISER

With the development of Fourier transform ion cyclotron resonance mass spectrometry (FT-ICR or FTMS) and the greater diversity and availability of lasers, a broad frontier of chemistry has opened. These two instruments are ideally suited for coupling. The FT-ICR with its high sensitivity and the ability to acquire a complete, high-resolution mass spectrum from a single laser pulse makes it a method of choice for laser desorption (LD) mass spectrometry. Also, the ability of the FT-ICR to store ions for long periods of time, along with the high flux densities available in continuous-wave (CW) and pulsed lasers, makes it ideal for the study of single and multiphoton/ion interactions.

A large volume of work has been performed by researchers in both laser desorption and photodissociation/photoionization studies, much of which is exemplified by other articles in this book. This review will focus on three main thrust areas in our research group: (1) multiphoton ionization of organic and organometallic compounds, (2) laser desorption of metal ions and the use of FT-ICR for the study of the gas-phase ion/molecule reactions they undergo, and (3) photodissociation studies of various organometallic ion species.

EXPERIMENTAL

The fundamental principles, instrumentation, and applications of ICR and FT-ICR have been discussed in detail in a number of excellent recent reports.[1-9] The experiments described herein have been carried out on two different FT-ICR instruments in our laboratory. The prototype Nicolet FTMS-1000 is a diffusion-pumped system utilizing a 15-in. electromagnet that is usually operated at 0.9 T. The 2-in. cubic cell has mesh transmit plates to allow light into the ion trap region (see Fig.

11.1). Also, one transmit plate has a variety of metal targets mounted on it for laser desorption studies. Windows allow introduction of a Quanta-Ray Nd:YAG laser beam (at the fundamental 1064 nm or frequency multiplied $\times 2$, $\times 3$, or $\times 4$ at 532, 356, and 266 nm, respectively) and the output from a 2.5-kW Hg-Xe arc lamp for laser desorption and photodissociation studies, respectively. The arc lamp is used either in conjunction with a Schoeffel 0.25-m monochrometer operating at 10-nm resolution or, alternatively, with cut-off filters to control the wavelength of light impinging upon the ions.

The second FT-ICR system is a Nicolet FTMS-2000 differentially pumped (diffusion pumps) dual-cell instrument. The dual-cell configuration permits ions to be generated in the high-pressure "source" cell and then transferred to the low-pressure "analyzer" cell for detection. The system utilizes a 3-T superconducting solenoid that, compared to the 0.9-T system, offers a variety of advantages including higher resolution, higher mass range, higher collision energies, and more efficient trapping. The system has also recently been equipped with a prototype external ion source that provides the versatility of generating ions either at the cell or well removed from the cell (Fig. 11.2).

The inlet system on the FTMS-1000 consists of three Varian precision leak valves and two General Valve Series 9 pulsed valves. The FTMS-2000 is equipped with two leak valves and two pulsed valves on both the analyzer and source sides. The pulsed valves allow the introduction of reagents at specified times during the pulse sequence, creating short bursts of high pressure ($\sim 10^{-5}$–10^{-4} torr) that allow reactions to take place.[10] Ions of interest can then be isolated by ion ejection tech-

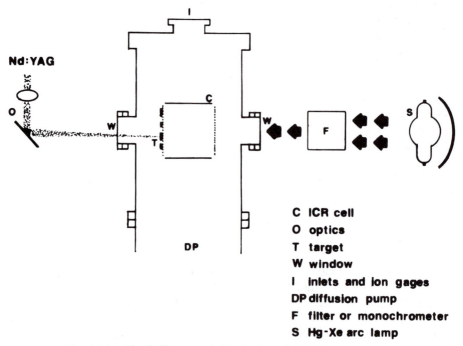

C   ICR cell
O   optics
T   target
W   window
I   inlets and ion gages
DP diffusion pump
F   filter or monochrometer
S   Hg-Xe arc lamp

Fig. 11.1. Block diagram of the single-cell FT-ICR instrument.

FT–ICR–MS with External Ion Source

A analyzer cell
S source cell
L lens support
T target
K airlock
V gate valve
R solids probe

W window
D dewar
G ion extraction grid
M superconducting magnet
I inlets and ion gages
DP diffusion pump
O optics

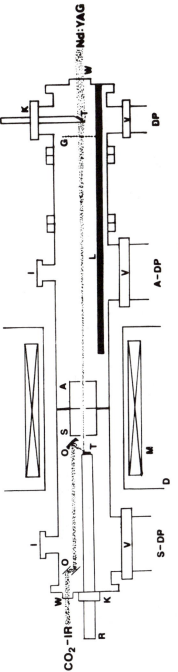

Fig. 11.2.   Block diagram of the dual-cell FT-ICR instrument.

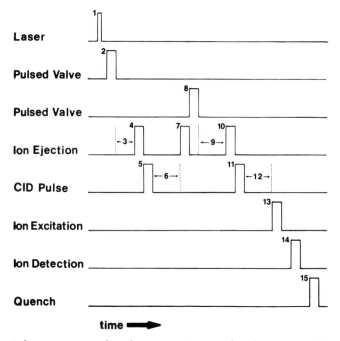

Fig. 11.3. Pulse sequence for the generation and subsequent CID studies of $NbFeC_6H_6^+$ (see text for details).

niques and subjected to a variety of other experimental steps including (1) high-resolution detection, (2) further reactions with other reagents, (3) irradiation from various light sources for photodissociation, or (4) collision-induced dissociation.

The pulse sequences used for the experiments are as varied as the different reactions studied. Indeed, it is one of the outstanding features of FT-ICR that completely different multistep (MS^n) experiments may be conducted with just the changing of a few computer-controlled parameters.[4] These are as follows: (1) Ion excitation—by applying at any selected time in an experimental sequence either a swept or single-frequency rf signal of variable amplitude and duration to the transmit plates, a range of ions or a particular mass ion may be excited for detection, ejected from the cell, or excited to higher energies prior to undergoing collision-induced dissociation (CID)[11] or collision-induced reaction.[12] (2) Delays—variable delays in a sequence allow time for ion–molecule reactions, photodissociation, or CID to take place. (3) Pulsed valves—as stated earlier, these allow the introduction of reagent or collision gases at any point in the pulse sequence. (4) Addressable pulsed devices—pulsed lasers, shutters, sources, etc. can be suitably timed to accomplish the desired result. Figure 11.3 shows a typical pulse sequence used in a specific case to study the CID products of $NbFeC_6H_6^+$ (Fig. 11.4) on the FTMS-1000 and is described in detail in the section on heteronuclear clusters. Pulse sequences for the FTMS-2000 are similar, but clearly more complex, involving source side pulses, ion transfer times, and analyzer side pulses. This configuration offers tremendous flexibility in the types of experiments that can be performed.

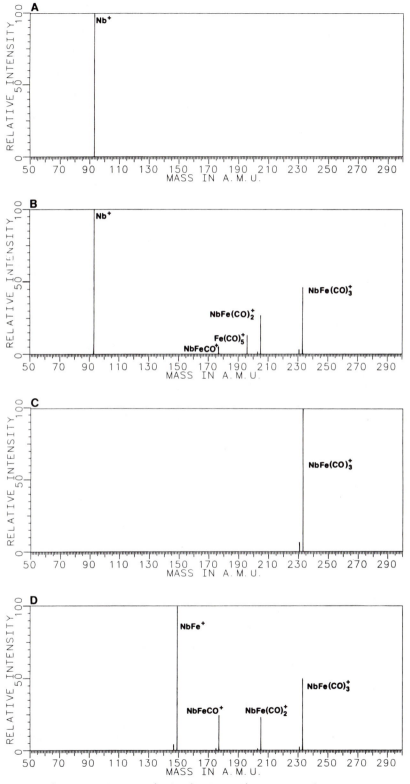

Fig. 11.4. Spectra obtained using MS⁴ sequence of Fig. 11.3.

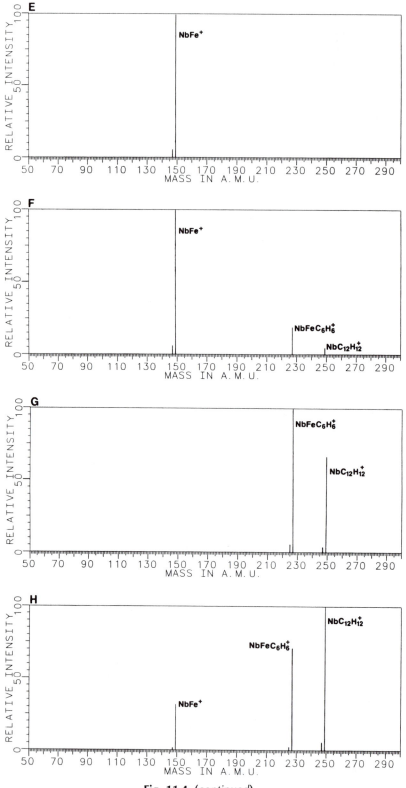

Fig. 11.4. (*continued*)

APPLICATIONS

Multiphoton Ionization

The feasibility and the promise of combining multiphoton ionization (MPI) with FT-ICR have been demonstrated in several laboratories.[13-16] Preliminary studies in our laboratory[14] were performed on the FTMS-1000 using the quadrupled beam of the Nd:YAG laser (266 nm, 4.7 eV). At this wavelength many aromatic and some organometallic compounds absorb and, therefore, can be ionized in sufficient quantity to produce a signal in the ICR cell. The molar absorptivity threshold for a molecule to yield a detectable number of ions was found to be $\sim 10^4$ L/mol-cm (or a cross section of ~ 0.17 Å^2) at pressures below 10^{-8} torr. Holes in the cell plates allowed the unfocused laser beam to pass through the cell without striking a metal surface, thus eliminating any interference from laser-desorbed metal ions. The laser power used was adjusted qualitatively to achieve the largest signal, typically 1–4 mJ per 4- to 7-nsec pulse. When $Cr(CO)_6$ was leaked into the cell and photoionized, only one photofragment, Cr^+, was observed, which is characteristic of MPI and demonstrated that no other ionization process was occurring.

The differences in molar absorptivity of two compounds frequently allow MPI to be used as a method for selective ionization in mixtures in which EI would ionize both compounds. A simple example is shown in Fig. 11.5 for a mixture of isopropyl sulfide, $[(CH_3)_2CH]_2S$, at 10^{-6} torr and perdeuterodiphenyl, $(C_6D_5)_2$, at a pressure of 2×10^{-7} torr. As can be seen, the EI spectrum shows peaks from both compounds, whereas using MPI, only the perdeuterodiphenyl is ionized in spite of its lower concentration. The sharp contrast in these two spectra is due to the large differences in molar absorptivity, $\sim 8 \times 10^3$ L/mol-cm (~ 0.13 Å^2) for perdeuterodiphenyl vs ~ 0 for isopropyl sulfide, which yields the excellent selectivity of MPI. In contrast the ionization potentials differ by only 0.1 eV and, hence, the difference in the spectra must be due to the differences in the molar absorptivities.

Another important advantage of MPI is that wavelength and power can be used to control the softness of ionization. Thus, varying the laser power used for MPI affects the amount of fragmentation observed. As shown in Fig. 11.6 for 2,6-diethylaniline, the fragment ion/parent ion ratio increases as the power is increased.

In trying to obtain thermodynamic and mechanistic information from ion–molecule reaction studies, it is important to know the degree of electronic and/or kinetic excitation that is imparted to the reactant ion during its formation. MPI of $Cr(CO)_6$ and $Mn_2(CO)_{10}$ were observed to produce "cool" ions (i.e., ions in their ground electronic state). Ground state Cr^+ is unreactive with methane, but excited state Cr^+ reacts with methane to generate $CrCH_2^+$ and H_2.[17,18] Cr^+ produced by MPI showed no reaction, in contrast to EI, which does yield a population of Cr^+ that reacts to form the carbene. Similarly Mn^+ produced by MPI was unreactive with isobutane, indicating that the ions were formed in their ground state.

These examples demonstrate several of the well-known advantages of MPI. Further advantages are obtained by coupling this technique to FT-ICR, the most important being the ability to collect a complete, high-resolution mass spectrum from each laser shot. Other MPI systems have utilized time-of-flight (TOF) or scan-

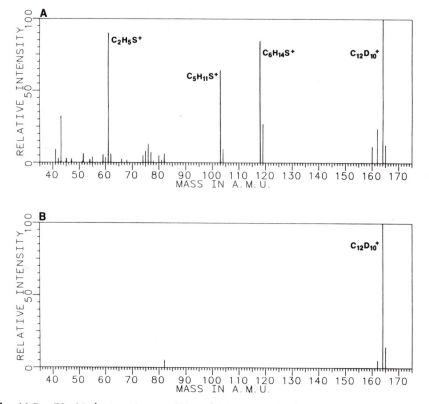

Fig. 11.5. 70 eV electron impact (A) and 266 nm multiphoton ionization (B) spectra of isopropyl sulfide and perdeuterodiphenyl (see text for details).

ning instruments (quadrupole or sector) that have certain inherent limitations. TOF, although capable of obtaining a complete mass spectrum with virtually unlimited mass range, is a much lower resolution technique than FT-ICR. Scanning instruments can provide high-resolution spectra but require multiple laser shots to produce an entire spectrum, which may not be possible in some experiments. For example, Gross and co-workers[15] have utilized MPI with FT-ICR as the detector for a gas chromatography (GC) system, a case where it is necessary to rapidly obtain an entire mass spectrum for each laser shot. The other key advantage of using FT-ICR with MPI has been touched on in the experimental section and results from the ability to trap and study the ions after their formation, whereby chemical and structural information may be obtained using CID, ion–molecule reactions, photodissociation, etc., as illustrated below.

Laser Desorption

Laser desorption FT-ICR has been applied to a wide variety of systems including biologically interesting molecules such as carboxylic acids, dipeptides, oligopeptides, porphyrins, oligosaccharides, glycosides, and nucleosides, as well as organometallics, clusters, and intractable polymers.[19-26] The research in our laboratory has focused on the use of laser desorption (LD) techniques as an excellent method to generate bare atomic metal ions and metal cluster ions.[4,27-29]

The versatility of the FT-ICR experiment allows many aspects of metal ion chemistry to be examined.[4,28] Novel organometallic ions may be synthesized in the gas phase environment of the ICR cell and their structures probed by a variety of

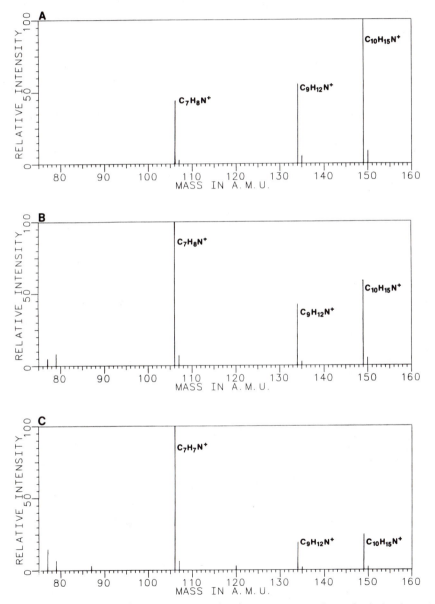

Fig. 11.6. Effect of MPI laser power on the fragmentation of 2,6-diethylaniline. (A) Unfocused beam, ~1 mJ/pulse. (B) Unfocused beam, ~2.5 mJ/pulse. (C) Focused beam, ~2.5 mJ/pulse.

experiments.[30-35] Kinetic data may be obtained by varying the reaction time and noting changes in ion intensity, yielding rate constant information.[36] Bond energies can be determined qualitatively by the observation of exothermic reactions[36-38] or quantitatively by measuring photodissociation thresholds.[39-44] Mechanisms are explored by ejecting potential intermediates and noting changes in relative peak intensities. The unusual variety of data that may be obtained from these experiments is the reason FT-ICR has emerged as one of the best techniques for examining gas-phase ion–molecule reactions.

The trapping of laser-generated metal ions in the FT-ICR cell requires no special pulse sequences or trapping potentials. The ease of trapping laser-desorbed ions was discovered in the early MPI experiments in which it was found that the beam had to pass through the cell and vacuum chamber to prevent the metal ion signal from interfering with the signal from the organics being examined (see preceding section). As is often the case in science, however, an unexpected experimental "problem" can open up an exciting new field of research and the use of laser desorption in our laboratory has greatly outweighed studies involving MPI. The laser desorption method of generating metal ions has proven to be far superior to other methods such as MPI and EI on volatile inorganic complexes, in that virtually any metal ion can be generated cleanly and simply by focusing the laser on a pure metal target.

Bare Atomic Metal Ions

The goals of this work have been threefold: determination of fundamental reaction mechanisms and trends in reactivity, comparison of gas-phase to solution-phase results, and development of metal ions as selective chemical ionization reagents. These goals have been pursued thus far in extensive studies of a wide variety of metal ions (Fig. 11.7) with an array of organic compounds. These reactions are too numerous to discuss in detail here and have been extensively examined in several recent reviews.[4,28,45,46]

A typical experimental sequence using FT-ICR for these studies consists of laser desorption to form the metal ions followed by pulses of reagent gases, delays for

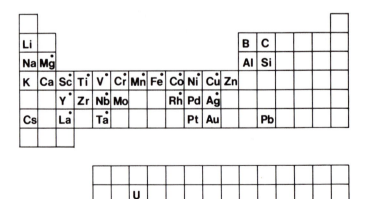

Fig. 11.7.   Representation of elements that have been laser desorbed in our laboratory; dot indicates extensive chemical characterization.

ion–molecule reactions to take place, single or swept frequency ion isolation events, CID pulses for ion structure determination, and finally a detection sequence. Routinely the collision gas (e.g., Ar) is kept at a static pressure while reagents are pulsed into the system. The pulsing of the reagent gases allows them to be removed from the system and thus prevents the possibility of further complicating reactions later in the experimental sequence. CID pulses are used not only to identify the structure of a particular ion, but often to generate a new ion of interest or an intermediate to the particular ion of interest.[47-49] Specifics of how these events are utilized will be covered in more detail in the examples in the following sections. Temporal control of both the neutral and ion concentrations is truly an outstanding feature of the FT-ICR.

Metal Clusters

The study of metal clusters is an area of great current interest.[50] Much of the emphasis has been on understanding the chemistry of clusters as models for the reactions that take place on the surfaces of catalysts. Of particular interest has been the effect of the size and composition of the metal cluster species. FT-ICR methods offer a powerful means for controlling and studying these two parameters. Several groups have been successful in forming silicon,[26] gold,[51] and carbon clusters[52] using laser desorption with FT-ICR. The research in our group has focused on both homonuclear and heteronuclear transition-metal clusters,[29,49,53-59] as briefly discussed below.

In addition to laser desorption, a number of other techniques have been used in conjunction with FT-ICR to generate and study cluster ions. These include direct electron impact on polynuclear carbonyl complexes,[60-62] sputtering sources,[63-66] and, more recently, supersonic beams.[67] The latter methods involve producing clusters in an external source and then injecting them into the analyzer cell in which all of the features of FT-ICR may be utilized.

Heteronuclear Clusters. The heteronuclear clusters studied in our laboratory have been mainly of the form MFe^+ with M = Rh, La, Y, Sc, Mg, Co, Co_2, V, Cu, and Nb.[42,49,53-59] These ions are produced by reacting $Fe(CO)_5$ with the trapped metal ion (produced using laser desorption) forming $MFe(CO)_n^+$. The latter species may be isolated and made to undergo CID producing MFe^+. The neutral carbonyl can be varied to produce other cluster series and, for example, $RhCo^+$ has been produced in a similar manner by reacting Rh^+ with $Co(CO)_3NO$ producing the intermediate $RhCo(CO)_nNO^+$ where n = 1 or 2.[68]

An example of the types of sophisticated experiments that may be performed is illustrated by the reaction sequence shown in Fig. 11.3 and the corresponding spectra A–H in Fig. 11.4 for $NbFe(C_6H_6)^+$. Here, Nb^+ is generated via laser desorption (step 1, spectrum A) and trapped in the analyzer cell. $Fe(CO)_5$ is pulsed in (step 2) and allowed to react (step 3, spectrum B). $NbFe(CO)_3^+$ is isolated with ion ejection rf pulses (step 4, spectrum C) after which it is accelerated (step 5) and allowed to undergo CID (step 6, spectrum D) with Ar, present at a static background pressure, and the $NbFe^+$ formed may be isolated (step 7, spectrum E). Next, cyclohexene is introduced to the cell (step 8) and allowed to react (step 9) forming $NbFeC_6H_6^+$ and $NbC_{12}H_{12}^+$ (spectrum F). The benzene complex may then be quasiisolated by ejecting lower mass ions (step 10, spectrum G), accelerated (step 11), and allowed to undergo CID (step 12). The dissociation products are then detected (steps 13 and 14, spectrum H) utilizing standard FT-ICR techniques. Similar sequences are used for studying other clusters.

The FT-ICR studies of the heteronuclear clusters have yielded a great deal of chemical information. As with the bare metal ions, they have been reacted with a great number of compounds including linear and cyclic alkanes and alkenes, alkynes, aldehydes, ketones, and alcohols. Again, the specifics of these reactions are beyond the scope of this review and have appeared in detail elsewhere.[59] In summary, the heteronuclear complexes in general show chemistry that is significantly different from that of either of the bare atomic metal ions. Activation of $C-H$ and $C-C$ bonds in hydrocarbons is a particular area of interest. The reactions are varied and complex and remain a vast area for future research.

Recently work in our group has expanded to doubly charged heteronuclear clusters.[69] $LaFe^{2+}$ has been produced in a method analogous to $LaFe^+$ by reacting La^{2+} with $Fe(CO)_5$. Only a few reactions have been studied. For example, $LaFe^{2+}$ reacts with ethane producing $LaFe(C_2H_4)^{2+}$ and $LaFe(C_2H_4)_2^{2+}$. This is similar to the reactions of $LaFe^+$ but different from La^{2+}. Theoretical and experimental work is currently underway involving this fascinating and unusual species to try and determine the nature of the bonding involved and to further characterize its chemical reactivity.

Homonuclear Clusters. Homonuclear clusters can be produced by direct laser desorption from metal and metal oxide surfaces. Recently we reported, for example, that laser desorption produces good yields of Zn_2^+ and Ag_3^+ from the metal oxides.[29] The ratio of cluster produced relative to the monomer was observed to vary inversely with the incident laser power. Surprisingly, using a pellet of an AgO and ZnO mixture enhanced Ag clustering, producing Ag_n^+ ($n = 1-9$) with no Zn^+ or $AgZn^+$ clusters apparent (Fig. 11.8). As has been observed previously, the odd-numbered clusters are enhanced. Collision-induced dissociation of Ag_5^+ produced Ag_3^+ exclusively, which could not be further dissociated (up to 100 eV). One interesting observation from these studies is that the species $Ag_3L_2^+$ and $Ag_5L_2^+$ (L = *sec*-

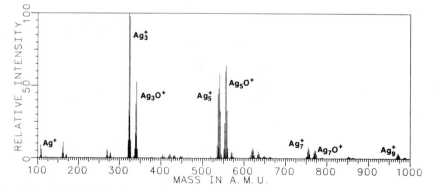

Fig. 11.8. LD-FT-ICR of an AgO–ZnO pellet showing enhanced clustering (see text for details).

butylamine) react with a third molecule of *sec*-butylamine through deamination and dehydrogenation, as opposed to simple attachment. Thus, ligands can play an important role in affecting the chemistry of the metal cluster. Again, this area presents a great expanse for ongoing and future studies.

Recently, we observed another interesting effect involving the formation of gold clusters from a pure metal target.[70] At low laser powers, only the monomer is observed. If the laser power is increased significantly, still only the monomer is observed. However, on lowering the power to the initial levels, enhanced clustering is observed forming Au_n^+ where $n = 2$ or 3 (Fig. 11.9). After several laser shots at the lower power, clustering is greatly reduced as the atomic ion Au^+ once again dominates. If the power is turned up again for one laser shot, and then lowered, clusters are again observed dissipating with each shot. This effect has been termed laser roughening, in analogy to electrochemical roughening, and has also been observed to enhance formation of Ag_3^+, Ag_5^+, Cu_2^+, Cu_3^+, Pd_2^+, and Pt_2^+ where usually only the monomer is observed.[70]

Doubly Charged Ions

Until recently it was believed that doubly charged metal cations would undergo rapid charge exchange as the only reaction pathway with simple organic molecules. Although from limited studies this appears to be true for the later transition metals, Tonkyn and Weisshaar recently reported that Ti^{2+} undergoes clustering reactions with methane and hydride abstraction with ethane in the high-pressure regime of a flowing afterglow.[71] These interesting results led us to look at the reactivity of some other doubly charged early transition metals, whose second ionization potentials are relatively low.

The metals we have examined to date include Nb,[38] Ta, Zr, and La.[72] These doubly charged ions are generated in much the same way as the singly charged ions in that a laser is focused on a high-purity block of the metal of interest. By varying the laser power and the position of laser impact on the metal surface, doubly charged metal ions can be formed in good yields.

Nb^{2+} was the first doubly charged metal ion examined in our group.[38] Because multiply charged ions are known to be formed by laser desorption with greater average kinetic energies than the singly charged ions,[73] Nb^{2+} was first thermalized by 1×10^{-5} torr Ar for 0.5 sec ($\sim$600 collisions). Nb^{2+} was then allowed to react with the reagent gas at $\sim$6 $\times$ 10^{-7} torr and a plot of natural log of the Nb^{2+} signal with time was found to give a good straightline fit to 2.5 half-lives (81% reacted), suggesting that the ions were predominantly in their ground states.

Nb^{2+} reacts with methane as shown in Reactions (11.1)–(11.3) and an example of the reaction sequence can be seen in Fig. 11.10. Although charge exchange is

$$52\%$$
$$\longrightarrow NbCH_2^{2+} + H_2 \tag{11.1}$$

$$7\%$$
$$Nb^{2+} + CH_4 \longrightarrow NbH^+ + CH_3^+ \tag{11.2}$$

$$41\%$$
$$\longrightarrow Nb^+ + CH_4^+ \tag{11.3}$$

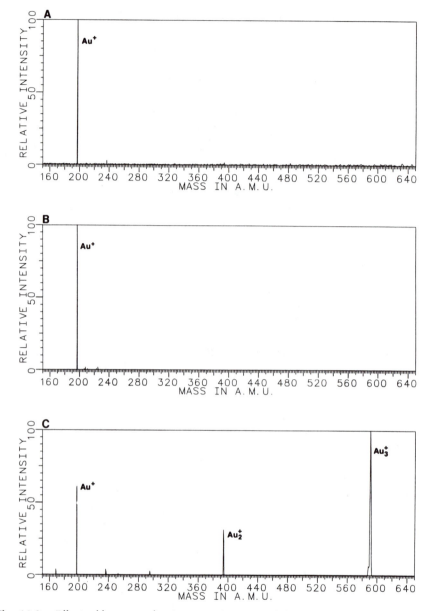

Fig. 11.9. Effect of laser roughening on a Au target. (A) Low power. (B) High power. (C) Low power after a single high-power shot.

observed to be a major pathway, there is also a surprising amount of oxidative addition/reductive elimination to produce the doubly charged carbene. Subsequent reactions of $NbCH_2^{2+}$ were monitored and, together with Reaction (11.1), yielded a value for $D^0(Nb^{2+}-CH_2) = 197 \pm 10$ kcal/mol.

The other doubly charged metals studied show similar reactions with methane. Ta^{2+} and Zr^{2+} both react primarily to form the doubly charged carbene with vary-

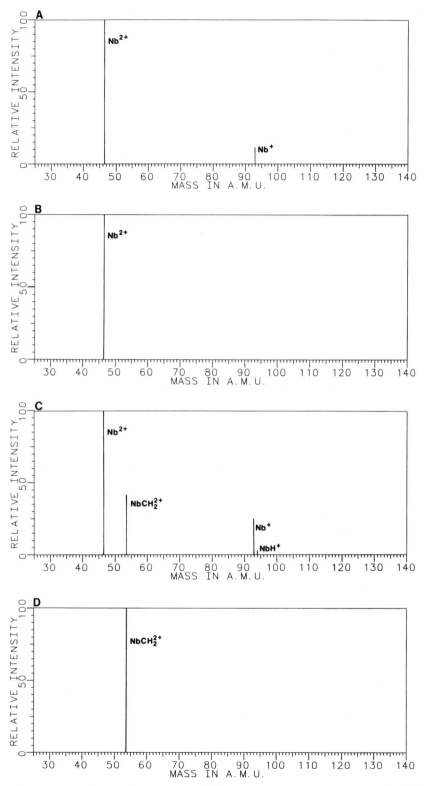

Fig. 11.10. Formation and reaction of Nb^{2+} with methane. (A) Formation of Nb^{2+} via LD. (B) Isolation of Nb^{2+}. (C) Reaction with methane. (D) Isolation of $NbCH_2^{2+}$.

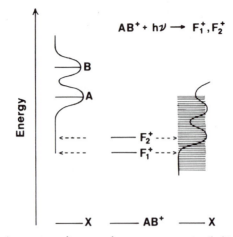

Fig. 11.11. Electronic energy diagram for spectroscopic (left) and thermodynamic (right) photodissociation thresholds (see text for details).

ing amounts of other charge-exchange products. La^{2+}, however, is unreactive with methane. The absence of any charge exchange is not surprising since the second ionization potential of La is lower than the first ionization potential of methane. Other studies currently underway indicate that C_2–C_4 alkanes also react with these doubly charged metals by pathways other than charge exchange, and this work is being pursued.

Photodissociation Studies

The ion-trapping capability of ICR is ideally suited for performing photodissociation studies. Such studies have been demonstrated to yield a wealth of information on ion structure, electronic states, and thermochemistry.[74] For an ion to undergo photodissociation, Reactions (11.4) and (11.5), first the ion must absorb a photon, second the photon must have sufficient energy to fragment a bond in the ion, and third the quantum yield for photodissociation must be nonzero.[75] The latter requirement simply means that dissociation must effectively compete with radiative relaxation.

$$AB^+ + \hbar\nu \begin{array}{c} \longrightarrow F_1^+ \\ \longrightarrow F_2^+ \end{array}$$

$$(11.4)$$

$$(11.5)$$

Photodissociation can be used to help determine ion structures in two ways: (1) the photodissociation spectrum gives an indirect measure of the UV-VIS absorption spectrum of the gas-phase ion and (2) the fragmentation pattern or the change in the fragmentation at low vs high photon energies can be interpreted in a manner similar to collision-induced dissociation. Photodissociation thresholds can be used to determine upper limits for bond energies and in some cases can give absolute bond energies. As shown in Fig. 11.11, if an ion has its first excited state above the energy necessary for bond rupture, then photodissociation gives only an upper limit as is the case for most organic ions. If, however, the first excited state lies fortuitously close to the thermodynamic threshold for dissociation or if there is a high density of excited states around the energy necessary to fragment a bond, then photodissociation can yield the absolute bond energy. Our studies have shown that in many cases an absolute metal ion–ligand bond energy can be obtained from photodissociation, because of the high density of states introduced by the metal center.[39-44] Table 11.1 provides a list of various bond energies determined in our laboratory and also compares our numbers to other values where available. In addition the spectrum of VFe$^+$ is shown in Fig. 11.12 and is illustrative of the type of data obtained. Finally, photodissociation has been suggested as holding promise for the activation of high molecular weight ions ($m/e > 1000$ amu) and has been demonstrated in a number of laboratories on polypeptide ions.[76]

Table 11.1. Bond energies

A^+-B	Photodissociation $D^0(A^+-B)$ (kcal/mol)	Literature (kcal/mol)
Sc^+-Fe	48 ± 5^a	49 ± 6^b
Ti^+-Fe	60 ± 6^a	
V^+-Fe	75 ± 5^c	
$V^+-C_6H_6$	62 ± 5^d	
$VC_6H_6^+-C_6H_6$	57 ± 5^d	
Cr^+-Fe	50 ± 7^a	
Fe^+-Fe	62 ± 5^a	$\sim 69^e$
Fe^+-O	68 ± 5^d	$68 \pm 3^{f,g}$
Fe^+-OH	73 ± 3^h	76 ± 5^i
Fe^+-S	61 ± 6^j	$65 \pm 5^d, 74 > x > 59^m$
Fe^+-S_2	48 ± 5^j	
FeS^+-S_2	49 ± 5^j	
$FeS_2^+-S_2$	49 ± 5^j	
$FeS_3^+-S_2$	43 ± 5^j	
$FeS_4^+-S_2$	38 ± 5^j	
Fe^+-NH	61 ± 5^k	
Fe^+-C	94 ± 7^l	89^n
Fe^+-CH	101 ± 7^l	115 ± 20^n
Fe^+-CH_2	82 ± 5^l	96 ± 5^f
Fe^+-CH_3	65 ± 5^d	$69 \pm 5^o, 68 \pm 4^f$
$Fe^+-(butadiene)$	48 ± 5^d	$45-60^p$
$Fe^+-(c-C_5H_6)$	55 ± 5^r	
$Fe^+-C_6H_6$	55 ± 5^d	58 ± 5^q
$FeC_5H_5^+-H$	46 ± 5^r	
Co^+-Fe	62 ± 5^a	66 ± 7^e
Co^+-OH	71 ± 3^h	
Co^+-S	62 ± 5^d	$74 > x > 59^m$
Co^+-C	90 ± 7^l	98^n
Co^+-CH	100 ± 7^l	
Co^+-CH_2	84 ± 5^l	$85 \pm 7^{f,s}$
Co^+-CH_3	57 ± 7^d	$61 \pm 4^{f,g,o}$
$Co^+-C_6H_6$	68 ± 5^d	$71 > x > 61^e$
Ni^+-Fe	64 ± 5^a	$<68 \pm 5^l$
Ni^+-S	60 ± 5^d	

Table 11.1. Bond energies (*continued*)

A^+-B	Photodissociation $D^0(A^+-B)$ (kcal/mol)	Literature (kcal/mol)
$Ni^+-2C_2H_4$	80 ± 5^d	74 ± 2^u
Cu^+-Fe	53 ± 7^a	
Nb^+-Fe	68 ± 5^a	
Nb^+-C	$>138^v$	
Nb^+-CH	145 ± 8^v	
Nb^+-CH_2	109 ± 7^v	112^w
$Nb^+-C_6H_6$	64 ± 3^b	
Rh^+-C	$>120^v$	164 ± 16^x
Rh^+-CH	102 ± 7^v	
Rh^+-CH_2	91 ± 5^v	94 ± 5^y
$Rh^+-C_6H_6$	66 ± 7^v	$>49^z$
La^+-C	102 ± 8^v	
La^+-CH	125 ± 8^v	
La^+-CH_2	106 ± 5^v	
Ta^+-Fe	72 ± 5^a	

[a]Ref. 42.

[b]L. M. Lech, Ph.D. Thesis, Purdue University 1988.

[c]Ref. 56.

[d]Ref. 41.

[e]Ref. 49.

[f]P. B. Armentrout, L. F. Halle, and J. L. Beauchamp, *J. Am. Chem. Soc.* **103**, 6501 (1981).

[g]P. B. Armentrout and J. L. Beauchamp, *J. Am. Chem. Soc.* **103**, 784 (1981).

[h]Ref. 39.

[i]E. Murad, *J. Chem. Phys.* **73**, 1381 (1980).

[j]Ref. 44.

[k]S. W. Buckner, J. R. Gord, and B. S. Freiser, *J. Am. Chem. Soc.* **110**, 6606 (1988).

[l]Ref. 40.

[m]T. J. Carlin, Ph.D. Thesis, Purdue University 1984.

[n]J. L. Beauchamp, private communication.

[o]L. F. Halle, P. B. Armentrout, and J. L. Beauchamp, *Organometallics* **1**, 963 (1982).

[p]D. B. Jacobson and B. S. Freiser, *J. Am. Chem. Soc.* **105**, 7484 (1983).

[q]D. B. Jacobson and B. S. Freiser, *J. Am. Chem. Soc.* **106**, 3900 (1984).

[r]Y. Huang and B. S. Freiser, *J. Am. Chem. Soc.*, in press.

[s]P. B. Armentrout and J. L. Beauchamp, *J. Chem. Phys.* **74**, 2819 (1981).

[t]D. B. Jacobson and B. S. Freiser, unpublished results.

[u]D. B. Jacobson and B. S. Freiser, *J. Am. Chem. Soc.* **105**, 7492 (1983).

[v]Ref. 43.

[w]Ref. 38.

[x]D. B. Jacobson, G. D. Byrd, and B. S. Freiser, *Inorg. Chem.* **23**, 553 (1984).

[y]D. B. Jacobson and B. S. Freiser, *J. Am. Chem. Soc.* **107**, 5870 (1985).

[z]G. D. Byrd and B. S. Freiser, *J. Am. Chem. Soc.* **104**, 5944 (1982).

FUTURE WORK

Hopefully, the above examples have served to illustrate the advantages of combining lasers with FT-ICR. In particular we have greatly benefited from this powerful combination in our studies of the structure and reactivity of laser-desorbed metal ions, an area still ripe for investigation. Future work in our laboratory will continue to focus on metal–ligand and metal–cluster species, determining more bond energies and ion structures to better understand the complex mechanisms of their gas-

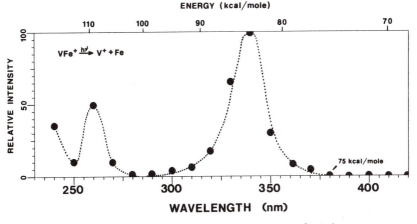

Fig. 11.12. Photodissociation spectrum of VFe^+.

phase ion–molecule reactions. Along these lines, we have recently put into service a CW Ar ion-pumped dye laser to aid in the photodissociation studies. This tunable source provides higher wavelength resolution and permits the study of multiphoton processes. Also, we have begun to use this source to examine photoinduced reactions and rearrangements, as models for photoinduced surface chemistry.[77,78]

As previously mentioned, we have recently added an external ion source to the FTMS-2000. As has been demonstrated in several other laboratories,[67,79–81] this will enable a multitude of sources such as FAB, [252]Cf, SIMS, and supersonic molecular beams to be interfaced. Using these sources we plan to study higher molecular weight clusters, biopolymers, and so forth. This work is in the preliminary stages, but we have been successful in producing metal ions using laser desorption in the external source and observing excellent ion signals in both the source and analyzer cells. Also, we hope to exploit the dual-cell arrangement of the FTMS-2000 to study much more complex reactions than those possible in the single-cell instrument. These advantages will be most utilized when conducting laser photodissociation experiments in which the ion to be dissociated can be transferred into a "clean" low-pressure cell, preventing it from reacting away and eliminating any possibility of interferences.

Currently there are limitations in both of our FT-ICR instruments in the number of rf pulses that are available. In particular it is difficult to go much above an MS^5 sequence with these limitations, even though in most instances there is an ample number of ions in the analyzer cell. The most promising solution to this problem is a new generation of software under commercial development that will utilize the stored waveform techniques (SWIFT) of Marshall and co-workers.[82] Using SWIFT, a single pulse of a specified rf waveform will take the place of what now requires multiple pulses. Thus, for example, a single SWIFT pulse can be formulated to simultaneously eject unwanted ions from the cell and excite for CID the desired ions. SWIFT alone would enable reaction schemes beyond MS^8 with the current software, but other changes will essentially make the schemes and the potential for studying new chemistry limitless!

ACKNOWLEDGMENT

Acknowledgment is made to the Division of Chemical Sciences in the Office of Basic Energy Sciences in the United States Department of Energy (DE-FG02-87ER13776) for supporting the transition-metal ion research and to the National Science Foundation (CHE-8612234) for continued support of FTMS methodology.

REFERENCES

1. M. L. Gross and D. L. Rempel, *Science* **226**, 261 (1984).

2. A. G. Marshall, *Acc. Chem. Res.* **18**, 316 (1985).

3. K.-P. Wanczek, *Int. J. Mass Spectrom. Ion Phys.* **60**, 11 (1984).

4. B. S. Freiser, *Talanta* **32**, 697 (1985).

5. D. A. Laude, Jr., C. L. Johlman, R. S. Brown, D. A. Weil, and C. L. Wilkins, *Mass Spec. Rev.* **5**, 107 (1986).

6. D. H. Russell, *Mass Spec. Rev.* **5**, 167 (1986).

7. *Anal. Chim. Acta* **178**(1) (1985).

8. *Int. J. Mass Spectrom. Ion Phys.* **72**(1 and 2) (1986).

9. M. V. Buchanan (ed.), *Fourier Transform Mass Spectrometry: Evolution, Innovation, and Applications* (ACS Symposium Series 359). American Chemical Society, Washington, DC, 1987.

10. T. J. Carlin and B. S. Freiser, *Anal. Chem.* **55**, 571 (1983).

11. R. B. Cody, R. C. Burnier, and B. S. Freiser, *Anal. Chem.* **54**, 96 (1982).

12. R. A. Forbes, L. M. Lech, and B. S. Freiser, *Int. J. Mass Spectrom. Ion Process.* **77**, 107 (1987).

13. M. P. Irion, W. D. Bowers, R. L. Hunter, F. S. Rowland, and R. T. McIver, Jr., *Chem. Phys. Lett.* **93**, 375 (1982).

14. T. J. Carlin and B. S. Freiser, *Anal. Chem.* **55**, 955 (1983).

15. T. M. Sack, D. A. McCrery, and M. L. Gross, *Anal. Chem.* **57**, 1290 (1985).

16. C. H. Watson, G. Baykut, M. A. Battiste, and J. R. Eyler, *Anal. Chim. Acta* **178**, 125 (1985).

17. R. B. Freas and D. P. Ridge, *J. Am. Chem. Soc.* **102**, 7129 (1980).

18. L. F. Halle, P. B. Armentrout, and J. L. Beauchamp, *J. Am. Chem. Soc.* **103**, 962 (1981).

19. (a) D. A. McCrery, E. B. Ledford, Jr., and M. L. Gross, *Anal. Chem.* **54**, 1435 (1982); (b) D. A. McCrery and M. L. Gross, *Anal. Chim. Acta* **178**, 91 (1985).

20. R. E. Shomo, II, A. G. Marshall, and C. R. Weisenberger, *Anal. Chem.* **57**, 2940 (1985).

21. C. E. Brown, P. Kovacic, C. E. Wilkie, R. B. Cody, and J. A. Kinsinger, *J. Polym. Sci. Polym. Lett. Ed.* **23**, 453 (1985).

22. C. L. Wilkins, D. A. Weil, and C. L. Ijames, *Anal. Chem.* **57**, 520 (1985).

23. C. S. Giam, R. R. Weller, and J. A. Mayernik, presented at ACS 194th Annual Meeting, New Orleans, LA, 1987.

24. R. B. Cody, J. A. Kinsinger, S. Ghaderi, I. J. Amster, F. W. McLafferty, and C. E. Brown, *Anal. Chim. Acta* **178**, 43 (1985).

25. R. B. Brown, D. A. Weil, and C. L. Wilkins, *Macromolecules* **19**, 1255 (1986).

26. W. D. Reents, Jr., M. L. Mandich, and V. E. Bondybey, *Chem. Phys. Lett.* **131**, 1 (1986).

27. R. B. Cody, R. C. Burnier, W. D. Reents, Jr., T. J. Carlin, D. A. McCrery, R. K. Lengel, and B. S. Freiser, *Int. J. Mass Spec. Ion Phys.* **33**, 37 (1980).

28. B. S. Freiser, *Anal. Chim. Acta* **178**, 137 (1985).

29. S. W. Buckner, J. R. Gord, and B. S. Freiser, *J. Chem. Phys* **88**, 3678 (1988).

30. T. C. Jackson, D. B. Jacobson, and B. S. Freiser, *J. Am. Chem. Soc.* **106**, 1252 (1984).

31. D. B. Jacobson and B. S. Freiser, *J. Am. Chem. Soc.* **106**, 3891 (1984).

32. D. B. Jacobson and B. S. Freiser, *J. Am. Chem. Soc.* **107**, 2605 (1985).

33. D. B. Jacobson and B. S. Freiser, *J. Am. Chem. Soc.* **107**, 7399 (1985).

34. T. C. Jackson and B. S. Freiser, *Int. J. Mass Spec. Ion Process.* **72**, 169 (1986).

35. C. J. Cassady and B. S. Freiser, *J. Am. Chem. Soc.* **108**, 2537 (1986).

36. L. Operti, E. C. Tews, and B. S. Freiser, *J. Am. Chem. Soc.* **110**, 3847 (1988).

37. S. W. Buckner and B. S. Freiser, *J. Am. Chem. Soc.* **109**, 4715 (1987).

38. S. W. Buckner and B. S. Freiser, *J. Am. Chem. Soc.* **109**, 1247 (1987).

39. C. J. Cassady and B. S. Freiser, *J. Am. Chem. Soc.* **106**, 6176 (1984).

40. R. L. Hettich and B. S. Freiser, *J. Am. Chem. Soc.* **108**, 2537 (1986).

41. R. L. Hettich, T. C. Jackson, E. M. Stanko, and B. S. Freiser, *J. Am. Chem. Soc.* **108**, 5086 (1986).

42. R. L. Hettich and B. S. Freiser, *J. Am. Chem. Soc.* **109**, 3537 (1987).

43. R. L. Hettich and B. S. Freiser, *J. Am. Chem. Soc.* **109**, 3543 (1987).

44. T. J. MacMahon, T. C. Jackson, and B. S. Freiser, *J. Am. Chem. Soc.,* **111**, 421 (1989).

45. J. Allison, *Prog. Inorg. Chem.* **34**, 627 (1986).

46. S. W. Buckner and B. S. Freiser, *Polyhedron* **7**, 1583 (1988).

47. T. J. Carlin, L. Sallans, C. J. Cassady, D. B. Jacobson, and B. S. Freiser, *J. Am. Chem. Soc.* **105**, 6320 (1983).

48. L. Sallans, K. Lane, R. R. Squires, and B. S. Freiser, *J. Am. Chem. Soc.* **105**, 6352 (1983).

49. D. B. Jacobson and B. S. Freiser, *J. Am. Chem. Soc.* **106**, 4623 (1984).

50. For recent reviews of cluster research, see (a) *Surf. Sci.* **156** (1985); (b) A. W. Castleman and R. G. Kesee, *Annu. Rev. Phys. Chem.* **37**, 525 (1986); (c) M. D. Morse, *Chem. Rev.* **86** (1986); (d) *The Physics and Chemistry of Small Clusters,* edited by P. Jena, B. K. Rao, and S. N. Khanna, Plenum Press, New York, 1987.

51. (a) D. A. Weil and C. L. Wilkins, *J. Am. Chem. Soc.* **107**, 7316 (1985); (b) M. Moini and J. R. Eyler, *Chem. Phys. Lett.* **137**, 311 (1987).

52. (a) S. W. McElvany, H. H. Nelson, A. P. Baronavski, C. H. Watson, and J. R. Eyler, *Chem. Phys. Lett.* **134**, 214 (1987); (b) S. W. McElvany, B. I. Dunlap, and A. O'Keefe, *J. Chem. Phys.* **86**, 715 (1987).

53. D. B. Jacobson and B. S. Freiser, *J. Am. Chem. Soc.* **106**, 5351 (1984).

54. D. B. Jacobson and B. S. Freiser, *J. Am. Chem. Soc.* **107**, 1581 (1985).

55. D. B. Jacobson and B. S. Freiser, *J. Am. Chem. Soc.* **108**, 27 (1986).

56. R. L. Hettich and B. S. Freiser, *J. Am. Chem. Soc.* **107**, 6222 (1985).

57. E. C. Tews and B. S. Freiser, *J. Am. Chem. Soc.* **109**, 4433 (1987).

58. Y. Huang and B. S. Freiser, *J. Am. Chem. Soc.* **110**, 387 (1988).

59. S. W. Buckner and B. S. Freiser, *Gas Phase Inorganic Chemistry,* D. Russell, Ed., Plenum, New York, 1988.

60. D. P. Ridge, in *Lecture Notes in Chemistry,* Vol. 31, edited by H. Hartmann and K.-P. Wanczek. Springer-Verlag, West Berlin, p. 140, 1982.

61. B. S. Larsen, R. B. Freas, and D. P. Ridge, *J. Phys. Chem.* **88**, 6014 (1984).

62. R. B. Freas and D. P. Ridge, *J. Am. Chem. Soc.* **106**, 825 (1984).

63. R. B. Freas, M. M. Ross, and J. E. Campana, *J. Am. Chem. Soc.* **107**, 6195 (1985).

64. R. B. Freas and J. E. Campana, *J. Am. Chem. Soc.* **107**, 6202 (1985).

65. T. F. Magnera, D. E. David, and J. Michl, *J. Am. Chem. Soc.* **109**, 936 (1987).

66. L. Hanley and S. L. Anderson, *Chem. Phys. Lett.* **122**, 410 (1985).

67. J. M. Alford, P. E. Williams, D. J. Trevor, and R. E. Smalley, *Int. J. Mass Spectrom. Ion Process.* **72**, 33 (1986).

68. Y. Huang, S. Buckner, and B. S. Freiser, in *The Physics and Chemistry of Small Clusters*, edited by P. Jena, B. K. Rao, and S. N. Khanna. Plenum, New York, 1987, p. 8.

69. Y. Huang and B. S. Freiser, *J. Amer. Chem. Soc.* **110**, 4435 (1988).

70. J. R. Gord, S. W. Buckner, and B. S. Freiser, *Chem. Phys. Lett.* **153**, 577 (1988).

71. R. Tonkyn and J. C. Weisshaar, *J. Am. Chem. Soc.* **108**, 7128 (1986).

72. T. J. MacMahon, S. W. Buckner, J. R. Gord, Y. Huang, and B. S. Freiser, unpublished results.

73. H. Kang and J. L. Beauchamp, *J. Phys. Chem.* **89**, 3364 (1985).

74. See, for example, L. R. Thorne and J. L. Beauchamp, Chap. 18, p. 41; R. C. Dunbar, Chap. 20, p. 129; P. S. Drzaic, J. Marks, and J. I. Brauman, Chap. 21, p. 167, in *Gas Phase Ion Chemistry: Ions and Light,* Vol. 3, edited by M. T. Bowers. Academic Press, New York, New York, 1984.

75. B. S. Freiser and J. L. Beauchamp, *J. Am. Chem. Soc.* **98**, 3136 (1976).

76. W. D. Bowers, S.-S. Delbert, R. L. Hunter, and R. T. McIver, Jr., *J. Am. Chem. Soc.* **106**, 7288 (1984).

77. J. R. Gord, S. W. Buckner, and B. S. Freiser, *J. Am. Chem. Soc.* **111**, 3753 (1989).

78. J. R. Gord and B. S. Freiser, *J. Am. Chem. Soc.* **111**, 3754 (1989).

79. R. T. McIver, Jr., R. L. Hunter, and W. D. Bowers, *Int. J. Mass Spectrom. Ion Process.* **64**, 67 (1985).

80. D. F. Hunt, J. Shabanowitz, R. T. McIver, Jr., R. L. Hunter, and J.E.P. Syka, *Anal. Chem.* **57**, 765 (1985).

81. P. Kofel, M. Allemann, Hp. Kellerhals, and K.-P. Wanczek, *Int. J. Mass Spectrom. Ion Process.* **65**, 97 (1985).

82. A. G. Marshall, T.-C. Wang, and T. L. Ricca, *J. Am. Chem. Soc.* **107**, 7893 (1985).

12

Lasers and Fourier Transform Mass Spectrometry: Applications with a YAG Laser

M. PAUL CHIARELLI and MICHAEL L. GROSS

The purpose of this chapter is to describe some applications of lasers coupled with Fourier transform mass spectrometry (FTMS). The chapter will be centered on applications of the neodymium:YAG laser because it produces radiation in the near infrared (1064 nm) and the output frequency can be readily multiplied, affording not only IR and UV-VIS resonant desorption but also multiphoton ionization. Other reviews, broader in scope, were recently published, and they describe the basic principles of FTMS operation and other applications.[1-4]

There are four features of FTMS that make it appealing for chemical analysis employing a laser: the multichannel advantage, the ability to store and manipulate ions, the high mass resolution, and the high upper mass limit. The Nd:YAG laser permits both laser desorption (LD) and multiphoton ionization (MPI), which are effective methods for generating ions. LD is now a proven method for getting large, fragile molecules into the gas phase. MPI offers control of ionization and fragmentation not possible with electron ionization (EI). Because an Nd:YAG laser is pulsed, ions cannot be formed continuously, and, therefore, it is preferred that the ions be detected simultaneously. This may be accomplished by making use of the multichannel advantage of FTMS. Scanning mass spectrometers, on the other hand, would sample only a small fraction of the mass range of the ions generated because of the short ion production time ($\sim 10^{-8}$ sec) in the source. Furthermore, because a FT mass spectrometer is capable of storing ions for several seconds, various experiments, such as collisional or photoactivation, may be executed with ions from one laser pulse, thus allowing for more information to be obtained. The utility of laser desorption is enhanced by the high mass resolution and upper mass limit of the FTMS. Mass resolution of 300 million for oxygen ions of m/z 16 has been demonstrated.[5] Theoretical calculations show that a conventional 5.0-cm cubic cell at a trapping potential of 1 V in a magnetic field of 13 T is capable of storing an ion having an m/z as large as 950,000.[2]

The applications of YAG lasers in FTMS can be divided into three areas: multiphoton ionization, desorption of nonvolatile molecules, and photodissociation of ions to obtain structural information. MPI in FTMS was first demonstrated by McIver and co-workers.[6] The utility of laser desorption was first recognized by

Freiser and co-workers,[7] and they used it for producing gas-phase metal ions to react with organic molecules. More recently, these workers employed the desorption of transition metal ions from solid targets concurrent with the MPI of Fe(CO)$_5$ to yield some novel dimetal ions (e.g., $CuFe^+$, $CoFe^+$, and VFe^+) whose reactivity with organic molecules can now be assessed.[8,9]

Since the original demonstration of LD-FTMS for nonvolatile organics and biomolecules,[10] the combination has been used to characterize polymers,[11] polypeptides,[12] oligosaccharides,[13] nucleosides,[14] and porphyrin metal complexes.[15] Studies have also been undertaken to ascertain mechanistic aspects of laser desorption. The extent of gas-phase cationization of sucrose by sodium as a function of laser power density was recently elucidated.[16] The lactim–lactam (keto–enol) equilibrium of the pyrimidine nucleosides shifts as a function of power density during laser desorption at a wavelength of 266 nm.[17] This photoinduced process influences the nature of the desorption ionization. These mechanistic studies will be discussed in greater detail in this chapter.

Photodissociation may prove more effective than collisionally activated dissociation (CAD) for inducing the fragmentation of large molecules. The limited translational energy excitation of ions, especially high mass ions, that is achievable in an FTMS cell and the inefficiency of energy deposition hinder the application of CAD in FTMS. Thus far, the multiphoton dissociation (MPD) of ions has been demonstrated by using lasers in the IR[18] and UV[19] wavelength range. In the former, continuous wave (CW) lasers were employed to fragment ions in the 400–1500 amu mass range. Photodissociation employing higher energy photons in the 266- to 308-nm wavelength range was used to form first and second generation fragment ions from a small peptide. Although photodissociation of large ions is in its infancy, significant progress is expected in the future.

GAS CHROMATOGRAPHY/MULTIPHOTON IONIZATION FTMS

Analysis of Polynuclear Aromatic Hydrocarbons (PAH)

Gas chromatography/mass spectrometry (GC–MS) has been employed to solve many problems involving the analysis of complex mixtures; however, there is a limit to the number of components that can be separated in a finite amount of time. The more complex the mixture the more probable it will be that there will be coeluting components. One approach to solving the problem is to employ a selective ionization technique, such as MPI, to increase analyte specificity. The advantages of FTMS already had been demonstrated for both GC–MS[20,21] and for MPI,[7] and it seemed logical in the planning stages of this research that the three-way partnership would have unique properties.[22]

MPI offers four features that make it appealing as an ion source for any mass spectrometer. (1) Ionization is wavelength dependent so analyte selectivity can be achieved. (2) In addition, when the photon energy corresponds to an allowed transition, a large enhancement in ion yield occurs such that 100% ionization efficiency may be achieved in the light beam.[23,24] (3) The fragmentation induced by MPI may be controlled to give mass spectra similar to those seen upon EI or to yield almost no fragmentation so that the whole sample response may be concentrated in the molecular ion. (4) The instrumentation involved is much less cumbersome than that of a vacuum-UV source needed to generate higher energy photons for single photon ionization.

The coupling of GC with MPI and FTMS requires a means of controlling the pressure, and a pulse valve interface was employed.[21] The application of a pulse valve prevents the degradation of resolution by allowing un-ionized GC eluent to be pumped away before ion detection at low pressure. Resolution is enhanced at the expense of the duty cycle. The laser was timed to fire when the eluent pressure was the greatest to maximize the amount of analyte ionized. The sequence of experimental events is given in Fig. 12.1.

Two test mixtures were chosen to evaluate the performance of GC–MPI-FTMS. The first mixture consisted of several polynuclear aromatic hydrocarbons, and the second was a commercially obtained gasoline sample. The results of the GC–MPI-FTMS determinations were contrasted with those obtained by low-energy EI GC–FTMS to evaluate the new features MPI lends to GC–FTMS.

A PAH mixture was employed as a test in the GC–MPI-FTMS determination because PAHs are known to ionize at 266 nm.[23,25] Moreover, PAHs are often toxic and carcinogenic[26] and occur as complex mixtures; thus, sensitive and selective analytical methods are needed. The EI and MPI spectra obtained for the five PAHs are similar; each is dominated by the molecular ion. However, chromatograms (Fig. 12.2) show that very different multiphoton ionization efficiencies pertain. For example, it was found that MPI is sensitive to a trace phenanthrene impurity in the anthracene, which could not be detected by using EI. If a wavelength of 310 nm instead of 266 nm could have been used, the anthracene may have been solely ionized in the presence of the coeluting phenanthrene. It is surmised that a tunable UV laser would allow the selective determination of a given class and possibly a

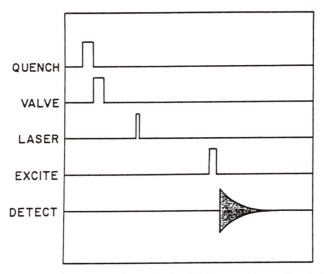

Fig. 12.1. Sequence of events in the GC-MPI-FTMS experiment. Reprinted with permission from Ref. 22.

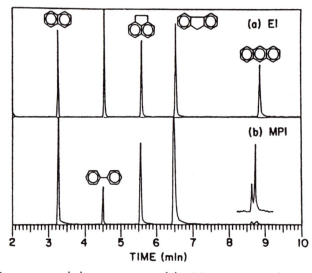

Fig. 12.2. Reconstructed chromatograms of the PAH mixture under (a) EI and (b) MPI conditions. Approximately 40 ng of each component was detected. The column temperature was programmed from 150 to 225°C at 10°C/min. Inset: chromatogram resulting from the injection of the anthracene stock solution, showing the impurity at 8.6 min. The minor differences in the retention times in (a) and (b) are due to small variations in the chromatographic conditions employed. Reprinted with permission from Ref. 22.

subclass of compounds. Cooling by jet expansion could add another opportunity for specificity.

Resolution of 87,000 for the naphthalene molecular ion of m/z 128 was demonstrated for a scan time of 1.96 sec. This time is too long for the narrow capillary peaks if one wishes to obtain 5–10 pts across the chromatographic profile to define the peak. Nonetheless with minor adjustments, mass resolution of 30,000 or 40,000 for ions in the 100–150 amu mass range may be routinely achieved even at magnetic fields of 1.2 T. Higher resolution is obtainable at higher magnetic fields, which in turn can be generated with superconducting magnets. Additional specificity is added by the high mass resolution.

To assess the sensitivity of GC–MPI-FTMS, three factors had to be determined. First, the absolute or instrument limit of detection (LOD) was established for the four PAHs. Second, the LODs obtained by MPI and EI were compared for differences, and finally the linear dynamic range was established for GC–MPI-FTMS. The LODs of the PAHs obtained in the EI determination are all similar, whereas the MPI determination gave LODs that were somewhat different because of different ionization efficiencies. The lowest LOD was obtained in the MPI mode for naphthalene: 8 pg injected on column gave a molecular ion peak with a signal-to-noise (S/N) ratio of 2. The LODs obtained for the other PAHs are similar in both the EI and MPI modes. The LODs are not improved for MPI of certain PAHs because the wavelength employed is not coincident with absorption maxima of these compounds. It is likely that detection limits in the femtogram range can be obtained with a tunable UV laser working at a wavelength of maximum absorption for each of these compounds.

The linear dynamic range of GC–MPI-FTMS was established by injecting different amounts of naphthalene onto the column and computing the complex area associated with the m/z 128 ion over the GC elution profile. The linear portion of the curve extends from 20 pg upward to 2.5 orders of magnitude until space charge effects cause deviation from linearity. Extension of the dynamic range may be achieved at the expense of LOD by cutting the laser power if it is known that larger amounts of sample are to be analyzed.

Once the model system determination was completed, a gasoline sample was characterized. Gasoline constitutes a more complex mixture, which contains hundreds of components, some of which are aromatic hydrocarbons that should easily ionize at a wavelength of 266 nm. The chromatograms generated by EI and MPI exhibit major differences. The GC components corresponding to light, aliphatic hydrocarbons are present in the EI but not in the MPI chromatogram because the aliphatics lack a resonance at 266 nm. MPI was found to accentuate the peaks of those components having large absorption cross sections at 266 nm. The components that have long elution times, such as the methyl- or ethylnaphthalenes (retention time > 20 min), are not detectable in the EI chromatogram but are quite evident in the MPI chromatogram.

Analysis of Alkene and Alkynes: Fe^+ CI

In the above application, specificity was obtained by using laser MPI in the primary ionization step of a resonant analyte and high mass resolution for detection. In the second example we present here, MPI is employed to generate reactive species that allow for selectively detecting nonresonant analytes and for obtaining structural information as well. The examples focus on the evaluation of Fe^+ CI for the analysis of alkene and alkyne isomers by using GC–FTMS.[27] Capillary GC is used to separate olefin mixture components that then react with Fe^+ in the FTMS cell. The Fe^+ is the reactive species, which is generated by multiphoton dissociation and ionization of $Fe(CO)_5$ [Eq. (12.1)]. The use of Fe^+ as a CI reagent gives molecular weight information for the hydrocarbon and permits double bond location through a specific oxidative addition to an allylic bond followed by a smaller neutral olefin loss (illustrated for 1-decene in Eqs. (12.2) and (12.3).

$$Fe(CO)_5 \xrightarrow{h\nu} Fe \xrightarrow{h\nu} Fe^+ \tag{12.1}$$

$$Fe^+ + 1\text{-}C_{10}H_{20} \longrightarrow Fe(1\text{-}C_{10}H_{20})^+ \longrightarrow (C_3H_6)Fe(C_7H_{14})^+ \tag{12.2}$$

$$(C_3H_6)Fe(C_7H_{14})^+ \longrightarrow Fe(C_7H_{14})^+ + C_3H_6 \tag{12.3}$$

The alternatives to pulsed GC–FTMS for analysis of olefin mixtures are MS/MS and GC–HRMS. If simple mixtures containing similar concentrations of olefin isomers are to be analyzed, then MS/MS is applicable. The use of MS/MS for congeners, however, is difficult because, for example, the ion that results from C_3H_6

loss from Fe(1-decene)$^+$ would have the same mass as Fe(heptene)$^+$. The application of GC–HRMS ensures congener separation before mass analysis. The overall combination of capillary column GC, which is necessary for separating isomeric alkenes, CI, and medium resolution MS, however, is difficult to implement with conventional sector mass spectrometers. At least medium resolution ($R \geq 5000$) is required because Fe(CO)$_x$ overlaps in mass with Fe(CH$_2$)$_{2x}$. Calibrating the accurate masses of CI-produced ions by using fast scanning MS is a difficult enterprise. An alternative is to use pulsed sources to produce only Fe$^+$ and avoid ionizing the organic compounds altogether. This may be achieved by the selective MPI of Fe(CO)$_5$ to yield predominantly Fe$^+$ or the direct LD of Fe$^+$ ions from a metal surface. Acquisition of full mass spectra of the products of the Fe$^+$–olefin reaction must be done on the chromatographic time scale, and this is done by utilizing both the ion storage (to permit the Fe$^+$ to react with the organics) and the multichannel (to permit simultaneous detection) features of FTMS.

The combination of pulsed GC–FTMS and Fe(I) CI simplifies and disposes of the requirements of both high-resolution tandem MS and GC–HRMS. There are several features of FTMS that are attractive and even unique for the analysis of olefin mixtures by metal ion CI. Low-resolution mass analysis is only needed for the method to be general. The pulsed YAG generates Fe$^+$ without ionizing the organic olefins.

The reactions of the isomeric decenes with Fe$^+$ are illustrative of the method for locating double bonds. Once Fe$^+$ is produced, it will insert in the allylic position of 1-decene causing it to lose propene [Eqs. (12.2) and (12.3)]. When 2- and 3-decene isomers react, butene and pentene are eliminated, respectively. Detection limits are in the low nanogram range.

A homologous series of 1-alkenes (C$_{10}$–C$_{13}$) was investigated to see whether or not longer chain alkenes could be analyzed by using this method. Low-energy EI is not optimal because the molecular ions are not sufficiently abundant. However, the Fe$^+$ CI experiment produces spectra of adducts that give both molecular weight and structural information for alkenes as large as decenes. Unfortunately, the reaction of larger alkenes such as 1-tetradecene with Fe$^+$ yields only the double propene loss ion Fe(C$_8$H$_{16}$)$^+$ in addition to the metal ion adduct [Eq. (12.4)]:

$$\text{Fe}^+ + 1\text{-C}_{14}\text{H}_{28} \xrightarrow[-\text{C}_3\text{H}_6]{} \begin{array}{c} \text{Fe}(\text{C}_{11}\text{H}_{22})^+ \\ \text{(not seen)} \end{array} \xrightarrow[-\text{C}_3\text{H}_6]{} \text{Fe}(\text{C}_8\text{H}_{16})^+ \qquad (12.4)$$

The high reactivity of Fe$^+$ leads to sequential oxidative addition and hinders the use of this method for locating double bonds in large alkenes (C$_{12}$ or greater). A possible alternative is to use metal ions of reduced reactivity in an attempt to slow the oxidative addition reaction so that a double bond-specific fragment may be observed.

EVALUATION OF TRANSITION METAL ION REACTIVITY

Subsequent studies[28] were undertaken to assess transition metal reactivity, and laser desorption was employed to generate directly univalent gas-phase metal ions from a metal surface. The reactions of nine transition metal ions with 1-pentene were investigated by using FTMS. All metals were laser desorbed from the appropriate metal surface, and relative rate constants were obtained. V^+ and Ti^+ react rapidly with 1-pentene to yield dehydrogenation products, whereas Fe^+, Cu^+, Ni^+, and Cu^+ react slightly slower to give products originating from $C-C$ allylic bond activation. Cr^+ and Mn^+ do not activate $C-C$ bonds, but condense slowly to yield $MC_5H_{10}^+$. Zn^+ undergoes charge exchange with 1-pentene in accord with the difference in the ionization energies of Zn and 1-pentene.

The surprising reactivity of Cu^+ appears to be general with alkenes, paralleling that of Fe^+.[29,30] This behavior is remarkable because the d shell of Cu^+ is filled so the ion is expected to behave more like Cr^+ ($3d^5$) than like Fe^+. It remains to be seen if Cu^+ will react with larger alkenes to yield double bond-specific fragment ions.

The results of Cu^+/1-pentene studies have prompted further investigation of Cu^+ reactivity with other organic substrates so that its reactivity may be better understood. Differences in the fragmentation of the Cu^+–nitrile adducts and those of Fe^+, Co^+, and Ni^+ yield evidence for a side-on Cu^+–nitrile complex, whereas the other metal ions form end-on complexes by interacting with the lone pair of electrons of the nitrile nitrogen.[31]

Reactivity and Specificity of Fe(I), Cr(I), and Mo(I) with Alcohols

Application of laser MPI/MPD was carried further to study the reactions of Fe^+, Cr^+, and Mo^+ with aliphatic alcohols of different chain length and degree of branching.[32] The focus of this investigation was turned more toward the nature of the organic substrate to enhance the analytical utility of MPI/MPD for transition metal CI. Concepts of reactivity and specificity were used successfully to correlate the insertion reactions of these metal ions with the alcohols. The aliphatic chain length, degree of branching of the alcohol, and the electronic configuration of the metal ion were found to influence reactivity.

The metal ion-induced fragmentations are accounted for by three types of reactions. Dehydration is preceded by metal ion insertion into a $C-O$ bond followed by a β-hydrogen shift and subsequent water loss. Insertion into $C-C$ bonds, particularly into the nonterminal bonds, is also followed by a β-hydrogen shift. Dehydrogenation is initiated by metal ion insertion into an $O-H$ or $C-H$ bond, although the former may not be thermodynamically favored, and is followed by another β-hydrogen shift to lead to loss of H_2.

Fe^+ is the most reactive of the metal ions toward larger alcohols, and its high reactivity can be ascribed to its electronic configuration (s^1d^6). The low reactivity of Cr^+ is also understood in terms of its ground state configuration. In fact, a more controlled reactivity of Cr^+ with Cr $(CO)_6$ occurs if Cr^+ is prepared by MPI-MPD at a wavelength of 266 nm.[33] The propensity for Mo^+ insertion into $C-H$ bonds followed by multiple dehydrogenations is not easily understood. $C-H$ bond insertions are the most energy demanding, and multiple dehydrogenations are the only reactions observed for Mo^+ and alcohols of four or more carbons. This high specificity is unusual for a metal ion so reactive. The higher reactivity of Mo^+ compared to Cr^+ is ascribed to its expanded size d orbitals.

The Cr^+ reaction specificity is generally high, and the nature of the reaction is determined by the chain length and the type of branching of the alcohol. The specificity of Mo^+ is generally high with short chain primary alcohols and decreases with increasing chain length. The specificity of Fe^+ is high only when the alcohols are highly branched or when they are not primary. The specificity of the reactions of Cr^+ and Fe^+ are principally controlled by thermodynamics. Weil and Wilkins[34] also used a CO_2 laser to desorb group II metal ions from their oxides and salts and studied their reactivity with alcohols.

Other research groups are involved in the evaluation of metal ion CI for chemical analysis. Of particular note is the group of Frieser, who has extensively used both MPI–MPD and LD to generate gas-phase metal ions. They studied the reactions of Cu^+ with ketones and esters[35] and the reactions of Fe^+ with ethers,[36] ketones,[37] and hydrocarbons.[38] Their research has been extended further to previously ignored metal ions such as Rh^+,[39] Y^+,[40] and La^+.[40]

LASER DESORPTION

Desorption with CO_2 Laser

Most investigations of laser desorption of organic molecules have employed pulsed CO_2 lasers. In fact, the first demonstration of LD-FTMS for nonvolatile organic molecules was with a CO_2 laser.[10] LD had been employed with a scanning MS previously,[41] but the pulsed nature of the laser is not highly compatible with these instruments. The first commercial LD instrument was a time-of-flight (TOF) MS,[42] and although these instruments possess the multichannel advantage necessary for LD, they suffer from poor mass resolution ($R < 1000$). Some improvements may be in the offing because a mass resolution of 13,000 for m/z 372 of crystal violet was recently achieved by pulsed secondary ion mass spectrometry (SIMS) and a reflectron TOF.[43] The FTMS possesses considerably higher mass resolution and also the multichannel advantage necessary to maximize the utility of LD.

The compounds used in the original FTMS demonstration experiments are representative of those employed in earlier characterizations with other mass spectrometers (e.g., organic acids, oligopeptides, and quaternary ammonium salts). Conventional ionization and/or volatilization techniques do not work well for these compounds.

A mass resolution of 10,000 was obtained in the original work for the succinate $(M - H)^-$ anion at m/z 117 at a pressure of 10^{-7} torr. The other carboxylic acids, adipic and citric acid, yield abundant $(M - H)^-$ anions as well. The LD mass spectra of ammonium salts are dominated by R_4N^+ ions. The dipeptide Gly-Tyr gives an $(M - H)^-$ in the negative ion mode and an $(M + H)^+$ in the positive ion mode, provided the sample was wetted with HCl before admitting to the vacuum.

Two major differences are seen in the spectra of these compounds when FTMS is contrasted with TOF.[44,45] First, no abundant Na^+ and K^+ signals were observed in the FTMS determinations. This is possibly due to desorption with kinetic energy and ion loss during the longer time scale involved in ion detection. Second, the fragment ions observed are generally less abundant in FTMS, and at all laser powers the molecular ion is dominant.

Desorption with YAG Laser

In the next sequence of investigations,[14] the optical arrangement was changed by replacing the CO_2 laser with an Nd:YAG laser. The adaptation of an Nd:YAG laser permits the photon energy to be multiplied by two, three, or four so that studies of MPI and resonant laser desorption can be done in the visible or UV. Although the use of a CO_2 laser does not allow the wavelength to be varied over a wide range, many research groups still pursue laser desorption studies with a CO_2 laser.[13,44,46] The success obtained in the initial demonstrations with the CO_2 laser prompted the investigation of more complex compounds, in particular, nucleosides, oligosaccharides, glycosides, and other oligopeptides.

The nucleosides constitute an important class of biological compounds. They are the primary constituents of DNA and RNA and are found in antibiotics as well. The laser desorption of uridine is typical of nucleosides in the negative ion mode.[14] The most prominent desorbed ions are the $(M - H)^-$ and $(B - H)^-$ ions, where B is the nucleic base. The latter ion is normally the most abundant. Both purine and pyrimidine nucleosides fragment to lose HCNO and to form the NCO^- ion. Nucleoside desorption requires high-power densities, about 10^8 W/cm². The precision of the spectra was found to be poor; that is, relative abundance ratios vary to within $\sim \pm 50\%$. The positive ion spectra of the nucleosides is described later in this chapter.[17]

Monosaccharides give abundant $(M - H)^-$ ions and fragment to lose H_2O and CH_2OH. Di- and trisaccharides do not yield an $(M - H)^-$ anion or any ion in the negative ion mode that is indicative of molecular weight. Their spectra are dominated by fragments resulting from glycosidic cleavages. If NaCl is added to the analyte, however, natriated molecular ions are desorbed by the laser. The $(M + Na)^+$ ions were found to be the most abundant at power densities three times those employed to desorb negative ions.

Both reducing and nonreducing sugars (e.g., sucrose and lactose, respectively) undergo cationization on desorption. The primary fragmentation of reducing sugars is glycosidic cleavage, whereas nonreducing sugars mainly undergo loss of CH_2O. The abundance of cationized ions was improved by turning the Q-switch off to lower the power density to $\sim 10^6$ W/cm² and to increase the energy delivered as a result of lengthening the laser pulse.

The success achieved with the oligosaccharides prompted the study of more complex glycosides. Quercitrin, rutin, and xanthorhamnin are closely related flavanoid glycosides, having a flavone bonded to a mono-, di-, and trisaccharide, respectively. These compounds were mixed with NaCl and desorbed at power densities in the range $1-5 \times 10^6$ W/cm². All three compounds yield abundant $(M + Na)^+$ ions. The most important fragment ions for quercitrin and rutin involve loss of CH_2O and a cleavage within the saccharide ring bonded to flavone. Xanthorhamnin gives spectra similar to quercitrin and rutin upon LD; the most important difference is the observation of a fragment ion resulting from the cleavage of the sugar–flavone bond with Na^+ retention on the trisaccharide.

In the negative ion mode, rutin and quercitrin give nearly identical spectra except the $(M - H)^-$ of the latter is of low abundance. Xanthorhamnin desorbs to give an $(M - H)^-$ ion and a charged flavone.

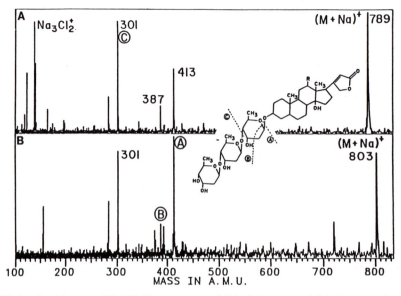

Fig. 12.3. Positive-ion LD-FTMS spectrum of (A) digitoxin and (B) digoxin obtained from one shot of the laser. Both samples were mixed with NaCl. Reprinted with permission from Ref. 14.

Digoxin and digitoxin, two steroidal glycosides, give abundant $(M + Na)^+$ ions when desorbed in the presence of NaCl, and the most abundant fragment ions result from cleavages in the saccharide chain, consistent with the fact that sodium attaches to the sugar portion of the molecule (see Fig. 12.3).

Erythromycin A is an antibiotic, composed of the aglycone erythronolide and the sugars cladinose and desosamine. The most abundant desorbed ion is $(M + Na)^+$. Fragment ions result from losses of cladinose, water, and both.

Once experimental conditions are understood, LD gives good results in FTMS for many classes of compounds.[10-15] The spectra are largely free of chemical noise, unlike fast atom bombardment (FAB) mass spectra, and detection limits are lower than those achievable by FAB. In spite of the advantages, certain obstacles stand in the way of widespread acceptance. First, desorption conditions must be adjusted for different classes of compounds, and that leads to difficulties in handling an unknown substance. Second, FAB has worked thus far more successfully for higher mass biomolecules than does LD. Third, there have been fewer investigations of LD, and thus it is less understood than FAB. Fourth, FT mass spectrometers are not as common as sector or quadrupoles. Finally, the advantage of LD, producing spectra of lower chemical noise, may be partially offset by new developments in FAB such as continuous-flow FAB.

MECHANISTIC STUDIES OF LASER DESORPTION

Recognizing that one of the obstacles for widespread application of LD is a lack of understanding to guide experiment design, we shifted our interest to mechanistic studies, which we believed would better serve the utility of LD. The remainder of the chapter is a survey of some of our attempts to understand the mechanism of LD.

Internal Energy Distributions of Laser-Desorbed versus FAB-Desorbed Ions

The decomposition pathways of isomeric and homologous phosphonium ions produced by LD and FAB are appropriate for establishing qualitatively the energy distributions associated with various desorption processes.[47] The *n-*, *sec-*, and isobutyltriphenylphosphonium ions give more fragmentation when desorbed by the laser at 1064 nm than when desorbed by FAB. It was also found that changing the power density of the laser offers more latitude in changing the abundance of fragment ions than does changing the beam flux of the atom gun in FAB. Fragmentation increases with successive laser shots as well, demonstrating that sample thickness also influences the fragmentation.

All three butyl isomers yield LD and FAB spectra that can be employed for compound identification, but complete compound-specific fragmentation does not occur on FAB or LD. This information can be obtained with collisional activation in both FT and sector mass spectrometers. The collisionally activated decompositions are examples of charge-remote fragmentations, a class of reactions that is useful for structure determination of fatty acids and related materials.[48,49] These reactions are often not initiated by LD and FAB desorption, but require higher energy inputs such as from collisional activation.

The internal energy distributions of ions desorbed by FAB or by the laser were compared by assessing both the abundance and the nature of the fragment ions. Phosphonium ions undergo two charge-proximate decompositions, which occur either as a simple cleavage to yield a triphenylphosphonium ion, $(C_6H_5)_3P^{+}$, or as a hydrogen rearrangement to yield a protonated phosphine, $(C_6H_5)_3PH^+$. The abundance ratio of these two ions is a measure of the relative internal energies of the decomposing ions sampled in both techniques. Lowering the internal energy of the fragmenting ions favors the formation of the protonated phosphine ion of m/z 263. This is in accord with the time-honored concept that rearrangements are favored for ions having low internal energies and that simple cleavages are more facile for ions having higher internal energy.

Under FAB desorption, the *n*-alkyltriphenylphosphonium ions prefer to undergo simple cleavage reactions. However, the total abundance of fragment ions is small relative to those produced by LD. These observations are evidence for an internal energy distribution centered below the fragmentation thresholds and a high-energy tail to accommodate the observation of low abundance fragment ions formed by simple cleavage.

On the other hand, laser-desorbed *n*-alkyltriphenylphosphonium ions preferably yield the rearranged $(C_6H_5)_3PH^+$ ions. This indicates that the internal energy distribution associated with laser-desorbed ions is wider and centered at a higher value relative to the distribution obtained by FAB, albeit with a less pronounced high-energy tail (see Fig. 12.4 for a schematic diagram).

It is emphasized that a mass resolution of 75,400 was demonstrated for the *n*-tetradecyltriphenylphosphonium ion at a pressure of 1×10^{-8} torr. Resolving pow-

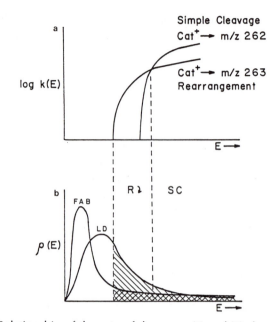

Fig. 12.4. (a) Relationship of the rate of decomposition, $k(E)$, for rearrangement and simple cleavage processes to the internal energy, E, of a decomposing phosphonium ion (Cat, cation). (b) Proposed schematic of the probability function, $\rho(E)$, describing the distribution of internal energies, E, of phosphonium ions formed by FAB and LD ionization. The regions labeled R and SC represent internal energies at which rearrangement or simple cleavage is dominant, respectively. Reprinted with permission from Ref. 47.

ers of this magnitude are currently impossible for time-of-flight and sector mass spectrometers to achieve. Moreover, the resolving power can be extended to high mass ions. Ijames and Wilkins[50] recently demonstrated in elegant fashion a resolving power of 60,000 for laser-desorbed ions from a synthetic polymer in the mass range of ~6000. This resolving power extrapolates to ~130,000 for an ion having the m/z of n-tetradecyltriphenylphosphonium ion at 1.2 T, in reasonable agreement with the value we observed in the phosphonium ion study.[47]

Extent of Gas-Phase Cationization of Laser-Desorbed Sucrose

As part of a second mechanistic study, we investigated the extent of gas-phase cationization of sucrose by sodium ions as a function of laser power density. The motivation arises from the results of the earlier determinations. In the previously discussed study of disaccharides, it was noted that the addition of NaCl to the sample is essential for observing a molecular ion. Furthermore, overall sensitivity increases in the positive ion mode for many compounds when NaCl is added. Thus, it is important to understand the mechanism of cationization.

Mechanisms of desorption were studied previously, and a brief summary of that work will be presented to lay the foundation for this study. Three principal mechanisms of desorption have been elucidated to date: thermal,[44] shock wave driven,[51] and resonant.[52] The former is found to predominate under low-power density conditions in which desorption proceeds as a function of temperature. The second is characterized at higher power densities in which little fragmentation is observed (unlike thermal), and the sample probe or substrate does not participate in the desorption process. Resonant desorption occurs when the analyte itself undergoes an electronic transition when irradiated by the laser; fragmentation is extensive near threshold, and there also appears to be no effect of the sample probe.

Gas-phase cationization of sucrose by potassium occurs when the two are thermally desorbed from separate, shielded sources in a quadrupole mass analyzer.[53] Extensive gas-phase cationization also occurs when potassium and sucrose are continuously desorbed from separate sources coincident with a 200 W/cm^2 IR laser.[54] We extended these studies to a pulsed YAG laser at power densities normally employed for analytical determinations.

A double-substrate, split probe was used to study the extent of gas-phase formation of the $(M + Na)^+$ of sucrose at power densities of 10^6 and 10^{10} W/cm^2.[16] The spectra obtained at these power densities are representative of the thermal and shockwave desorption mechanisms, respectively. NaCl was loaded on a 47% transmittant tungsten mesh, which was mounted above a smooth copper probe. The distance of separation was the thickness of the mesh, 30 μm, and the laser spot size was approximately 0.5 mm in diameter.

A series of control experiments were executed to establish meaningful limits and to assess the amount of cationization that would constitute evidence of gas-phase cationization. The most significant control involved desorbing sucrose alone from the copper probe with the unloaded mesh above it. At both power densities mentioned, the amount of cationized adduct is insignificant.

When both sucrose and NaCl were loaded on the probe and mesh, respectively, strong evidence for gas-phase cationization was obtained for a power density of 10^6 W/cm^2, but none at 10^{10} W/cm^2 for 30-μm separation. Gas-phase cationization occurs even at a separation of 1 mm at 10^6 W/cm^2, approaching the separation used in the experiment cited in Ref. 54.

The extent of gas-phase cationization is consistent with the mechanisms of desorption believed to predominate at the two power densities employed. Time-resolved studies[44] show that alkali metal ion desorption is coincident with the highest temperatures generated in the sample probe and that neutral desorption is sustained over a longer period of time. Higher temperatures are obtained in the tungsten mesh relative to the copper, and sodium desorption will be prolonged relative to sucrose desorption from the copper. This temperature differential enhances the probability of gas-phase cationization and compensates for the fact that the mesh is not 100% transmitting.

Laser desorption at 10^{10} W/cm^2 proceeds without interaction from the substrate (sample probe) and, thus, is not driven by the temperature. The lack of broadening present in peaks observed in a TOF mass spectrometer for laser-desorbed ions indicates that desorption times are much shorter than at lower power densities. A theoretical calculation indicates that ion formation must occur within 11 μm of the probe surface in the TOF experiment.[16,55] The selvedge or intermediate region between the solid state and the gas phase is not as diffuse, and the material ejected during desorption must be largely condensed (microparticulate). Thus, there is little probability of gas-phase cationization.

Resonant Desorption of Nucleosides

The interactions of the analyte and metal ions in the solid state also affect the degree of cationization. This was demonstrated in a study of the low-power LD of nucleosides. The desorption/ionization characteristics of four nucleosides were studied as a function of laser energy and wavelength.[17] The pyrimidine nucleosides, thymidine and uridine, were found to exhibit very different desorption characteristics relative to the purine nucleosides adenosine and guanosine (see structures 1–4) especially at a wavelength of 266 nm and near the desorption threshold.

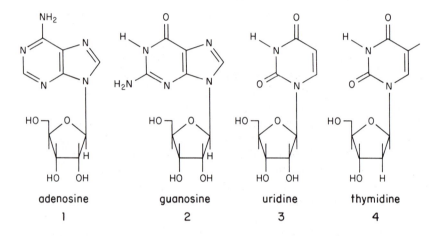

| adenosine | guanosine | uridine | thymidine |
| 1 | 2 | 3 | 4 |

Under near-threshold resonant desorption conditions ($\sim 10^8$ W/cm^2 and a wavelength of 266 nm), doubly sodiated molecular ions $(M - H + 2Na)^+$ are predominantly desorbed from pyrimidines whereas purines yield singly sodiated base ions $(B + Na)^+$. As the power density is increased, the doubly sodiated base fragment ion, $(B - H + 2Na)^+$, becomes the most abundant for guanosine, whereas the pyrimidines eventually desorb as $(M + Na)^+$ ions (Figs. 12.5 and 12.6).

At 1064 nm, the nucleosides desorb in the power density range of 1–5×10^5 W/cm^2. Near the desorption threshold at 1064 nm, guanosine displays desorption characteristics similar to those near threshold at 266 nm, but at higher power density, the monosodiated base, $(B + Na)^+$, remains the most abundant ion. The other nucleosides desorb like guanosine at onset, but at higher powers, the $(M + Na)^+$ peak dominates. The doubly sodiated ions never exhibit large abundances.

One possibility for the dominant formation of doubly sodiated molecular ions from the pyrimidines at the 266-nm desorption onset is a lowering of the pK_a in the S$_1$ state, as is known to occur for thymine.[56] This increase in acidity enhances the probability of proton loss and makes doubly sodiated ion formation more facile. This mechanism was previously proposed as a driving force in resonant laser desorption,[9,57] but no systematic investigations have been undertaken to prove the hypothesis. The model compounds, 1-naphthoic acid and 2-naphthol, were analyzed separately and in mixtures to assess evidence of excited state proton transfer by comparison of $(M - H)^-$ ion abundances in the negative ion mode at wavelengths of 266, 532, and 1064 nm. These two compounds were chosen because 2-naphthol undergoes a decrease in pK_a from 9 to 2 in the S_1 state, whereas 2-naphthoic acid undergoes an increase in pK_a from 6 to 7 in the S_1 state. No evidence for excited state chemistry was obtained.

Because excited state proton transfer is not apparent in the 2-naphthol/1-naphthoic acid system, excited state proton transfer is an unlikely cause for the pyrimidine behavior near the desorption onset at 266 nm. If the imide hydrogen does not become labile via an intermolecular process to permit exchange with Na^+, then an intramolecular process may be responsible for the hydrogen lability. Quantum mechanical calculations[58,59] indicate that the lactim (enol) form of uracil and thymine is favored in the S_1 state and the lactam (keto) form is favored in the ground state. Guanine and adenine favor the lactam form in both S_0 and S_1 states. Furthermore, luminescence studies of two isomers that are locked in the lactim

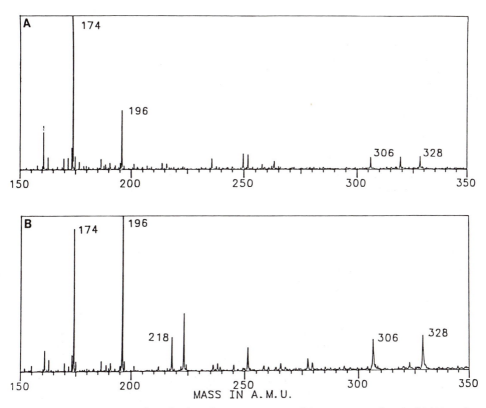

Fig. 12.5. Guanosine desorbed under resonant conditions at near threshold (A) and at higher power (B). The ion of m/z 306 is $(M + Na)^+$ and the ion of m/z 328 is $(M - H + 2Na)^+$.

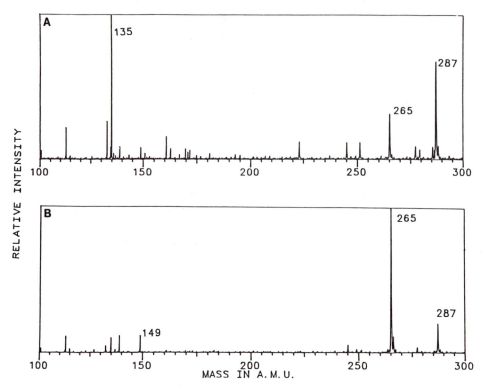

Fig. 12.6. Thymidine desorbed under resonant conditions at near threshold (A) and at higher power (B). The ion of m/z 265 is $(M + Na)^+$ and m/z 287 is $(M - H + 2Na)^+$.

(2,4-dimethoxypyrimidine) and the lactam (N_1,N_3-dimethyluracil) forms show that luminescence occurs from the lactim form. This is evidence that the pyrimidine nucleosides are converted to the lactim form when excited at 266 nm[60] [Eq. (12.5)].

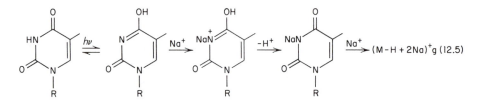

The mechanism of $(M - H + 2Na)^+$ formation can be envisioned as shown in Eq. (12.5). Once the pyrimidine is excited, the imide proton vacates its position on the nitrogen and moves to a carbonyl oxygen, making it possible for a sodium ion to associate with the imide nitrogen. Following deexcitation, the nitrogen will associate more strongly with the sodium ion, and the hydrogen released from the oxygen is blocked by the sodium so it cannot return to the imide nitrogen. Further support for this mechanism is seen in the desorption characteristics of adenosine (structure **1**), which has no imide nitrogens and gives no doubly sodiated ions on laser desorption.

Other evidence indicates that it is the triplet rather than the singlet state that is responsible for the mechanism of proton loss. The ratios of di- to monosodiated ions decrease by a factor of 10 and 4 for thymidine and uridine, respectively, as the laser energy is raised from 5 to 6.5 mJ. These decreases are consistent with the lifetimes of the T_1 states in solution: 25 and 2 μsec for thymidine and uridine, respectively.[61] The longer lifetime triplet state shows greater probability of disodiation. Furthermore, phosphorescence is enhanced greatly in the solid state when the imide nitrogen is deprotonated.[60]

Turning to another matter, we find that, unlike the laser desorption of sucrose, there is no evidence for formation of $(M - H + 2Na)^+$ or $(M + Na)^+$ in the gas phase from nucleosides at a wavelength of 1064 nm and at a power density of $1–5 \times 10^5$ W/cm^2. The reason nucleosides do not undergo gas-phase cationization is that they tend to self-associate in the solid state, and that may be carried over to the gas phase. Vapor pressure osmometry[62] was employed to show that nucleosides will self-associate and oligomerize in solution. In the solid state, nucleosides "stack" on the purine or pyrimidine ring systems, and sugar units are staggered to minimize ribosyl repulsion.[62] TOF and quadrupole MS experiments provided evidence for the cluster desorption of guanosine (G); broadening of flight times indicates that a $(6G + K)^+$ cluster decomposes in the flight tube when laser desorbed.[63] These clusters are not observed in the FT instrument presumably because of the longer time scale for detection. Sucrose, on the other hand, is used as a standardizing agent for vapor pressure osmometry[61] and viscosity measurements[64] because it has no tendency to self-associate even at high concentrations. These arguments are consistent with the property of sucrose, but not nucleosides, to desorb as an independent unit and undergo gas-phase cationization.

Reproducibility of LD/FTMS

One of the issues that stands in the way of widespread use of LD as an analytical method is the alleged lack of reproducibility of the spectra. Here we show that if care is taken, the measurement of relative abundances can be reasonably precise.

Two independent investigations were undertaken to assess the reproducibility of laser desorption. In the first study,[65] the nature of the copper probe surface (rough or smooth) and sample deposition techniques (electrospray[66] or pipetting) were varied. n-Propyltriphenylphosphonium bromide and NaCl/sucrose were employed as model systems because they represent two different classes of desorbing compounds. The former exists as an ion in the solid and desorbs directly whereas sucrose requires cationization for a molecular ion to be observed. Ion abundance ratios were compared to assess reproducibility. The most reproducible ratio gave an RSD of $\pm 14\%$ for the abundance ratio of two abundant sucrose fragment ions when the solution was electrosprayed onto a "smooth" probe surface. It was reasoned that the smooth surface gave better uniformity and that electrospray provided a more homogeneous codeposition of NaCl and sucrose and, therefore, a more consistent codesorption. In general, however, the desorption mass spectra of the phosphonium ions (an ion in the solid state) are more precise than those of sucrose.

In a second study of reproducibility, a resonant wavelength of 266 nm was

Table 12.1.   Ion abundance ratios[a]

	(m/z 174)/(m/z 196)		(m/z 306)/(m/z 328)	
	Electrospray	Pipet	Electrospray	Pipet
2 mJ	1.6 ± 0.2	2.2 ± 0.2	1.00 ± 0.14	0.98 ± 0.05
4 mJ	0.94 ± 0.10	0.65 ± 0.03	0.81 ± 0.08	0.48 ± 0.02

[a]Ratios of the monosodiated to disodiated bases, $(m/z\ 174)/(m/z\ 196)$, and molecular ions, $(m/z\ 306)/(m/z\ 328)$, desorbed at 2 and 4 mJ and a wavelength of 266 nm from a mixture of guanosine and an equal weight of NaCl deposited by electrospray and pipetting. N = eight determinations.

employed to desorb guanosine from a smooth copper probe surface. The resonant desorption of amino acids and dipeptides at a wavelength of 266 nm is known to yield more reproducible results than nonresonant desorption when TOF detection is used.[52] LD was carried out at pulse energies of 2 and 4 mJ ($1-5 \times 10^8$ W/cm^2) when the sample, containing an equal weight of NaCl, was electrosprayed or pipetted onto the probe, depositing 1.5 and 10 μg of sample on the probe, respectively. The ion abundance ratios of mono- to disodiated base and molecular ions are compared in Table 12.1 to assess reproducibility.

Significant improvement in reproducibility was found under resonant conditions for LD-FTMS at a wavelength of 266 nm when compared to the nonresonant conditions of an earlier study. Every measured ion abundance ratio is more precise than the most precise ratio obtained in the earlier study. The results also indicate that pipetting is the better sample deposition technique when resonant desorption is used in LD-FTMS. One reason for the better precision obtained for pipetting is that more ions are sampled ($\sim$45% more than when electrospray is used) because, when dealing with a set amount of material, more sample is laid down on the probe when pipetting ($\sim$85% of the sample is lost to the probe under our electrospray conditions).

The improved precision in the resonant desorption mode suggests that the use of a resonantly absorbing matrix such as purine or pyrimidine derivative may enable quantitative analysis by LD-FTMS.

SUMMARY

This chapter is a survey of the applications of lasers with FTMS as have been demonstrated at the University of Nebraska. The work has been focused on both MPI and on laser desorption.

The motivation for research in MPI-FTMS has been twofold. One purpose is to demonstrate the high selectivity and low detection limits that can be achieved by using direct ionization of resonant analytes with a YAG laser combined with an FT mass spectrometer. The other purpose is to show the utility of metal ions formed by MPI/MPD as highly specific chemical ionization reagents.

The work in laser desorption was focused on compound characterizations initially and more recently has been turned to mechanistic studies. The latter have been centered on the advantages of resonant desorption at 266 nm, the nature of gas-phase cationization, and reproducibility of LD-FTMS.

Note added in proof: Recent work by M. Karas and F. Hillenkamp serves to demonstrate that UV-laser desorption with a Nd-YAG-laser is capable of producing gas-phase molecule ions for proteins having molecular masses in excess of 200,000 dalton. Mass analysis has been by time-of-flight mass spectrometry. It is expected that FTMS should also play an important role in the mass analysis of these high molecular weight materials.

ACKNOWLEDGMENT

This chapter was prepared with support by the U.S. National Science Foundation, Midwest Center for Mass Spectrometry (CHE 8620177).

REFERENCES

1. A. G. Marshall, *Acc. Chem. Res.* **18,** 316 (1985).

2. M. L. Gross and D. L. Rempel, *Science* **226,** 261 (1984).

3. D. H. Russell, *Mass Spectrom. Rev.* **5,** 167 (1986).

4. D. A. Laude, C. L. Johlman, R. S. Brown, D. A. Weil, and C. L. Wilkins, *Mass Spectrom. Rev.* **5,** 107 (1986).

5. M. Bamberg and K.-P. Wanzcek, presented at 35th ASMS Conference on Mass Spectrometry and Allied Topics, Denver, CO, May 24–29, 1987.

6. M. P. Irion, W. D. Bowers, R. L. Hunter, F. S. Rowland, and R. T. McIver, Jr., *Chem. Phys. Lett.* **93,** 375 (1982).

7. R. C. Burnier, T. J. Carlin, W. D. Reents, Jr., R. B. Cody, R. K. Lengel, and B. S. Frieser, *J. Amer. Chem. Soc.* **101,** 7127 (1979).

8. D. B. Jacobsen and B. S. Frieser, *J. Amer. Chem. Soc.* **107,** 1581 (1987).

9. E. C. Tews and B. S. Frieser, *J. Amer. Chem. Soc.* **109,** 4432 (1987).

10. D. A. McCrery, E. B. Ledford, Jr., and M. L. Gross, *Anal. Chem.* **54,** 1437 (1982).

11. R. S. Brown, D. A. Weil, and C. L. Wilkins, *Macromolecules* **19,** 1255 (1986).

12. C. L. Wilkins and R. S. Brown, in *Mass Spectrometry in the Analysis of Large Molecules,* edited by C. J. McNeal. John Wiley, Chichester, 1986.

13. M. L. Coates and C. L. Wilkins, *Anal. Chem.* **59,** 197 (1987).

14. D. A. McCrery and M. L. Gross, *Anal. Chim. Acta* **178,** 91 (1985).

15. R. S. Brown and C. L. Wilkins, *Anal. Chem.* **58,** 3196 (1986).

16. M. P. Chiarelli and M. L. Gross, *Int. J. Mass Spectrom. Ion Process.* **78,** 37 (1987).

17. M. P. Chiarelli and M. L. Gross, *J. Phys. Chem.,* in press.

18. C. H. Watson, G. Baykut, and J. R. Eyler, *Anal. Chem.* **59,** 1133 (1987).

19. W. D. Bowers, S.-S. Delbert, and R. T. McIver, Jr., *Anal. Chem.* **58,** 969 (1986).

20. E. B. Ledford, Jr., R. L. White, S. Ghaderi, C. L. Wilkins, and M. L. Gross, *Anal. Chem.* **52,** 2450 (1980).

21. T. M. Sack and M. L. Gross, *Anal. Chem.* **55,** 2419 (1983).

22. T. M. Sack, D. A. McCrery, and M. L. Gross, *Anal. Chem.* **57,** 1290 (1985).

23. M. Seaver, J. W. Hudgens, and J. J. DeCorpo, *Int. J. Mass Spectrom. Ion Phys.* **34,** 159 (1980).

24. D. M. Lubman and M. N. Kronick, *Anal. Chem.* **54,** 660 (1982).

25. G. Rhodes, R. B. Opsal, J. T. Meek, and J. P. Reilly, *Anal. Chem.* **55,** 280 (1983).

26. A. Bjorseth, A. J. Dennis (eds.), *Polynuclear Aromatic Hydrocarbons: Chemistry and Biological Effects.* Battelle Press, Columbus, OH, 1980.

27. D. A. Peake, S.-K. Huang, and M. L. Gross, *Anal. Chem.* **59,** 1557 (1987).

28. D. A. Peake and M. L. Gross, *J. Am. Chem. Soc.* **109,** 600 (1987).

29. D. A. Peake, M. L. Gross, and D. P. Ridge, *J. Am. Chem. Soc.* **103**, 4307 (1984).
30. D. B. Jacobsen and B. S. Frieser, *J. Am. Chem. Soc.* **105**, 4784 (1983).
31. B. Lebrilla, T. Drewell, and H. Schwarz, *Organometallics* **6**, 2450 (1987).
32. S.-K. Huang, R. W. Holman, and M. L. Gross, *Organometallics* **5**, 1857 (1986).
33. S.-K. Huang and M. L. Gross, *J. Phys. Chem.* **89**, 4422 (1985).
34. D. A. Weil and C. L. Wilkins, *J. Am. Chem. Soc.* **107**, 7316 (1985).
35. R. C. Burnier, G. D. Byrd, and B. S. Frieser, *Anal. Chem.* **52**, 1641 (1980).
36. R. C. Burnier, G. D. Byrd, and B. S. Frieser, *J. Am. Chem. Soc.* **103**, 4360 (1981).
37. D. B. Jacobsen and B. S. Frieser, *J. Am. Chem. Soc.* **105**, 5197 (1983).
38. G. D. Byrd, R. C. Burnier, and B. S. Frieser, *J. Am. Chem. Soc.* **104**, 3565 (1982).
39. G. D. Byrd and B. S. Frieser, *J. Am. Chem. Soc.* **104**, 5944 (1982).
40. Y. Huang, M. B. Wise, D. B. Jacobsen, and B. S. Frieser, *Organometallics* **6**, 346 (1987).
41. R. A. Bingham and P. L. Salter, *Anal. Chem.* **48**, 1735 (1980).
42. K.-D. Kupka, F. Hillenkamp, and C. H. Schiller, *Adv. Mass Spectrom.* **8A**, 935 (1979).
43. E. Nichius, T. Heller, H. Feld, and A. Benninghoven, Presented at the 35th ASMS Conference on Mass Spectrometry and Allied Topics, Denver, CO, May 24–29, 1987.
44. R. J. Cotter and J.-C. Tabel, *Anal. Chem.* **56**, 1662 (1984).
45. H. J. Heinen, S. Meier, H. Vogt, and R. Wechsung, *Adv. Mass Spectrom.* **8A**, 942 (1979).
46. R. E. Shomo, A. G. Marshall, and R. P. Lattimer, *Int. J. Mass Spectrom. Ion Process.* **72**, 209 (1986).
47. D. A. McCrery, D. A. Peake, and M. L. Gross, *Anal. Chem.* **57**, 1181 (1985).
48. N. J. Jensen, K. B. Tomer, and M. L. Gross, *J. Am. Chem. Soc.* **107**, 1863 (1985).
49. J. Adams, L. J. Deterding, and M. L. Gross, *Spectrosc. Int. J.* **5**, 199 (1987).
50. C. F. Ijames and C. L. Wilkins, *J. Am. Chem. Soc.* **110**, 2687 (1988).
51. B. Linder and V. Seydel, *Anal. Chem.* **57**, 895 (1985).
52. M. Karas, D. Bachman, and F. Hillenkamp, *Anal. Chem.* **57**, 2935 (1985).
53. G.J.Q. van der Peyl, K. Isa, J. Haverkamp, and P. G. Kistemaker, *Org. Mass Spectrom.* **16**, 416 (1981).
54. R. Stoll and F. W. Röllgen, *Z. Naturforsch.* **37A**, 9 (1982).
55. F. Hillenkamp, personal communication (1986).
56. J. F. Ireland, in *Advances in Physical Organic Chemistry* edited by V. Gold and D. Bethell. Academic Press, London, 1976.
57. C. D. Porter and D. M. Hercules, *Anal. Chem.* **58**, 25 (1986).
58. V. J. Danilov, *Biofizika* **12**, 540 (1967).
59. N. K. Kochetkov and E. I. Badovskii, *Organic Chemistry of Nucleic Acids,* Part B. Plenum Press, London, 1972.
60. J. W. Longworth, R. O. Rahn, and R. G. Shulman, *J. Chem. Phys.* **45**, 2930 (1966).
61. C. Salet, R. Bensasson, and R. S. Becker, *Photochem. Photobio.* **30**, 325 (1979).
62. A. D. Broom, M. P. Schweizer, and P.O.P. T'so, *J. Am. Chem. Soc.* **89**, 3612 (1967).
63. E. D. Hardin and M. L. Vestal, *Anal. Chem.* **53**, 1492 (1981).
64. J. Eisinger and A. A. Lamola, in *Excited States of Proteins and Nucleic Acids,* edited by R. F. Steiner and I. Weinryb. Plenum Press, London, 1971.
65. D. A. McCrery and M. L. Gross, *Anal. Chim. Acta* **178**, 105 (1985).
66. C. J. McNeal, R. D. MacFarlane, and E. L. Thurston, *Anal. Chem.* **54**, 336 (1979).

13

Lasers Coupled with Fourier Transform Mass Spectrometry

LYDIA M. NUWAYSIR and CHARLES L. WILKINS

Fourier transform mass spectrometry (FTMS) has evolved as a versatile analytical tool capable of ultrahigh resolution, accurate mass measurement, and low-pressure chemical ionization. In addition to its applicability for structure analysis, it also facilitates study of positive- and negative-ion gas-phase ion–molecule reactions. Many excellent review articles summarizing the numerous analytical, environmental, biological, and physical applications of FTMS have been published.[1-5] Because of its wide mass range, FTMS permits analysis of otherwise refractory substances. However, to realize this potential, a method of volatilizing and ionizing high mass compounds is needed. A number of techniques that enable the ionization of non-volatile, thermally labile molecules have been developed in recent years.[6-10] For FTMS, best resolution is obtained when low cell pressures are maintained. Because FTMS detects and records all masses at once, pulsed sources are compatible. Laser desorption is ideally suited for use with FTMS and is particularly well suited to analysis of high mass materials.

For many organic compounds, laser desorption FTMS yields primarily molecular weight information. For low molecular weights (hundreds or less), collision-activated dissociation (CAD) can be used to produce structurally informative ions. However, when masses are greater (above several thousand daltons), CAD is much less effective for fragmenting ions under low energy FTMS conditions. In these cases, photodissociation using a second laser pulse (or pulses) may provide an alternative.

Work in our laboratory has been devoted to the development of laser desorption-FTMS (LD-FTMS) for analytical application to biomolecules, polymers, porphyrins, and other high-mass, nonvolatile and/or thermally labile compounds. The first part of this chapter includes a basic introduction to laser desorption theory and reviews research done in our laboratory using LD-FTMS. FTMS theory will not be covered in this chapter because it is discussed elsewhere in the book. The second part reviews briefly the photodissociation technique and describes results obtained using photodissociation to obtain structural information from laser-desorbed ions. Research done in other laboratories purposely has been omitted, not because it is less important, but because the intention of the present chapter is to review and summarize work done at the University of California, Riverside. It is suggested that the reader refer to one of the general articles previously cited for a more comprehensive overview of FTMS.

Laser desorption and laser desorption/photodissociation studies are performed in our laboratory using Nicolet FTMS-1000 and Nicolet FTMS-2000 instruments. A block diagram of the FTMS-1000 laser system is shown in Fig. 13.1a. This instrument utilizes a 3-T superconducting magnet and a 2-in. cubic single-section cell with a vacuum system that maintains a base pressure of $\sim 2 \times 10^{-9}$ torr. Output from a Tachisto 215G pulsed CO_2 laser is directed into the cell through a ZnSe window and focused onto the tip of a direct insertion probe by a ZnSe or KBr lens. Power density at the probe tip is estimated to be between 10^6 and 10^8 W/cm^2 for a spot size of about 1 mm^2, with a pulse width between 40 and 100 nsec. Laser desorption/photodissociation experiments are performed using a Lambda Physik EMG 201 MSC excimer laser operating at either 193 nm (ArF) or 308 nm (XeCl). The excimer laser beam is directed into the FTMS-1000 through a quartz window in the vacuum flange by a series of beam-steering mirrors and reflected into the cell by an Al-MgF mirror mounted at a 45° angle on the cell assembly. The beam is unfocused. This experimental arrangement was designed after other configurations failed to produce consistent results. In the previous design, the excimer laser beam was directed through a hollow direct insertion probe and focused into the center of the cell by a lens mounted near the probe tip. Beam power density was high enough to cause desorption from the analyzer cell front and back trap plates. This caused uncertainties regarding the origin of ions observed following excimer laser irradiation. Also, both focused and unfocused portions of the beam came in contact with ions in the cell, resulting in potentially different interactions with different ions. The experimental design shown in Fig. 13.1a eliminated these inconsistencies and was used to obtain most of the FTMS-1000 laser desorption/photodissociation results.

The FTMS 2000 has a differentially pumped dual $1\frac{7}{8}$-in. cubic cell and maintains base pressures of $\sim 2 \times 10^{-8}$ torr in the source cell and $\sim 2 \times 10^{-9}$ torr in the analyzer cell. This instrument utilizes a 7-T superconducting magnet and is equipped with a computer-controlled solids probe. The CO_2 laser beam is directed into the source cell via a series of beam steering mirrors and is focused onto the tip of the solids probe by an off-axis paraboloid mirror. A 2-mm conductance limit connects the two cells and allows ions to be transferred into the analyzer cell following desorption into the source cell (Fig. 13.1b). A quartz window on the analyzer vacuum flange allows the excimer laser beam to pass into the analyzer cell for photodissociation experiments. Use of this cell configuration, as well as higher magnetic field, allows for ultrahigh resolution. Improved low-frequency noise characteristics (together with use of the higher field) extend the mass range of this FTMS-2000 beyond that obtained with the FTMS-1000.

Solid samples typically are prepared by dissolving them in suitable solvents and then depositing the resulting solution onto the stainless-steel probe tip. Solvent is allowed to evaporate, leaving a thin film of sample material. Salts added to enhance cationization are either dissolved in the solvent with the sample or deposited onto the probe tip prior to or following sample deposition. Alternatively, potassium salts

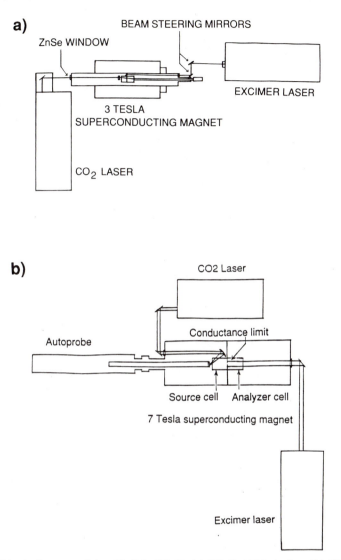

Fig. 13.1. Block diagram of the LD/PD-FTMS. (a) FTMS-1000 design. (b) FTMS-2000 design.

are intimately mixed with sample and pressed into the grooves of the probe tip to form a semitransparent disk of material. For samples that are not soluble or that form gels in solution, double-sided tape is used to affix finely ground sample material to the probe tip.

Collision-activated dissociation (CAD) was performed for a number of peptide samples to increase the extent of fragmentation and aid in sequence analysis. For these studies, water vapor was present at a constant pressure of $\sim 9 \times 10^{-8}$ torr throughout the course of the experiment.

LASER DESORPTION BACKGROUND AND BASIC THEORY

The desire to study thermally labile and high mass compounds by mass spectrometry prompted development of the laser desorption technique. Alternatively, polar substituents of nonvolatile samples can be chemically derivatized to produce compounds with higher vapor pressures amenable to electron ionization or chemical ionization mass spectrometry. However, the limitations of this procedure become apparent as the mass and number of polar groups increase. Field desorption is another method used to analyze solid samples, but affords only moderate reproducibility. Flash pyrolysis is frequently used to create fragment ions reproducibly, but provides no molecular weight information for thermally unstable compounds. Secondary ion mass spectrometry (SIMS), fast atom bombardment (FAB), and plasma desorption are more recent techniques developed concurrently with laser desorption and have been utilized successfully for the analysis of high mass organic compounds and biomolecules.[7,9,10]

Numerous experiments have been performed in attempts to understand the nature of the laser desorption process.[11-15] Difficulties arise as a result of the many experimental parameters that can be manipulated. Power density, wavelength, pulse duration, substrate, matrix, and sample preparation can all affect the mass spectrum and such changes may suggest different desorption mechanisms. One notable feature frequently observed in laser desorption spectra of organic molecules is production of molecular ions by cationization or protonation rather than electron abstraction. In the negative ion mode, deprotonation rather than electron attachment usually is observed. Radical ions rarely are generated.

For thin sample films, laser power densities of approximately 10^8 W/cm^2, and pulse durations between nanoseconds and microseconds, it is generally accepted that laser desorption occurs via a fast heating of the substrate, resulting in thermionic emission of ions from the hot center of the laser spot and vaporization of intact neutral sample molecules further away at lower temperatures. Gas-phase reactions in the resultant plasma of ions and neutrals above the surface can produce cationized or protonated molecular ions.[13,16] Metal-attached ions can result from reactions with alkali salts added to the sample, or present as contaminants in either the sample or the spectrometer. Cationization arising from laser ablation of metal substrates (e.g., Fe$^+$ ions) also has been observed. Desorbed neutrals can be protonated by fragment ions or residual solvent. Use of acidic solvents enhances the formation of protonated molecular ions.

As long as heating is rapid, evaporation of intact sample molecules rather than thermal degradation is favored.[17,18] It is generally found that use of higher power densities at the surface results in more fragmentation and that laser wavelength is important only for determining threshold desorption values (i.e., the minimal laser energy needed to desorb sample molecules). Samples that exhibit strong absorptions at the desorbing wavelength tend to have lower threshold desorption values. The qualitative aspects of spectra are influenced relatively little by laser wavelength. However, some variations in relative ion abundances are observed for spectra obtained using a wavelength corresponding to an absorption band in the sample, as compared with spectra obtained with a laser wavelength in which the sample does not absorb.[15]

SAMPLES STUDIED BY LD-FTMS

Analysis of small amounts of complex biomolecules is one of the most challenging analytical problems. Laser desorption Fourier transform mass spectrometry (LD-FTMS) is used in our laboratory as a fast and sensitive method for volatilizing and detecting small amounts of large, thermally labile biomolecules. Oligopeptides are one example.

Table 13.1 lists some of the peptides we have analyzed by LD-FTMS. Besides those listed, numerous di-, tri-, and tetrapeptides, as well as the 20 naturally occurring amino acids have been examined. Figure 13.2 is the laser desorption mass spectrum of bradykinin and is typical of the LD-FTMS spectra obtained from oligopeptides.[19] Abundant protonated molecular ions are observed (m/z 1061) as well as fragment ions arising from cleavages along the peptide chain. For the most part, abundant fragment ions correspond to A and Z sequence ions. Figure 13.3 is the laser desorption mass spectrum of angiotensin I.[20] $[M + H]^+$ (m/z 1297) predominates. Some higher mass fragment ions are also detected.

These results prompted the development of a method for sequencing peptides using LD-FTMS.[21] Pattern recognition and use of accurate mass measurements are the main techniques on which the methodology is based. The sequences of moderate sized oligopeptides (13–15 amino acid residues) can be determined by interpreting the fragmentation pattern in their mass spectra. Both positive- and nega-

Table 13.1. Oligopeptides successfully analyzed by LD-FTMS

Oligopeptide	Number of residues	MW
Arg-Pro-Lys-Pro-Gln-Gln-Phe-Phe-Gly-OH	9	1103.6
Angiotensin I	10	1295.7
Angiotensin II	8	1045.5
Bacitracin A	11	1421.7
Bradykinin	9	1059.6
Lys-Bradykinin	10	1187.7
Met,Lys-Bradykinin	11	1318.7
Bombesin	14	1618.8
Gramicidin D	15	1881.1
Gramicidin S	10 (cyclic)	1140.7
Leu-Enkephalin	5	555.6
Melanocyte-stimulating hormone (α-MSH)	13	1663.8
Neurotensin	13	1671.9
Renin substrate tetradecapeptide (porcine)	14	1757.9
Pentaalanine	5	373.1
Pentaglycine	5	298.0
Somatostatin	14 (cyclic)	1636.7
Tetraphenylalanine	4	550.1
Trp-Met-Asp-Phe-NH$_2$	4	633.2

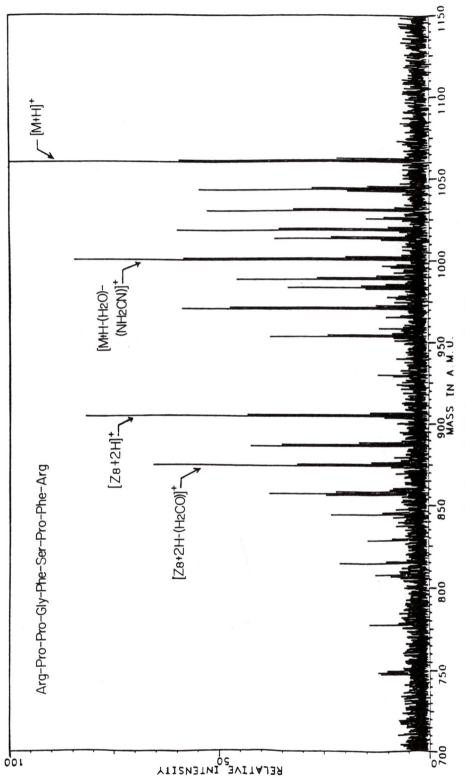

Arg–Pro–Pro–Gly–Phe–Ser–Pro–Phe–Arg

$[M+H]^+$

$[M+H-(H_2O)-(NH_2CN)]^+$

$[Z_8+2H]^+$

$[Z_8+2H-(H_2CO)]^+$

Fig. 13.2. LD-FTMS spectrum of bradykinin.

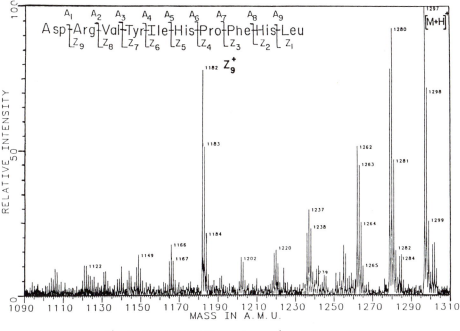

Fig. 13.3. LD-FTMS spectrum of angiotensin I.

tive-ion spectra are employed in the identification process. Accurate mass measurements are used to confirm the identity of ions. For peptides that do not yield abundant fragment ions on laser desorption, CAD is used to induce fragmentation.

Table 13.2 summarizes the types of ions observed in positive- and negative-ion mass spectra of oligopeptides as a result of laser desorption or CAD.[22] To distinguish specific amino acids from others, the laser desorption spectra of the 20 common amino acids were measured and used as a guide for interpreting peptide spectra. For example, a spectrum that contains ions corresponding to losses of 17, 42, and 60 mass units from one specific ion suggests an arginine is present. The relative positions of the amino acid residues can be deduced by careful analysis of the mass differences between fragment ions and the molecular ions. To determine the molecular weight of a peptide, $[M + H]^+$ and $[M - H]^-$ are located in the positive- and negative-ion spectra. From this information, the molecular weight can be inferred. $[M + Na]^+$ and $[M + K]^+$ (recognized by the 16 amu mass difference) frequently also are found in the positive-ion spectra and can be used to confirm molecular weight assignments. Next, series ions are identified. For example, ions of the same mass in both the positive- and negative-ion spectra correspond to B series fragments. Ions in the positive-ion spectrum differing by 2 mass units from ions in the negative-ion spectrum indicate Y, Z, or C series fragments. After labeling fragments according to their series, mass differences from one ion in a series to the next are used to determine the identity of the amino acid residue at that position in the peptide. Figures 13.4 and 13.5 are representative positive- and negative-ion spectra of gramicidin D covering the mass range 1600 to 1950 and gramicidin S from 300

Table 13.2. Oligopeptide fragment ions observed in LD-FTMS[a,b]

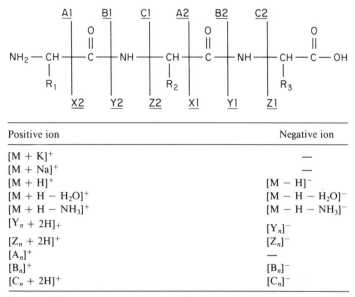

Positive ion	Negative ion
$[M + K]^+$	—
$[M + Na]^+$	—
$[M + H]^+$	$[M - H]^-$
$[M + H - H_2O]^+$	$[M - H - H_2O]^-$
$[M + H - NH_3]^+$	$[M - H - NH_3]^-$
$[Y_n + 2H]_+$	
	$[Y_n]^-$
$[Z_n + 2H]^+$	$[Z_n]^-$
$[A_n]^+$	—
$[B_n]^+$	$[B_n]^-$
$[C_n + 2H]^+$	$[C_n]^-$

[a]A, B, and C ions result when the charge is retained at the N-terminal fragment. X, Y, and Z ions result when the charge is retained at the C-terminal fragment.

[b]Nomenclature for oligopeptide fragment ions is based on conventions outlined in Ref. 22.

to 1200 illustrate the methodology previously outlined (middle mass ranges are omitted for brevity). Blind testing of various peptide samples was performed in our laboratory to determine the advantages and limitations of the LD-FTMS method and how it compares with more conventional methods of analysis.[21]

Carbohydrates are another class of compounds we have studied by LD-FTMS. Maltooligosaccharides, polysaccharides, glycoalkaloids, steroid glycosides, and antibiotics have been examined and the results compared with those obtained from alternative mass spectrometric analyses.[23–26]

Maltooligosaccharides are composed of D-glucose units joined by α-(1-4) linkages. Their laser desorption mass spectra show characteristic fragmentation patterns.[23] Six distinct series of fragment ions are observed. These can be represented in the form $[(M + K)^+ - (162)_n - X]^+$, where M is the mass of the molecular ion, K is the mass of potassium, 162 corresponds to the mass of glucose, n is the number of glucose units, and X is the mass of the ring fragment left on the saccharide chain after cleavage (Table 13.3). The most abundant fragment ions appear to arise from cleavages within the glucose rings rather than from cleavages at the glycosidic linkages, hence the variable X in the series equation. Figure 13.6 compares the LD-FTMS spectra of maltotetraose, maltoheptaose, and starch desorbed from KBr matrices. As the chain length of the molecule increases, the relative abundance of the potassium-attached molecular ion decreases; eventually, only short chain length fragment ions (1 to 8 glucose units) are observed (e.g., starch). Also observed is a decrease in the abundance of a and c series ions and an increase in the abundance of f series ions as the chain length of the molecule increases.

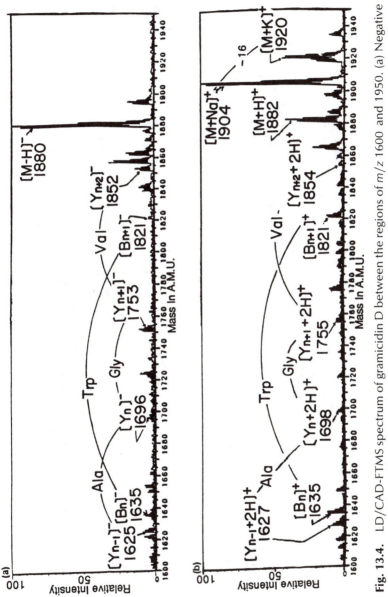

Fig. 13.4. LD/CAD-FTMS spectrum of gramicidin D between the regions of m/z 1600 and 1950. (a) Negative-ion spectrum; (b) positive-ion spectrum.

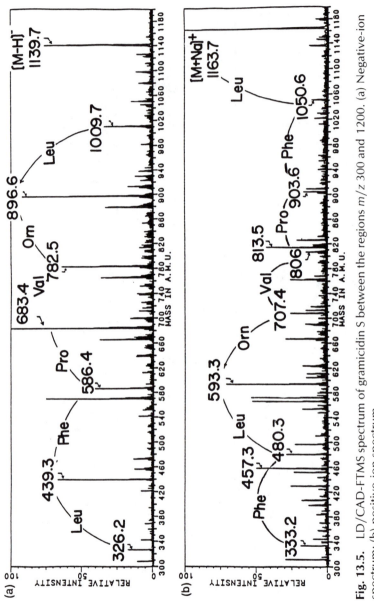

Fig. 13.5. LD/CAD-FTMS spectrum of gramicidin S between the regions m/z 300 and 1200. (a) Negative-ion spectrum; (b) positive-ion spectrum.

Table 13.3. Ion series observed in the laser desorption Fourier transform mass spectra

Maltooligosaccharides: $[(M + K)^+ - (162)_n - X]^+$

A: $X =$	0
b: $X =$	18
c: $X =$	60
d: $X =$	76
e: $X =$	120
f: $X =$	136

Polysaccharides: $[(162)_n + X + K]^+$

A: $X =$	0
B: $X =$	42
C: $X =$	60
G: $X =$	74
J: $X =$	88
K: $X =$	90
L: $X =$	102
M: $X =$	104
Q: $X =$	144
R: $X =$	148

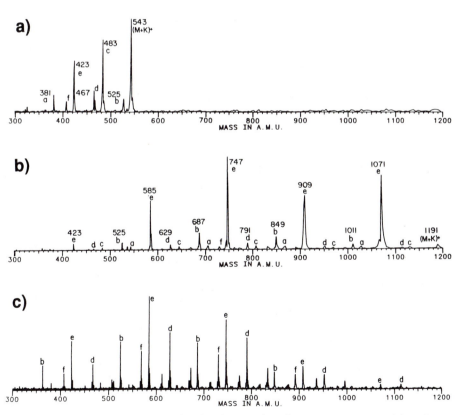

Fig. 13.6. LD-FTMS spectrum of maltooligosaccharides in KBr matrices. (a) Maltotetraose. (b) Maltoheptaose. (c) Starch. Note: labeling is explained in text and in Table 13.3 and is different from the polysaccharide labeling scheme.

Because distinct fragmentation patterns were obtained for the maltooligosac-charides, polysaccharides also were studied to determine whether or not LD-FTMS would produce similar characteristic fragmentation patterns.[24] Because of the structural complexity of these samples, as well as their high molecular weights (2,000–>500,000 amu) and thermal instability, a discrete and reproducible series of fragment ions would greatly aid in structural characterization. Samples studied include the polyhexoses (locust bean gum, white dextrin, dextran, cellulose, starch, and xanthan gum), a polyglucoseamine (chitin), and polygalactoses (agar and agarose).

Positive-ion spectra of the polyhexoses desorbed in the presence of KBr exhibit extensive fragmentation, but with a definite and reproducible pattern. Series of ions seem to originate from fragmentations both between and within the hexose rings and are of the form $[(162)_n + X + K]^+$ where 162 is the mass of a hexose ring, n is the number of rings, X is an integer corresponding to the mass of a ring fragment left on the saccharide chain after cleavage, and K is the mass of potassium (Table 13.3). Dextran, cellulose, and starch have glucose as their repeating unit and differ only in the way these units are linked. Nevertheless, their fragmentation patterns differ greatly in both the types and abundances of series ions present (Fig. 13.7). For dextran, the F series ions are the most abundant in the spectrum. Less abundant A, K, and Q series ions are also observed. For cellulose, both A and F series ions are abundant, with lower abundance fragments arising from D, G, L, and Q series ions. The starch spectrum contains abundant ions representing A, D, F, and M series and less abundant ions from the J and R series.

Similar behavior was observed for the other polyhexoses, chitin, agar, and agarose. Their spectra also contain unique fragmentation patterns. The regularity of the fragmentation patterns for all polysaccharides studied and the reproducibility of the spectra (compare the spectra of starch in Fig. 13.6 and Fig. 13.7) indicate that cleavages are highly specific and should be useful for structural analysis.

In contrast with the behavior of maltooligosaccharides and polysaccharides, positive-ion LD-FT mass spectra of glycoalkaloids and steroid glycosides desorbed in the presence of KCl contain predominantly cationized molecular ions.[25] These compounds are comprised of straight or branched sugar chains attached to steroid moieties. Gitoxin and lanatoside A contain three and four sugars, respectively, arranged in a linear chain. α-Solanine, α-tomatine, and digitonin contain three, four, and five sugars arranged in a branched chain. It appears that the presence of the steroid group may stabilize potassium-attached molecular ions against fragmentation. On the other hand, negative-ion spectra yield fragment ions resulting from loss of one or two sugar groups from $[M + Cl]^-$ and $[M - H]^-$. Comparisons of the negative-ion spectrum of gitoxin with the corresponding negative-ion spectrum of α-solanine, and the spectrum of lantoside A with that of α-tomatine suggest that the degree of branching in the sugar chain is the main factor influencing the extent of fragmentation. In both cases, more fragment ions are observed for the compounds with branched sugar chains. Chain length also may influence fragmentation; for example, the spectrum of α-tomatine shows more fragment ions than the spectrum of α-solanine, and the digitonin spectrum shows even more fragment ions than either.

Some antibiotics contain sugar moieties as part of their structures. We have studied a number of such antibiotics that have approximately cyclic structures,

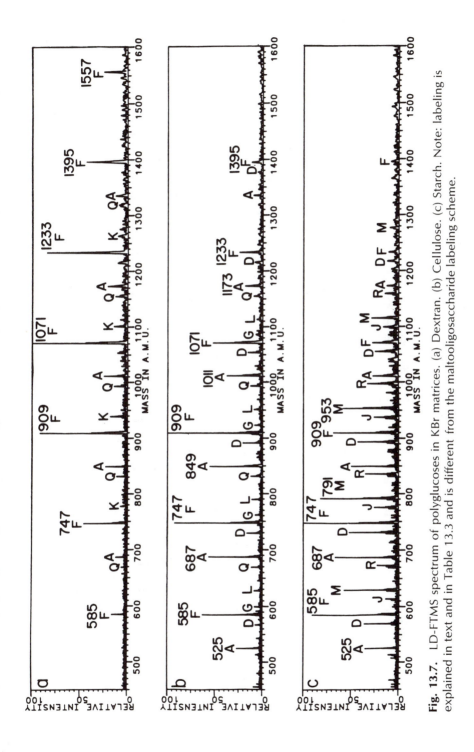

Fig. 13.7. LD-FTMS spectrum of polyglucoses in KBr matrices. (a) Dextran. (b) Cellulose. (c) Starch. Note: labeling is explained in text and in Table 13.3 and is different from the maltooligosaccharide labeling scheme.

except for certain aromatic systems that are composed of fused rings.[26] Peptide or sugar moieties may or may not be present as branching groups. Generally, the presence of a branching group and the complexity of its structure govern the extent of fragmentation observed in the spectrum. The polyenes filipin, nystatin, and amphotericin B have the same core structure. Nystatin and amphotericin B also contain sugar branching groups, and their spectra are characterized by the presence of many fragment ions. In contrast, the spectrum of filipin, with no branching groups, is dominated by cationized molecular ions. Ristocetin contains six sugars branching from a peptide core structure. The spectrum contains only fragment ions resulting from loss of all six sugars. This can be compared with the spectrum of mithramycin, a fused-ring core structure with two-sugar and three-sugar branches. Here, fragment ions resulting from sequential loss of sugars from $(M + K)^+$ are observed, as well as cationized molecular ions, indicating that the stability of the core structure also affects the extent of fragmentation of these compounds.

Nucleosides and nucleotides are another class for which routine analysis is hindered by their nonvolatility and thermal instability. For this reason, the systematic study of 4 nucleosides and 15 related nucleotides was undertaken to evaluate the utility of LD-FTMS for their analysis.[27] Nucleosides are composed of a nitrogenous base derived from either purine or pyrimidine linked to ribose or deoxyribose. Nucleotides have phosphoric acid attached to ribose. Samples studied included adenosine, guanosine, cytidine, and uridine and their mono-, di-, and triphosphates, as well as the oligonucleotides derived from the dimers, tetramers, and hexamers of adenosine and deoxyadenosine.

Nucleoside positive-ion LD-FTMS spectra are characterized by abundant cationized or protonated molecular ions. The major fragment ions are cationized or protonated base, resulting from cleavage of the base–sugar bond. Nucleotide spectra are generally dominated by the same base ion. In negative-ion spectra, deprotonation of the base occurs. Molecular ions are not detected in either the positive- or negative-ion spectra. Instead, cleavage of the phosphate groups occurs, yielding cationized or protonated nucleoside in the positive-ion spectrum and deprotonated nucleoside in the negative-ion spectrum. This cleavage also results in inorganic phosphate ions that increase in relative abundance as the number of phosphate groups on the parent nucleotide increases. Even though molecular ions are not produced, a nucleotide can still be identified by the base ion produced and evaluation of the abundance and type of inorganic phosphate ions present. The dinucleotide adenylyladenosine does yield deprotonated molecular ions in the negative-ion spectrum as well as fragment ions resulting from base–sugar bond cleavage and phosphate–sugar bond cleavage. However, molecular ions are not observed in the spectra of the tetramer and hexamer oligonucleotides of deoxyadenosine. Rather, these spectra reveal predominantly inorganic phosphate and base ions.

The LD-FTMS of underivatized steroids proved to be a viable alternative to conventional electron ionization mass spectrometry.[28] LD spectra of underivatized cholestane, pregnane, androstane, and related steroid hormones were qualitatively similar to EPA/NIH reference spectra but displayed evidence of less thermal decomposition; molecular ions were abundant and fewer low mass fragment ions were observed.

Ascorbic and isoascorbic acids and their sodium and potassium salts were analyzed by LD-FTMS in an effort to understand unusual results reported in an earlier study.[29] The negative ion spectra of these compounds were dominated by [M − H]− ions, and contained dimer ions when longer delay times between desorption and observation were employed. Fragment ions resulting from loss of both OH and the side chain were also observed. Spectra of the sodium and potassium salts did not contain m/z 41 ions, as reported in the earlier study. Positive ion spectra contained [M + H]+ ions, dimer ions, and some fragment ions but did not contain the previously reported m/z 95 ions. The results obtained provided no support for the earlier conclusion that the alkali salts exist as electroneutral species, because abundant molecular ions were detected in both positive- and negative-ion spectra.

LD-FTMS was used to characterize a series of synthetic linear epoxypoly-acenes.[30] Attempts to record their mass spectra using electron ionization resulted in extensive decomposition, and limited solubility made them difficult to analyze by FAB. In contrast, laser desorption FTMS yielded abundant molecular ions and some fragment ions in the negative-ion spectra. The positive-ion spectra contained ions resulting from sequential CO losses from the molecular ions. Accurate mass measurements were performed and the results were consistent with the proposed structures.

The nonvolatility and high mass ranges of polymer mixtures make laser desorption FTMS ideally suited for their study. LD-FTMS can provide molecular weight distributions and structural information as well as allow direct determination of polymer additives. A survey study of high mass compounds demonstrated the potential of this technique for analysis of polymers.[31] In this study, a poly(ethylene glycol) polymer with average molecular weight 3350 (PEG 3350) and Krytox 16140 (a polyperfluorinated ether) were analyzed. The positive-ion spectrum of the PEG sample contained an envelope of peaks corresponding to potassium-attached oligomer ions. A second envelope superimposed over the first corresponded to sodium-attached oligomer ions. The Krytox negative-ion spectrum contained an ion envelope extending to m/z 6825 that corresponded to F-[CF(CF$_3$)CF$_2$O]$_n^+$ with n = 41.

As a result of the success of the earlier study, a more detailed analysis of a series of polymers was undertaken to evaluate the general utility of LD-FTMS for such analyses on a routine basis and to compare results with other methods.[32] Positive-ion spectra were measured for a series of polar polymers and one nonpolar polymer. Spectra generally were characterized by abundant ions corresponding to sodium or potassium attachment to the individual oligomers with very little fragmentation. Figure 13.8 is the positive-ion spectrum of poly(ethyleneimine) (PEI) 1200 mixed with KBr and is typical of the spectra obtained for all polymers. Potassium-attached oligomer ions dominate the spectrum. Very minor peaks corresponding to loss of NH$_3$ are also observed 17 amu lower than the molecular ion envelope. This loss of NH$_3$ for PEI is analogous to loss of H$_2$O from the higher mass poly(ethylene glycol) polymers (PEG 3350 and 6000).

In the same study, molecular weight distributions (MWD) were calculated and compared to values provided by the manufacturers, obtained by more classical methods. Mass spectral determinations of MWD values can be influenced by excessive fragmentation and mass discrimination. In addition, for high masses, isotopic distributions must be taken into consideration. Polymer spectra obtained in this study were characterized by minimal fragmentation, assisting the calculation of

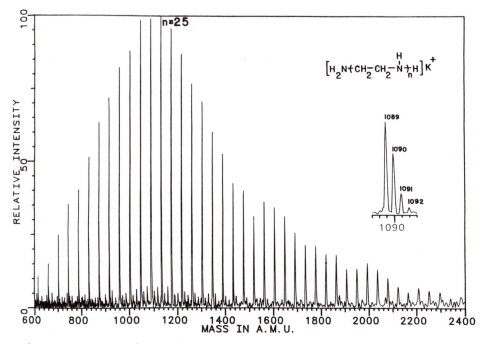

Fig. 13.8. LD-FTMS low-resolution spectrum of poly(ethyleneimine) 1200 in a KBr matrix. Inset, high resolution, $n = 24$ oligomer ions.

MWD. Furthermore, because FTMS does not suffer from mass discrimination, and has the added advantage that the resolution can be manipulated by varying the observation time of the signal (i.e., number of data points Fourier transformed), it can provide accurate MWD. For MWD calculations, spectra were generated with low resolution to allow all isotopic distributions for a given ion to be displayed as one peak, and to enhance the signal-to-noise ratio. The spectrum of PEI 1200 in Fig. 13.8 illustrates this point nicely, showing the low-resolution spectrum obtained for MWD calculations and the high-resolution inset for $n = 24$ obtained by Fourier transform of more data points. MWD values were in good agreement with nominal molecular weight designations of the manufacturers. For PEI samples, a comparison with other mass spectral techniques revealed LD-FTMS to be superior since all other studies gave substantially lower number-average molecular weights than the nominal molecular weight provided by the manufacturer, whereas LD-FTMS results were in good agreement.

A more recent study compared the LD-FTMS spectra of a series of alkoxylated pyrazole and hydrazine polymers with EI, FAB, and SIMS spectra of the same compounds.[33] The spectra were characterized by abundant potassium-attached oligomer ions, with little or no fragmentation observed for the laser power used. Fragmentation was more obvious for the EI, FAB, and SIMS spectra. MWD values were calculated and higher results were obtained for LD-FTMS spectra than for spectra obtained by the other methods. This was consistent with earlier comparisons of PEI molecular weight determinations.[32]

A comparison of LD-FTMS and FAB analysis of polymer additives showed that the LD-FTMS technique was superior for all samples studied.[34] LD-FTMS spectra provided predominantly molecular weight information. In contrast, the FAB spectra revealed extensive fragmentation or could not be obtained at all. LD-FTMS provided rapid direct analysis of the polymer additive Irganox 1010 in dry mixes and films with polyethylene. Spectra were also obtained by extracting the additive, which was present at the 100 ppm level, from a commercial resin.

We have recently reported high-resolution LD-FTMS of high mass organic ions.[35] Figure 13.9 contains the high mass region of a PEG 8000 spectrum obtained on the FTMS-2000 instrument and displays K^+ attached oligomer ions up to m/z 9700. This is the highest mass range LD-FTMS spectrum obtained to date. Figure 13.10 is the high mass region of the unapodized spectrum of poly(propylene glycol) (PPG) 4000, demonstrating unprecedented mass resolution of 60,000 at m/z 5922.

Polymers are now routinely used as calibration compounds for the FTMS instruments in our laboratory.[19] Mass calibration is accomplished by fitting the measured masses of sodium- or potassium-attached oligomers to their theoretical masses. Normally, PEG samples with ions ranging from m/z 200 to 2000 are used because of the ease with which spectra of good quality and high signal-to-noise ratio can be obtained.

Porphyrins and metalloporphyrins are another category of high-molecular-weight, nonvolatile compounds that are well suited for analysis by LD-FTMS. Because previous studies by other methods suggested LD-FTMS might not be applicable to certain biomolecules,[36,37] LD-FTMS was applied to the analysis of

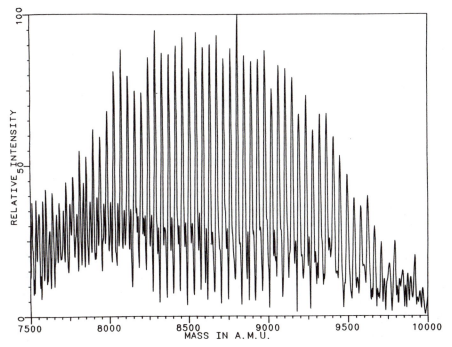

Fig. 13.9. LD-FTMS spectrum of poly(ethylene glycol) 8000 in a KCl matrix showing m/z 9700 ions.

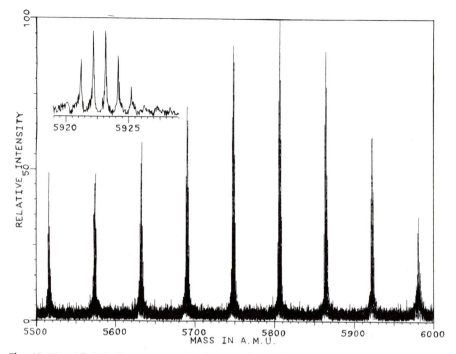

Fig. 13.10. LD-FTMS spectrum of poly(propylene glycol) 4000 in a KCl matrix. Resolution = 60,000 at m/z 5922 (full width at half height).

chlorophyll a and chlorophyll b.[38] Positive-ion spectra of these samples desorbed in the presence of KBr yielded abundant $[M + K]^+$ for both chlorophyll a and chlorophyll b, as well as fragment ions resulting from cleavages involving the phytyl chain. Chlorophyll b spectra in the absence of KBr showed more fragmentation. Ions corresponding to $[M]^+$ were also observed, establishing that at least some of the molecular ions have lifetimes orders of magnitude greater than those reported earlier.

A follow-up study analyzed a series of synthetic porphyrins and metalloporphyrins.[39] All of the porphyrins produced abundant molecular ions or alkali-attached molecular ions. Fragment ions resulted from cleavage of substituents from the main porphyrin ring; their relative abundance could be controlled by adjustment of the laser power employed for laser desorption. Accurate mass measurement of the positive-ion molecular ion region of tetrakis(4-ferrocenyl-phenyl)porphyrin was accomplished with 1.07 ppm accuracy and verified the molecular formula. Metalloporphyrin positive-ion spectra were dominated by $[M]^+$ ions and ions resulting from loss of counterion, and generally were less dependent on the laser desorption power used. Negative-ion spectra contained $[M - H]^-$ ions as well as structurally significant fragment ions resulting from cleavage of porphyrin substituents.

More recently, we have investigated porphyrin behavior upon excimer laser irradiation of ions derived from CO_2 laser desorption. These results, as well as other photodissociation studies, are discussed in the following section.

LASER DESORPTION/PHOTODISSOCIATION BACKGROUND

Photodissociation of ions with mass spectrometric detection is a technique often used to probe the spectroscopy, chemical reactivity, and reaction kinetics of small molecules. Theoretical discussions as well as numerous applications of photodissociation and mass spectrometry can be found in any of a number of review articles.[40-43] In recent years, there has been much interest in the potential use of photodissociation for structural elucidation of larger molecules such as porphyrins and peptides.[44-46] In our laboratory, studies intended to investigate the utility of laser desorption/photodissociation FTMS (LD/PD-FTMS) for analysis of biomolecules and other high mass, intractable species are underway.

Often, CO_2 laser desorption produces mass spectra containing molecular ions but relatively few fragment ions. In those cases, CAD sometimes can be used to enhance analytical information. In an FTMS cell CAD is performed by translationally exciting ions, causing them to collide with a collision gas also present in the cell with sufficient energy to fragment. Too little excitation results in insufficient energy for fragmentation on collision and too much excitation ejects the ions from the cell. For higher mass ions (thousands of daltons), it becomes increasingly difficult to impart enough energy to cause fragmentation without ejection. A further complication is the fact that larger molecules can accommodate more energy before dissociation occurs (due to the large number of vibrational modes); also, energy is transferred less efficiently due to the greater disparity in the masses of analyte and collision gas. These factors contribute to the inefficiency of low-energy CAD for higher mass ions. For such samples, photodissociation provides an attractive alternative.

Efficient photodissociation requires that the ions being irradiated must absorb photons that are sufficiently energetic to cause fragmentation (assuming a one-photon process and a nonzero quantum yield for photodissociation; see Refs. 42 and 43). We have taken two different approaches to satisfying these criteria. The first involves using a laser with a wavelength at which the underivatized sample absorbs. Peptides have been photodissociated in this way by using 193-nm radiation to induce fragmentation involving the carbonyl and aryl chromophores.[45,46] We have employed 308-nm radiation for photodissociation of various compounds. If the underivatized sample does not absorb, attachment of metal ions can permit photodissociation using 308-nm excitation. This approach is the subject of recent investigations in our laboratory.

Instrument designs for both the FTMS-1000 and FTMS-2000 were described earlier and are diagrammed in Fig. 13.1. The timing diagram in Fig. 13.11 depicts the sequence of events for typical LD/PD-FTMS experiments. First, a quench pulse rids the cell(s) of any residual ions from past experiments. This is followed by the CO_2 laser event for desorption of sample. Immediately after firing the CO_2 laser, desorbed ions are transferred from the source cell to the analyzer cell in the FTMS-2000. For the FTMS-1000, with its single-cell configuration, this step is omitted. A delay allows neutrals to be pumped away (more important for the FTMS-1000 experiment), and ejection sweeps isolate ions of interest. Next, the excimer laser is fired at 60-80 Hz for periods of up to hundreds of milliseconds. Daughter ions are

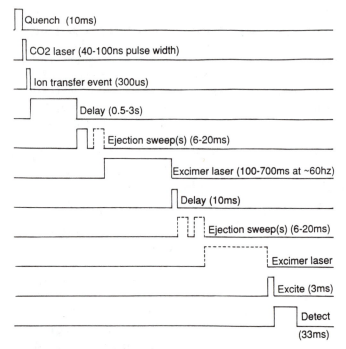

Fig. 13.11. Laser desorption/photodissociation pulse diagram for LD/PD-FTMS experiments.

excited and detected, or, alternatively, certain daughter ions may be isolated by further ejection sweeps for photodissociation into granddaughter ions.

SAMPLES STUDIED BY LD/PD-FTMS

In our laboratory, photodissociation has been performed using 308-nm radiation. Samples that have been photodissociated fall into two categories: those that exhibit absorption maxima around 308 nm (303–312 nm) for their solution UV spectra, and those that do not. Rescinnamine is a typical example of a sample that absorbs 308-nm radiation in solution, and photodissociates as gas-phase ions. The daughter ions arise from cleavages of bonds linking ring systems and the resulting spectra are easily interpreted. Other samples that fall into this category include rhodamine B, noscapine, (+)-amethopterin, brucine, calycanthine, and distamycin.[47] For these compounds, absorptions occur at 308 nm as a result of their general conjugated structures, rather than one particular structural feature as is the case for the carbonyl and aryl chromophores of the peptides photodissociated with 193-nm irradiation. A few compounds do not photodissociate, in spite of the presence of absorption bands at 308 nm in their solution UV spectra. One reason may be that on volatilization and ionization, absorbances shift. Solvent effects, not present for gas-phase ions, also contribute to differences between solution and gas-phase spectral characteristics.

Photodissociation sometimes produces detailed information regarding energetics and molecular structure. The $(2M + K)^+$ and $(2M + Na)^+$ ions of the alkaloid, noscapine, photodissociate to yield $(M + K)^+$ and $(M + Na)^+$ ions revealing that the alkali-bound dimer is a more weakly bound complex than the cationized molecular ions. Distamycin, an antibiotic, produces enough fragment ions upon CO_2 laser desorption to make photodissociation unnecessary as a primary structural tool. However, photodissociation of the laser-desorbed m/z 273 fragment ions produces only daughter ions of m/z 149, although both m/z 149 and 151 ions are present in equal abundance in the original laser desorption spectrum. Further photodissociation of the m/z 149 ions results in the loss of CO, producing granddaughter ions of m/z 121 and verifying ion assignments.

For samples that do not absorb 308-nm radiation (in solution), photodissociation can be accomplished by first attaching a chromophore, either by derivatization or by *in situ* metal ion attachment. Because derivatization is time consuming and not readily applicable to all types of compounds, as well as being difficult for samples in limited quantity, we have investigated the attachment of metal ions. It is known that metal ion complexes absorb broadly and photodissociate readily in the ultraviolet and visible regions (see, for example, Ref. 43).

Metalloporphyrins were chosen for initial studies because they contain attached metal ions. Nonmetalated porphyrins are used for comparisons. The LD-FTMS spectrum of manganese tetraphenylporphyrin chloride (MnTPPCl) is shown in the top of Fig. 13.12 and the corresponding photodissociation spectrum of the isolated

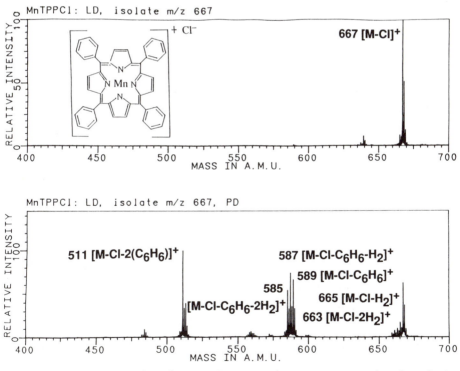

Fig. 13.12. LD-FTMS and LD/PD-FTMS spectra of manganese tetraphenylporphyrin chloride. Top, LD-FTMS. Bottom, LD/PD-FTMS.

$[M - Cl]^+$ ions is shown in the bottom. In this case, photodissociation yields ions resulting from loss of one and two phenyl substituents (m/z 589 and 511, respectively), as well as ions resulting from dehydrogenation. Granddaughter ions with m/z 511 are produced by photodissociation of the isolated daughter ions with m/z 589 (Fig. 13.13). Iron tetraphenyl porphyrin chloride and chromium tetraphenylporphyrin chloride also have been examined by LD/PD-FTMS and yield similar results. Photodissociation of other porphyrins is not as consistent. $[M + K]^+$ ions, produced from the laser desorption of dimethylaminotetraphenylporphyrin, photodissociate, even though only an alkali metal is present, and yield ions resulting from cleavage of a dimethylaminophenyl substituent. This spectrum does contain valuable structural information not present in the LD-FTMS spectrum. LD-FTMS of 4-fluorotetraphenylporphyrin yields $[M]^+$ ions and no fragment ions. Attempts to photodissociate these $[M]^+$ ions were unsuccessful. *In situ* incorporation of iron into the porphyrin, accomplished by depositing the porphyrin on an $FeCl_3$ substrate followed by laser desorption, results in abundant $[M + Fe]^+$ ions as well as $[M + FeCl]^+$ ions. Photodissociation of these ions results in sequential loss of two fluorophenyl substituents (Fig. 13.14). In this case, attachment of the metal ion

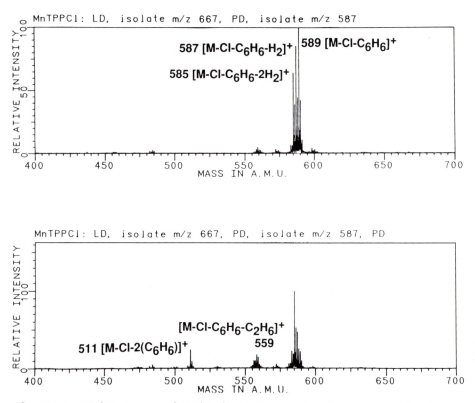

Fig. 13.13. LD/PD-FTMS and LD/PD/PD-FTMS spectra of manganese tetraphenylporphyrin chloride. Top, LD/PD-FTMS with ejection sweeps to isolate daughter ions at m/z 587. Bottom, LD/PD/PD-FTMS of daughter ions at m/z 587.

4-FTPP + Fe: LD, isolate m/z 740

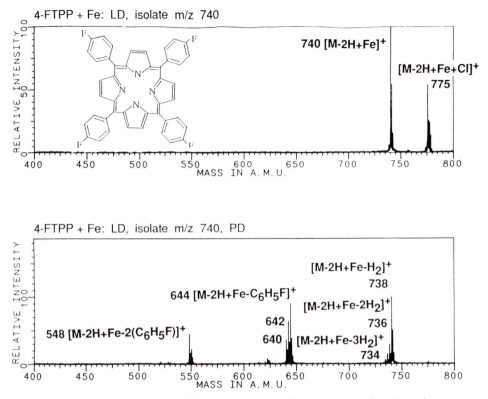

Fig. 13.14. LD-FTMS and LD/PD-FTMS spectra of 4-fluorotetraphenylporphyrin on an $FeCl_3$ substrate. Top, LD-FTMS. Bottom, LD/PD-FTMS.

facilitated photodissociation and, as a result, structurally informative ions were produced.

We are presently investigating relative photodissociation efficiencies as a function of attached metal and are exploring the potential applicability of LD/PD-FTMS to metal-attached biomolecules (e.g., oligopeptides and saccharides).

LASER-ASSISTED ION–MOLECULE REACTION STUDIES

Finally, as mentioned at the outset, FTMS is an extremely valuable tool for the study of gas-phase ion–molecule reactions. Of interest in the present context is the applicability of some of the techniques developed for analytical laser desorption studies to investigations of gas-phase metal ion chemistry. One of our contributions to this field was the observation that abundant metal ions could be generated directly by CO_2 laser ablation from their oxides or salts.[48] Using this technique, we have investigated the gas-phase reactions of gold, copper, and silver with alcohols[48] and the reactions of gold with hydrocarbons and alkyl halides.[49] In a recent study, similar reactions of aluminum, gallium, and indium with a variety of organic reactants were also investigated.[50] In the latter case, the potential utility of these metal ions as chemical ionization reagents was evaluated.

ACKNOWLEDGMENTS

We gratefully acknowledge support for our research from the National Institutes of Health (GM-30604), the National Science Foundation (CHE-85-19087), and the Shell Development Company.

REFERENCES

1. A. G. Marshall, in *Mass Spectrometry in the Health and Life Sciences* (Anal. Chem. Symp. Series Vol. 24), edited by A. L. Burlingame and N. Castagnoli Jr. Elsevier, Amsterdam, 1985, pp. 265–286.

2. N.M.M. Nibbering, *Adv. Mass. Spectrom.* 417 (1985).

3. D. A. Laude, C. L. Johlman, R. S. Brown, D. A. Weil, and C. L. Wilkins, *Mass Spectrom. Rev.* **5,** 107 (1986).

4. D. H. Russell, *Mass Spectrom. Rev.* **5,** 167 (1986).

5. D. H. Russell and M. E. Castro, in *Mass Spectrometry in Biomedical Research,* edited by S. J. Gaskell. Wiley, Chichester, 1986, pp. 313–338.

6. H. R. Schulten and W. D. Lehmann, *Mikrochim. Acta* **2,** 113 (1978).

7. R. Roepsdorf and B. Sundqvist, in *Mass Spectrometry in Biomedical Research,* edited by S. J. Gaskell. Wiley, Chichester, 1986, pp. 269–285.

8. M. L. Vestal, in *Mass Spectrometry in the Health and Life Sciences* (Anal. Chem. Symp. Series Vol. 24), edited by A. L. Burlingame and N. Castagnoli Jr. Elsevier, Amsterdam, 1985, pp. 99–118.

9. K. L. Rinehart, Jr., in *Mass Spectrometry in the Health and Life Sciences* (Anal. Chem. Symp. Series Vol. 24), edited by A. L. Burlingame and N. Castagnoli Jr. Elsevier, Amsterdam, 1985, pp. 119–148.

10. D. J. Colton, D. A. Kidwell, and M. M. Ross, in *Mass Spectrometry in the Analysis of Large Molecules,* edited by C. J. McNeal. John Wiley, Chichester, 1986, pp. 13–48.

11. F. Hillenkamp, in *Ion Formation from Organic Solids,* edited by A. Benninghoven. Springer-Verlag, Berlin, 1983, pp. 190–205.

12. M. A. Posthumus, P. G. Kistemaker, and H. L. C. Meuzelaar, *Anal. Chem.* **50,** 985 (1978).

13. G.J.Q. van der Peyl, K. Isa, J. Haverkamp, and P. G. Kistemaker, *Int. J. Mass Spectrom. Ion Phys.* **47,** 11 (1983).

14. B. Lindner and U. Seydel, *Anal. Chem.* **57,** 895 (1985).

15. M. Karas, D. Bachmann, and F. Hillenkamp, *Anal. Chem.* **57,** 2935 (1985).

16. G.J.Q. van der Peyl, J. Haverkamp, and P. G. Kistemaker, *Int. J. Mass Spectrom. Ion Phys.* **42,** 125 (1982).

17. R. D. Macfarlane and D. F. Torgerson, *Science* **191,** 920 (1976).

18. R. J. Beuhler, E. Flanagan, L. J. Greene, and L. Friedman, *J. Am. Chem. Soc.* **96,** 3990 (1974).

19. C. L. Wilkins and C.L.C. Yang, *Int. J. Mass Spectrom. Ion Process.* **72,** 195 (1986).

20. R. S. Brown and C. L. Wilkins, in *Fourier Transform Mass Spectrometry: Evolution, Innovation and Applications* (ACS Symposium Series No. 359), edited by M. V. Buchanan. American Chemical Society, Washington DC, 1987, pp. 127–139.

21. L. C. Yang, *Laser Desorption Fourier Transform Mass Spectrometric Studies of Oligopeptides.* University of California Riverside, 1988. *Diss. Abstr. Int. B* 1988.

22. P. Roepstorff and J. Fohlman, *Biomed. Mass Spectrom.* **11,** 601 (1984).

23. M. L. Coates and C. L. Wilkins, *Biomed. Mass Spectrom.* **12,** 424 (1985).

24. M. L. Coates and C. L. Wilkins, *Anal. Chem.* **59,** 197 (1987).

25. M. L. Coates and C. L. Wilkins, *Biomed Environ. Mass Spectrom.* **13,** 199 (1986).

26. M. L. Coates, *Laser Desorption/Fourier Transform Mass Spectrometry of Saccharides and Glycosides.* University of California, Riverside, 1987. *Diss. Abstr. Int. B* 1988.

27. L. M. Nuwaysir and C. L. Wilkins, *Laser Desorption-Fourier Transform Mass Spectrometry of Nucleosides and Nucleotides.* Presented at the 1987 Pittsburgh Conference & Exposition on Analytical Chemistry and Applied Spectroscopy, March 9–13, Atlantic City, NJ. Paper No. 851.

28. E. T. Fung and C. L. Wilkins, *Biomed. Environ. Mass Spectrom.,* **15,** 609 (1988)

29. P. Coad, R. A. Coad, C.L.C. Yang, and C. L. Wilkins, *Org. Mass Spectrom.* **22,** 75 (1987).

30. L. L. Miller, A. D. Thomas, C. L. Wilkins, and D. A. Weil, *J. Chem. Soc., Chem. Commun.* 661 (1986).

31. C. L. Wilkins, D. A. Weil, C.L.C. Yang, and C. F. Ijames, *Anal. Chem.* **57,** 520 (1985).

32. R. S. Brown, D. A. Weil, and C. L. Wilkins, *Macromolecules* **19,** 1255 (1986).

33. L. M. Nuwaysir and C. L. Wilkins, *Anal. Chem.* **60,** 279 (1988).

34. C. L. Johlman and C. L. Wilkins, *Laser Desorption/Fourier Transform Mass Spectrometry for the Analysis of Polymer Additives.* Presented at the 35th ASMS Conference on Mass Spectrometry and Allied Topics, May 24–29, 1987, Denver, Colorado. Paper No. 785.

35. C. F. Ijames and C. L. Wilkins, *J. Am. Chem. Soc.,* **110,** 2687 (1988).

36. B. T. Chait and F. H. Field, *J. Am. Chem. Soc.* **106,** 1931 (1984).

37. J. C. Tabet, M. Jablonski, R. J. Cotter, and J. E. Hunt, *Int. J. Mass Spectrom. Ion Process.* **65,** 105 (1985).

38. R. S. Brown and C. L. Wilkins, *J. Am. Chem. Soc.* **108,** 2447 (1986).

39. R. S. Brown and C. L. Wilkins, *Anal. Chem.* **58,** 3196 (1986).

40. R. C. Dunbar, in *Gas Phase Ion Chemistry,* Vol. 2, edited by M. T. Bowers. Academic Press, New York, 1979, pp. 182–220.

41. K. Levsen, *Adv. Mass Spectrom.* **359,** 1985.

42. B. S. Freiser, *Anal. Chim. Acta* **178,** 137 (1985).

43. B. S. Freiser, in *Fourier Transform Mass Spectrometry: Evolution, Innovation, and Applications* (ACS Symposium Series No. 359), edited by M. V. Buchanan. American Chemical Society, Washington DC, 1987, pp. 155–174.

44. E. K. Fukuda and J. E. Campana, *Anal. Chem.* **57,** 952 (1985).

45. W. D. Bowers, S-S. Delbert, and R. T. McIver, Jr., *Anal. Chem.* **58,** 972 (1986).

46. D. F. Hunt, J. Shabanowitz, and J. R. Yates III, *J. Chem. Soc., Chem. Commun.* 548 (1987).

47. L. M. Nuwaysir and C. L. Wilkins, *Two Color Laser Experiments Using Fourier Transform Mass Spectrometry.* Presented at the 1988 Pittsburgh Conference & Exposition on Analytical Chemistry and Applied Spectroscopy, February 22–26, New Orleans, LA. Paper No. 1050.

48. D. A. Weil and C. L. Wilkins, *J. Am. Chem. Soc.* **107,** 7316 (1985).

49. A. K. Chowdhury and C. L. Wilkins, *J. Am. Chem. Soc.* **109,** 5336 (1987).

50. A. K. Chowdhury and C. L. Wilkins, *Int. J. Mass Spectrom. Ion Process.* **82,** 163 (1988).

14

Applications of Laser Desorption–Fourier Transform Mass Spectrometry to Polymer and Surface Analysis

ROBERT B. CODY, ASGEIR BJARNASON, and DAVID A. WEIL

The use of lasers with Fourier transform mass spectrometry (FTMS)[1-8] is well established for a variety of applications. High-power pulsed lasers have been extensively used to generate atomic metal ions for gas-phase ion–molecule reaction studies.[9-16] Multiphoton ionization has been combined with FTMS[17,18] and used for both surface analysis[19] and as an efficient and selective ionization source for GC-FTMS.[20] Both single and multiphoton photodissociation have been employed for structural and thermochemical studies.[10,21-23] Photodissociation has been proposed as an alternative to collisional activation for MS/MS of large ions.[24] Recently, sequencing of oligopeptides up to about 2000 u has been successfully demonstrated by using a tandem quadrupole-FTMS instrument.[25]

The most widely used analytical application of lasers in Fourier transform mass spectrometry to date has been for desorption and ionization of thermolabile and involatile substances. Following the first report by McCrery et al.[26] on the use of combined laser desorption-Fourier transform mass spectrometry (LD-FTMS) to desorb and ionize organic compounds, a wide variety of applications have been reported. LD-FTMS has been applied to the analysis of synthetic organic compounds,[27-29] organometallic compounds,[27,30-34] compounds of biological interest,[27,34-43] drugs,[27,39,44-48] and surface and inorganic samples.[39-49] The largest number of published applications of LD-FTMS has been for the analysis of polymers and polymer additives.[27,34,50-61] From these applications, it is apparent that laser desorption has proven to be an extremely useful tool for determining the composition and molecular weight distribution for low-molecular-weight polymers (i.e., those having molecular weight distributions below 10,000 u), and for examining the nature and effect of polymer additives.

In this report, we describe applications of laser desorption to a number of analytical problems. We put particular emphasis on polymer applications, since we have found this area to be one in which LD-FTMS is the analytical method of choice. Applications to surface and inorganic analysis are also described, along with a brief summary of work being done to develop and apply laser microprobe instruments for surface analysis, based upon FTMS technology.

EXPERIMENTAL

The Dual-Cell LD-FTMS Interface

Since the basic principles of Fourier transform mass spectrometry and the dual-cell geometry are described in the literature,[82] we will not discuss them here, but we will concentrate on the laser desorption interface. The dual-cell geometry is used because it provides a simple and flexible solution to the requirement for low-pressure analysis to maintain high resolution in FTMS.

The design of a laser desorption source for a Fourier transform mass spectrometer with a dual-cell geometry must satisfy several objectives. The source must provide a "soft ionization" method for organic analysis, must be compatible with the dual-cell, and must not interfere with other modes of operation, such as solids probe work and GC–MS. Although the laser desorption source is intended primarily for laser desorption/ionization of organic compounds, it is also applicable to inorganic compounds, or to "bulk analysis" of a variety of samples, such as semiconductors and minerals. A schematic diagram of the laser desorption interface and the dual-cell geometry is given in Fig. 14.1. The laser desorption source has a small spot size at the focal point (30–150 μm), but it is not intended as a laser microprobe. A pulsed CO_2 laser[62] having an output of 100 mJ to 2 J/pulse at the 10.6-μm wavelength is used. The pulse duration is 40 nsec at half height, but with a "tail" extending 1 μsec. This same type of laser has been employed by Cotter[63] with good success for laser desorption of organic and biological compounds.

Reflective optics are used, rather than transmissive optics, since reflective optics are less susceptible to laser damage. The materials used for reflective optics (aluminum oxide-coated mirrors) are also easier to obtain and less expensive than those for transmissive optics (zinc selenide) at the operating wavelengths for the CO_2 laser. Further, the reflective optics improve focusing, and permit the use of source-side optics, which means that the laser need not be focused through the ana-

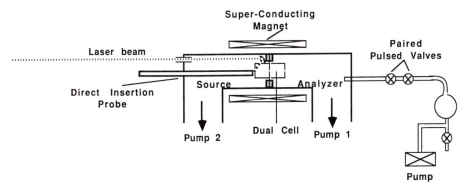

Fig. 14.1. Schematic diagram of the vacuum system used in these experiments.

lyzer cell into the source cell. The resulting smaller spot size increases available power density at the surface.

Since microprobe characteristics are not the objective for this interface, optics are designed to give a focal diameter of approximately 30–150 μm at the target. These dimensions are only an approximation, since the 33° angle of incidence of the laser causes the shape of the beam at the target to be an ellipse. The power density of the laser at the surface ranges from 10^5 to 10^8 W/cm^2, depending on the amount of light passed through an adjustable iris attenuator. The amount of light available to be focused on the target may be controlled with the iris attenuator, or by varying the high voltage setting for the laser. This allows the use of the same laser for organic laser desorption or for desorption of inorganics or polymers.

The laser beam is focused onto the target, which is held in place by inserting it into the end of the probe tip. The probe is positioned so that the laser focal point falls off-axis on the target. In this way, the probe can be rotated to expose a fresh target surface for successive laser shots. Seventy-two target positions may be accessed through software control of the probe position; smaller adjustments are possible with manual positioning.

Mass analysis may be performed either in the source cell or the analyzer cell. The laser target is aligned with the conductance limit to permit transfer of laser-desorbed ions from the source to the analyzer cell for high-resolution analysis. A micrometer adjustment is provided to adjust the target position for samples of different thickness, to keep the laser focal point constant.

Experimental Event Sequences for Laser Desorption

A typical event sequence for laser desorption experiments is given in Fig. 14.2. Since the principles of FTMS are described in detail elsewhere in the literature,[82] we will provide only a brief description here.

The event sequence begins with quench events to clear the cell of any ions remaining from previous experiments. As the laser is triggered (100 μsec), the source cell trapping plate is grounded to permit ions to enter the source cell. For experiments with analyzer-cell detection, the conductance limit may also be grounded during this event and during the (optional) ion transfer time following the laser trigger.

Following a variable delay period, two ion ejection events are provided to remove unwanted ions and thereby increase the dynamic range. Another delay follows; the variable delays permit the operator to sample the ion population at different pressures or reaction periods. For source-cell experiments, these delays may be several seconds long to permit desorbed gasses to be pumped away before the detection period.

After the last delay period, ions are excited into coherent motion with a swept-frequency ("chirp") or stored-waveform inverse Fourier transform (SWIFT)[64,65] excitation. The image currents produced on the receiver plates by the coherently moving ions are then detected, digitized, and mathematically converted into the mass spectrum.

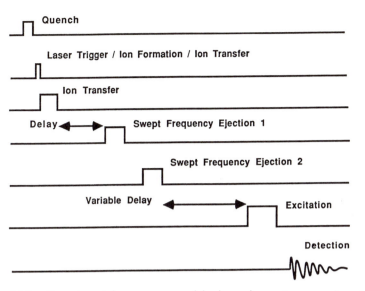

Fig. 14.2. Experimental sequence used for laser desorption experiments.

APPLICATIONS

Types of Ions Observed in LD-FTMS

The nature of an LD-FTMS spectrum depends very strongly on the nature of the sample and on its surface characteristics. The subject of the mechanisms of ion formation in laser desorption is too extensive to consider in this chapter. However, the discussions that follow may be better appreciated if the reader is aware of the types of ions that are observed in LD-FTMS.

Some compounds (such as aromatic compounds or some organometallics) produce stable molecular ions. For polar compounds (such as saccharides and antibiotics) deposited on the surface in a thin film, cation attachment is frequently observed, giving rise to $[M + K]^+$ or $[M + Na]^+$ ions. Cations may come from trace impurities present on the target surface, from the sample itself, or from dopants such as KBr, KCl, or NaCl deliberately added to the sample. Small cluster ions such as K_2Cl^+ or KBr_2^-, are often observed, as are cation-bound dimers from the sample ($[2M + Na]^+$, etc.). Cation-attached fragments are also observed. Atomic cations (K^+, Na^+) often make up a large percentage of the total ion current by themselves.

Although it is less common, proton transfer is sometimes observed, resulting in the formation of $[M + H]^+$ ions. Proton transfer occurs by sample–sample interactions or when certain dopants such as NH_4Br are added.

If the compound has a structure that allows it to delocalize an added electron, M^- peaks may be observed in the negative ion spectrum. These are commonly observed for compounds such as porphyrins and polynuclear aromatics. More frequently, dissociative electron attachment results in fragment ions in the negative ion spectrum. In many cases, $[M - H]^-$ ions provide molecular weight informa-

Table 14.1. Summary of types of ions observed in LD-FTMS

Compound class	Molecular		Quasimolecular ions formed with dopants $[M + K]^+, [M + Na]^+$
	Cations M^+	Anions M^- or $[M - H]^-$	
Nucleotides	No	Yes	Yes
Peptides	No	Yes ($[M - H]^-$)	Yes
Polysaccharides	No	No	Yes
Oligosaccharides	No	Yes ($[M - H]^-$)	Yes
Glycoconjugates	No	Maybe	Yes
Porphyrins	Yes	Yes	No
Additives	Compound dependent		
Organometallics	Yes	Yes	
Salts	Yes	Yes	
Carbonyls	Yes	Yes	
Polymers	Compound dependent (Table 14.2)		

tion. Occasionally, if alkali salts have been used as dopants, anion attachment (e.g., $[M + Br]^-$) occurs at much lower intensity than the corresponding cation attachment.

Other types of ions may be observed under various circumstances. Some very polar compounds will give ions such as $[M + 2K - H]^+$ or $[M + K + Na - H]^+$. Doubly charged cations or anions will produce clusters (see discussion of calcium stearate in the following section on polymer additives). Some doubly charged ions may undergo reduction to produce singly charged ions: we have reported an example for an Ni^{2+} organometallic complex that produces $[M - H]^+$ ions resulting from a one-electron reduction.[6] Stable doubly charged ions are less common, but occur for some compounds (aromatics, in particular). Reactions of laser-desorbed organic ions are not common under the conditions employed, even if source-side detection is used (detection under relatively higher pressure conditions). However, laser-generated metal ions tend to react quickly, and are best detected under the lowest pressure conditions possible (using the analyzer cell). Carbon clusters may be observed under certain conditions and will be described in detail in a following section. A summary of types of ions observed in LD-FTMS is given in Table 14.1.

Laser Desorption Analysis of Polymers

As evident from the many publications on the subject, one of the major applications of LD-FTMS is polymer analysis. The first reported observation of a polymer molecular weight distribution by LD-FTMS was for a sample of the antiherpes drug acyclovir, which was dispersed in poly(ethylene glycol) having an average molecular weight of 400 u.[27] While the negative-ion spectrum of the drug showed an $[M - H]^-$ ion, the positive-ion spectrum was dominated by sodium and potassium-attached molecular ions for the poly(ethylene glycol) oligomers. Since that time, many studies of polymers by LD-FTMS have been reported; extensive work in this area has been conducted by Wilkins (U.C. Riverside) and co-workers, and Brown (Medical College of Wisconsin). A summary of polymer types studied by LD-FTMS is presented in Table 14.2.

The reasons to choose LD-FTMS as a method for polymer analysis are many. It has been suggested that LD-FTMS provides accurate molecular weight distributions, with somewhat less fragmentation than other ionization methods.[53,59,66] Further, it is often possible to obtain spectra for "difficult" polymers, such as those that are insoluble, or otherwise difficult to handle; laser desorption analysis need not be limited to thin films.[50] One of the most appealing characteristics of laser desorption is that there is a good chance to obtain information from almost any sample as long as it can be mounted on the laser target. This means that laser desorption can extend mass spectrometric analysis of polymers to "materials" problems (including *in situ* polymer and rubber additive analysis) without extensive sample pretreatment.

Characteristics of LD-FTMS for polymer analysis can best be appreciated by considering a few representative examples. One of the first polymer classes to be examined by LD-FTMS is the polyglycols: poly(ethylene glycol) (PEG) and poly(propylene glycol) (PPG).[27,34,53] These polymers are readily soluble in organic solvents such as acetone and methanol. If KBr is added as a dopant, the positive-

Table 14.2. Summary of polymer types studied by LD-FTMS

Polymer	Cations (including $[M + H]^+$, $[M + K]^+$, $[M + Na]^+$)	Positive M^+	Negative M^-	Dominant fragments	Clusters	References
PEG	Yes	No	(Weak)	No	No	27, 34, 53
PPG	Yes	No	(Weak)	No	No	53
PEGME	Yes	No	(Weak)	No	No	53
PEI	Yes	No	—	—	No	53
Polystyrene	Yes	No	Yes	No	No	53, 86
Polycaprolactonediol	Yes	No	—	No	No	53
PPP	No	Yes	(Weak)	No	No	50, 51, 54
PDMS	No	Yes	No	Some	No	86
PMMA	Yes	No	No	Yes	No	52
PPS	No	Yes (cyc.)	No	Yes (lin.)	No	56
Fluorinated HC	No	Yes	No	Yes	No	86
Polybutadiene	Yes	No	No	Some	No	86
Polyaniline	No	?	No	Rearrange	No	56
Polythienylene	No	Yes	No	No	+High m/z	54, 56
Poly-Ph-pyrrolylene	No	No	No	Positive and negative	No	55
Poly-Me-pyrrolylene	No	Yes	No	No	No	55
Polyselenienylene	No	No	(Weak)	No	No	55
Polyperfluoroether	No	No	Yes	No	No	34
Alkox. pyrazoles	Yes	No	—	—	No	59
Alkox. hydrazine	Yes	No	—	—	No	59
Polyester	No	No	No	Yes	+High m/z	86
Polyvinylphenol	Yes		$[M - H]^-$			56
Polypyrene	No	No	—	Yes	Positive	56
Brom. PPP	No	No	No	No	Positive and negative	54
Hydrocarbon wax	No	Yes	No	Yes	Yes	86

ion spectra are characterized by a distribution of $[M + K]^+$ ions. Such cation attachment is required for high-quality spectra of these types of polymers. If a dopant is not supplied, pseudomolecular ions are still observed, resulting from cationization by potassium and sodium as trace impurities. Doping increases the ion current, and simplifies the spectra by controlling which cation is available for attachment. A spectrum of a poly(ethylene glycol) is shown in Fig. 14.3.

An interesting example may be found in the analysis of a poly(dimethylsiloxane) sample, believed to have an average molecular weight of 2000 u. For LD-FTMS analysis, this sample was dissolved in methanol and deposited as a thin film on the laser target without any added dopant. The positive-ion spectrum (Fig. 14.4) shows a distribution of molecular ions separated by 74 u (Me$_2$SiO). The result is surprising: the molecular ion distribution extends up to m/z 8000, far beyond the expected average molecular weight distribution. We speculate that previous methods used to determine the molecular weight of this particular sample must have been in error.

It is only fair to note that laser desorption is not the only approach that has been reported for examining poly(dimethylsiloxane)s. Early analyses of poly(dimethylsiloxane)s by Gardella and Hercules (using a laser microprobe/time-of-flight instrument)[67] and Briggs (using SIMS)[68] showed primarily low mass fragments, with no molecular ion distributions. More recently, Bletsos, Hercules, and co-workers have shown SIMS spectra for a poly(dimethylsiloxane) sample with

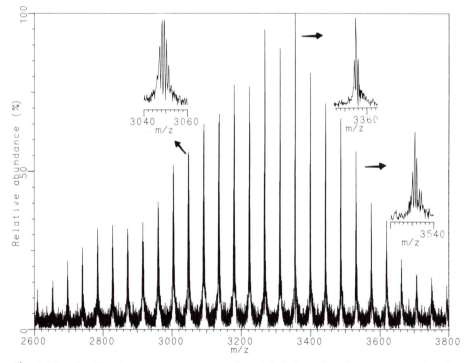

Fig. 14.3. Positive-ion mass spectrum of poly(ethyleneglycol), average molecular weight (3350) acquired at unit mass resolution.

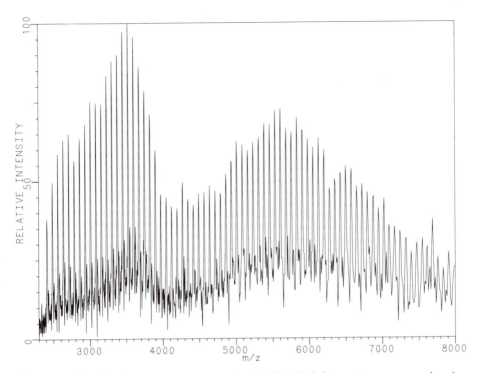

Fig. 14.4. Positive ion mass spectrum of poly(dimethylsiloxane), average molecular weight (2000).

ions in the range 1200–4010 u.[69] This latter result was accomplished by placing a nickel grid on the surface of the sample to reduce charge buildup.

Hydrocarbon waxes are particularly difficult to analyze by most mass spectrometric methods. Laser desorption was used to examine a "hard wax" sample (partially unsaturated long-chain hydrocarbons). Direct ionization of the sample by laser desorption does not provide good results; the spectrum is weak, and contains many fragments. However, the spectrum can be improved by turning on the electron beam after the laser has been fired. Electron impact ionization of the desorbed neutrals ("postionization") produces excellent results (Fig. 14.5A). Molecular ions for alkenes and dienes are evident as series of peaks having the general formulas $C_nH_{2n-2}^+$ and $C_nH_{2n}^+$, respectively (Fig. 14.5B and C). The remaining peaks consist of alkyl fragments ($C_nH_{2n+1}^+$) and carbon-13 isotope peaks.

Fragments and Carbon Clusters

An unusual effect of laser desorption on certain polymers involves the formation of high-mass carbon clusters. This was first reported by Brown, who observed a series of peaks separated by 24 u in the high-mass region of the negative-ion spectrum of a dehydrocoupled aromatic polymer.[51] Subsequently, positive-ion clusters were observed in the LD-FTMS spectrum of poly(2,5-thieneylene) and of the product from dehydrocoupling of pyrene.[56] This latter product shows positive-ion clusters containing more than 400 carbons. Clusters with masses up to about 3000 u

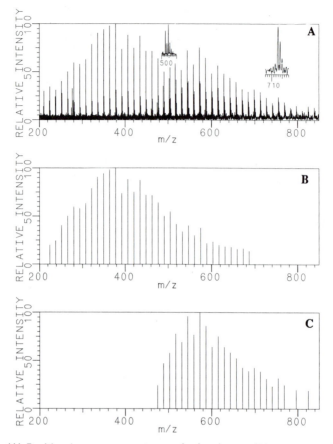

Fig. 14.5. (A) Positive ion mass spectrum of a hard wax. (B) Low mass alkene series present in the hard wax mass spectrum. (C) Low mass diene series present in the hard wax mass spectrum.

have recently been reported by Mauney[70] for laser microprobe/time-of-flight analysis of carbonaceous materials. Creasy and Brenna[71] have also reported high-mass carbon clusters from polyimide, using a laser microprobe/FTMS (described later in this chapter). These clusters are similar to those observed in laser desorption of graphite.[13,72–79] However, it is likely that different mechanisms are involved in producing clusters observed using different instruments. For example, the stable C_{60} cluster reported by Smalley and co-workers[72–75] for desorption of clusters from graphite is not observed in the clusters formed from polymers with a CO_2 laser. In contrast, such stable clusters ("magic numbers") are observed by Creasy and Brenna with the FTMS microprobe using a UV laser. Mauney reports that carbon clusters were not observed from direct ionization of graphite under his conditions, yet these clusters are observed by Creasy and Brenna.

Figure 14.6 shows the high-mass portion of a positive-ion LD-FTMS spectrum of an aromatic polyester. Since this sample was insoluble in various organic solvents, pellets of the material were frozen in liquid nitrogen and pulverized. Particles of the resulting fine powder that adhered to the target were examined by laser desorption. Peaks differing by 24 u appear over the mass range 1000–4000 u. Low-mass fragments (not shown) also appear, and result from fragmentation or pyrolysis of the polymer. It is interesting to note that several other samples of polyesters having similar chemical compositions were examined; only one showed carbon clusters of this type. We do not have enough information at this point to speculate on the mechanism of carbon cluster formation and how it may be related to specific chemical properties or sample handling conditions. It appears that carbon cluster formation is favored for samples in thick layers (or in this case, larger particles) rather than in thin films.

Large polymers (e.g., those with molecular weights above 10,000) have not been shown to be amenable to molecular weight distribution determination by LD-FTMS at its current state of development, although recent work by Wilkins and Ijames shows very promising results for polymers near 10,000 u.[60] However, it is often possible to obtain structural information about large polymers from low-mass fragments produced by laser pyrolysis, or laser-induced fragmentation of the polymer. Applications of laser pyrolysis/mass spectrometry of polymers were first described by Kistemaker and Meuzelaar.[80,81] As an example of laser-induced fragmentation of a large polymer, the high-mass (>200 u) negative-ion laser desorption spectrum of polytetrafluoroethylene (Teflon) is shown in Fig. 14.7. The low-mass fragment ions were ejected from the cell to enhance the detection of high-mass fragment ions. Teflon typically contains about 1000 CF_2 units, giving it a molecular

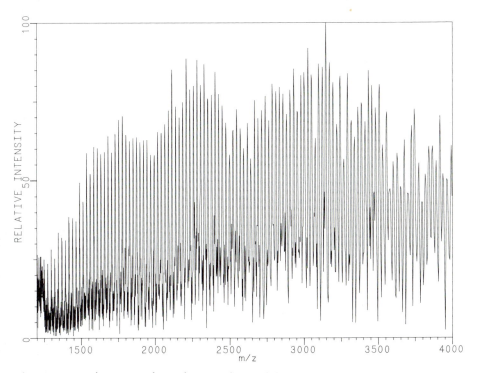

Fig. 14.6. High-mass carbon clusters obtained from an intractable high-molecular-weight polyester.

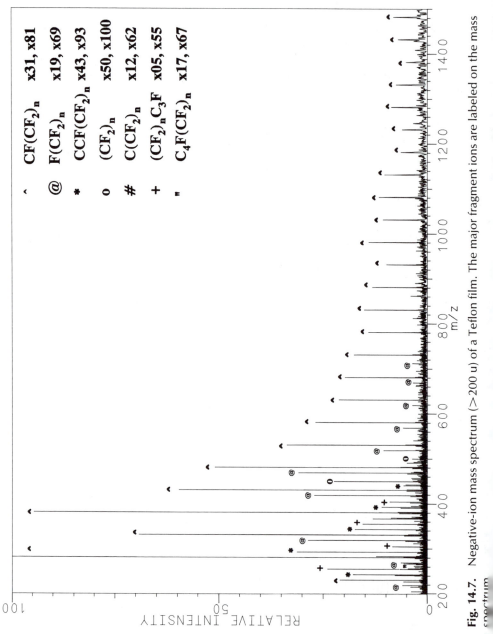

Fig. 14.7. Negative-ion mass spectrum (>200 u) of a Teflon film. The major fragment ions are labeled on the mass spectrum.

Legend:

^	$CF(CF_2)_n$	x31, x81
@	$F(CF_2)_n$	x19, x69
*	$CCF(CF_2)_n$	x43, x93
o	$(CF_2)_n$	x50, x100
#	$C(CF_2)_n$	x12, x62
+	$(CF_2)_nC_3F$	x05, x55
"	$C_4F(CF_2)_n$	x17, x67

weight of about 50,000 u. The LD-FTMS spectrum consists of fluoroalkyl fragments, with several series of peaks that differ by CF_2 units. These are marked on the spectrum and are similar to the results recently reported by Bletsos and co-workers using SIMS techniques.[69] Thus, even though the polymer has too large a molecular weight to be characterized by intact molecular ions, the fragmentation pattern is distinctive and informative.

Related Analyses: Carbon Clusters from Asphaltenes

Other analytical problems exist that have similarities to polymer analysis. For example, laser desorption may be used to obtain molecular weight distributions for the complex petroleum mixtures that are referred to as "asphaltenes." Positive- and negative-ion laser desorption spectra of an asphaltene sample are shown in Fig. 14.8A and B. Note that carbon clusters are observed in the negative-ion spectrum above m/z 900.

Polymer Additives

In addition to characterization of the polymers themselves, an important application of LD-FTMS may be analysis of polymer additives. A wide variety of compounds such as dyes, surfactants, antioxidants, and plasticizers are frequently additives in commercial polymeric products. Determining which additives are present, and in what quantity, is an important question for the polymer industry both for quality control and in the analysis of competitive products.

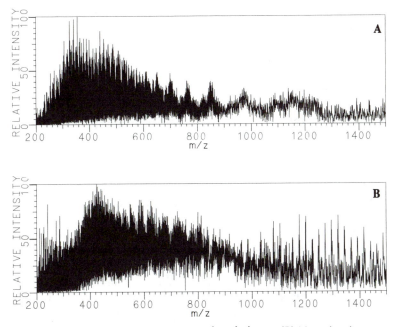

Fig. 14.8. (A) Positive-ion mass spectrum of asphaltene. (B) Negative-ion mass spectrum of asphaltene.

Some preliminary work has been done in this area. Marshall and Hsu have shown that dyes in polymethylmethacrylate (PMMA) can be identified by LD-FTMS at lower levels than by reflectance infrared spectroscopy.[58] Johlman et al. have reported that LD-FTMS gave more informative spectra than those obtained by fast atom bombardment (FAB) for a number of polymer additives.[61] We have also examined a number of polymer additives in our own laboratory as part of our activity in polymer analysis.

One aspect of polymer additive analysis that makes it a challenging problem is that the chemical nature of various additive types differs widely, ranging from salts, such as zinc or calcium stearate, to polymers [such as poly(ethyleneglycol)]. Laser desorption seems to offer an ionization method that is effective for a wide variety of compound types, and that maybe applicable to direct analysis of additives in the polymer, without prior extraction. Laser desorption spectra of a number of polymer additives of widely varying chemical classes are shown in Fig. 14.9.

Salts such as zinc or calcium stearate pose problems for mass spectrometry because of the ionic nature of the compounds and because it is difficult to deal with doubly charged zinc and calcium cations. Laser desorption is a useful approach to the analysis of compounds of this type. For salts, we expect to see clusters or fragments that produce singly charged ions. In Fig. 14.10, we see positive- and negative-ion spectra for calcium stearate. In the positive-ion spectrum, we see an ion resulting from loss of one stearate anion, leaving a net single positive charge (323 u, or $[CaSt]^+$), and also a peak due to loss of CO at 295 u. At high mass we see cluster ions that are assigned to $[Ca_2St_3]^+$ and successive carbonyl losses (one per stearate). The negative-ion spectrum shows the stearate anion (283 u), and loss of a carbonyl (255 u). Clusters at high mass are assigned to $[CaSt_3]^+$ and carbonyl losses for each stearate. Cluster formation of this type seems to be characteristic of many surfactants, and is enhanced for higher laser power densities and/or higher sample concentrations.

As an example of the application of laser desorption to a manufacturing problem, Fig. 14.11 shows a laser-desorption spectrum of an additive claimed to be "sorbitol tristearate." The compound was believed to be a pure sample of sorbitol, with stearates substituted at three of the hydroxyl groups. The potassium-attached molecular ion is observed at 1020 u, corresponding to attachment of three stearate groups to sorbitol. The series of peaks in the range 850 up to 1020 u consist of loss of H_2O from the molecular ion, further H_2 losses from the $[M + K - H_2O]^+$ peak, and series of successive C_2H_4 losses from each of these species. Peaks at higher mass show that the compound was not pure, but contained other substitutions. The peak at 1286 u results from substitution of an additional stearate to a fourth hydroxyl group, and the peak at 1552 u results from a fifth stearate addition. Peaks at 1258 and 1524 u may be assigned to substitution of a palmitate (C_{16}) group rather than a stearate group (C_{18}) for the tetra- and penta-substituted sorbitols. Indeed, other substitutions of palmitates may be present, but are difficult to distinguish from successive C_2H_4 loss peaks without more information. In this case, laser desorption provides a quick method for screening the polymer additive, verifying its composition, and determining the presence and nature of impurities in the sample.

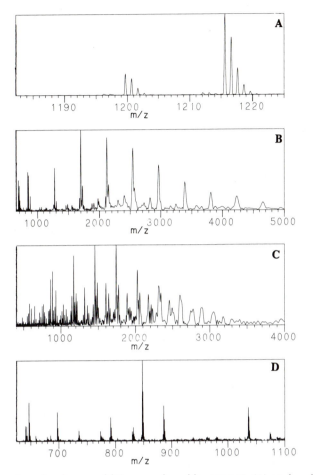

Fig. 14.9. A series of polymer additives analyzed by LD-MS. (A) Molecular ion region of Irganox 1010, [M + K]$^+$ showing the high-resolution capabilities of LD-FTMS. (B) A low-resolution positive-ion mass spectrum of Spinuvex A36, showing a molecular weight distribution of the oligomer (442 u) from $n = 2$ to $n = 11$. (C) A low resolution positive-ion mass spectrum of Tinuvin 622, showing a molecular weight distribution of the oligomer (283 u) from $n = 2$ to $n = 11$. The major peaks correspond to potassium cation attachment with loss of MeOH (from OCH$_3$ end group). (D) A positive-ion mass spectrum of Santostab PEPQ, with quasimolecular ions [M + H]$^+$ at m/z 1036 and [M + K]$^+$ at m/z 1074.

Surface and Bulk Analysis

Characterization of the surface and bulk properties of materials is an important application of LD-FTMS. Laser desorption allows us to examine materials that are otherwise difficult or impossible to study by mass spectrometric methods. FTMS offers high resolution, elemental compositions (from accurate mass measurements), and the possibility of obtaining further information by MS/MS methods such as collisional activation, photodissociation, or selective reactions. An early

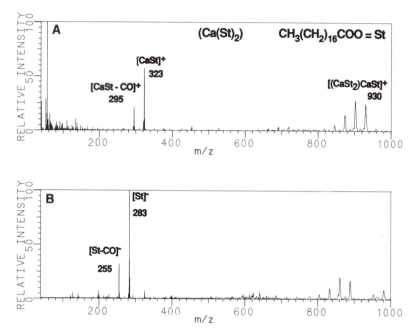

Fig. 14.10. (A) Positive-ion mass spectrum of calcium stearate. (B) Negative-ion mass spectrum of calcium stearate.

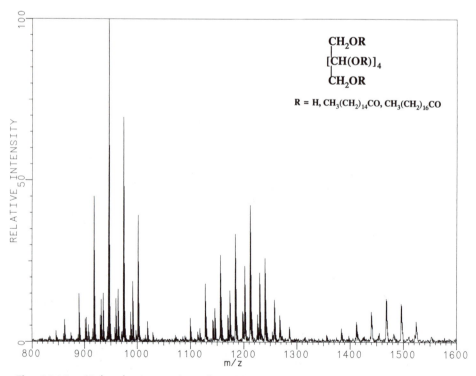

Fig. 14.11. Molecular ion region of positive-ion mass spectrum of tri-, tetra-, and penta-substituted stearate and palmitate polymer additive.

report of applications of LD-FTMS to surface and bulk analysis was presented in 1985 in which Cody and co-workers described the identification of fluorinated lubricants on magnetic storage media and analysis of lead, uranium, and rare earths in the mineral zircon.[39] Cody and Kinsinger reported the identification of a surface stain on a copper part by LD-FTMS.[82] Recently, McIver and co-workers have reported surface characterization using low laser power and postionization of desorbed neutrals[19] by electron impact or multiphoton ionization.

An important problem for surface characterization involves distinguishing interferences such as oxides from atomic ions and their dimers and multiply charged ions. To separate these by mass spectrometry requires high resolution. An example is the analysis of silicon wafers used in the semiconductor industry. We examined two samples: a "control wafer" and a sample believed to have phosphorus-containing impurities. The results are shown in Fig. 14.12. The control wafer shows primarily Si^+ and $SiOH^+$ on the surface, along with some trace potassium. In contrast, the presence of phosphorus in the "impure" sample is obvious from the strong PO^+ peak, and, to a lesser extent, a P^+ peak. An expanded view of the region between 55 and 57 u shows five major peaks at m/z 56, detected at a resolving power of 7000. The negative-ion spectrum of the control wafer (not shown) also indicates the presence of carbon in the wafer with a series of carbon cluster peaks C_n^+, with n ranging from 4 to 12.

FTMS Laser Microprobe Instruments

The analytical utility of LD-FTMS for surface analysis could be greatly increased by adding sample positioning and viewing capabilities, and having optics that would permit analysis of very small (1–2 μm) spots. At least three instruments have been built with this goal in mind.

Multipurpose Laser Microprobe/FTMS. Two research groups at IBM have modified the Nicolet FTMS-2000 dual-cell Fourier transform mass spectrometer for laser desorption analysis with small-spot viewing optics. The first such instrument was designed at the Materials Laboratory of the General Products Division of IBM at San Jose, California, by C. Hignite and S. Ghaderi in conjunction with Pacific Precision Laboratories of Canoga Park, California and Nicolet Instruments. This instrument uses a variation on the laser interface described previously. A focusing lens and flat mirror are substituted for the parabolic mirror to provide better sample-viewing optics. Sample viewing is accomplished with a video camera using the same optical path as the laser beam. A fiber optic bundle and external light source provide sample illumination. Magnification is 70×, and a capability is provided for photographing the sample surface. The same CO_2 laser used for the standard FTMS-2000 is presently employed for ionization, and provides a spot size of approximately 50 μm with an operating wavelength of 10.6 μm. To reduce the spot size to 10 μm, they plan to replace the CO_2 laser with an Nd:YAG laser.

In practice, the sample is positioned using a combination of x,y rotation and in–out (z-axis) movement, which allows any spot on the surface to be targeted. A crosshair is used to indicate the laser target. Once the desired spot has been targeted, the laser is fired and a spectrum obtained. Both direct laser ionization and laser desorption with electron impact postionization are employed to examine a

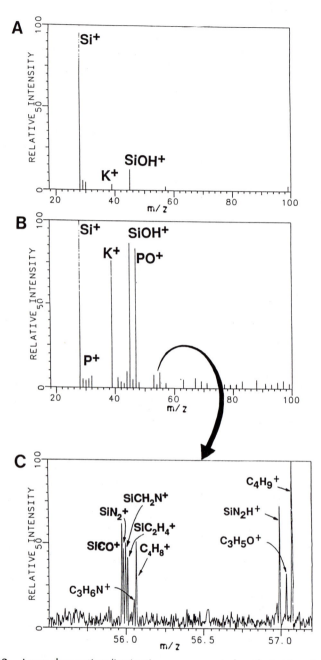

Fig. 14.12. Laser desorption/ionization mass spectral analysis of surface contaminants on silicon wafers. (A) Control wafer shows a relatively simple spectrum of peaks corresponding to Si^+, $SiOH^+$, and a trace of K^+. (B) The contaminated wafer shows a more complex spectrum, with much higher levels of $SiOH^+$, and K^+, plus P^+, PO^+ and peaks at m/z 56 and m/z 57. (C) An expanded plot clearly shows the different components at m/z 56 and 57. Accurate mass measurements at high resolution permit the classification of the species present.

variety of samples, including organic and inorganic contaminants on electronic components, as well as polymers and lubricants. Typical spectra are shown in Fig. 14.13, which shows two spectra of contaminated disk sliders (from magnetic storage media). Figure 14.13A shows niobium contamination on the disk slider, whereas Fig. 14.13B shows tin contamination that has been picked up by the slider from the disk surface.

A similar instrument[83] has been built by J. T. Brenna at IBM in Endicott, New York (also in collaboration with Pacific Precision Laboratories and Nicolet). This instrument uses an Nd:YAG laser operating at the fourth harmonic (266 nm) or the second harmonic (532 nm). A special interface was constructed using an IBM PC-AT to mediate between the trigger generated by the computer on the mass spectrometer and special trigger requirements of the laser. Fiber optics and a parabolic mirror are used to illuminate the target surface, and a flat mirror and focusing lens are used for both the Nd:YAG laser beam and the sample viewing optics. The laser spot size for this instrument is 5–8 μm. Sample positioning is accomplished with a combination of the standard probe rotational motion and a specially constructed sliding sample stage controlled by a plunger assembly.

This instrument is used to examine and identify large polymers on surfaces. The characteristic ions formed on laser irradiation with this instrument are somewhat different than those for the standard instrument using a CO_2 laser, due to higher power densities and/or the laser wavelengths. Extensive fragmentation frequently occurs; poly(ethyleneglycol)s show only low-mass fragments and no cationized molecular ions under these conditions. Therefore, the term "laser ablation" is preferred to "laser desorption" in describing this instrument.

Since the polymers of interest (e.g., polyimides) have molecular weights too large to characterize by mass spectrometry, this instrument is being used to identify "fingerprint" patterns in the fragmentation patterns of different varieties of these polymers. In particular, Creasy and Brenna have found that observed carbon cluster patterns are sufficiently different that they may be used to recognize specific polyimides.[84] This work has also led to a better understanding of the laser ablation process, which has important implications for laser drilling. An example of the type of spectrum obtained under these conditions is shown in Fig. 14.14, which shows a spectrum of polyimide obtained using 532-nm ablation. Note the presence of carbon clusters at higher masses.

The spatial resolution and spot size for these general purpose microprobe/ FTMS instruments are slightly less than is available for commercial laser microprobe systems using time-of-flight detection. However, both of these instruments permit viewing and manipulation of the laser target while preserving other modes of operation of the Fourier transform mass spectrometer for high resolution, heated direct probe work, CI, MS/MS, GC–MS, etc.

A Dedicated Laser Microprobe/FTMS. A different design has been used in a dedicated laser microprobe instrument designed and built by Prof. J. F. Muller at the University of Metz, France, in collaboration with Nicolet Instruments.[85] This instrument uses an excimer laser charged with KrF (wavelength = 249 nm) with a pulse energy of 35 mJ delivered in 20 nsec. The energy at the surface is approximately 70 μJ, with a power density of 2×10^9 W/cm^2. In contrast with the two

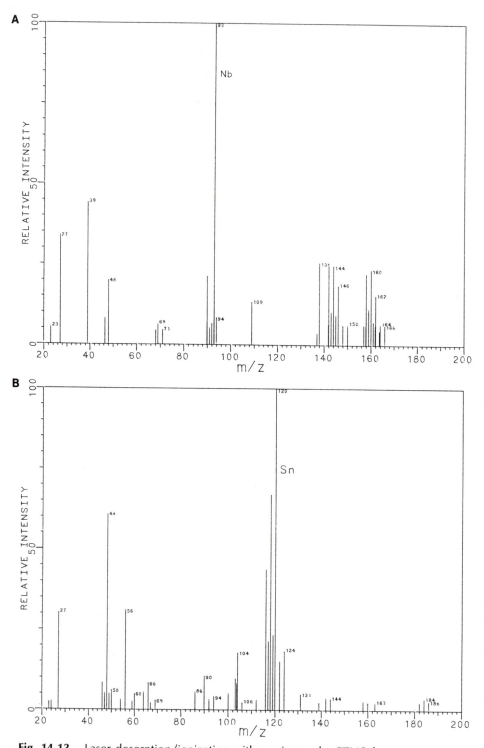

Fig. 14.13. Laser desorption/ionization with a microprobe FTMS for contamination identification on electronic media, using a pulsed CO_2 laser. (A) Positive-ion mass spectrum of niobium contamination on a disk slider. (B) Positive-ion mass spectrum of tin contamination on a particulate disk. (Reprinted, by permission of S. Ghaderi.)

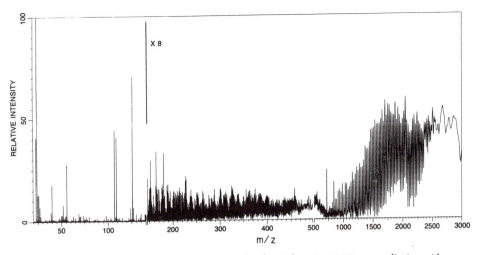

RELATIVE INTENSITY

X 8

m/z

Fig. 14.14. A positive-ion mass spectrum of polyimide using 532-nm radiation. Above mass 140, the relative intensity scale is multiplied by 8. The mass scale changes at mass 140 and 500. (Reprinted, with permission, from Ref. 84.)

previously described instruments, the laser beam travels parallel to the magnetic field, and is focused through the conductance limit onto the sample stage by a quartz lens. This permits the laser to be focused to a circular spot (rather than an oval spot as in the multipurpose microprobe/FTMS instruments). The minimum spot diameter obtainable with the optics is 3–4 μm. Sample illumination is accomplished by use of a flexible polymer lightpipe and an external light source. An image of the sample stage (20× magnification) is viewable through an endoscope. Coarse and fine sample x, y, and z positioning are accomplished with a new sample introduction probe that has fine micromanipulators to move the sample stage.

To test the high resolution and accurate mass capabilities of the instrument, a glass sample was prepared from SiO_2 and PbO with 10% impurities: 2% chromium, 2% cobalt, 2% tin, 1% neodymium, 1% samarium, 1% europium, and 1% gadolinium. The positive-ion mass spectrum shows both metal and metal oxide cations. Under high-resolution conditions (resolution >26,000 over a 35 u mass range) the doublets at 158 and 160 u can be identified as NdO^+ and Gd^+ (Fig. 14.15).

CONCLUSION

Laser desorption combined with Fourier transform mass spectrometry has been shown to be a useful approach to the analysis of polymers and surfaces. Polymer analysis by LD-FTMS may involve direct determination of the molecular weight distributions for oligomers with molecular weights approaching 10,000 u. For larger polymers, structural information may be obtained from fragment ions. Also attractive is the fact that LD-FTMS may be used to characterize polymer additives such as surfactants or antioxidants.

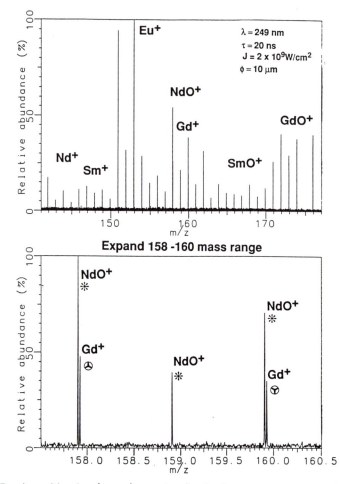

Fig. 14.15. A positive-ion laser desorption/ionization mass spectrum of a rare earth oxide obtained on a prototype laser microprobe FTMS, with an excimer laser ($\lambda = 249$ nm, pulse energy 3.5 mJ in 20 ns), under high-mass resolution conditions. Expansion of 158–160 mass range demonstrates the separation of NdO^+ and Gd^+. (Reprinted, with permission, from Ref. 85.)

For surface and inorganic analysis, LD-FTMS offers the advantage of high resolution, which is important for separating interferences such as oxides from atomic ions. The development of laser microprobe optics for use with Fourier transform mass spectrometry has been described and promises to extend the analytical utility of LD-FTMS to small spot surface analysis.

ACKNOWLEDGMENTS

The authors would like to thank Sahba Ghaderi and Tom Brenna of the IBM Corporation and Prof. J. F. Muller of the University of Metz, France for providing information on the laser microprobe systems and for permission to present these results. In addition we would like to thank Mark Johnston, Pat Brown, and Brad Spencer for their assistance in preparing the manuscript.

REFERENCES

1. M. B. Comisarow and A. G. Marshall, *Chem. Phys. Lett.* **25,** 28 (1974).
2. M. B. Comisarow and A. G. Marshall, *Chem. Phys. Lett.* **26,** 489 (1974).
3. A. G. Marshall, *Chem.* **18,** 316 (1985).
4. D. A. Laude, Jr., C. L. Johlman, R. S. Brown, D. A. Weil, and C. L. Wilkins, *Mass Spec. Rev.* **5,** 107 (1986).
5. M. Johnston, *Spectroscopy* **2**(2), 14 (1987).
6. M. Johnston, *Spectroscopy* **2**(3), 14 (1987).
7. M. L. Gross and D. L. Rempel, *Science* **226,** 261 (1984).
8. M. V. Buchanan (ed.), *Fourier Transform Mass Spectrometry: Evolution, Innovation, Applications.* American Chemical Society, Washington, DC, 1987.
9. B. S. Freiser, *Anal. Chim. Acta* **178**(1), 137 (1985).
10. B. S. Freiser, *Talanta* **32,** 697 (1985).
11. M. L. Mandich, W. D. Reents, Jr., and V. E. Bondybey, *J. Phys. Chem.* **90,** 2315 (1986).
12. W. D. Reents and V. E. Bondybey, *Chem. Phys. Lett.* **125**(4), 324 (1986).
13. S. W. McElvany, W. R. Creasy, and A. O'Keefe, *J. Chem. Phys.* **85**(1), 632 (1986).
14. J. M. Alford, P. E. Williams, D. J. Trevor, and R. E. Smalley, *Int. J. Mass Spec. Ion Process.* **72,** 33 (1986).
15. D. A. Weil and C. L. Wilkins, *J. Am. Chem. Soc.* **107,** 7316 (1988).
16. A. K. Chowdhury and C. L. Wilkins, *Int. J. Mass Spec. Ion Process.* **82,** 163 (1988).
17. T. J. Carlin and B. S. Freiser, *Anal. Chem.* **55,** 955 (1983).
18. M. P. Irion, W. D. Bowers, R. L. Hunter, F. S. Rowland, and R. T. McIver, Jr., *Chem. Phys. Lett.* **93,** 375 (1982).
19. M. G. Sherman, J. R. Kingsley, J. C. Hemminger, and R. T. McIver, Jr., *Anal. Chim. Acta* **178**(1), 79 (1985).
20. T. M. Sack, D. A. McCrery, and M. L. Gross, *Anal. Chem.* **57,** 1290 (1985).
21. C. H. Watson, G. Baykut, M. A. Battiste, and J. R. Eyler, *Anal. Chim. Acta* **178**(1), 125 (1985).
22. C. H. Watson, G. Baykut, and J. R. Eyler, *Anal. Chem.* **59,** 1133 (1987).
23. W. J. van der Hart, L. J. de Koning, N.M.M. Nibbering, and M. L. Gross, *Int. J. Mass Spec. Ion Process.* **72,** 99 (1986).
24. W. D. Bowers, S.-S. Delbert, and R. T. McIver, Jr., *Anal. Chem.* **58,** 969 (1986).
25. D. F. Hunt and J. Shabanowitz, presented at the 35th Annual Conference on Mass Spectrometry and Allied Topics, Denver CO, 1987.
26. D. A. McCrery, E. B. Ledford, Jr., and M. L. Gross, *Anal. Chem.* **54,** 1435 (1982).
27. R. E. Hein and R. B. Cody, Presented at the 31st Annual Conference on Mass Spectrometry and Allied Topics, Boston, MA, May 8–13, 1983.
28. L. L. Miller, A. D. Thomas, C. L. Wilkins, and D. A. Weil, *J. Chem. Soc., Chem. Commun.* 661 (1986).
29. D. A. McCrery, D. A. Peake, and M. L. Gross, *Anal. Chem.* **57,** 1181 (1985).
30. R. S. Brown and C. L. Wilkins, *Anal. Chem.* **58,** 3196 (1986).
31. M. B. Comisarow, D. P. Fryzuk, and R. B. Cody, unpublished results.
32. D. A. Weil, C. F. Ijames, R. S. Brown, M. L. Coates, C. L. Yang, and C. L. Wilkins,

Presented at the 33rd ASMS Annual Conference on Mass Spectrometry and Allied Topics, May 26–31 1985, San Diego, CA, collected abstracts, p. 183.

33. L. Dahl, R. B. Cody, and A. Bjarnason, unpublished results.

34. C. L. Wilkins, D. A. Weil, C.L.C. Yang, and C. F. Ijames, *Anal. Chem.* **57**, 520 (1985).

35. R. B. Cody, I. J. Amster, and F. W. McLafferty, *Proc. Natl. Acad. Sci. U.S.A.* **82**, 6367 (1985).

36. M. L. Coates and C. L. Wilkins, *Biomed. Mass Spectrom.* **12**, 424 (1985).

37. D. A. McCrery and M. L. Gross, *Anal. Chim. Acta* **178**(1), 91 (1985).

38. O. Barbera, J. F. Sanz, J. Sanchez-Parareda, and J. A. Marco, *Phytochemistry* **25**(10), 2361 (1986).

39. R. B. Cody, J. A. Kinsinger, Sahba Ghaderi, I. J. Amster, F. W. McLafferty, and C. E. Brown, *Anal. Chim. Acta* **178**, 43 (1985).

40. R. S. Brown and C. L. Wilkins, *J. Am. Chem. Soc.* **108**, 2447 (1986).

41. C. L. Wilkins and C.L.C. Yang, *Int. J. Mass Spec. Ion Process.* **72**, 195 (1986).

42. Z. Lam, M. B. Comisarow, G.G.S. Dutton, D. A. Weil, and A. Bjarnason, *Rapid Commun. Mass Spectrom.* **1**(5), 83 (1987).

43. M. L. Coates and C. L. Wilkins, *Anal. Chem.* **59**, 197 (1987).

44. C. E. Brown, S. C. Roerig, V. T. Berger, R. B. Cody, and J. M. Fujimoto, *J. Pharm. Sci.* **74**, 821 (1985).

45. M. L. Coates and C. L. Wilkins, *Biomed. Environ. Mass Spectrom.* **13**, 19 (1986).

46. R. E. Shomo, II, A. G. Marshall, and C. R. Weisenberger, *Anal. Chem.* **57**, 2940 (1985).

47. R. E. Shomo, A. G. Marshall, and R. P. Lattimer, *Int. J. Mass Spec. Ion Process.* **72**, 209 (1986).

48. M. P. Chiarelli and M. L. Gross, *Int. J. Mass Spec. Ion Process.* **78**, 37 (1987).

49. D. A. Weil and R. E. Hein, Presented at the 35th ASMS Annual Meeting on Mass Spectrometry and Allied Topics, May 24–29, 1987, Denver, CO, collected abstracts, pp. 783–784.

50. C. E. Brown, P. Kovacic, C. A. Wilkie, R. B. Cody, and J. A. Kinsinger, *J. Polym. Sci., Part C Polym. Lett.* **23**, 453 (1985).

51. C. E. Brown, P. Kovacic, C. A. Wilkie, J. A. Kinsinger, R. E. Hein, S. I. Yaniger, and R. B. Cody, *J. Polym. Sci., Part A: Polym. Chem.* **24**, 255 (1986).

52. C. E. Brown, C. A. Wilkie, J. Smukalla, R. B. Cody, and J. A. Kinsinger, *J. Polym. Sci., A: Polym. Chem.* **24**, 1297 (1986).

53. R. S. Brown, D. A. Weil, and C. L. Wilkins, *Macromolecules* **19**, 1255 (1986).

54. C. E. Brown, P. Kovacic, C. A. Wilkie, R. B. Cody, R. E. Hein, and J. A. Kinsinger, *Synthet. Metals* **15**, 265 (1986).

55. C. E. Brown, P. Kovacic, R. B. Cody, Jr., R. E. Hein, and J. A. Kinsinger, *J. Polym. Sci., C: Polym. Lett.* **24**, 519 (1986).

56. C. E. Brown, P. Kovacic, K. J. Welch, R. B. Cody, R. E. Hein, and J. A. Kinsinger, *J. Polym. Sci., Part A: Polym. Chem.* **26**, 131 (1987), in press.

57. C. E. Brown, P. Kovacic, K. J. Welch, R. B. Cody, R. E. Hein, and J. A. Kinsinger, *Arab. J. Sci. Engineer.*, **13**, 163 (1988).

58. A. T. Hsu and A. G. Marshall, *Anal. Chem.* **60**, 932 (1987).

59. L. M. Nuwaysir and C. L. Wilkins, *Anal. Chem.* **60**, 279 (1988).

60. C. F. Ijames and C. L. Wilkins, *J. Am. Chem. Soc.* **110**, 2687 (1988).

61. C. L. Johlman, C. L. Wilkins, and M. Youssefi, Presented at the 35th ASMS Conference on Mass Spectrometry and Allied Topics, May 24–29, 1987, Denver, CO, collected abstracts, pp. 785–786.

62. Tachisto model 216 pulsed CO_2 laser.

63. R. J. Cotter, *Anal. Chem.* **56**, 485A (1984).

64. L. Chen, T.-C.L. Wang, T. L. Ricca, and A. G. Marshall, *Anal. Chem.* **59,** 449 (1987).

65. R. B. Cody, R. E. Hein, and S. D. Goodman, *Rapid Commun. Mass Spectrom.* **1,** 99 (1987).

66. R. J. Cotter, J. P. Honovich, J. K. Olthoff, and R. P. Lattimer, *Macromolecules* **19,** 2996 (1986).

67. J. A. Gardella, Jr. and D. M. Hercules, *Fres. Z. Anal. Chem.* **308,** 297 (1981).

68. D. Briggs, *Surf. Interface Anal.* **9,** 391 (1986).

69. I. V. Bletsos, D. M. Hercules, J. H. Magill, D. vanLeyen, E. Niehuis, and A. Benninghoven, *Anal. Chem.* **60,** 938 (1988).

70. T. Mauney, *Anal. Chim. Acta* **195,** 337 (1987).

71. W. R. Creasy and J. T. Brenna, *Chem. Phys.* **126,** 453 (1988), submitted.

72. Q. L. Zhang, S. C. O'Brien, J. R. Heath, Y. Liu, R. F. Curl, H. W. Kroto, F. K. Tittel, and R. E. Smalley, *J. Phys. Chem.* **90,** 525 (1986).

73. J. R. Heath, S. C. O'Brien, Q. L. Zhang, Y. Liu, R. F. Curl, H. W. Kroto, F. K. Tittel, and R. E. Smalley, *J. Am. Chem. Soc.* **107,** 7779 (1985).

74. J. R. Heath, Q. L. Zhang, S. C. O'Brien, R. F. Curl, H. W. Kroto, and R. E. Smalley, *J. Am. Chem. Soc.* **109,** 359 (1987).

75. S. C. O'Brien, J. R. Heath, R. F. Curl, and R. E. Smalley, *J. Chem. Phys.* **88,** 220 (1988).

76. R. E. Rohlfing, D. M. Cox, and A. Kaldor, *J. Chem. Phys.* **81,** 3332 (1984).

77. A. O'Keefe, M. M. Ross, and A. P. Baronavski, *Chem. Phys. Lett.* **130,** 17 (1986).

78. S. W. McElvaney, H. H. Nelson, A. P. Baronavski, C. H. Watson, and J. R. Eyler, *Chem. Phys. Lett.* **134,** 214 (1987).

79. R. D. Knight, R. A. Walch, S. C. Foster, T. A. Miller, S. L. Mullen, and A. G. Marshall, *Chem Phys. Lett.* **129**(4), 331 (1986).

80. H.L.C. Meuzelaar, P. G. Kistemaker, and M. A. Posthumus, *Biomed. Mass Spectrom.* **1,** 312 (1974).

81. P. G. Kistemaker, J. H. Boerboom, and H.L.C. Meuzelaar, *Dynamic Mass Spectrom.* **4,** 139 (1976).

82. R. B. Cody and J. A. Kinsinger, in *Fourier Transform Mass Spectrometry,* edited by M. V. Buchanan. American Chemical Society, Washington, DC, 1987, pp. 59–80.

83. J. T. Brenna, W. R. Creasy, W. McBain, and C. Soria, *Rev. Sci. Instrum.* **59,** 873 (1988).

84. J. T. Brenna, private communication.

85. M. Pelletier, G. Krier, J. F. Muller, R. M. Johnston, and D. A. Weil, *Rapid Commun. Mass Spectrom.* **2,** 146 (1988).

86. A. Bjarnason, R. B. Cody, D. A. Weil, unpublished results.

15

Laser Photodissociation on a Tandem Quadrupole Fourier Transform Mass Spectrometer

JEFFREY SHABANOWITZ and DONALD F. HUNT

Fourier transform mass spectrometers have tremendous potential for the analysis of large biomolecules[1-3] because (1) they record the masses of all ions in the spectrum simultaneously and sample is not continuously consumed during the time needed to generate a mass spectrum, (2) they function as ion storage devices that permit accumulation of ions produced in low abundance from small amounts of sample, (3) they facilitate direct analysis of mixtures by using ion ejection techniques,[2-3] and (4) they appear to be ideally suited for laser photodissociation experiments.[4-6]

One disadvantage of Fourier transform mass spectrometers is that they must be operated at a pressure less than 10^{-8} torr in the analyzer to prevent collisions between ions and neutral gas molecules from interfering with the mass measurement process. This requirement has made it difficult to use Fourier transform instruments in conjunction with volatile liquid matrices employed in the highly successful particle-bombardment ionization techniques for biological molecules. Problems associated with high gas flow are circumvented in a tandem quadrupole Fourier transform mass spectrometer (QFTMS), because sample introduction and ionization take place in a differentially pumped ion source, and only the ions of interest are transferred through the fringing fields of a superconducting magnet and into the ion cyclotron resonance (ICR) cell for mass analysis.[7-9] The tandem quadrupole Fourier transform mass spectrometer has demonstrated that ions generated from picomole amounts of large oligopeptides can be trapped and mass analyzed when injected into the ICR cell from an external ion source.[10]

Laser photodissociation has received considerable attention from physical chemists but has only recently shown promise as an analytical technique in mass spectrometry.[11-13] Photodissociation of ions in conventional magnetic sector and quadrupole instruments does not usually proceed in high yield because the interaction time of the ion and photon beams is too short to facilitate efficient excitation and fragmentation. This is to be contrasted with the situation in the Fourier transform ion cyclotron resonance mass spectrometer and the ion trap mass spectrometer in which ions can be selected and stored in orbits, a few millimeters in diameter, for several minutes.[6,14,15]

The laser chosen for photodissociation of oligopeptides on the tandem QFTMS is an ultraviolet excimer laser. Excitation at 193 nm (argon fluoride, 6.42 eV) is used because this wavelength corresponds closely to that for absorption by the amide bond. Amino acids are linked via amide bonds in the backbone of the oligopeptide structure. In contrast, a pulsed TEA-CO_2 (10.6 μm) laser is used for photodissociation of the saccharide and glycerol oligomers, because these molecules do not contain an amide linkage and, therefore, are transparent to light from the UV excimer laser.

INSTRUMENTATION

The tandem quadrupole Fourier transform mass spectrometer, shown schematically in Fig. 15.1, has been described in detail elsewhere.[7,8] In brief, the front end of the instrument consists of a standard ion source and lens system that is evacuated differentially by a cyrogenic pump. All sample introduction and ionization modes are compatible with the configuration of this ion source. For large biomolecules, the ionization method of choice is fast atom bombardment or liquid secondary ion mass spectrometry (LSIMS). Sample plus a suitable liquid matrix are placed on the gold-plated tip of a stainless-steel probe. After evacuation to 10^{-3} torr, the probe is positioned into the ion source in which sample and matrix are sputtered by a high energy Cs^+ ion beam (up to 10 keV). Secondary ions produced from both sample and matrix are then extracted and transported through the fringing fields of a 7-T superconducting magnet via long quadrupole rods. A second cryopump is used to differentially evacuate the first set of quadrupoles (25 cm long) from the much longer (86.25 cm) second set of quadrupoles. The latter are employed to guide the ions into the ICR cell. The longer quadrupoles and the cell are maintained at a vacuum better than 10^{-9} torr by a third and final cryopump. The electronics that drive the two quadrupoles have been modified to operate in the rf-only mode at a frequency of 870 kHz. The rf-only mode allows the system to function with a high-mass cutoff well above 10 kDa and, more importantly, a low-mass cutoff that can be set anywhere below mass 3 kDa. The transmitter and receiver plates of the ICR cell are constructed of highly polished stainless-steel plates, 7.75 × 2.8 cm. The trapping plates, 2.8 × 2.8 cm, contain 1.75-cm^2 square holes, which are covered with 90% transmission nickel mesh. These holes function as entrance and exit apertures. The laser beam for photodissociation is brought in from the backside of the magnet through an appropriate window in the rear vacuum flange and enters the ICR cell via trapping plate 2.

EXPERIMENTAL

Sample analysis for biopolymer sequencing on the tandem quadrupole Fourier transform mass spectrometer is performed by adding 0.5–1.0 μL of 5% acetic acid containing peptide or saccharide at the 10–100 pmol/μL level to 0.5 μL of a matrix

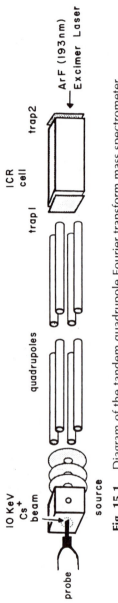

Fig. 15.1. Diagram of the tandem quadrupole Fourier transform mass spectrometer.

of glycerol:monothioglycerol (1:3) on the sample probe. The glycerol oligomer spectra are obtained by just adding glycerol to the probe. A positive voltage of 1.5 V is applied to the sample probe after it has been inserted into the ion source. The sample matrix is then bombarded with a 4-msec pulse of 10-keV Cs^+ projectiles. Ions sputtered into the gas phase by the above process are extracted from the ion source and focused into the quadrupoles. The latter are operated in the rf-only mode with an offset potential of -20 V and a low-mass cut-off of 0.8 kDa. Under these conditions, the quadrupoles pass all ions above m/z 800 and reject most of the abundant low-mass ions produced from both the sample and liquid matrix. If allowed to enter the cell, these low-mass ions would quickly exceed the capacity of the ICR cell and impair the ability of the instrument to detect weaker signals at the high-mass end of the spectrum. For the stachyose and the glycerol oligomer analysis similar low-mass cut-offs at 0.6 and 0.35 kDa, respectively, were employed. To contain the ions within the cell, a potential of $+3$ V is applied to trapping plates 1 and 2. The receiver and transmitter plates of the ICR cell are maintained at 0 V potential.

After the above parameters are set, mass spectra are acquired using a series of pulses controlled by the data system. First, a quench pulse applies potentials of $\pm$ 10 V to trapping plate 2, which is used to clear the cell of all charged particles. The second, or beam pulse, lowers the voltage on the extractor lens of the Cs^+ gun and allows 10-keV Cs^+ ions to impact on the sample matrix for 4 msec. The same pulse places a rf potential on the quadrupole rods and lowers the dc potential on trapping plate 1 to 0 V. This process allows efficient transport of the ions through the fringing fields of the magnet and into the ICR cell. At this point, a sweep-out rf pulse applied to the transmitter plates of the cell can be used to remove unwanted ions. After a 10-msec delay, the data system can also trigger a 10-nsec pulse of light from the UV-excimer laser to photodissociate the trapped ions. In the next step, 10 msec later, an rf pulse of 35 V, peak-to-peak, is applied to the transmitter plates for 2.5 msec. This pulse contains a range of frequencies that accelerates all ions of all masses trapped in the cell and causes them to move coherently at their characteristic cyclotron frequencies. Ion image currents induced by these orbiting ions on the ICR cell receiver plates are monitored for 8–64 msec, depending on the amount of resolution desired, amplified, and then digitized. Data from 20 to 200 such experiments are acquired, summed together, and the resulting time-averaged signal is then converted to the desired mass spectrum by using Fourier transform analysis. The total experiment time to produce the required data is seconds, and the total sample ionization or consumption time is less than a second.

APPLICATIONS

Examples of laser photodissociation mass spectra recorded on glycerol oligomers, stachyose, and oligopeptides are now discussed. Fragmentation observed in the laser photodissociation of the oligopeptides is discussed according to the nomenclature rules proposed by Roepstorff and Fohlman.[16]

1. The effect of laser photodissociation on glycerol oligomers is illustrated in Fig. 15.2. Sputtering of neat glycerol [$HOCH_2CH(OH)CH_2OH$], molecular mass 92

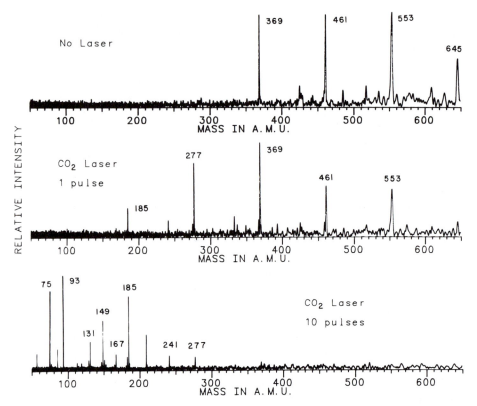

Fig. 15.2. Pulsed CO_2 laser effect on dissociation of protonated glycerol oligomers.

Da, with high-energy Cs^+ ions produces proton-bound oligomers corresponding to $(glycerol)_nH^+$. The top spectrum shows ions of this type for the glycerol tetramer, pentamer, hexamer, and heptamer. A single pulse of infrared CO_2 radiation (10.6 μm, 0.12 eV) causes only minimal dissociation of these ions as shown in the center spectrum. Substantial fragementation does occur, however, if the ions are exposed to 10 consecutive pulses of laser radiation as shown in the bottom spectrum. Laser photodissociation of the protonated oligoglycerols generates the protonated trimer, dimer, and monomer of glycerol, as well as fragment ions formed by loss of water molecules from these oligomers.

2. A typical spectrum resulting from laser photodissociation of the oligosaccharides is shown in Fig. 15.3. In this example, stachyose (molecular mass 666 Da) is ionized by cationization and the resulting $(M + Na)^+$ ion is subjected to photodissociation. Ten pulses of radiation from the CO_2 laser on the pseudomolecular ion at m/z 689 results in fragmentation that occurs by two complementary pathways. Successive losses of dehydromonosaccharide (162 Da) from the sodiated molecule $(M + Na)^+$ account for the ions at m/z 527 and 365. The signals at m/z 163, 325, and 487 correspond to oxonium ions derived from mono-, di-, and trisaccharide units, respectively. Similar results were obtained on a maltohexose of molecular mass 990 Da.

3. Shown in Fig. 15.4 is the sequence derived from the laser photodissociation

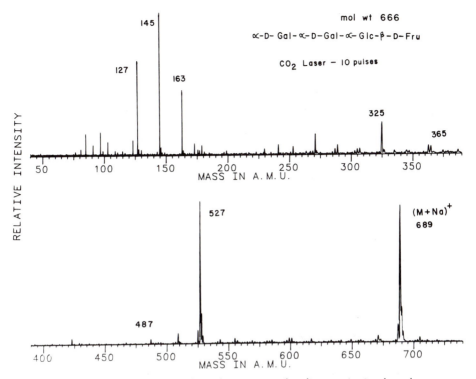

Fig. 15.3. Pulsed CO_2 laser photodissocation of sodium cationized stachyose.

mass spectrum generated from $(M + H)^+$ ions of a 15-residue tryptic peptide (M_r 1771). This particular peptide was obtained by digestion of beef spleen, purple acid phosphatase, a glycoprotein ($M_r \sim 38,000$) of unknown structure presently being studied in our laboratory.[17] A sweep-out pulse was employed to eject most of the ions below m/z 1600 from the cell and then the remaining ions were exposed to a single pulse of radiation at 193 nm from an argon fluoride excimer laser. Data from 50 such experiments on the same 10 pmol sample were acquired and summed in less than 20 sec to produce the data shown in Fig. 15.5. All mass assignments are within 0.2 amu of the expected values.

Assignment of an unambiguous sequence to the tryptic peptide is facilitated greatly by comparing the laser photodissociation spectrum in Fig. 15.4 to that obtained on a 10-pmol sample of the corresponding methyl ester. The $(M + H)^+$ ion for the latter derivative is shifted to a higher mass by 56 amu, a number con-

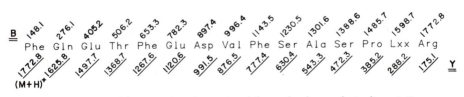

Fig. 15.4. Amino acid sequence determined from the laser photodissociation mass spectrum on 10 pmol of a tryptic peptide.

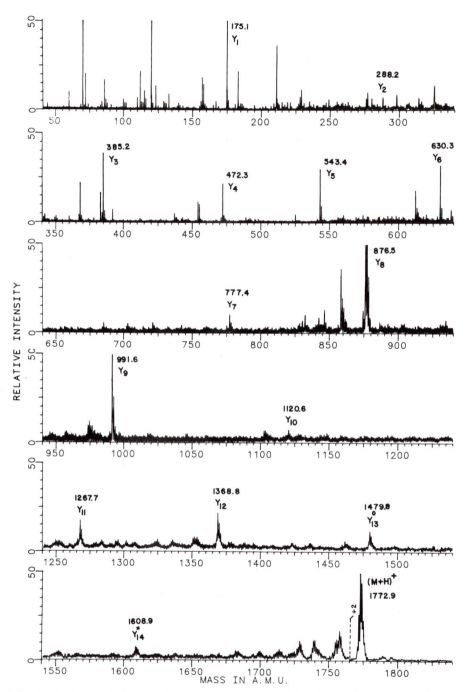

Fig. 15.5. Laser photodissociation mass spectrum of the (M + H)$^+$ ion derived from 10 pmol of a tryptic peptide. (Republished with permission of Ref. 17.)

sistent with the presence of four carboxylic acid groups in the molecule. All major fragment ions in Fig. 15.5 also shift to higher mass by multiples of 14 in the methyl ester spectrum and are thus assigned as being type Y, containing the COOH terminus of the molecule. Mass differences between the observed type Y_n fragment ions allow the sequence in Fig. 15.4 to be assigned to this sample.

In Fig. 15.4, the residue designated Lxx indicates that Leu and Ile cannot be differentiated by the present methodology. Ions in the spectrum at m/z 1479.8 and 1608.9 result from a loss of water and ammonia, respectively, from ions of type Y containing Glu and Gln at the NH_2 terminus. Mass analysis of peptides shortened by one amino acid as a result of three separate Edman degradation steps confirmed the identity of the three NH_2-terminal residues of the proposed sequence.

4. The photosynthetic membranes of green plants contain phosphoproteins that are thought to control the distribution of excitation energy between the two photosystems of the light harvesting chlorophyll complex. In order to identify and locate these phosphorylation sites, a tryptic digest of photosystem core particles was employed. The proteolytic digest yielded blocked phosphopeptides that could not be sequenced by classical Edman techniques.[18] Shown in Fig. 15.6 is the laser photodissociation mass spectrum obtained from 30 pmol of one of these phosphopeptides. The $(M + H)^+$ ions at m/z 1386.4 stored in the ICR cell were exposed to a single 10-nsec pulse of irradiation at 193 nm from an ArF excimer laser. Low-mass fragment ions observed at m/z 44.1, 74.1, 86.1, 87.1, 98.1, 112.1, 120.1, and 157.1 suggest the presence in the peptide of alanine, threonine, leucine, asparagine, N-acetylphosphothreonine, phenylalanine, and arginine, respectively. The appearance of an abundant fragment ion at an m/z value 98 Da below that of the $(M + H)^+$ ion (loss of phosphoric acid) suggests the presence of phosphate in the sample. Assignment of a sequence to the unknown peptide was facilitated by knowledge that the sample was generated by the action of trypsin on the parent protein. Trypsin digestion produces peptides that should contain either lysine or arginine at the COOH terminus. The signal at m/z 175.1 in the photodissociation spectrum indicates a series of Y-type ions with arginine as the COOH-terminal residue. The complete sequence of amino acids in the peptide is obtained from the m/z values of signals due to other members of the type Y series. Ions of this type result from cleavage of the various amide bonds in the peptide and retention of the charge on the COOH-terminal half of the molecule. Ions of type Y are labeled in Fig. 15.6 and also on the bottom of the structure in Fig. 15.7.

Assignment of the NH_2-terminal residue as N-acetylphosphothreonine is made from the mass difference of 223.1 Da observed between the $(M + H)^+$ ion and the fragment ion of type Y formed by loss of a single residue from the NH_2 terminus of the peptide. Note that signals due to Y_5 and Y_7 both occur at m/z values 101.0 Da above Y_4 and Y_6, respectively. If the NH_2 terminus was phosphorylated, the observed mass increment would be expected to be 181.0 Da (83.0 Da if loss of phosphoric acid accompanied formation of these Y-type ions). That this is not the case supports the assignment of the NH_2-terminal threonine as the site of phosphorylation. Additional support for this assignment comes from the fact that all ions of type B observed in the spectrum lose phosphoric acid and, therefore, occur 98.0 Da lower than expected.

5. The final example illustrates the use of laser photodissociation mass spec-

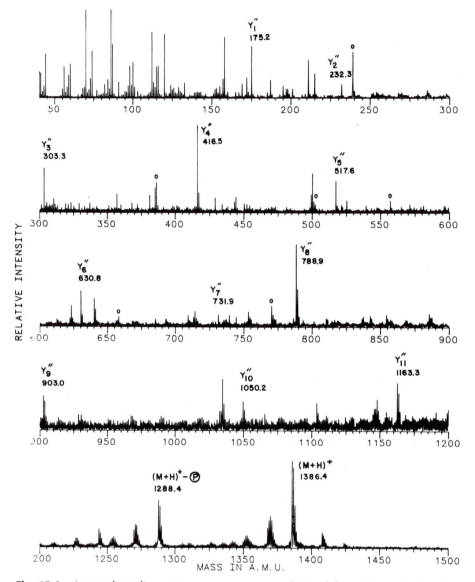

Fig. 15.6. Laser photodissociation mass spectrum obtained from 30 pmol of a phosphopeptide isolated from a tryptic digest. (Republished with permission of Ref. 18.)

trometry to elucidate an amino acid sequence at the protein binding site for the plant growth hormone, cytokinin. In this particular study, [^{14}C]2-azido-N^6-benzyladenine (AzBA) (Fig. 15.8) was chosen as a radiolabeled photoaffinity ligand to probe the cytokinin binding site since [N^6]-benzyladenine is a synthetic cytokinin with a high affinity for CBF-1. Radiolabeled ligand was covalently attached to the protein and the resulting sample was then cleaved into a collection of oligopeptides by the enzyme, endoproteinase Glu-C. Separation of this mixture by a combination

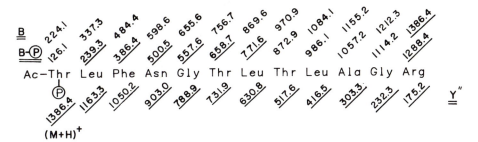

Fig. 15.7. Amino acid sequence of 30 pmol of a phosphorylated tryptic peptide determined by laser photodissociation.

of reversed-phase and anion-exchange high-performance liquid chromatography afforded a singly labeled 12-residue peptide.[19]

Mass spectra recorded on the labeled peptide and its corresponding methyl ester showed abundant $(M + H)^+$ ions at m/z 1605.1 and 1661.1, respectively. The observed mass shift of 56 amu indicates the presence of four carboxyl groups and, therefore three acidic amino acids in the molecule. Shown in Fig. 15.9 is the laser photodissociation mass spectrum recorded on 10 pmol of the peptide methyl ester. Fragment ions at m/z 70.1, 110.1, and 120.1 indicate the presence of histidine, proline, and phenylalanine in the peptide.

The search for a series of fragment ions having m/z values corresponding to the loss of one or more residues from the COOH terminus of the peptide, type B ions, is facilitated by the knowledge that the labeled peptide was generated from the protein by endoproteinase Glu-C digestion and therefore should contain a glutamic

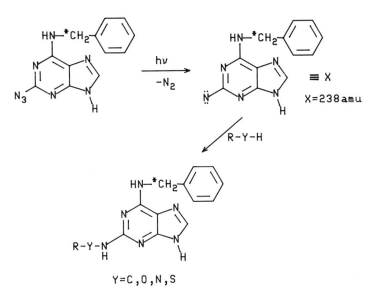

Fig. 15.8. Structure and activation mechanism for the radiolabeled photoaffinity label (AzBA) used to probe the CBF-1 binding site.

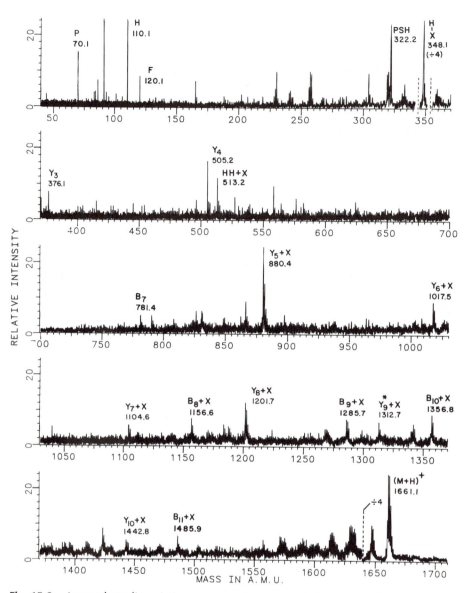

Fig. 15.9. Laser photodissociation mass spectrum recorded on 10 pmol of photoaffinity labeled peptide methyl ester. (Republished with permission of Ref. 19.)

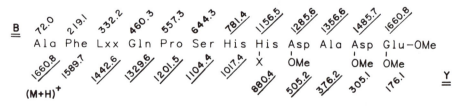

Fig. 15.10. Amino acid sequence of 10 pmol of photoaffinity labeled peptide methyl ester determined by laser photodissociation.

acid residue at the COOH terminus. The loss of glutamic acid affords the fragment ion at m/z 1485.9. Additional B-type fragment ions corresponding to the loss of aspartic acid, alanine, and aspartic acid from the COOH terminus are also observed in the spectrum at m/z 1356.8, 1285.7, and 1156.6, respectively. Additional sequence information is obtained from a search for fragment ions derived from the parent molecule by loss of one or more amino acid residues from the NH_2 terminus, type Y ions. Strong signals at m/z 376.1 and 505.2 in the spectrum are assigned to Y_3 and Y_4, respectively. The next ion in the series appears at m/z 880.4, and corresponds to the addition of an AzBA-modified histidine residue. The modified histidine (with the typical loss of the α-carbonyl group) also appears by itself at m/z 348.1. Additional Y-type ions appear at 1017.5, 1104.6, 1201.7, 1312.7, and 1442.8 and specify the addition of histidine, serine, proline, glutamine, and either leucine or isoleucine to the proposed sequence.

The remaining mass difference (218 Da) between the last observed Y-type ion and the $(M + H)^+$ ion can correspond to only one of two dipeptide combinations, methionyl–serine or alanyl–phenylalanine. To differentiate these possibilities, one cycle of Edman degradation was performed on the sample and the loss of 71 Da from the peptide specified the NH_2-terminal residue as alanine and, therefore, alanyl–phenylalanyl is the correct combination. The complete sequence is shown in Fig. 15.10. Values of m/z for the expected B- and Y-type ions are all within 0.2 mass units of those measured from the mass spectrum.

These examples of biopolymer ions analyzed by laser photodissociation on a tandem quadrupole Fourier transform mass spectrometer demonstate that it is possible to obtain structural sequence information using only picomole amounts of material. The methodology described is well suited for amino acid sequence analysis of peptides containing blocked NH_2-termini, posttranslationally modified residues, and unusual amino acids.

In summary, laser photodissociation is an efficient and versatile means of generating fragment ions indicative of structure. The tandem quadrupole Fourier transform mass spectrometer offers a fast and simple means of selecting, detecting, and analyzing trapped ions with high sensitivity and resolution. Laser photodissociation in combination with the tandem quadrupole Fourier transform mass spectrometer appears to be a powerful method for the structural characterization of biopolymers.

ACKNOWLEDGMENTS

This research was supported by grants from NIH (GM 37537), NSF (CHE-8618780), and CIT (BIO-87-006), as well as instrument development funds from the Monsanto Company.

REFERENCES

1. D. H. Russell, *Mass Spectrom. Rev.* **5**, 167 (1986).

2. R. B. Cody, Jr., I. J. Amster, and F. W. McLafferty, *Proc. Natl. Acad. Sci. U.S.A.* **82**, 6367 (1985).

3. M. L. Gross and D. L. Rempel, *Science* **226**, 261 (1984).

4. W. D. Bowers, S.-S. Delbert, R. L. Hunter, and R. T. McIver, Jr., *J. Am. Chem. Soc.* **106**, 7288 (1984).

5. W. D. Bowers, S.-S. Delbert, and R. T. McIver, Jr., *Anal. Chem.* **58**, 969 (1986).

6. D. A. Laude, Jr., C. L. Johlman, R. S. Brown, D. A. Weil, and C. L. Wilkins, *Mass Spectrom. Rev.* **5**, 107 (1986).

7. D. F. Hunt, J. Shabanowitz, R. T. McIver, Jr., R. L. Hunter, and J.E.P. Syka, *Anal. Chem.* **57**, 765 (1985).

8. D. F. Hunt, J. Shabanowitz, J. R. Yates, III, R. T. McIver, Jr., R. L. Hunter, J.E.P. Syka, and J. Amy, *Anal. Chem.* **57**, 2728 (1985).

9. R. T. McIver, Jr., R. L. Hunter, and W. D. Bowers, *Int. J. Mass Spectrom. Ion Process.* **64**, 67 (1985).

10. D. F. Hunt, J. Shabanowitz, J. R. Yates, III, N.-Z. Zhu, D. H. Russell, and M. Castro, *Proc. Natl. Acad. Sci. U.S.A.* **84**, 620 (1987).

11. A. L. Burlingame, T. A. Baillie, and P. J. Derrick, *Anal. Chem.* **58**, 174R (1986).

12. A. L. Burlingame, D. Maltby, D. H. Russell, and P. T. Holland, *Anal. Chem.* **60**, 306R (1988).

13. M. J. Welch, R. Sams, and E. White, V, *Anal. Chem.* **58**, 890 (1986).

14. C. H. Watson, G. Baykut, and J. R. Eyler, *Anal. Chem.* **59**, 1133 (1987).

15. J. N. Louris, J. S. Brodbelt, and R. G. Cooks, *Int. J. Mass Spectrom. Ion Process.* **75**, 345 (1987).

16. P. Roepstorff and J. Fohlman, *Biomed. Mass Spectrom.* **11**, 601 (1984).

17. D. F. Hunt, J. Shabanowitz, and J. R. Yates, III, *J. Chem. Soc., Chem. Commun.* 548 (1987).

18. H. Michel, D. F. Hunt, J. Shabanowitz, and J. Bennett, *J. Biol. Chem.* **263**, 1123 (1988).

19. A. C. Brinegar, J. E. Fox, G. Cooper, A. Stevens, C. R. Hauer, J. Shabanowitz, and D. F. Hunt, *Proc. Natl. Acad. Sci. U.S.A.* **85**, 5927 (1988).

16

Resonant Two-Photon Ionization Spectroscopy of Biological Molecules in Supersonic Jets Volatilized by Pulsed Laser Desorption

DAVID M. LUBMAN and LIANG LI

Mass spectrometry (MS) remains one of the most powerful means of chemical analysis based on exact mass identification of molecular species. One of the most active areas within this field has been the development of new and more versatile ionization sources. For identification purposes soft ionization methods are often desirable in which only the molecular or parent ion is obtained for mass analysis. However, an additional parameter for identification and structural analysis can be based upon the ability to produce fragment ions that are characteristic of the structure of that species. Numerous methods have been devised to produce soft ionization including chemical ionization (CI), fast atom bombardment (FAB), plasma desorption (PD), field ionization (FI), and secondary ion mass spectrometry (SIMS). Other methods such as electron bombardment or collision-induced dissociation are capable of generating extensive fragmentation.

In this article we discuss laser-induced multiphoton ionization (MPI) as an alternative ionization source with unique properties for mass spectrometry. MPI is an optical ionization technique in which the electronic absorption of a molecule can provide unique identification and discrimination of molecules in a mixture by wavelength selectively producing ions in a mass spectrometer. In addition, laser MPI is a versatile ionization source capable of producing either soft or hard ionization by simply controlling the laser beam intensity and frequency. In particular, in our work MPI has been explored as an ionization source for controlling the fragmentation of labile biomolecules in a time-of-flight mass spectrometer (TOF-MS) for structural analysis. More recently laser ionization spectroscopy of these molecules in supersonic jet expansions has proved to be a means of identification based upon the sharp spectral features that result under the ultracooling conditions provided by this method. The goal of our laboratory has been to explore these capabilities of the MPI technique and to extend the method to compounds of biomedical and pharmaceutical importance ranging from small neurotransmitters to larger peptides.

MULTIPHOTON IONIZATION

The multiphoton ionization technique (MPI) depends upon the absorption of several photons by a molecule on irradiation with an intense visible or ultraviolet light source.[1-49] When the laser frequency is tuned to a real intermediate electronic state, the cross section for ionization is greatly enhanced. This technique is known as resonance-enhanced multiphoton ionization (REMPI). A molecule will ionize only if the sum of the energy of the photons absorbed exceeds the ionization potential of the molecule. When the laser wavelength is not tuned to a real electronic state, the probability for MPI is negligible. Thus, although ions are produced as the final product for detection in mass spectrometry (MS), the ion current obtained depends on the ionization cross section of the molecule at the selected laser frequency. Thus, the ionization cross section reflects the absorption–excitation spectrum of the intermediate state. The truly unique property of MPI spectroscopy is that it can be used as a means of achieving spectral selection of a compound prior to mass analysis.

There are several means by which MPI can be induced as shown in Fig. 16.1. The MPI method that has found most extensive application in analytical chemistry and is used throughout our work is resonant two-photon ionization (R2PI).[23-47] In

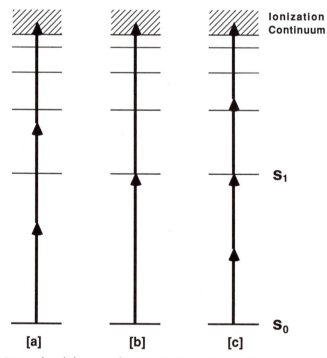

Fig. 16.1. Energy level diagram showing MPI transitions: (a) nonresonant MPI, (b) resonant two-photon ionization, (c) two-photon resonant REMPI.

this process, one photon excites a molecule to an excited electronic state (i.e., $S_0 \rightarrow S_1$), and a second photon ionizes the molecule (see Fig. 16.1b). Thus, the sum of the two photon energies must be greater than the ionization potential of the molecule for R2PI, although the two photons may have either the same or different frequencies. Since most organic species have ionization potentials between 7 and 13 eV, R2PI can be achieved using near-UV pulsed laser sources. Thus, high peak power broadly tunable Nd:YAG and excimer-pumped frequency-doubled dye lasers in the UV region serve as versatile sources for selective resonant two-photon ionization of molecules.

Other MPI processes such as two-photon resonant ionization (TPRI) or, more generally, n-photon resonant ionization are also possible. In these processes, at least two or more photons are needed to reach the first resonance (see Fig. 16.1c). Because at least one photon is nonresonant and interacts with a very short-lived virtual state ($<10^{-15}$ sec), the efficiency for ionization in these processes is far less than that achieved in R2PI. Totally nonresonant ionization is also possible, however, very high laser power is needed to drive this very inefficient process (Fig. 16.1a). In addition, the wavelength selectivity that is unique to the MPI process as an ionization source for mass spectrometry is lost.

R2PI has several important attributes for mass spectrometry in addition to its potential for selectivity. In particular, R2PI can be used to control the fragmentation pattern obtained. This method can provide very efficient soft ionization of molecules yielding the molecular ion with little or no fragmentation.[17-21,23-30,33-36] Although the full capability of this property has not been explored, soft ionization generally appears to occur at modest laser energies ($<10^6$ W/cm^2) in a wide range of organic species. In small aromatics such as aniline,[23,25,28] benzene,[8,16,21] naphthalene,[21,28,30] and other larger but rigid aromatics such as polynuclear aromatic hydrocarbons,[28-30,37] strong molecular ion signals with no fragmentation can be routinely produced. In more recent studies, relatively soft ionization has resulted from R2PI of fragile biological and thermally labile molecules as discussed in more detail later in this chapter.[37-46] This process simplifies the mass spectrum considerably and raises the possibility of identification according to molecular weight. Of course, some compounds, such as certain inorganic complexes [e.g., Fe(CO)$_5$] and some organometallics, are quite fragile and fragment even at low power due to photodissociation by the first photon before MPI can occur[11]; however, these systems are the exception rather than the rule.

The R2PI technique is an extremely efficient means of producing molecular ions. Often ionization of up to several percent of the seed molecules present in a molecular beam can be produced without fragmentation within the intersection of the laser beam and the molecular beam while the laser is on.[13,14,21,29,30,33] In several cases such as aniline and naphthalene, ionization efficiencies ranging from 10 to 100% have been estimated.[21,23,26,48] The high efficiency of R2PI contrasts sharply with that of the electron impact source that typically ionizes less than $\frac{1}{10^4}$ of those molecules that enter the ionization chamber of a mass spectrometer. Also, soft ionization with an electron beam can be achieved only at low beam energies at which the ionization cross section is small resulting in a loss in ionization efficiency of several orders of magnitude.

The ultimate limits to the efficiency of laser R2PI are fundamental considerations, such as the absorption cross section of the molecule at a particular wavelength and the radiationless transition rate (generally due to internal conversion), that is, the rate at which energy leaks out of S_1 before the second photon can induce ionization. Thus, molecules with groups that induce radiationless transitions, such as chlorinated and brominated groups on aromatic rings, will generally exhibit less efficient ionization than their unsubstituted counterparts. Theories that model the competing processes in MPI and discuss the effect of substitution on the efficiency of R2PI have appeared in the literature.[24,35]

Multiphoton ionization is also capable of providing extensive fragmentation either by increasing the laser power density or as a function of the wavelength chosen for ionization. Zandee and Bernstein, for example, have demonstrated that C^+ is the predominant ion obtained from resonance-enhanced multiphoton ionization fragmentation of benzene when a laser power density of $\sim 10^9$ W/cm^2 is used at a wavelength of 391.4 nm.[17-20] As the laser power density is varied, the fragmentation pattern will correspondingly alter where different ratios of carbon fragments including $C_6H_n^+$, $C_5H_n^+$, $C_4H_n^+$, $C_3H_n^+$, $C_2H_n^+$ and CH_n^+ will be observed. Similar results have been observed in MPI as a function of laser power in the near-UV (300–260 nm) by numerous other investigators.[10,14,21,25-38] The fragmentation process generally occurs through the ladder switching mechanism for most organic molecules using a 10-nsec laser pulse.[49] In this mechanism the molecular ion is initially produced by R2PI or by one of the other MPI processes outlined previously. As the laser power is increased subsequent absorption of additional photons may occur, resulting in excitation to a state that dissociates, producing ionic (and neutral) fragments. The ionic fragments may absorb subsequent photons, producing yet smaller ionic fragments.

Thus, the R2PI technique has great versatility as an ionization source for mass spectrometry in that either soft ionization or very extensive fragmentation via successive up-pumping can be produced. This can be accomplished by simply changing the laser output power. In either case ionization occurs with very high efficiency. In addition, the laser source can produce fragments with high appearance potentials with relative ease by increasing the laser energy[39] and, in some cases, can produce unusual fragmentation not readily observed in electron impact (EI)[39].

It should be noted that the fragmentation patterns produced by laser ionization are initiated by light typically ~ 300 nm in wavelength (i.e., much larger than the dimension of the molecule), so that the transitions occurring are Franck–Condon controlled. According to the Franck–Condon principle, electronic transitions occur rapidly with respect to the nuclear motions so that an electronic transition involves a change to a vibrational state in a new electronic configuration in which the positions and momenta of the nuclei are essentially the same as in the initial state. Thus, light-induced transitions occur vertically without a change in the internuclear distance and only transitions to certain states will be readily allowed. Also because of spin conservation from molecule to ion, fewer states are accessible than by EI. This is true because ionization by 70-eV electron impact corresponds to radiation of ~ 0.15 nm energy, so that strong perturbation of the molecule occurs and many more states are accessible with the potential for much more interesting fragmentation for analysis. Thus, the laser fragmentation patterns may be similar, although different than those generated by EI due to the different selection rules that are operating.

An alternative spectroscopic means of detection is laser-induced fluorescence (LIF).[50] In this technique a narrow-band laser is scanned and excites the absorption profile of a resonant electronic state. The total integrated emission is then monitored as a function of the wavelength. This is a particularly sensitive method for molecules that do fluoresce, however, many large polyatomic molecules have negligible or very low quantum yields and of course LIF cannot produce ions for mass spectrometry. However, the ionization and detection of nonfluorescent molecules can be induced by R2PI with a laser at moderate UV photon flux. The inability to detect fluorescence from these so-called "dark" molecules is due to rapid radiationless transitions that cause relaxation before emission can occur from this short-lived excited state manifold.

Using R2PI even molecules with very short lifetimes will ionize efficiently. Indeed, molecules such as azulene and bromobenzene that have picosecond or subpicosecond lifetimes can be easily ionized by increasing the rate of up-pumping by increasing the laser output power, thereby overcoming the radiationless transition rate.[21,24,35] Thus, R2PI serves as a means of generating an optical-absorption spectrum of nonfluorescent compounds, for which a fluorescence excitation spectrum obviously cannot be measured.

PULSED LASER DESORPTION/VOLATILIZATION

The R2PI method to date has been applied mainly to systems that are reasonably volatile. The main problem with extending the R2PI method to significant biological systems and other large molecules in the gas phase is that they are nonvolatile or thermally labile, that is, they decompose on heating and have very low vapor pressures at low temperatures. A number of methods are being developed to incorporate such molecules into supersonic jet expansions for introduction into a mass spectrometer. The method discussed here is pulsed laser desorption in which a high-powered pulsed infrared laser is used to induce rapid heating that desorbs molecules from a surface before they have time to decompose.[51,52] In the desorption process both ions and neutrals are formed in a ratio that depends to a first approximation on the surface temperature induced by the laser.[53] The actual number of ions produced in the desorption process is generally quite small compared to the large number of neutral species that result at power densities below 10^8 W/cm^2, unless precharged ions or organic salts are present.[54,55]

Most investigators to date have focused their attention on the small percentage of ions produced during the desorption event. Consequently, most studies reported in the literature involve the operation of lasers for desorption at relatively high power density ($>10^8$ W/cm^2). An alternate means of vaporizing biological molecules involves using a two-step process in which a low-powered laser is used to desorb mainly neutrals from a surface[37-46] with subsequent ionization by R2PI. The result is a much larger sample available for ionization as compared to the generation of ions in direct laser desorption mass spectrometry (LDMS). In addition, the desorption and ionization processes are now separated in time and space. Thus, the laser energy for desorption and ionization are now independent of one another. By controlling the output power of the ionization laser, R2PI can be used to produce mass spectral patterns that range from molecular ions to extensive fragmentation. Also, the desorbed neutrals can be entrained into a pulsed supersonic flow and then into a mass spectrometer in which the sample can be studied by R2PI spectroscopy and wavelength selectivity can now be achieved.

EXPERIMENTAL METHODOLOGY

Apparatus

The strategy behind the experiment described in this chapter involves entraining the neutral species produced by pulsed laser desorption into a supersonic jet expansion. These species are subsequently ionized by laser R2PI in a time-of-flight mass spectrometer (TOF-MS). The experimental setup used in this work is shown in Fig. 16.2. The apparatus consists of a supersonic beam time-of-flight mass spectrometer.[39] The experiment is oriented such that a pulsed supersonic molecular beam crosses the acceleration region of the TOF device perpendicular to the flight tube. A laser beam orthogonal to both ionizes the sample within the acceleration region of the mass spectrometer.

The TOF-MS is a simple linear device designed after that of Wiley and

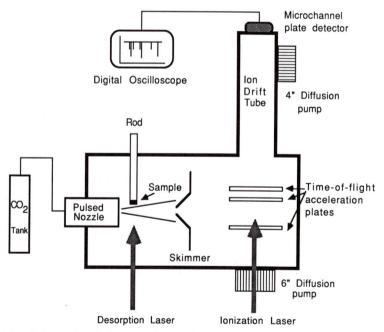

Fig. 16.2. Schematic of the experimental apparatus. (Reproduced with permission from Ref. 69.)

McLaren,[56] which has been modified for supersonic beam injection. The TOF-MS is mounted vertically in a 6-in. stainless-steel cross that is pumped by a 6-in. diffusion pump. The acceleration region is enclosed by a liquid N_2-cooled cryoshield that contains a conical skimmer to skim the molecular beam and ports to allow the laser beam to enter this region. This enclosure introduces a conductance limit to pumping by the main pump so that a 4-in. liquid N_2-baffled diffusion pump provides a pressure of $<10^{-5}$ torr in the flight tube. The combination of the cryoshield and differential pumping significantly reduces the background ionization over that observed in a single-stage system.[37,39]

The supersonic beam source uses a stainless-steel pulsed valve (R. M. Jordan Co., Grass Valley, CA) that is based on the magnetic repulsion principle.[57-60] This nozzle provides gas pulses of $\sim$55 μsec FWHM at "choked flow" so that at a 10-Hz laser repetition rate the valve is on only 550 μsec/sec. Because we are using pulsed laser desorption and excitation sources, it makes sense to pulse the gas injection to achieve an improvement in duty cycle of nearly 2000-fold. The use of this valve allows a small pumping apparatus to be used with a large orifice (500 μm) for high on-axis density. In a continuous expansion, an orifice of $\sim$40 μm would be needed to maintain a pressure of $<2 \times 10^{-5}$ torr in the TOF-MS with a 1 atm argon expansion. The resulting decrease in on-axis density drops as D^2, where D is the orifice diameter, with a corresponding decrease in sensitivity.

The use of a short beam pulse increases the proportion of sample probed by the laser. Since the jet velocity is $\sim$4 $\times$ 10^4 cm/sec and the pulse width is 55 μsec, the beam spread in space is $\sim$2 cm so that a 4-mm laser beam probes $\sim$20% of the available molecules. A longer gas pulse would obviously result in a smaller fraction of molecules being probed and the inherent detection capabilities become less favorable. In addition, this source has been designed so that "choked flow" or unobstructed hydrodynamic flow occurs so that the maximum throughput is achieved during the pulse for maximum sensitivity. The nozzle-to-excitation distance is 16 cm, which is limited by the constraints of our particular apparatus. The shortest distance that places the jet expansion in its "free flow" or collisionless region by the time the skimmer is reached is preferable since the density decreases on-axis as $1/x^2$ (x = distance from the orifice). The jet must reach the free-flow region before the skimmer or else shock waves will destroy the jet expansion.

One important feature of the use of supersonic jets is a marked improvement in the resolution of the TOF device. The Boltzmann energy spread of molecules in the acceleration region generally limits the width of the ion packets at the detector so that a mass resolution of $\sim$150–200 may be observed in a TOF device. However, in an ideal supersonic jet all the molecules are traveling in the same direction with the same velocity so that the energy spread is minimized, and a resolution of up to 1600 has been obtained, limited only by our 5-nsec laser pulse.[28] This resolution, of course, holds only in the absence of large energy release or if severe fragmentation does not occur where spatial effects may limit the mass resolution.

Desorption Method

In our experiments,[37-40] a thick coat of sample that consists of hundreds of monolayers is desorbed by using a CO_2 laser at $\sim$10–40 mJ/pulse (Quanta Ray-Exc-1).

This contrasts to the direct laser desorption (LD) experiments in which generally a thin layer of material several monolayers thick is rapidly desorbed and ionized by using a high-powered pulsed CO_2 laser ~1 J in energy.[61-63] Because we are interested in neutral species rather than ions, significantly lower desorption energies can be used. In this process, we believe that an initial desorption occurs from the surface that sends a shock wave through the solid causing material to desorb from the sample into the gas phase. This mechanism is strongly enhanced in these thick samples in which most of the bonding in the sample is intermolecular and is due to weak van der Waals forces that can be easily disrupted. Although the molecules near the surface are probably physisorbed, the bonding is still weak relative to a situation in which chemisorption occurs. These surface-adsorbed molecules also constitute only a small fraction of the species in the total sample. Considering the low laser energy used in these weakly bound molecules, it appears that mainly neutrals are generated in this method.

Sample introduction into the supersonic expansion was performed by desorbing the compound of interest from the surface of an $\frac{1}{8}$-in.-diameter rod of machineable Macor® that was found to produce the least amount of decomposition in the desorption process. In our initial laser R2PI mass spectrometry experiments, the sample was deposited on the face of the rod by dissolving the compound in benzene or methanol and coating the surface with the use of a spatula. Rather large (milligram) samples were employed in our preliminary studies to demonstrate the general feasibility of the method rather than to demonstrate quantitative analysis. A typical sample may consist of many hundreds of monolayers. By controlling the laser energy, several monolayers can be desorbed on each laser pulse. The rod was situated 4 mm from the molecular beam axis and ~4 mm from the nozzle orifice. This situation does not provide optimal internal cooling; however, such cooling was not required for the laser R2PI mass spectrometry experiments.

The key to the desorption method used for the R2PI jet-cooled spectroscopy experiments is the ability to obtain shot-to-shot desorption stability. This is achieved by dissolving the material of interest in a high viscosity fluid such as glycerol or silicone diffusion pump fluid. Generally ~100 μg of sample is used in these experiments and the glycerol matrix causes the sample to form a very even, thin layer from which pulse-to-pulse stability appears quite good (i.e. somewhat less than ±5%). The rate of desorption can then be controlled by adjusting the laser power on the surface so that total desorption can be produced in seconds or over an extended length of time (i.e., more than a half hour). In addition, the glycerol matrix serves as a heat sink which may prevent thermally induced decomposition. Thermal decomposition can be produced if the thick samples used absorb the laser radiation in the desorption process.

Cooling Optimization

To perform R2PI jet-cooled spectroscopy by using the pulsed desorption method for entrainment in the supersonic jet expansion, the cooling must be optimized because the sample is no longer expanded with the carrier gas through the nozzle but instead is introduced into the jet expansion outside the nozzle orifice. Thus, a critical parameter for obtaining optimal rovibronic cooling is the position of the ceramic rod with respect to the jet expansion. There are two dimensions to consider here: (1) the distance of the rod from the molecular beam axis and (2) the distance of the rod from the nozzle orifice. In the first case, the best cooling was observed when the rod was as close to the supersonic beam as possible without destroying the jet. This was performed in our work by including a volatile sample such as aniline in the jet and moving the rod toward the beam until the molecular ion peak intensity started to decrease. As the rod was moved further, generally a dramatic decrease occurred as the rod began to interfere with the jet. In the second case, the distance between the center of the rod and the nozzle in our experiments was ~5.5 mm. This distance provided excellent cooling, however, if this distance was decreased, a significant decrease in signal was observed with no noticeable increase in cooling. This appeared to occur because at shorter distances at which the carrier density was very high, it was difficult to efficiently penetrate the beam without causing shock waves. At a longer expansion distance at which the carrier density was lower, sample penetration appears to be much more efficient. The use of CO_2 versus Ar or He provides an increased number of collisions at a longer expansion distance for enhanced collisional cooling.[64]

The desorption was performed using a pulsed CO_2 laser at 10.6 μm. The IR beam was softly focused with a 10-cm focal length biconvex germanium lens to a ~2–3 mm spot for desorption, although the focus was adjusted for each sample molecule to optimize our results. The amount of sample desorbed per pulse depends on the power density on the surface and on the properties of the sample, in particular, the melting point. An estimate of the desorption power density on the surface can be made by measuring the beam energy using a power meter (Scientech Model 365) and from the known beam temporal profile.[65] In the case of catechol, a low melting point compound (mp = 106°C), the desorption was so efficient that a low laser power was used in the experiments (~1 $\times$ 10^5 W/cm^2), whereas for tyrosine (mp = 325°C) a power density of nearly 5 $\times$ 10^6 W/cm^2 was used. By carefully controlling the desorption laser power the sample can be made to last for an extended period of time for spectroscopic scans. In several spectroscopic scans in our work a 100-μg sample has lasted for ~0.5 hr at a repetition rate of 10 Hz. This is equivalent to ~3 $\times$ 10^{13} molecules or ~0.1 monolayers desorbed per CO_2 laser pulse.

Laser Ionization

The laser ionization (R2PI) was performed using the frequency doubled dye output from a Quanta-Ray DCR-2A Nd:YAG pumped dye laser system. For the R2PI mass spectrometry experiments, the fourth harmonic of the Nd:YAG laser at 266 nm can also be used as the ionization source. The 6-mm output beam was collimated with a combination 30-cm focal length positive lens and a 10-cm focal length negative lens to produce a laser beam 2–3 mm in diameter. The power density was adjusted to obtain the desired signal level, which depended on the amount of material entrained in the jet and the efficiency of ionization of each particular compound.

The actual sequence of events was controlled by several delay generators in which the pulsed CO_2 laser fires first to produce desorption followed by the pulsing of the valve. The two events were synchronized in time so that the desorbed plume was entrapped into the jet expansion of CO_2 and carried into the acceleration region of the TOF-MS. The flight time of the jet from the pulsed valve to the region was ~ 300 μsec and the laser was therefore set to pulse as the gas pulse arrived at this point. Laser R2PI was produced and a LeCroy 9400 digital oscilloscope was used to record the mass spectrum. The wavelength spectrum was obtained by using an SRS 250 gated integrator to monitor only the molecular ion as the dye laser was scanned.

APPLICATIONS

Neurotransmitters and Other Small Biological Species

In our laboratory, we have used the pulsed laser desorption/laser R2PI technique to study the mass spectrometry of various labile compounds. These include a number of important classes of biological molecules such as catecholamines, indoleamines and their metabolites,[39,40] vitamins, amino acids, neuroleptic drugs,[39] nucleoside bases,[82] and small peptides ranging from di- to hexapeptides.[38,65] In particular, we have been interested in the detection of catecholamines and their metabolites because their selective detection would be of great utility in the investigation of neuroblastomas and other neurogenic tumors, Parkinson's disease, and psychological stress.[66,67] For example, neuroblastomas affect nerve structures involved in synthesizing catecholamines, so this condition can be diagnosed by the concentration of metabolites in the urine. The quantitative and qualitative determination of these compounds is of importance for cases in which high concentration values for the metabolites homovanillic acid and vanilmandelic acid in urine are indicative of an unfavorable prognosis and rapid metastasis. At present, simple chemical tests are used to determine total catecholamine concentration or the presence of several metabolites, but the degree of selectivity available is unsatisfactory for the type of exact identification needed for the analytes present. A more selective means of specifically monitoring each catecholamine metabolite would be highly desirable.[66] In addition, the lability of these compounds even under mild temperature conditions makes this an important test of the capabilities of any given technique for analyzing other fragile biological species.

The key result using the pulsed laser desorption/laser R2PI method is that generally molecular ions with little or no fragmentation were observed at modest laser input energy at 280 nm ($<7 \times 10^5$ W/cm^2). There are some exceptions such as epinephrine and norepinephrine, which undergo facile cleavage even at the threshold laser ionization energy used to obtain an observable signal, so that production of only the molecular ion is not observed. R2PI of 3-methoxytyramine, dopamine, DOPA, and tyrosine produced only the molecular ion at 280 nm using laser beam energies below 2–3 mJ. However, even a small increase in energy of the 280-nm ionizing beam, such as was used to obtain the results in Fig. 16.3, significantly enhances the intensity of the molecular ion while only moderately promoting bond scission in the alkylamine fragment. Similar results were observed for several indoleamines and related compounds. In the case of tryptophan and tryptamine,

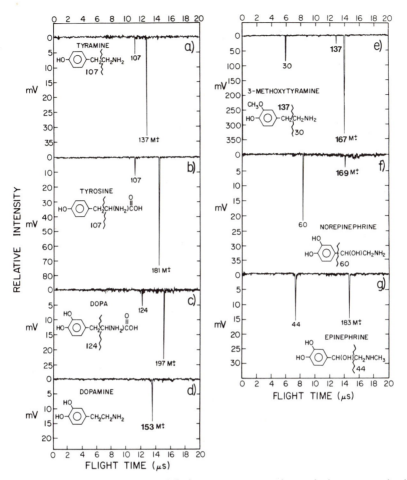

Fig. 16.3. Laser desorption time-of-flight mass spectra of catecholamines and related derivatives. R2PI was performed at (a) 280 nm, (b) 280 nm, (c) 266 nm, (d) 280 nm, (e) 280 nm, (f) 266 nm, and (g) 280 nm. (Reproduced with permission from Ref. 39.)

for example, a strong molecular ion peak was observed, although it was difficult to prevent simple β-cleavage even at low input ionizing energies as observed for the catecholamines.

The effect of ionization wavelength on mass spectral fragmentation patterns was studied for several of the catecholamines and indoleamines as illustrated in Fig. 16.4 for tryptamine. In all cases studied there is no significant change in the type of fragments observed as the ionization wavelength decreases. In fact, most of the fragment ions can be produced at any wavelength at which the molecule absorbs simply by increasing the ionizing laser power. However, the threshold power density required for the appearance of fragments with a relatively high appearance potential decreases as the lasing frequency increases. In addition, substantial changes in the relative fragment ion abundance are observed as a function of wavelength at the same laser beam intensity as demonstrated in Fig. 16.4. This occurs as the laser frequency increases since the excitation proceeds high into the

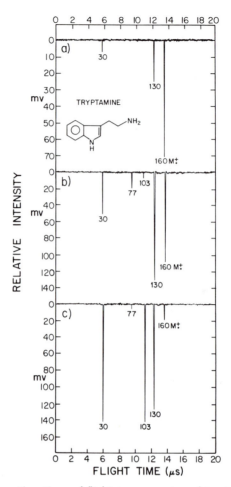

Fig. 16.4. Laser desorption time-of-flight mass spectra of tryptamine obtained as a function of wavelength: (a) 266 nm, (b) 245 nm, (c) 222 nm. (Reproduced with permission from Ref. 39.)

ground ionic state at which the energy may be greater than the appearance potential for the formation of these fragment ions. As the laser frequency increases further, the probability for fragmentation will thus increase.

An important point common to all spectra observed throughout this work is that the laser-induced mass spectra exhibit fragment ions that are similar to those obtained in conventional electron impact spectra. This result was not unexpected as other investigators have reported this to be the case.[68] However, the great advantage of laser MPI is that it provides a greater degree of control over the number of fragments observed and their relative abundance. By simple variation of the laser energy, it is often possible to obtain only molecular ions with high efficiency or diagnostic fragments of relatively high appearance potential. The mass spectra are frequently simpler than those obtained in EI because the selection rules involved are different. However, using MPI, fragments whose production are energetically unfavorable can be produced in higher abundance by successive photon pumping as the laser power is increased.

Although the main fragments observed in the laser-induced mass spectra shown here are those that are expected, generally fewer fragments are observed than in EI and these fragments are characteristic of the structure of the molecule. This is demonstrated for the catecholamines at 280 or 266 nm (see Fig. 16.3) in which the fragmentation pattern of the catecholamines is dominated by two competing mechanisms when a relatively low energy ($<10^6$ W/cm^2) is used. These mechanisms, which are similar to those of electron beam ionization, involve either simple cleavage of the benzylic $C_\alpha-C_\beta$ bond to form a stable even electron fragment ion or rapid charge migration to the amine nitrogen and subsequent β-cleavage to this atom.[68] Following charge migration, bond scission probably occurs via expulsion of an aralkyl radical, again resulting in formation of an even electron cation.

Another important point common to all the spectra observed in these experiments is the absence of cationization. This phenomenon appears to be ubiquitous in laser desorption experiments in which ionization and desorption are performed in a single step with the use of a high laser power density ($>10^8$ W/cm^2).[61-63] However, in the experiments performed here cationization was never observed. In previous work[37] pulsed laser desorption with a CO_2 laser followed by R2PI using a UV laser beam was performed within the acceleration region of the TOF-MS. The use of the pulsed CO_2 laser alone at the power densities used in these experiments produced essentially no ions but only neutrals that could be subsequently ionized with the UV source. The power density of the pulsed CO_2 laser used here is typically at least one to two orders of magnitude lower than that used in LDMS experiments, to enhance the production of neutrals over ions.

Peptides

This technique has been recently used to systematically study small peptides by R2PI.[38,69] In particular, di- and tripeptides with an absorbing aromatic group such as tryptophan, tyrosine, and phenylalanine have been studied. These compounds generally absorb radiation at either 280 or 266 nm through the $\pi-\pi^*$ transition of the aromatic moiety and subsequently are ionized by R2PI. Each of the aromatic groups is known to have a strong origin absorption in the region around 280 nm. It must be emphasized that R2PI is a spectroscopic ionization tool since it depends on the presence of an intermediate absorbing electronic state. Thus, it appears that the presence of small linear amino acids such as glycine, alanine, and leucine serve as substituent groups on the aromatic center that may shift the absorption center or change the absorption cross section of the aromatic group, although they do not absorb the radiation directly themselves. However, liquid-phase spectra show that the basic origin absorption transition still occurs between 260 and 290 nm in most of the peptides studied. Thus, in these studies the 280- or 266-nm wavelength is absorbed by most of the small peptides. Although 245- and 222-nm wavelengths were tested as ionization sources, they were generally found to be less effective than the longer wavelengths. We suspect that as the lasing frequency increases, the molecule is excited higher into the S_1 manifold and the radiationless transition rate increases dramatically, thus reducing the R2PI efficiency.

The key result observed in the R2PI of these peptides[38,69] is that generally molecular ions are observed at 266 nm at modest laser intensity if the molecules have achieved optimal cooling in the jet expansion. If full cooling is not obtained, then higher laser energy is needed for ionization and some fragmentation may be difficult to prevent. In addition, as the laser power is increased, simple fragmentation characteristic of the structure of the molecule is observed.

In our early R2PI experiments on dipeptides[38] the rovibronic cooling was incomplete. Therefore, under the cooling conditions obtained in our present experiments an attempt was made to adjust the conditions to check whether this might alter the mass spectral intensity pattern.[69] In the case of glycyltyrosine the ionizing wavelength was adjusted to a sharp resonance at 280.70 nm in the cold jet expansion. On the origin band resonance relatively low power ($\sim$0.1 mJ) could be used for ionization because of the increased absorption efficiency of the process. Figure 16.5 illustrates the mass spectral pattern as the laser power density is varied over the range of 8×10^5 to 2×10^5 W/cm^2. At high laser power density extensive fragmentation results although a dominant M$^+$ peak appears. As the laser power decreases the fragmentation is reduced substantially, although even at the lowest energy used it was difficult to observe only M$^+$ without some minor fragmentation. Nevertheless, by exciting this dipeptide on a strong resonance under the ultracooling conditions of the jet, the M$^+$ peak can be obtained with minimal fragmentation due to the relatively low laser energy used, although, of course, at the expense of sensitivity. This was also found to be true for numerous other dipeptides studied such as alanyltyrosine and prolyltyrosine.[69]

In recent work[70] this technique has been extended to pentapeptides such as met- and leu-enkephalin, which are important neurotransmitters in the brain. By using high laser power ($P = 1.5 \times 10^7$ W/cm^2) at 266 nm, extensive fragmentation could be induced in these compounds for structural analysis. The fragmentation induced under these conditions resulted in all the classical A, B, C and X, Y, Z cleavage products with formation of characteristic acylium and aldimine ions as well as β-cleavage at the aromatic moiety. In addition, a substantial molecular ion, M$^+$, is observed in every case and $(M - OH)^+$ or $(M - COOH)^+$ ions are often also observed. In every case only M$^+$ and not MH$^+$ as in fast atom bombardment mass spectrometry (FABMS) is observed. The result is that the fragmentation generated by R2PI/MPI would be expected to be very different from that generated by FABMS as indeed is observed. The laser-induced mass spectrum of leu-enkephalin is shown in Fig. 16.6.

In the case of leu-enkephalin (Fig. 16.6b) cleavage occurs at the $-CO-NH$ peptide bond with a resulting series of acylium ions at m/z values 278, 221, and 164. In addition, at 28 mass units lower than the acylium ions the corresponding acyl-immonium ions are observed due to loss of CO at m/z values of 397 and 136. A second degradation pathway begins from the N-terminal end and results in formation of aldimine ions at m/z values of 391, 335, and 277 in the laser-induced fragmentation of leu-enkephalin. Such aldimine ions are often difficult to produce using electron impact ionization. In the R2PI/MPI-induced mass spectra these ions are formed by bond cleavage of the $-CO-NH$ bond in the molecular cation with

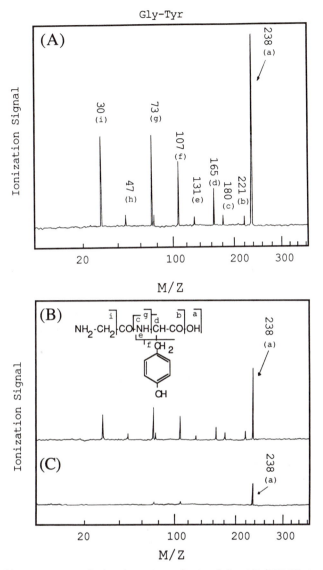

Fig. 16.5. Mass spectra of glycyltyrosine obtained by LD/REMPI technique. The power density at λ = 280.70 nm for the ionization laser was (A) 8×10^5 W/cm², (B) 4×10^5 W/cm², and (C) 2×10^5 W/cm². (Reproduced with permission from Ref. 69.)

a resulting even-electron ion. No hydrogen migration is observed as in FABMS[71] except for m/z 335 in leu-enkephalin in which a hydrogen has been added to the Y-cleavage product to form an odd-electron ion.

A series of fragments due to cleavage at every N−C bond is observed for leu-enkephalin in which charge retention can occur on the N or C fragment. Charge retention on the N results in a series of fragments at m/z values of 179, 236, and 441 whereas charge retention on the C results in ions at m/z 376, 319, and 262. The formation of the fragment at m/z 441 formed by cleavage of the N−C bond

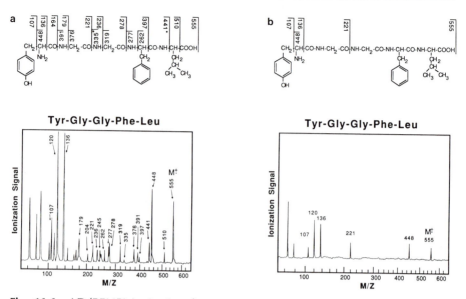

Fig. 16.6. LD/REMPI ionization–fragmentation pattern at λ = 266 nm of (a) leu-enkephalin at high laser ionization energy ($P = 1.5 \times 10^7$ W/cm²) and (b) at low ionization energy ($P = 1.0 \times 10^6$ W/cm²). (Reproduced with permission from Ref. 70.)

with retention of the charge at the N-containing fragment involves the addition of one H to form an odd-electron fragment so that C′ is formed. However, the fragments at m/z 179 and 236 result in even-electron fragments with no migration of the H. In addition, simple β-cleavage of the aromatic side chain in tyrosine occurs with charge retention on either fragment resulting in an m/z of 107 if the charge remains on the hydroxybenzyl ion or in an m/z of 448 if the charge remains on the remaining fragment. Also, a strong peak is observed at an m/z of 120 due to an internal fragment resulting from the formation of an immonium ion characteristic of the phenylalanine group.[71]

It should be noted that when the laser energy is lowered to ~1–2 × 10⁶ W/cm² the molecular ion can be obtained with reduced fragmentation as demonstrated in Fig.16.6a for leu-enkephalin. The ability to reduce fragmentation will be of importance in analysis of complicated mixtures. However, this occurs at the expense of sensitivity by an order of magnitude or more.

One extremely important problem in biomedical analysis is the discrimination of isomeric peptides. This is demonstrated in using R2PI at 266 nm ($P = 1.5 \times 10^7$ W/cm²) for the tripeptide Gly-Gly-Trp, Gly-Trp-Gly, and Trp-Gly-Gly in Fig. 16.7.[70] In each case a strong M⁺ ion is clearly observed as well as an ion at $m/z = 130$ u due to simple β-cleavage at the tryptophan moiety. Although the preferred cleavage under the conditions of the experiment appears to be β-cleavage with a resulting acylium ion fragment, in each isomer a different fragment pattern is obtained depending on the initial structure of the molecule. For example, the ion at M − 74 (244 u) is present in configuration 2 and 3 due to the cleavage as indicated. However, the same cleavage will not provide this ion when as in configura-

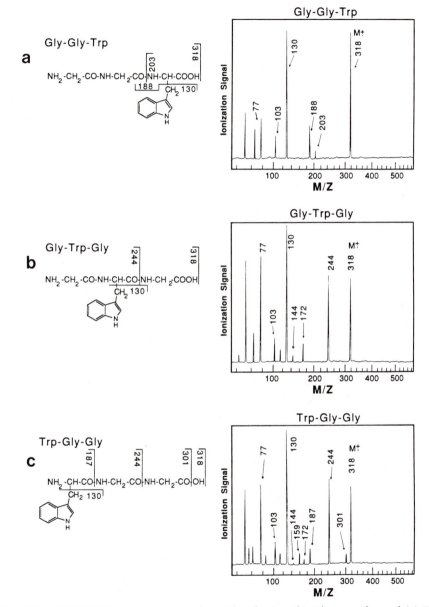

Fig. 16.7. LD/REMPI mass spectra of (a) Gly-Gly-Trp, (b) Gly-Trp-Gly, and (c) Trp-Gly-Gly at 266 nm ($P = 1.5 \times 10^7$ W/cm²). (Reproduced with permission from Ref. 70.)

tion 1 in Fig. 16.7, the tryptophan is on the other side of the bond. In general, the fragments obtained from the three species are quite different.

One truly unique feature of the fragmentation induced by laser R2PI/MPI is the ability to distinguish the isomeric amino acid pair leucine and isoleucine that cannot be easily distinguished based upon the fragmentation induced by classical methods.[71] This had been demonstrated previously by Zare and co-workers for PTH-leucine and PTH-isoleucine amino acids.[41] In our work[65] this has been illustrated for several CBZ-oligopeptides containing leucine and isoleucine as shown in Fig. 16.8, where the CBZ group provides an absorbing center at 266 nm for this linear peptide. At above $m/z = 100$ the mass spectra for Leu-containing and Ile-containing peptides are indistinguishable. However, at m/z below 100 these peptides can be distinguished by observing the intensity differences between the $m/z = 43$ and $m/z = 57$ ion. It was found that the $m/z = 43$ ion is of much greater relative abundance than the $m/z = 57$ ion for the Leu-containing peptide, whereas the opposite is true for the Ile-containing peptides. This is the result of different cleavages induced in the side chains of these amino acids. The $-CH(CH_3)_2$ fragment is more stable for leucine based upon the structure of the side chain whereas $-CH(CH_3)CH_2CH_3$ is the stable fragment produced from the isoleucine side chain. This is also demonstrated for the case of the isomers of CBZ-Leu, CBZ-Ile, and CBZ-Nle. For CBZ-Leu an $m/z = 43$ ion of much greater relative abundance is observed compared to the $m/z = 57$ ion; for CBZ-Ile an $m/z = 57$ ion is observed that is of relatively greater abundance than the $m/z = 43$ ion. In either case, the differences involve the presence of minor features in the mass spectrum that can be observed since there is no matrix background that would interfere at these low masses as in FABMS. In the case of CBZ-Nle, ions are observed at $m/z = 43$ and $m/z = 57$, which are of relatively low abundance compared to the other ions in the spectrum. Such isomeric discrimination has also been demonstrated for CBZ-(α)-lysine, CBZ-(ϵ)-lysine, and CBZ-glutamine.[65]

Important features of this method include sensitivity and the ability to perform quantitative measurements. To provide an estimate of the sensitivity of this method, successive dilutions of several different peptides in methanol:benzene (1:1) were prepared. Using a microliter syringe the sample was placed in a 1-mm-diameter hole bored into the rod surface. The desorption laser was focused to a ~1-mm beam and the power density raised to ~10^7 W/cm^2 so that desorption was complete within several laser pulses. The resulting R2PI molecular ion signal was monitored over the desorption period and a plot of signal versus concentration provided an estimate of the lower limit of detection with a signal-to-noise (S/N) ratio of 2 as shown in Table 16.1. The detection limit achieved in our work, which is typically tens of picograms, is not nearly as low as the picogram detection limit achieved for PTH-amino acids by Zare and co-workers.[41] This difference probably reflects the loss of molecules during the entrainment process as compared to Ref. 41 in which no beam entrainment is used.

The apparent detection limit appears to increase as the size of the peptide increases. A detection limit for CBZ-amino acids was typically 50–100 pg whereas for CBZ-tripeptides a detection limit of 500 pg–1 ng resulted. As the peptide size increases further, the detection limit becomes higher as in met-enkephalin where only 250 ng of sample can be detected. The detection limit appears to be affected

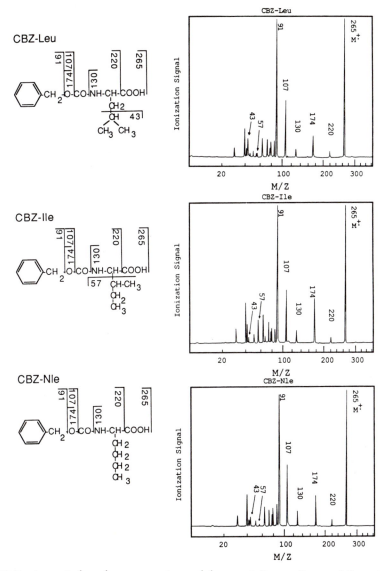

Fig. 16.8. Laser-induced mass spectra and fragmentation patterns of three isomers: CBZ-Leu, CBZ-Ile, and CBZ-Nle at 266 nm ($P = 1.5 \times 10^7$ W/cm^2). (Reproduced with permission from Ref. 65.)

mainly by the laser-induced ionization efficiency. The latter will be influenced by the absorption coefficient of the molecule and the competing radiationless processes. These will depend on the structure of the particular molecule and thus sensitivity in this technique may be expected to vary significantly with the sample. However, in nonrigid peptide chains there are many active modes that can promote internal conversion, thus resulting in a decreased ionization efficiency. The entrainment efficiency of the sample into the jet and the resulting jet cooling may also

Table 16.1. Detection limit for CBZ-derivatized compounds using R2PI at 266 nm[a]

Compound	Detection limit (pg)
CBZ-Ala	125
CBZ-Leu	56
CBZ-Pro	125
CBZ-Gly-Ala	333
CBZ-Leu-Gly	200
CBZ-Pro-Leu	250
CBZ-Gly-Gly-Ile	500
CBZ-Gly-Gly-Ala	300
CBZ-Ile-Gly-Gly	1000

[a] $P = 1.5 \times 10^7$ W/cm^2.

affect the detection limit, although the relative contribution of these processes in limiting the sensitivity is still unknown.

It should be noted that similar work has been carried out by Grotemeyer et al.[43-45,73] who are using a high resolution V-shaped reflectron TOF device. They have desorbed neutrals into a supersonic jet of Ar by using a pulsed CO_2 laser and performed REMPI with an excimer-pumped dye laser. This group has been able to obtain a mass resolving power of at least 6500 at $m/z = 96$ by using this technique.[72] In one study, they were able to ionize the decapeptide angiotensin I and obtain only the molecular ion at $m/z = 1295$ [43] by using 271.23-nm radiation at a power density of 1×10^6 W/cm^2. By increasing the laser power to 1×10^8 W/cm^2, they were able to generate fragments characteristic of the structure of the peptide (see Scheme 1 in Ref. 43). Fragmentation mainly due to bond breaking around the carbonyl groups of the peptide chain resulted in efficient formation of acylium ions. More recently, this group has extended the technique to significantly larger molecules such as insulin,[73] where the spectrum of insulin displayed a prominent molecular ion signal at m/z 5927. In addition, group peaks for the A-chain at m/z 2334 and for the B-chain at m/z 3895 appeared.

Nucleosides

The use of laser desorption/laser R2PI is particularly appropriate for the study of nucleosides. This group of molecules has presented a great challenge to mass spectrometry due to the polarity and thermal lability of these molecules. Most of the simple pyrimidine nucleosides and some of the purine nucleosides, such as adenosine, can be thermally vaporized and subsequently analyzed by electron impact (EI) or chemical ionization (CI) in a mass spectrometer. However, more polar nucleosides such as cytidine and guanosine and most of the complex natural nucleosides as well as all the nucleotides are not amenable to thermal vaporization without degradation. This includes a significant class of nucleosides, since in transfer RNA (tRNA) there are a variety of modified nucleosides present in addition to the four normally found in RNA. Modified nucleosides also occur in ribosomal RNAs,

DNA, and eukaryotic messenger RNA. The function of these modified nucleosides is known to be important and more than 60 have been isolated and studied.[74,75]

These thermally labile nucleosides and nucleotides have been studied by a number of techniques. Chemical derivatization has been used as a means of volatilizing these compounds[76] using conversion to trimethylsilyl derivatives. However, this process may result in chemical artifacts and uncertainties in the mass spectrum and for a small amount of sample may be inappropriate if conversion yields are low.[75] In addition, EI ionization (70 eV) of these compounds provides very extensive fragmentation with generally a very low relative abundance of the molecular ion peak, M^+, for identification.[77,78] More recently fast atom bombardment (FAB) has been used as a means of volatilizing and ionizing these compounds.[79,80] Using this method enhanced molecular ion peaks, MH^+, $M - H^-$, were observed in the positive and negative modes, respectively, although generally the base fragments BH_2^+ and B^- were the dominant ions. Field desorption (FD) and field ionization (FI) have also been applied to these compounds with relative soft ionization resulting, although BH^+ was generally found to be the most abundant ion.[81]

In our work,[82] the laser desorption/laser R2PI method was used to study adenosine, guanosine, and cytosine as well as their various modified nucleosides. The results for various adenosine analogs are shown in Table 16.2 at $\lambda = 266$ nm and $P = 8 \times 10^6$ W/cm^2. The salient feature of the mass spectra observed under these conditions is an intense molecular ion peak without the presence of cationization. In addition, there is generally one principal fragment observed due to bond cleavage at the sugar moiety. This fragment denoted, BH^+, (B is the molecular weight of the free base minus 1) corresponds to an ion formed from the base portion of the nucleoside with transfer of one hydrogen atom from the sugar moiety. This process is also observed in EI mass spectra of these compounds.[75,77] In addition, two characteristic fragments due to cleavage of the sugar moiety are observed, which are denoted as S_1 and S_2, respectively, as shown in Fig. 16.9. These fragment ions are particularly useful for structural analysis and easily allow the ability to pinpoint whether there is modification on the 2'-oxygen of a nucleoside. As Table

Table 16.2. Mass spectral fragmentation patterns of adenosine analogs

	Compound	M	S_1	S_2	BH	119	108	Others
1	Adenosine	267(43)	178(19)	164(33)	135(100)	119(29)	108(38)	
2	2'-Deoxyadenosine	251(42)	162(15)	164(23)	135(100)	119(17)	108(45)	
3	N^6-Methyladenosine	281(51)	192(17)	178(24)	149(100)	119(40)	108(24)	
4	N^6-Methyldeoxyadenosine	265(40)	176(17)	178(18)	149(100)	119(30)	108(23)	
5	1-Methyladenosine	281(42)	192(16)	178(18)	149(100)			
6	N^6,N^6-Dimethyladenosine	295(100)	206(7)	192(14)	163(50)	119(16)	108(5)	148(38),134(69),121(20)
7	3'-o-Methyladenosine	281(50)	178(5)	164(32)	135(100)	119(16)	108(19)	147(11)
8	1,N^6-Ethenoadenosine	291(99)			159(100)			132(21),105(39)
9	2,6-Diaminopurine-2'-deoxyribosine	266(61)	177(7)		150(100)		108(18)	134(10)
10	N^6-Furfuryladenosine	347(88)			215(100)	119(62)	108(50)	
11	N^6-(2-Isopentenyl)adenosine	335(100)			203(47)	119(73)	108(39)	188(43), 160(34), 135(46)
12	Puromycin aminonucleoside	294(77)	206(9)	192(35)	163(100)	119(23)	108(77)	222(9), 148(38), 134(47), 121(18), 114(6)

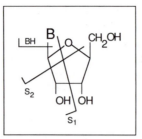

Fig. 16.9. Major ions observed for the nucleosides using laser desorption/laser ionization technique at 266 nm: molecular ion M⁺, base plus one hydrogen BH⁺, base and parts of sugar plus one hydrogen S_1^+ and S_2^+, and fragments from BH⁺. (Reproduced with permission from Ref. 82.)

16.2 indicates, the mass spectrum of 2′-deoxyadenosine shows a strong molecular ion peak m/z 251 and a fragment BH⁺, m/z 135. The mass difference between M⁺ and BH⁺ suggests that it is a deoxyribose. Two fragments, m/z 162 and m/z 164, corresponding to the S_1 and S_2 notation, respectively, can be used to locate the deoxy position (i.e. the 2′-position in this case). The formation of S_1 and S_2 is similar to EI in which an intramolecular rearrangement of the H from the sugar to the base occurs during fragmentation.[75,77] These S_1 and S_2 fragments are easy to detect using R2PI/MPI as opposed to FABMS in which their presence may be obscured due to matrix background.[79]

   One of the advantages of using lasers as an ionization source for studying nucleosides over the other ionization techniques mentioned is the ability to control the fragmentation pattern of the mass spectrum by simply changing the ionization laser power density. In particular, Fig. 16.10 shows the dependence of the fragmentation on the laser power density for N^6, N^6-dimethyladenosine. At 1×10^6 W/cm² only the molecular ion is obtained with no fragmentation, although in most of the nucleosides studied generally a BH⁺ fragment accompanies M⁺ even at the lowest energy used. As the laser power is increased, extensive fragmentation is obtained. In Fig. 16.10A extensive fragmentation of N^6, N^6-dimethyladenosine is obtained at 8×10^6 W/cm². The molecular ion is the dominant peak in the mass spectrum accompanied by the base, BH⁺, fragment at m/z 163. The S_1, S_2 fragments are also present in relatively low abundance at m/z 206 and m/z 192, respectively. The m/z 148 ion is the result of CH_3 expulsion from BH⁺. In addition, m/z 119 and m/z 108, which are characteristic of the adenine base, are observed.

Spectroscopic Experiments

The ultimate experiment for achieving selectivity would be if the pulsed laser desorption–entrainment into supersonic expansions could provide ultracold molecules with sharp spectral features in R2PI spectroscopy for identification of biological species. Indeed, this has been achieved for several groups of molecules including tyrosine and its derivatives,[69] tryptophan and its derivatives,[83,84] and various catecholamine metabolites.[84]

Figures 16.11 and 16.12 show cold jet-cooled R2PI spectra of tyrosine and its analogs[69,85] obtained using pulsed laser desorption with subsequent entrainment in the jet expansion. In each of these cases the molecular ion with either no or minimal fragmentation results from the R2PI process. Each of these spectra is the result of monitoring only the molecular ion as a function of wavelength using a gated integrator. In essence, each figure represents a mass-selected wavelength spectrum.

Figure 16.11B shows a jet-cooled spectrum of tyramine (mp = 163°C) obtained by pulsed laser desorption and entrainment into a jet of CO_2. Tyramine results as a side-product of tyrosine metabolism in the human body and is important for detection and identification since its presence can be particularly toxic in combination with certain antidepressant drugs.[86] In previous work, a spectrum of tyramine was obtained by heating in a hot pulsed oven at 100°C with subsequent expansion in Ar carrier.[87] In this case, substantial pyrolysis of the sample occurred with all but a fraction of the sample converted to a polymeric tar. There were also several decomposition products observed in the mass spectrum for this relatively labile molecule, although a wavelength spectrum was obtained by monitoring only

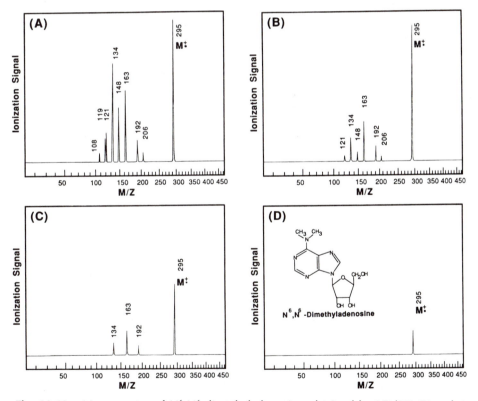

Fig. 16.10. Mass spectra of N^6,N^6-dimethyladenosine obtained by LD/REMPI technique at 266 nm. The ionization laser power density was (A) 8×10^6 W/cm^2, (B) 4×10^6 W/cm^2, (C) 2×10^6 W/cm^2, and (D) 1×10^6 W/cm^2. (Reproduced with permission from Ref. 82.)

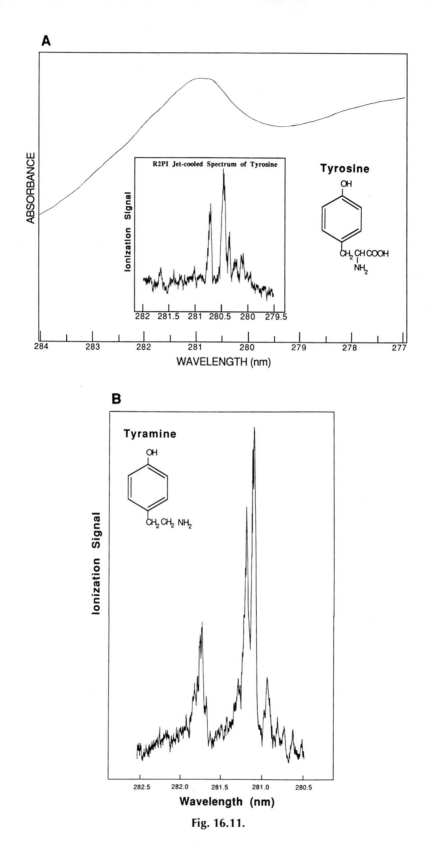

A

ABSORBANCE

R2PI Jet-cooled Spectrum of Tyrosine

Ionization Signal

282 281.5 281 280.5 280 279.5

Tyrosine

OH

CH₂CHCOOH
NH₂

284 283 282 281 280 279 278 277

WAVELENGTH (nm)

B

Tyramine

OH

CH₂CH₂NH₂

Ionization Signal

282.5 282.0 281.5 281.0 280.5

Wavelength (nm)

Fig. 16.11.

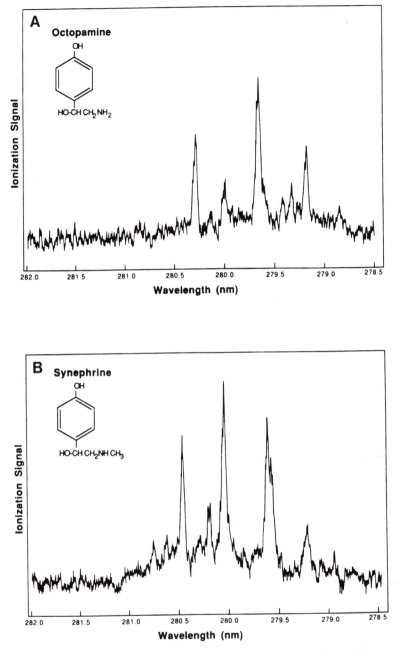

Fig. 16.12. R2PI spectra of (A) octopamine and (B) synephrine, obtained by monitoring the molecular ion in a TOF-MS using the pulsed laser desorption method for sample introduction. (Reproduced with permission from Ref. 69.)

←

Fig. 16.11. Resonant two-photon ionization (R2PI) spectra of (A) tyrosine and (B) tyramine, obtained by monitoring the molecular ion in a TOF-MS using the pulsed laser desorption method for sample introduction into the jet expansion. (Reproduced with permission from Ref. 69.)

the molecular ion at $m/z = 137$. Using the pulsed laser desorption method, volatilization of tyramine was accomplished with no apparent decomposition or pyrolysis products.[69] The spectrum obtained by pulsed laser desorption appears considerably sharper and the peaks better resolved than that obtained by direct heating. In particular, the peak at 281.75 nm can be resolved in our present study whereas sufficient cooling was not observed in the spectrum obtained by direct heating.[69,87]

Figure 16.11A shows a jet-cooled spectrum of tyrosine obtained by pulsed laser desorption–volatilization.[69] In our present setup we cannot use direct heating since the melting point of this fairly labile molecule is 325°C, which is beyond the capability of our present pulsed valve injection system. Nevertheless, using the pulsed laser desorption method we obtain sharp cold features for tyrosine in a jet expansion of CO_2. Since the tyrosine and tyramine structures are similar, their spectra are, as expected, also very similar. However, using the supersonic jet method the two spectra are distinctly unique so that unambiguous spectroscopic identification of each of these compounds is possible in conjunction with mass identification in our mass spectrometer.

Figure 16.12A and B show R2PI spectra of octopamine and synephrine[69] obtained by monitoring only the molecular ion as a function of wavelength. These two compounds are very similar in structure, the difference being an extra $-CH_3$ group attached to the linear substituent chain. Nevertheless, each spectrum has distinctive sharp spectral features that are shifted by several tenths of a nanometer relative to each other, but that can clearly be resolved by the laser. Thus, electronic spectroscopy serves as a sensitive probe revealing small differences in molecular structure. These two compounds are important products of tyrosine metabolism in the human body and can be clearly distinguished using jet spectroscopy from the original tyrosine precursor.

It should be noted that in the case of simpler substituted aromatics such as toluene or p-cresol, the origin region is a sharp band with no additional vibronic bands observed until there is significantly higher energy. However, in the spectra of these nonvolatiles there are several bands observed in the origin region. We attribute this to the presence of multiple rotational isomers in which several similar but different structures, which differ by the rotation of a group on the extended chain in space, are stable in the jet expansion. This phenomenon has been observed previously by Ito and co-workers[88] for several meta-substituted phenols and by Dunn et al.[89] for isomers of dihydroxybenzenes. Most recently rotational isomerization has been reported by Levy and co-workers[90] for indole-based compounds such as tryptophan, tryptamine, and 3-indoleacetic acid, which contain side chains similar to the tyrosine analogs reported in this work.

Argon was initially used in these experiments as the jet carrier gas since enhanced rotational cooling should be obtained as compared to a polyatomic carrier such as CO_2. However, we have found that increased penetration and enhanced signal levels are obtained in CO_2 over Ar by desorbing at longer distances (~ 5.5 mm) from the nozzle. The increased collisional rate at longer expansion distance for CO_2 than for the monatomic carriers provides a sufficient number of collisions at this point (i.e., 5.5 mm)[69] so that excellent cooling is still obtained. This occurs provided the beam is probed at a distance sufficient to be in the "free-flow" region in which the terminal Mach number and the optimal cooling are achieved. A number of other carrier gases were studied in this work including N_2 and air as well as mixtures of these gases with CO_2. However, CO_2 carrier was found to provide the optimal cooling.

The efficiency of the desorption–entrainment process into the jet expansion is unknown; however, some estimate of the efficiency can be made. If a typical signal on the tyramine 0–0 transition is 30 mV with a 10-ns FWHM peak width through a 50-Ω termination and the dual microchannel plate detector has an estimated gain of 5×10^6 at the given working voltage, then ~ 10 ions are detected per pulse. Assuming a 0.1% ionization efficiency and that a 3-mm laser beam intersects $\sim 15\%$ of the gas pulse[28] then $\sim 10^5$ neutrals are entrained in the beam per pulse. This is only $\frac{1}{10^8}$ of the estimated number of molecules desorbed per pulse. Nevertheless, a remarkably small amount of sample is used (~ 100 μg) to obtain the spectra shown here.

FUTURE PROSPECTS

Laser-induced REMPI exhibits many properties that make it a highly desirable addition to the arsenal of ionization methods currently available for mass spectrometry. The enhanced selectivity provided by REMPI in combination with supersonic jet methods has great potential for solving discrimination problems in analysis of biological isomers and in complicated mixture analysis in which conventional methods may not be sufficient. The development of widely usable applications in this area will depend upon the advent of broadly tunable high-powered pulsed lasers in the near-UV and UV-visible regions. The requirement of broad tunability analogous to that possible in present spectrophotometer systems will require the further development of tunable solid-state lasers.

The second unique property of REMPI is the ability to control the degree of fragmentation based upon laser power, where either soft ionization or extensive fragmentation may be induced. This may have great potential in peptide and nucleotide analysis for solving problems in identification and structural analysis of these compounds. The potential of this method will depend on a basic understanding of ionization–fragmentation processes and on developing various methods for controlling the fragmentation. This may include a knowledge of the dependence of the fragmentation mechanisms on laser power, pulse length, frequency, and molecular structure. In addition, multicolor laser REMPI methods may be used to enhance the characteristic fragmentation patterns obtained. Also, the limits of the soft ionization capabilities of R2PI must be further investigated.

Other developments that will be needed to make REMPI a widely accepted technique include the compilation of a broad data base that presently does not exist. This encompasses two aspects: (1) a data base for REMPI spectroscopy of each molecule so that wavelengths for selectivity purposes can be identified, and (2) a data base for fragmentation patterns under various conditions. Nevertheless, the unique properties of REMPI as a highly selective and sensitive method for mass spectrometric analysis will hopefully spur the growth of this base as various investigators continue to adapt laser-induced REMPI to their accepted experimental procedures.

ACKNOWLEDGMENTS

I would like to thank Roger Tembreull, Liang Li, Ho Ming Pang, and Chung Hang Sin for their various contributions to this work. We acknowledge financial support of this work under NSF Grant CHE 8419383 and NSF Grant DMR 8418095 for acquisition of the Chemistry and Materials Science Laser Spectroscopy Laboratory. David M. Lubman is a Sloan Foundation Research Fellow.

REFERENCES

1. P. M. Johnson, M. R. Berman, and D. Zakheim, *J. Chem. Phys.* **62**, 2500 (1975).

2. G. Petty, C. Tai, and F. W. Dalby, *Phys. Rev. Lett.* **34**, 1207 (1975).

3. P. M. Johnson, *J. Chem. Phys.* **62**, 4562 (1975). D. H. Parker, J. O. Berg, and M. A. El-Sayed, *Advances in Laser Chemistry* (Springer Series in Chemical Physics). Springer-Verlag, New York, 1978, p. 320.

4. R. P. Rava and L. Goodman, *J. Am. Chem. Soc.* **104**, 3815 (1982).

5. G. C. Nieman and S. D. Colson, *J. Chem. Phys.* **68**, 5656 (1978).

6. M. B. Robin and N. A. Kuebler, *J. Chem. Phys.* **69**, 806 (1978).

7. V. S. Antonov, I. N. Knyazev, V. S. Letokhov, V. M. Matjiuk, B. G. Moshev, and V. K. Potapov, *Opt. Lett.* **3**, 37 (1978).

8. U. Boesl, H. J. Neusser, and E. W. Schlag, *Z. Naturforsch.* **33a**, 1546 (1978).

9. K. Krogh-Jespersen, R. P. Rava, and L. Goodman, *Chem. Phys.* **44**, 295 (1979).

10. G. J. Fisanick, T. S. Eichelberger, IV, B. A. Heath, and M. B. Robin, *J. Chem. Phys.* **72**, 5571 (1980).

11. M. A. Duncan, T. G. Dietz, and R. E. Smalley, *Chem. Phys.* **44**, 415 (1979).

12. U. Boesl, H. J. Neusser, and E. W. Schlag, *J. Chem. Phys.* **72**, 4327 (1980).

13. R. Frueholz, J. Wessel, and E. Wheatley, *Anal. Chem.* **52**, 281 (1980).

14. M. Seaver, J. W. Hudgens, and J. J. DeCorpo, *Int. J. Mass Spectrom. Ion Processes.* **34**, 159 (1980).

15. P. M. Johnson, *Acc. Chem. Res.* **13**, 20 (1980), and references cited therein.

16. J. P. Reilly and K. L. Kompa, *J. Chem. Phys.* **73**, 5468 (1980).

17. L. Zandee, R. B. Bernstein, and D. A. Lichtin, *J. Chem. Phys.* **69**, 3427 (1978).

18. L. Zandee and R. B. Bernstein, *J. Chem. Phys.* **70**, 2574 (1979).

19. L. Zandee and R. B. Bernstein, *J. Chem. Phys.* **71**, 1359 (1979).

20. R. B. Bernstein, *J. Phys. Chem.* **86**, 1178 (1982), and references cited therein.

21. D. M. Lubman, R. Naaman, and R. N. Zare, *J. Chem. Phys.* **72**, 3034 (1980).

22. C. D. Cooper, A. D. Williamson, J. C. Miller, and R. N. Compton, *J. Chem. Phys.* **73**, 1527 (1980).

23. T. G. Dietz, M. A. Duncan, M. G. Liverman, and R. E. Smalley, *Chem. Phys. Lett.* **70**, 246 (1980).

24. T. G. Dietz, M. A. Duncan, M. G. Liverman, and R. E. Smalley, *J. Chem. Phys.* **73**, 4816 (1980).

25. C. T. Rettner and J. H. Brophy, *Chem. Phys.* **25**, 53 (1981).

26. U. Boesl, H. J. Neusser, and E. W. Schlag, *Chem. Phys.* **55**, 193 (1981).

27. J. E. Wessel, D. E. Cooper, and C. M. Klimcak, *Laser Spectroscopy for Sensitive Detection,* edited by J. A. Gelbwachs. *Proc. Soc. Photo-Opt. Instrum. Engineers* **286**, 48 (1981).

28. D. M. Lubman and M. N. Kronick, *Anal. Chem.* **54**, 660 (1982).

29. T. M. Sack, D. A. McCrery, and M. L. Gross, *Anal. Chem.* **57**, 1290 (1985).

30. G. Rhodes, R. B. Opsal, J. T. Meek, and J. P. Reilly, *Anal. Chem.* **55**, 280 (1983).

31. M. P. Irion, W. D. Bowers, R. L. Hunter, F. S. Rowland, and R. T. McIver, Jr., *Chem. Phys. Lett.* **93**, 375 (1982).

32. T. J. Carlin and B. S. Freiser, *Anal. Chem.* **55**, 955 (1983).

33. R. Tembreull and D. M. Lubman, *Anal. Chem.* **56**, 1962 (1984).

34. D. A. Gobeli, J. J. Yang, and M. A. El-Sayed, *Chem. Rev.* **85**, 529 (1985).

35. R. Tembreull, C. H. Sin, P. Li, H. M. Pang, and D. M. Lubman, *Anal. Chem.* **57**, 1186 (1985); *Anal. Chem.* **57**, 1084 (1985).

36. R. Tembreull, T. M. Dunn, and D. M. Lubman, *Spectrochim. Acta* **42A**, 899 (1986).

37. R. Tembreull and D. M. Lubman, *Anal. Chem.* **58**, 1299 (1986).

38. R. Tembreull and D. M. Lubman, *Anal. Chem.* **59**, 1003 (1987).

39. R. Tembreull and D. M. Lubman, *Anal. Chem.* **59**, 1082 (1987).

40. R. Tembreull and D. M. Lubman, *Appl. Spectrosc.* **41**, 431 (1987).

41. F. Engelke, J. H. Hahn, W. Henke, and R. N. Zare, *Anal. Chem.* **59**, 909 (1987).

42. J. H. Hahn, R. Zenobi, and R. N. Zare, *J. Am. Chem. Soc.* **109**, 2842 (1987).

43. J. Grotemeyer, U. Boesl, K. Walter, and E. W. Schlag, *Org. Mass Spectrom.* **21**, 595 (1986); *Org. Mass Spectrom.* **23**, 388 (1988).

44. J. Grotemeyer, U. Boesl, K. Walter, and E. W. Schlag, *Org. Mass Spectrom.* **21**, 645 (1986).

45. J. Grotemeyer, U. Boesl, K. Walter, and E. W. Schlag, *J. Am. Chem. Soc.* **108**, 4233 (1986).

46. D. M. Lubman, *Anal. Chem.* **59**, 31A (1986).

47. J. W. Hager and S. C. Wallace, *Anal. Chem.* **60**, 5 (1988).

48. J. H. Brophy and C. T. Rettner, *Opt. Lett.* **4**, 337 (1979).

49. U. Boesl, H. J. Neusser, and E. W. Schlag, *J. Chem. Phys.* **72**, 4327 (1980).

50. R. N. Zare and P. J. Dadigian, *Science* **185**, 739 (1974).

51. M. A. Posthumus, P. G. Kistemaker, H.L.C. Meuzelaar, and M. C. Ten Noever de Brauw, *Anal. Chem.* **50**, 985 (1978).

52. R. J. Cotter, *Anal. Chem.* **52**, 1767 (1980).

53. S. Datz and E. H. Taylor, *J. Chem. Phys.* **25**, 389 (1956).

54. P. G. Kistemaker, G.J.Q. van der Peyl, and J. Haverkamp, in *Soft Ionization Biological Mass Spectrometry,* edited by H. R. Morris. Heyden & Son, London, 1981.

55. R. J. Conzemius and J. M. Capellen, *Int. J. Mass Spectrom. Ion Processes.* **34**, 197 (1980).

56. W. C. Wiley and I. H. McLaren, *Rev. Sci. Instrum.* **26**, 1150 (1955).

57. W. R. Gentry and C. F. Giese, *Rev. Sci. Instrum.* **49**, 595 (1978).

58. R. L. Byer and M. D. Duncan, *J. Chem. Phys.* **74**, 2174 (1981).

59. M. G. Liverman, S. M. Beck, D. L. Monts, and R. E. Smalley, *J. Chem. Phys.* **70**, 192 (1979).

60. R. J. Rorden and D. M. Lubman, *Rev. Sci. Instrum.* **54**, 641 (1983).

61. D. A. McCrery and M. L. Gross, *Anal. Chim. Acta* **91**, 178 (1985).

62. C. L. Wilkins, D. A. Weil, C.L.C. Yang, and C. F. Ijames, *Anal. Chem.* **57**, 520 (1985).

63. M. L. Coates and C. L. Wilkins, *Anal. Chem.* **59**, 197 (1987).

64. D. M. Lubman, C. T. Rettner, and R. N. Zare, *J. Phys. Chem.* **86**, 1129 (1982).

65. L. Li and D. M. Lubman, *Appl. Spectrosc.* **42**, 411 (1988).

66. R. Robinson, *Tumors That Secrete Catecholamines.* John Wiley, Chichester, 1980.

67. L. Boniforti, G. Citti, O. Lostia, and C. Lucarelli, in *Recent Developments in Mass Spectrometry in Biochemistry, Medicine and Environmental Research,* Vol. 12, edited by A. Frigiderio. Elsevier, Amsterdam, 1983, p. 25.

68. H. Budzikiewicz, C. Djerassi, and D. H. Williams, *Interpretation of Mass Spectra of Organic Compounds.* Holden-Day, San Francisco, 1964, p. 63.

69. L. Li and D. M. Lubman, *Appl. Spectrosc.* **42**, 418 (1988); *Rev. Sci. Instrum.* **59**, 557 (1988); Appl. Spectrosc., 43, 543 (1989).

70. L. Li and D. M. Lubman, *Anal. Chem.* **60**, 1409 (1988).

71. K. Biemann and S. A. Martin, *Mass Spec. Rev.* **6**, 1 (1987).

72. R. Frey, G. Weiss, H. Kaminski, and E. W. Schlag, *Z. Naturforsch.* **40a**, 1349 (1985).

73. J. Grotemeyer and E. W. Schlag, *Org. Mass Spectrom.* **22**, 758 (1987).

74. J. A. McCloskey, *Mass Spectrometry in Biomedical Research,* edited by S. J. Gaskell. John Wiley, New York, 1986, p. 75.

75. J. A. McCloskey and S. Nishimura, *Acc. Chem. Res.* **10**, 403 (1977).

76. Y. Sasaki and T. Hashizume, *Anal. Biochem.* **16**, 1 (1966).

77. S. J. Shaw, D. M. Desiderio, K. Tsuboyama, and J. A. McCloskey, *J. Am. Chem. Soc.* **92**, 2510 (1970).

78. K. Biemann and J. A. McCloskey, *J. Am. Chem. Soc.* **84**, 2005 (1962).

79. F. W. Crow, K. B. Tomer, M. L. Gross, J. A. McCloskey, and D. E. Bergstrom, *Anal. Biochem.* **139**, 243 (1984).

80. D. J. Ashworth, W. M. Baird, C. Chang, J. D. Ciupek, K. L. Bush, and R. G. Cooks, *Biomed. Mass Spectrom.* **12**, 309 (1985).

81. H. K. Mitchum, F. E. Evans, J. P. Freeman, and D. Roach, *Int. J. Mass Spectrom. Ion Processes.* **46**, 383 (1983).

82. L. Li and D. M. Lubman, *Int. J. Mass Spectrom. Ion Processes,* **88**, 197 (1989).

83. J. R. Cable, M. J. Tubergen, and D. H. Levy, *J. Am. Chem. Soc.* **109**, 6198 (1987).

84. L. Li and D. M. Lubman, *Anal. Chem.* **60**, 2591 (1988).

85. L. Li and D. M. Lubman, *Anal. Chem.* **59**, 2538 (1987).

86. A. Burger, *A Guide to the Chemical Basis of Drug Design.* John Wiley, New York, 1983.

87. R. Tembreull and D. M. Lubman, *Appl. Spectrosc.* **41**, 43 (1987).

88. A. Oikawa, H. Abe, N. Mikami, and M. Ito, *J. Phys. Chem.* **88**, 5181 (1984).

89. T. M. Dunn, R. Tembreull, and D. M. Lubman, *Chem. Phys. Lett.* **121**, 453 (1985).

90. Y. D. Park, T. R. Rizzo, L. A. Peteanu, and D. H. Levy, *J. Chem. Phys.* **84**, 6539 (1986).

17

Analysis of Organic Molecules on Surfaces: Laser Desorption Spectroscopy

MATTANJAH S. de VRIES, HEINRICH E. HUNZIKER, and H. RUSSEL WENDT

Detection and identification of organic molecules on surfaces are very general analytical problems with ever increasing technological applications. In the electronics industry, for example, ever smaller amounts of surface contaminations are becoming important. Although many compounds may be present in the gas phase at low vapor pressures, sometimes as low as 10^{-8} torr, they often become important only by their presence on a surface. Moreover, although their low concentrations make them hard to detect in ambient air, they can often be concentrated on a suitable surface for analysis. As a consequence, surface analysis is a fundamentally important challenge. Other examples can be found in every area in which low vapor pressure materials need to be analyzed. These include the study of biological materials, certain pharmaceuticals, geophysical samples, or small organic particles. Volatile samples can be analyzed by the powerful combination of gas chromatography and many mass spectrometric techniques. However, solid compounds generally need an initial step of solution or vaporization, which may alter the compounds to be studied.

Many surface analytical techniques are available, each with its own merits and limitations. The most important ones are summarized in Table 17.1. This table is by no means intended to be exhaustive. Many techniques, such as surface-enhanced Raman scattering (SERS) or electron stimulated desorption (ESD) are omitted. The ones listed are those that are most routinely used for surface analysis. Often a combination of approaches will be required to gain all the necessary information on a problem. In general, these techniques involve heating or exposure to electrons or ions, conditions under which most organic molecules are not stable. The possibilities for obtaining chemical structure information about large molecules are limited mainly to those techniques in which the molecule is first desorbed. The first challenge, then, is to desorb the molecule without fragmenting or otherwise altering it. The same challenge applies for the detection phase, which often involves ionization. The merits of laser desorption in this respect will be discussed, together with various detection schemes.

Table 17.1. Commonly used surface analytical techniques

Technique	Advantage	Limitation	Ref.
Auger	Small areas (less than 0.1 μm). Elemental composition, some chemical information from peak shifts, sensitive to 0.01 monolayer	No molecular structure information. Electron beam damage possible	31
EDX (energy dispersed X rays)	Small areas. Inside electron microscope	Only elemental composition. Poor for hydrogen, carbon	32
EELS (electron energy loss spectroscopy)	Vibrational spectroscopy. High sensitivety	Limited resolution compared to optical spectroscopy	33
EXAFS/XANES	Can reveal local bond lengths		
IR spectroscopy	Can reveal functional groups	Limited sensitivity	34
LEED (low-energy electron diffraction)	Long range order. Crystalline structure	No molecular information	35
RBS (Rutherford backscattering)	Quantitative on absolute scale	Damage of polymers from ion bombardment. Background from lighter substrates. Decreasing resolution for higher masses	36
Static SIMS (secondary ion mass spectroscopy)	Mass information	Fragmentation of larger molecules, inconsistent ion yields	37
TPD (temperature-programmed desorption)	Probes bond strength with surface	Chemistry (decomposition) occurs during heating	
UPS/XPS (UV/X-ray photoelectron spectroscopy)	Elemental composition, some binding information from peak shifts, sensitive to 0.01 monolayer	No molecular structure information. Electron beam damage possible	31

LASER-INDUCED DESORPTION

In much of the work in which desorption is used with mass spectrometry, secondary ions are detected that are formed by the same mechanism that also causes the desorption. This is true for techniques such as fast atom bombardment (FAB) or secondary ion mass spectrometry (SIMS), usually involving keV projectiles, as well as for heavy ion-induced mass spectrometry or (^{252}Cf) plasma desorption, with MeV exposures. Also laser desorption mass spectrometry (LDMS) often refers to the situation in which the ions are formed at the surface or in the field of the desorption laser. Most of the early LDMS work uses laser powers at which extensive fragmentation takes place, as well as damage to the sample.[1] In those applications the laser ion source is used mainly to obtain elemental characterization of the sample. Nevertheless, even with these techniques, it is sometimes possible to observe high mass ions, including unfragmented parent molecules. Examples are the detection of polymer ions with masses up to 4500 amu by SIMS[2] or cationization of biological molecules.[3] It was speculated early on by MacFarlane that to observe parent ions the time scale for heating would have to be shorter than that for vibrational excitation of the bond that would lead to fragmentation.[4] This idea was discussed by Hall for laser desorption[5] in terms of rates $k = \theta v \exp(-E/RT)$ in which E

represents an activation energy, v a preexponential factor, and Θ the surface coverage. If the activation energy and the preexponential factor are larger for desorption than for reaction on the surface, then desorption will be preferred over reaction at high heating rates. This implies that is is possible to achieve desorption rather than chemistry at the surface, provided the heating rate is sufficiently fast. Recently, Zare and Levine reached the same conclusion with a different theoretical picture.[6] In their description of laser desorption the laser energy excites the phonons of the surface, the frequencies of which reasonably match the frequency v of the bond between the molecule and the surface, whereas the frequencies v' of the molecular vibrations are much higher. Consequently, the bond to the surface acts as a bottleneck for the flow of energy into the molecule. As a result, in this nonequilibrium picture, it is once again possible to desorb molecules without fragmentation. The heating rate has to be high enough, such that the time for desorption is small compared to $\exp(v'/v)/v$. It has been shown experimentally that molecules can be desorbed intact and with relatively little internal energy at high heating rates, whereas they decompose at low heating rates.[7] In general, the rates required to desorb molecules without fragmentation are of the order of 10^{11} K/sec, which corresponds to a 1000 K temperature jump in a typical 10-nsec laser pulse.

LASER PARAMETERS

Unfortunately, it is impossible to reliably measure temperatures at nanosecond time scales and submicron distance scales. Consequently, common wisdom is derived from models based on thermal properties of bulk materials.[8] Typically the peak temperature is reached within the laser pulse whereas the decay of the temperature with time depends on the material of the substrate. The diffusion of energy into the bulk is of the order of microseconds for metals whereas thermally insulating materials can stay at locally elevated temperatures for longer times. Desorption does not follow the temperature excursion since it is a strongly exponential process. For metal substrates it is possible to adjust the laser power such that desorption takes place only in the peak of the temperature excursion, making the time of desorption shorter than the laser pulse. As demonstrated in Fig. 17.1, an increase in desorption laser power of a factor of 1.2 can cause a 10-fold increase in desorption rate. We also find that often there is a relatively narrow range of laser powers between the limits of threshold for desorption on the one hand, and ablation on the other. Another consideration in this respect is the intensity profile of the laser beam. Hot spots can give rise to enormously higher desorption rates. A careful model study of spatial desorption profiles has been reported by Brand and George for applications on studies of thermal diffusion.[9]

A final laser parameter that needs to be discussed is the wavelength. Three considerations are important: (1) the photons need to be absorbed by the substrate, (2) the adsorbate should be transparent to the laser, and (3) the gas-phase molecules should not undergo photochemistry in the laser field. The requirements are not always compatible. For instance, to comply with (3) a long wavelength is preferable, however, certain substrates, such as metals, generally absorb better at shorter wavelength. With higher absorbtivity higher temperatures can be reached. When focus-

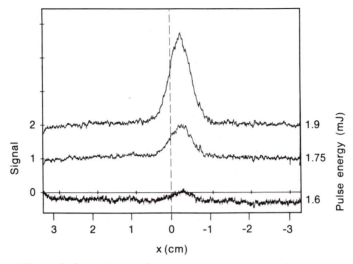

Fig. 17.1. Effect of desorption pulse energy on amount of perylene molecules, desorbed from a gold surface by 248-nm, 20 nsec laser pulse. Detection by LIF imaging as described in the text. Peak integrals are proportional to the number of molecules that are desorbed and entrained in the beam.

ing is desired to a spot of the order of a micron, UV wavelengths are also called for. As for the adsorbate, in the case of thin coverages it will generally be sufficiently transparent. On the other hand, in the case of multilayer coverage it may be preferable to choose a wavelength that is absorbed by the adsorbate. In that situation the dynamics of heating and desorption may be different,[10] and for mixed coverages matrix effects may play a role.[11]

APPLICATIONS

Laser Desorption Transfer (LDT)

Given the premise that with laser desorption large molecules can be desorbed intact, it is possible to create a very simple transfer device as shown in Fig. 17.2. A desorption laser beam traverses a substrate that is transparent for the laser and that is mounted as close as possible above a sample. The gap between substrate and sample is evacuated to about 0.05 torr. The distance between the sample and the substrate is made significantly less than the mean free path, which is approximately given by: L (cm) $= 5 \times 10^{-3}/P$ (torr). As the laser heats the sample, molecules are desorbed and travel in straight lines to the unheated substrate, where they stick. In the process spatial information is roughly preserved. This technique can be applied when analysis of molecules on the sample itself is difficult. An example of such a situation is IR analysis, for which the background from a substrate often limits the sensitivity for overlayers. LDT can transform a problem that is prohibitive in reflection IR to one that is feasible in transmission IR. Figure 17.3 shows an IR spectrum of material transferred onto an NaCl window from a ferrite sample,

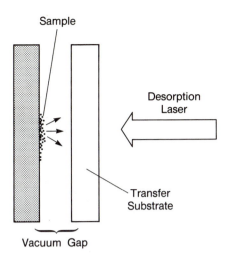

Fig. 17.2. Schematic of apparatus for laser desorption transfer.

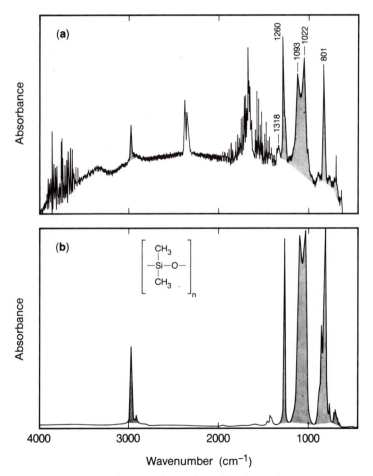

Fig. 17.3. (a) IR spectrum of material transferred by laser desorption onto NaCl window. (b) Reference spectrum of dimethylsiloxane, shown for comparison.

which is black in the infrared. The IR signature of dimethylsiloxane can be clearly recognized in this spectrum. To test the quantitative capability of LDT we evaporated a film of coronene on stainless steel with a thickness of 5 $\mu g/cm^2$ as measured with a quartz microbalance. The material was transferred with a single laser shot at 200 mJ/cm^2 from an Nd:YAG laser onto an NaCl substrate. Subsequent IR analysis showed that 90% of the material had been transferred. This implies that for application to IR analysis LDT can be as sensitive as transmission FT-IR. Assuming a cos^2 θ angular distribution for the desorbing molecules, the lateral spatial distribution can be preserved with a resolution of the order of the gap between sample and substrate. It may be noted that with a commercial FT-IR microscope a spatial resolution of 50 μm can be obtained. Other applications of LDT include other forms of analysis, such as laser desorption mass spectrometry for samples that cannot be disassembled to be placed in a vacuum apparatus themselves.

Jet Cooling

Two problems are inherent to any approach in which vaporization and ionization occur at the same time. First, at laser power levels well below those that would cause damage to the sample, orders of magnitude more neutrals than ions are generated. Second, lack of separate control of desorption and ionization steps prevents their separate optimization and complicates interpretation, since ionization is influenced by matrix effects, ion–molecule reactions, and plasma effects. Separate postionization as an improvement to laser desorption has taken a number of different forms. Electron impact[12] and chemical ionization[13] have been used for structure elucidation, laser-induced fluorescence (LIF) has been applied for the determination of internal state distributions,[14] and multiphoton ionization with different intensities has been used for controlled fragmentation,[15,16] or simply for improved detection efficiency.[17–21] Another attractive possibility is to make use of optical spectroscopy by resonance-enhanced multiphoton ionization (REMPI). This requires cooling as molecules are desorbed with at least the initial surface temperature. As is well known, cooling of internal degrees of freedom of molecules can be achieved by entrainment in a supersonic expansion.[22] The combination of laser desorption with jet cooling has been demonstrated for biomolecules and other complex organic species in at least two laboratories.[15,20,23–25]

A typical experimental arrangement, which we are using, is shown in Fig. 17.4. A desorption laser, usually 248 or 193 nm, at 2 mJ/pulse, focused onto a 1-mm-diameter spot, is directed at a 3-mm-wide sample bar, which is mounted directly in front of a pulsed nozzle. The distance of the sample with respect to the symmetry axis of the nozzle can be varied, as will be discussed below. The drive gas is usually He or N$_2$ with backing pressures from 1 to 5 atm. The beam is skimmed by a 1 × 4-mm slit-shaped skimmer after which it is crossed by the ionization laser beams, either from a dye laser or from an excimer laser or—for two-color multiphoton ionization (MPI)—from both. The ions are extracted at right angles in a linear time-of-flight (TOF) mass spectrometer of the Wiley-McLaren type.[26]

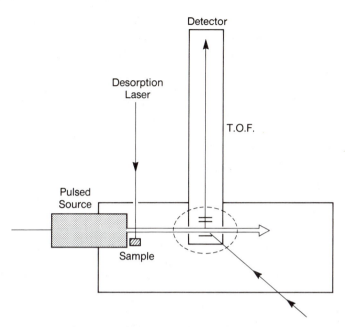

Fig. 17.4. Schematic of apparatus for laser desorption jet cooling.

Sensitivity

Any analytical application requires a discussion of sensitivity. In the case of laser desorption jet cooling the probability P of detecting a molecule on the surface is given by

$$P = P_d P_e P_i P_t$$

where P_d, the fraction of molecules desorbed, can in many cases approach unity. P_e is the fraction of desorbed molecules that is entrained and brought into the field of the ionizing laser. This factor is the most difficult to predict and will be discussed separately in the next paragraph. It will be shown that values of P_e of the order of 1% can be achieved. P_i is the ionization efficiency, which depends on the molecule, and can, in favorable cases, approach unity, while being less than 1% in unfavorable cases. The cooling helps significantly by concentrating the internal population in the lowest states. P_t is the transmission of the mass spectrometer which for TOF can be of the order of 10%. With these rough estimates the total detection efficiency P can be of the order of 10^{-5}, implying that as few as a million molecules should be detectable. This corresponds to 0.01 fmol. To test the sensitivity we have made calibrated samples by evaporating films of coronene through a mask, creating 1.5-mm-diameter circular spots with a thickness of 5 μg/cm^2. This corresponds to 100 ng of material in the spot, or 330 pmol. To ensure uniform exposure to the desorption laser a 3-mm-diameter laser beam was used. The material was desorbed with 248 nm and ionized with 193 nm. By recording the total signal over 100,000 laser shots and extrapolating down to the detection limit we found that in the present setup, which is not fully optimized, less than a femtomole of material can be detected. This may be compared with arrangements in which the desorbed molecules are ionized directly above the sample. In that situation, because of the angular and velocity distributions of the desorbed molecules, the overlap with the ionization laser is also not better than a few percent at best. As a result, similar detection sensitivities are obtained, without the added benefit of spectroscopic distinction. The importance of the additional capability of optical spectroscopy may be illustrated by noting that the optical resolution, when expressed as $v/\Delta v$, is of the order of 10,000, which is an order of magnitude better than the mass resolution ($m/\Delta m$) from TOF mass spectrometry. The beam also improves the mass resolution to some extent by reducing the spread in velocity components in the time-of-flight direction.

Spatial Distributions

Because only a small portion of the jet can be sampled, it is obviously important to optimize the fraction of adsorbate entrained in the jet that will appear in the detection volume. The size of the detection volume is limited by the nozzle/skimmer geometry and by the fact that, for a reasonable mass resolution, the range along the TOF direction (x axis) over which ions are formed has to be restricted. For our configuration this leads to a volume that is 1 mm in the x direction. By using a 1 $\times$ 4-mm slit-shaped skimmer and 9-mm-wide laser beam, we obtain a 1 $\times$ 4 $\times$ 9 mm^3 detection volume. However, at the given distance of 57 mm (114 nozzle diameters) from the orifice, the FWHM of the density in a free He jet is 67 mm.[22] Consequently, only a very small portion of the expansion is sampled by the detection laser. In order to optimize the sensitivity, it is desirable to confine the concentration of desorbed matter as much as possible to the jet axis. Furthermore the time spread should be limited in order to limit spreading along the nozzle beam direction (z axis). To learn how to achieve this optimization, and to estimate the optimal fraction, we have measured concentration profiles of entrained perylene vapor and explored their dependence on the adjustable parameters of the experiment.[27] This was done by replacing the MPI-TOF detection in our apparatus by LIF imaging detection as shown in Fig. 17.5. Concentration profiles of perylene vapor entrained in the jet were measured in an xy plane located at z = 57 mm (114 nozzle diameters) from the orifice. At this distance the expansion is collision free and the profile has reached its final form. A 6-mm-diameter probe beam traversed the xy plane in the x direction. The probe was a 6-nsec, 20-mJ pulse of 355-nm radiation derived from a Q-switched, tripled Nd:YAG laser. A camera lens (f 1.8) was used to form an image of the fluorescent probe beam trace on the 3 $\times$ 25-mm detector area of an optical multichannel analyzer. Profiling in the y direction was accomplished with a mechanism that translated a right angle prism inside the vacuum chamber together with the lens and multichannel detector outside. Profiling in the z direction was effectively achieved by varying the delay time between desorption and probe laser. Further details of the experiment are given in Fig. 17.5.

Results. The spatial distributions of perylene concentration produced by our experimental arrangement are ridge shaped, with the ridgeline running parallel to the sample plane. Vertical sections have a Gaussian shape for He, are very narrow

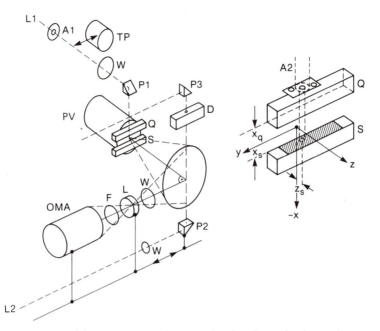

Fig. 17.5. Diagram of the experimental setup. The detail at right shows the coordinate system and the coordinates for laser desorption. Abbreviations: A1, 3.2-mm aperture; A2, 1.0-mm apertures (3); D, graphite beam dump with alignment hole; F, Schott GG400 filter; L, camera lens; L1, 248-nm excimer laser; L2, 355-nm Nd:YAG laser; OMA, optical multichannel detector; P1, P2, P3, right angle prisms; PV, pulsed valve; Q, UV-fused silica rod; S, sample rod with perylene film; TP, thermopile; W, vacuum chamber windows.

compared with the FWHM of the jet density, and are remarkably independent of delay time, source pressure, desorption geometry, and horizontal position. Figure 17.6 shows sequences of vertical concentration profiles recorded as a function of delay between desorption and detection time. The shape of the profile is approximately Gaussian, and its FWHM is about 7 mm independent of delay time. Although the early portion of the signal appears well above the beam axis, the center of the profile moves toward and below the axis as the delay is increased. The width of the horizontal profiles depended on the distance of the desorption spot from the nozzle. Data were collected with 1-mm spots, with their center 0.7, 1.7, and 2.7 mm away from the orifice, respectively. The average widths observed are close to the widths of the straightline projections of the desorption spots onto the observation plane, using tangents drawn from the edge of the jet orifice.

The major effects of other experimental variables can be summarized as follows:

1. *Effect of Sample Position.* The distance between sample surface and jet axis turns out to be critical for the vertical position of the concentration maximum. It can be made to arrive with its peak above, at, or below the jet axis by moving the sample up or down. This parameter serves as an important

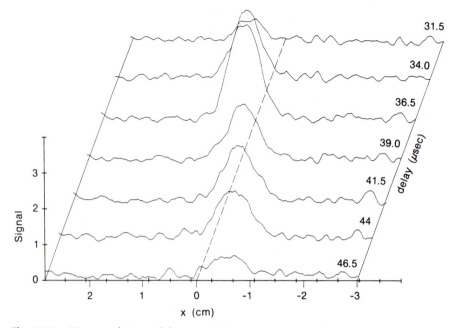

Fig. 17.6. Time evolution of the vertical concentration profiles. See text for details.

control for adjustment to let the main portion of the desorbed material pass the skimmer.

2. *Effect of Source Pressure.* Reservoir pressure of the jet expansion is another parameter that strongly affects the position of the concentration profile. Data were taken with backing pressures from 1 to 11.2 atm. With increased pressure the amplitude of the signal decreased and the vertical position of the concentration maximum moved down. Thus, backing pressure can also be used for aiming the desorbed material at the skimmer (it should be remembered that the backing pressure can also affect the cooling).

3. *Effect of Source Gas.* Some experiments in Ar were carried out to see whether the narrow, vertical profiles observed in He were related to the large perylene/He molecular mass ratio (252:4). The profiles in Ar are much broader and asymmetrical, particularly at lower pressure. If the vertical widths were determined by simple diffusion we would expect them to be larger in He. Since the opposite is observed, it seems more likely that a gas-dynamic separation effect is responsible, which increases with molecular mass ratio and thus is more effective in He.

A common feature in all our time profiles is the downward motion of the concentration maximum with delay time. We speculate that the downward motion reflects portions of desorbed material injected into the expansion with different velocities. The stopping distance depends on the velocity component in the z direction and on the density of the beam, which is higher closer to the nozzle. Similar arguments based on density effects would also qualitatively explain why vertical profiles at equal delay times are shifted down on the x axis when the source pressure is increased. A shorter penetration depth at higher density will increase the amount of vaporized material recondensing on the sample surface, which may explain the decrease of the total signal at higher source pressure.

Conclusions. Narrow concentration profiles, particularly in He, combined with the ability of aiming them, are a considerable advantage for applications of the entrainment technique. The fraction of entrained material that is extracted through a skimmer centered on the jet axis can be made much larger than the fraction of expansion gas passing through the skimmer. To optimize this fraction the profile maximum should be approximately on axis at the delay time corresponding to peak concentration. The two preferred control parameters for achieving this are sample height and source pressure. A second control surface above the jet orifice can also be used. Although this has some advantage in producing still narrower concentration profiles, the disadvantages may be more significant: interference with the expansion causes reduced final velocity and higher final temperature, and constant transparency of the second surface for the desorption beam may be hard to maintain due to coating by desorbed material. We have estimated the fraction of entrained material that can be extracted as a packet of convenient size for laser detection by using the measured spatial and time distributions. Because of the elongated shape of the two-dimensional profile it is advantageous to use a slit skimmer; we are employing one with a 4×1-mm cross section. This slit will pass about 8×10^{-4} times the total gas flow at the 57 mm sampling distance used in this work.[13] The fraction of entrained perylene passing through the skimmer is estimated by approximating the perylene flux, $F(t)$, as

$$F(t) \, dx \, dy = pv(x,t)h(y) \, dx \, dy$$

where $v(x,t)$, the time-dependent vertical profile, and $h(y)$, the horizontal profile, are given by the measurements, and p is a proportionality constant. We find that under the best conditions 1–1.5% of the entrained material can be contained in a packet 9 mm long in the z direction, and intercepted by a laser beam of that width, traveling in the y direction. Of course this fraction will vary with skimmer size, nozzle-to-skimmer distance, and volume chosen for the interaction region.

The horizontal fraction captured by the skimmer can be increased by moving the desorption spot further from the nozzle and by decreasing its diameter, thereby narrowing the horizontal concentration profile. Because distance from the nozzle will affect the degree of cooling, this parameter must be optimized depending on the application. Precise alignment of the desorption spot along the jet axis is extremely important.

Cooling

For analytical applications, besides sensitivity, specificity is important. The ultimate goal of laser desorption jet cooling is the ability to recognize desorbed molecules by their spectroscopy, or to selectively ionize certain molecules in the presence of many others by tuning to a unique resonance. It is therefore important to diagnose the degree of cooling that can be obtained. Ideally, for this purpose one would like to use a molecule with known spectroscopy as a thermometer. Unfortunately there is a general problem, in that the gas-phase spectroscopy of most invo-

Indole,desorbed

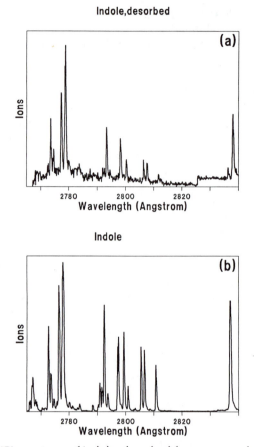

Indole

Fig. 17.7. (a) REMPI spectrum of indole, desorbed from porous glass and entrained in nitrogen. The baseline change at 282.5 nm is an experimental artifact. (b) REMPI spectrum of indole, seeded in nitrogen.

latile molecules of interest is unknown. This is the case simply because there is no other technique that allows such molecules to be studied cold and in the gas phase.

As an alternative, we have studied molecules with a vapor pressure sufficiently high to allow them to be seeded in a jet expansion for comparison. To achieve such comparison it is necessary to be able to desorb relatively high vapor pressure compounds from a substrate as well as seeding them in a beam. A suitable substrate is fritted glass, which can be saturated with the compound. The fritted glass acts as a wick, ensuring the presence of a film on the surface. Figure 17.7a shows the REMPI spectrum of indole, desorbed from fritted glass and entrained in a N_2 expansion with 5 atm backing pressure through a 0.5-mm orifice. The line at 283.8 nm is the $S_1(0-0)$ transition.[28] As a rule, better cooling was obtained with entrainment in N_2 than with entrainment in He. The reason for this is not obvious. One might speculate that since with N_2 final temperatures are reached at a larger distance from the nozzle, the influence of the surface on the expansion is smaller. This would suggest that the present geometry causes hydrodynamic effects that are not completely understood. The spectrum of desorbed indole can now be compared with that of indole, seeded in the same expansion, as shown in Fig. 17.7b. Similar results were obtained by Hager and Wallace by two-color REMPI of jet-cooled, seeded indole.[28] In comparing these spectra, relative peak intensities are not always meaningful because both desorption and ionization are strongly intensity dependent processes and, as a result, signals are influenced by shot-to-shot variations of both lasers. With that proviso, the agreement between the seeded and desorbed REMPI spectra is very good, showing that with desorption in front of the nozzle cooling can be achieved, similar to that obtainable in a seeded beam. In either case the linewidth is of the order of 0.03 nm, which is also the order of the ^{13}C isotope shift in the spectrum. Figure 17.8 shows two TOF spectra obtained with a difference of 0.025

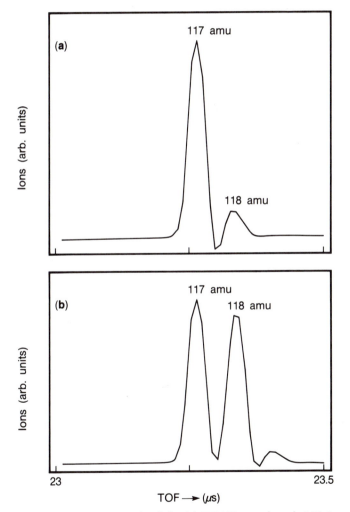

Fig. 17.8. Detail of TOF spectra of indole. (a) REMPI wavelength 238.8 nm resonant with $S_1(0-0)$ band in parent molecule, (b) REMPI wavelength resonant with the same transition in the ^{13}C isotope. This wavelength is 0.025 nm to the blue of that in (a).

nm in laser wavelength. One is resonant with the parent molecule, whereas the other is resonant with the ^{13}C isotope.

The same conclusion about cooling can be reached by inspecting Fig. 17.9, which shows the $S_1(0-0)$ transition of methyl benzoate, cooled in an expansion of N_2 at 5 atm. There is no significant difference in the linewidth, which is approximately 0.05 nm, whether the molecule is seeded or desorbed. These spectra were obtained by two-color REMPI, with 193 nm for the ionization laser. Single-color MPI, in this case, is frustrated by the fact that the intermediate excited state rapidly internally converts to a lower triplet state, as shown by Tomioka et al.[29] The same authors also showed that the singlet state becomes longer lived when methyl benzoate forms a complex with water, which apparently stabilizes the intermediate state. Figure 17.10 shows a two-color REMPI spectrum of methyl benzoate, seeded

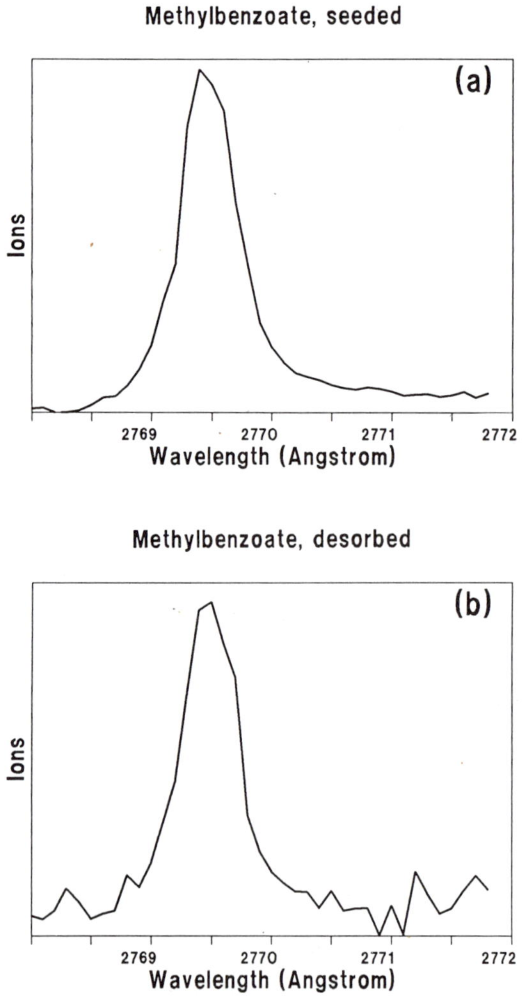

Fig. 17.9. Two color (dye + 193 nm) REMPI spectrum of methyl benzoate. (a) Seeded, (b) desorbed from fritted glass.

Methylbenzoate/Water

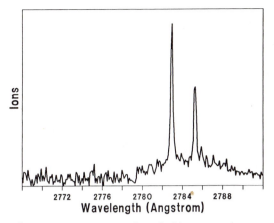

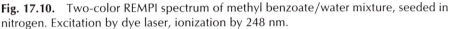

Wavelength (Angstrom)

Fig. 17.10. Two-color REMPI spectrum of methyl benzoate/water mixture, seeded in nitrogen. Excitation by dye laser, ionization by 248 nm.

in nitrogen together with water. The second laser wavelength in this case was 248 nm, which is enough to ionize the singlet state, while being insufficient to ionize the lower triplet state.

Complications with Jet Spectroscopy

Since every analytical technique has unique capabilities as well as problems, it is appropriate to discuss complications that can play a role with laser desorption jet spectroscopy. A number of those are demonstrated by the example of methyl benzoate, discussed in the previous paragraph.

Spectroscopy. While it can justly be argued that laser desorption jet cooling is an excellent new tool for the spectroscopic study of complex molecules, this also implies at the same time that cold spectra of such molecules are so far practically unknown. There are no fingerprint spectra available for identifying unknown species. Studies of model compounds for comparison will be required in many cases.

Two-Color Ionization. As demonstrated by the example of methyl benzoate, not every molecule lends itself easily to resonance-enhanced MPI. Preferably the resonant state is the first excited state, because at higher energy levels the density of states increases and the lifetime of the intermediate state may become very short, making it more difficult to obtain unique, molecule-specific information. A problem presents itself when the intermediate state is less than halfway between the ground state and the ionization potential (IP). In that case the molecule cannot be ionized by a single additional photon of the same wavelength and the laser fluence necessary for two-photon ionization of the intermediate state becomes too large, causing fragmentation.[30] This is the case either when the first excited state is too low, as with many symmetric aromatics, or when the lifetime of the excited state is too short. Sometimes this problem can be circumvented by two-color MPI, with the energy of the second photon, $h\nu_i$, sufficiently large to carry the molecule over the ionization potential. As pointed out by Hager and Wallace,[28] by this method REMPI can often be made more sensitive and more selective. However, there is still an inherent problem because the second photon has a higher energy than the resonant one, while many molecules have increased absorption coefficients toward the blue, sometimes by orders of magnitude. The result is competition of two photon ionization from the ground state by $2h\nu_i$. The spectra of Fig. 17.9 could be obtained only by strongly reducing the intensity of the ionization laser, and by delaying it with respect to the dye laser to allow for buildup of population in the long-lived triplet state.

Complex Formation. It is well known that in supersonic expansions clusters and complexes are easily formed. Figure 17.10 shows an example in which complex formation causes a significant shift in the wavelength of the REMPI spectrum. This may be a problem when the weakly bound complex falls apart in the ionization process, causing it to be detected at the original molecular mass. This could be confusing, particularly with attempts to distinguish similar species, such as isomers or isotopes, by small wavelength shifts. Complexes can easily be formed from other molecules or their fragments desorbed from the same surface. Figure 17.11 shows a TOF spectrum of coronene, mass 300, desorbed with 248 nm and ionized by 193 nm. The spectrum shows satellite peaks at higher masses, separated by 25 mass units. Apparently these are C_2H fragments created by the desorption laser. On the other hand, complex formation can also offer opportunities when used intentionally to alter spectra, or, for instance, by attaching a known chromophor to otherwise

Coronone TOF spectrum

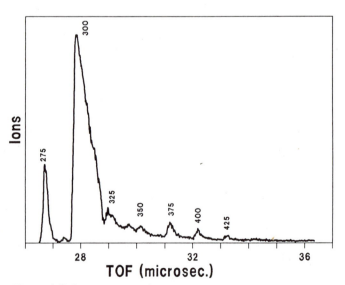

Fig. 17.11. Time-of-flight spectrum of coronene, desorbed with 248 nm and ionized with 193 nm. The masses of several peaks are indicated in amu.

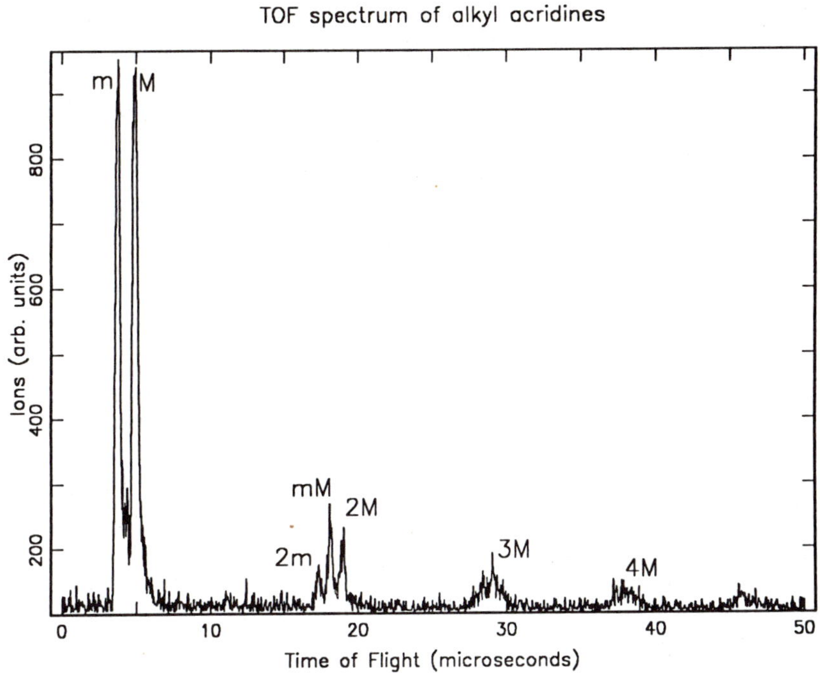

TOF spectrum of alkyl acridines

Fig. 17.12. Time-of-flight spectrum of mixture of long-chain acridines, desorbed from gold, showing cluster formation. M indicates heptadecylacridine, m indicates penta-decylacridine. The numbers refer to the number of basic units in the cluster.

hard to detect molecules. Similarly, clusters could be used to confirm identification of parent masses, as opposed to the fragmentation that is used in ordinary mass spectrometry. Figure 17.12 is an example of this. It shows a TOF spectrum with dimers and trimers of a mixture of heptadecylacridine and pentadecylacridine desorbed from a gold surface into an He jet.

SUMMARY

Identification of complex organic molecules on surfaces with sensitivity as well as specificity offers a new challenge, both to the analytical and to the surface science community. Neither routine analytical techniques nor standard surface diagnostic tools are particularly suitable for this task. Laser desorption appears to be a good approach, as large molecules can be brought into the gas phase without fragmentation. Subsequent detection can be done in a variety of ways, including IR analysis after transfer, optical spectroscopy after cooling, or high-resolution mass spectrometry. For the latter Fourier transform ion cyclotron resonance mass spectrometry (FTMS, FTICR) appears especially promising, particularly for the identification of unknowns. Perhaps the main limitation of that approach is the small dynamic range. This limitation may make it difficult to detect a small amount of a given compound in the presence of large amounts of others. On the other hand, for that kind of challenge, jet cooling with REMPI can be a very suitable approach, as it is possible in principle to selectively ionize specific molecules. This is particularly significant since we have shown the great sensitivity that can be obtained with careful attention to the entrainment dynamics. Since the precise spectroscopy of complex molecules is often uncharted terrain, the most suitable MPI scheme may have to be worked out for each individual case. Once this task is accomplished, the combined selectivity of mass and optical spectroscopy offers great promise, particularly for detecting molecules in mixtures with high sensitivity as well as specificity.

ACKNOWLEDGMENTS

The authors are greatly indebted to P. Arrowsmith for participation in jet cooling experiments and to D. Saperstein for his contributions to the Laser Desorption Transfer work.

REFERENCES

1. R. Conzemius and J. M. Capellen, *Int. J. Mass Spectrom. Ion Phys.* **34**, 197 (1980).

2. I. V. Bletsos, D. M. Hercules, J. H. Magill, D. van Leyen, E. Niehuis, and A. Benninghoven, *Anal. Chem.* **60**, 938 (1988).

3. M. A. Posthumus, P. G. Kistemaker, H.L.Z. Meuzelaar, and M. C. Ten Noever de Brauw, *Anal. Chem.* **50**, 985 (1978).

4. C. J. MacNeal, *Anal. Chem.* **54**, 43A (1982).

5. R. B. Hall, *J. Phys. Chem.* **91**, 1007 (1987).

6. R. N. Zare and R. D. Levine, *Chem. Phys. Lett.* **136**, 593 (1987).

7. D. Burgess, Jr., R. Viswanathan, I. Hussla, P. C. Stair, and E. Weitz, *J. Chem. Phys.* **79**, 5200 (1983).

8. J. F. Ready, *Effects of High-Power Laser Radiation.* Academic Press, New York, 1971.

9. J. L. Brand and S. M. George, *Surf. Sci.* **167**, 341 (1986).

10. K. Domen and T. J. Chuang, *Phys. Rev. Lett.* **59**, 1484 (1987).

11. M. Karas, D. Bachmann, and F. Hillenkamp, *Anal. Chem.* **57**, 2935 (1985).

12. A. T. Hsu and A. G. Marshall, *Anal. Chem.* **60**, 932 (1988).

13. J. Amster and J. Hemminger, private communication.

14. D. Burgess, Jr., D. A. Mantell, R. R. Cavanagh, and D. S. King, *J. Chem. Phys.* **85**, 3123 (1986).

15. J. Grotemeyer, U. Boesl, K. Walter, and E. W. Schlag, *Org. Mass Spectrom.* **21**, 645 (1986).

16. J. Grotemeyer, U. Boesl, K. Walter, and E. W. Schlag, *Org. Mass Spectrom.* **21**, 595 (1986).

17. N. Winograd, J. P. Baxter, and F. M. Kimock, *Chem. Phys. Lett.* **88**, 581 (1982).

18. J. H. Hahn, R. Zenobi, and R. N. Zare, *J. Am. Chem. Soc.* **109**, 2842 (1987).

19. F. Engelke, J. H. Hahn, W. Henke, and R. N. Zare, *Anal. Chem.* **59**, 909 (1987).

20. D. M. Lubman, *Anal. Chem.* **59**, 31A (1987).

21. C. H. Becker and K. T. Gillen, *Anal. Chem.* **56**, 1671 (1984).

22. D. H. Levy, *Science* **214**, 263 (1981).

23. R. Tembreull and D. M. Lubman, *Anal. Chem.* **59**, 1003 (1987).

24. R. Tembreull and D. M. Lubman, *Anal. Chem.* **59**, 1082 (1987).

25. R. Tembreull and D. M. Lubman, *Appl. Spectrosc.* **41**, 431 (1987).

26. W. C. Wiley and I. H. McLaren, *Rev. Sci. Instrum.* **26,** 1150 (1955).

27. P. Arrowsmith, M. S. de Vries, H. Hunziker, and H. R. Wendt, in press.

28. J. W. Hager and S. C. Wallace, *Anal. Chem.* **60,** 5 (1988).

29. Y. Tomioka, H. Abe, N. Mikami, and M. Ito, *J. Phys. Chem.* **88,** 2263 (1984).

30. P. M. Johnson and C. E. Otis, *Annu. Rev. Phys. Chem.* **32,** 139 (1981).

31. D. Briggs and M. P. Shea (eds.), *Practical Surface Analysis by Auger X-Ray Photoelectron Spectroscopy,* John Wiley, New York, 1983.

32. K.F.J. Heinrich, *Electron Beam X-Ray Microanalysis.* van Nostrand Reinhold, New York, 1981.

33. R. F. Willis, A. A. Lucas, and G. D. Mahan, in *The Chemical Physics of Solid Surfaces and Heterogeneous Catalysis,* Vol. 2, edited by D. A. King and D. P. Woodruff. 1983, pp. 59–164.

34. J. H. van der Maas, *Basic Infrared Spectroscopy.* Heyden & Son, 1972.

35. J. B. Pendry, *Low Energy Electron Diffraction Application to Determination of Surface Structure.* Academic Press, New York, 1974.

36. Wai-Kan Chu, J. W. Mayer, and M. A. Nicolet (eds.), *Backscattering Spectrometry.* Academic Press, New York, 1978.

37. A. Benninghoven (ed.), *International Conference on Ion Formation from Organic Solids.* Springer-Verlag, New York, 1986.

18

Multiphoton Ionization Spectroscopy of Biologically Related Molecules in a Supersonic Molecular Beam

THOMAS R. RIZZO and DONALD H. LEVY

Understanding how biological molecules interact with light is of central importance in many biochemical systems.[1-3] Detailed investigations of the excited electronic states of biological molecules are essential to developing fundamental descriptions of photobiological processes and providing new analytical tools to probe molecular structure or assist in molecular identification. Although spectroscopic studies of *small molecules in the gas phase* can provide information on their electronic structure, vibrational frequencies, directions of their electronic transition moments, and, in many cases, their precise geometry,[4] broadening of the features in the electronic spectra of *large molecules in the condensed phase* severely limits the amount of spectral information available. In fluid solution, electronic spectra of polyatomic molecules are typically broad and structureless due to the superposition of transitions from a large number of thermally populated states and to interactions with the solvent. Studies performed in the environment of low-temperature glasses are free from the complications due to thermal congestion, however, the microscopic heterogeneity of the amorphous media causes considerable spectral broadening.[5] Electronic spectra in crystals at liquid helium temperature can in many instances become quite sharp and provide detailed electronic structure information, although the effect of crystal packing forces on the molecular conformation and interactions of the molecule with crystal lattice vibrations add additional complications.[6-8] If, however, large molecules are studied in the gas phase in the absence of solvent, the degree of detail obtainable from spectroscopic studies can be similar to that for small volatile molecules.[9] As this chapter describes, it is now becoming possible to extract detailed information about the electronic structure and excited state photophysics of biologically related molecules by performing spectroscopic studies in the cold isolated environment of a supersonic molecular beam.[10,11]

Since the early work of Kantrowitz and Grey[12] and Kistiakowsky and Slichter,[13] supersonic molecular beams have been used extensively both for investigations of chemical reaction dynamics[14] and spectroscopic studies of small molecules,[9] and over the past 10 years this approach has revolutionized the field of molecular electronic spectroscopy. The advantages of this technique arise from the cooling that results when a gas expands into a vacuum through a hole that is large with respect to the mean free path of the gas. The expansion process converts the random kinetic energy of the gas to directed mass motion and results in a narrowing of the molecular velocity distribution. Using a monatomic gas, temperatures of 1 K can routinely be achieved.[9] Since the cooling process is much less efficient with polyatomic molecules, one typically "seeds" a small percentage of the polyatomic in a high pressure of monatomic carrier. At sufficiently low concentrations, the velocity of the larger molecule is accelerated to that of the carrier and experiences a similar degree of cooling. In addition, the internal degrees of freedom of the molecule begin to come to equilibrium with the translational degrees of freedom through collisions in the early part of the expansion. Temperatures associated with the rotational and vibrational motions are typically 0.5 K and 20–50 K, respectively,[9] and it is the combination of these low internal temperatures and the isolated environment of the molecular beam that has important consequences for molecular spectroscopy. The cooling in the supersonic expansion simplifies the spectrum by putting almost all the molecules into their lowest vibrational state and lowest few rotational states, and the collisionless environment of the molecular beam removes spectral broadening due to solvent interactions.[9,15]

Several types of information can result from studies of the electronic spectroscopy of biologically related molecules in the cold, isolated environment of a supersonic molecular beam. Splittings of the sharp spectral features of the ultracold molecules carry information on their stable conformations,[10] which can critically test the predictions of molecular structure calculations.[16–18] The sensitivity of the spectral features of a particular chromophore to its local chemical environment[10] could possibly be exploited to detect specific types of nearest-neighbor interactions in small biopolymers. One can imagine developing a data base of spectral shifts of a spectroscopic probe due to nearest-neighbor residues that would then be used to help determine primary structure in small peptides or nucleic acids. For molecules the size of individual nucleotide bases or amino acids, rotationally resolved spectra can determine the precise molecular geometry and the direction of the electronic transition moment,[19,20] information that is important for understanding structure-activity relationships in pharmacologically active small molecules[16] and for calculations of circular dichroism.[1,21] Highly resolved spectroscopic data can also reveal underlying dynamical processes such as electronic energy transfer,[1] intramolecular proton[22,23] or electron migration,[22,24] and exciplex formation in the isolated molecule,[25,26] and these can be compared with similar processes in the condensed phase to elucidate the role of solvent interactions.

Although the use of supersonic molecular beams has revolutionized the way people do molecular spectroscopy,[27,28] this appraoch is still somewhat limited in its scope by the requirement that a molecule possess sufficient vapor pressure to be entrained in a high pressure of carrier gas and expanded into a vacuum chamber in high enough concentrations to permit spectroscopic detection. Unfortunately, most biologically related molecules at room temperature do not meet this requirement and when heated to increase their vapor pressure they thermally decompose. Until recently, this entire class of molecules has not been available for study by this powerful spectroscopic technique.[10,11]

The problem of nonvolatility of biological compounds is one that analytical chemists have dealt with for some time. For the past 10 years an intense amount of investigation has been aimed at extending mass spectrometric techniques to nonvolatile biologically related molecules,[29,30] and much of this effort has been focused toward using mass spectrometers as detectors for liquid chromatographs. The basic problems encountered in liquid chromatography/mass spectrometry (LC/MS), those of sample volatilization and solvent removal, are quite similar to those encountered in producing molecular beams of isolated biological molecules. Several recent articles review the variety of techniques that have been applied to this problem.[31,32] The experimental requirements for forming molecular beams of nonvolatile substances are somewhat more stringent than those of mass spectrometry, however. The sample must not only be volatilized, but entrained in a carrier gas and expanded into a vacuum chamber without undergoing condensation or thermal decomposition. Residual traces of solvent used to transport the sample will cause large clusters to form in the expansion process, and hence must be completely removed.

In our work at the University of Chicago, we discovered that aromatic amino acids (whose spectroscopy we were particularly interested in) could be volatilized and injected into a supersonic expansion using relatively simple thermal techniques if care was taken in the manner in which the heat is applied. Although this did not turn out to be the generally applicable technique we sought, it worked well for the individual amino acids and their derivatives that were of immediate interest to us. This chapter summarizes the results of our spectroscopic studies of the amino acid tryptophan and some tryptophan analogs in the cold isolated environment of a supersonic molecular beam using laser and mass spectrometric techniques.[10,11,33,34]

Because of the possibility that these large, thermally labile molecules may fragment during the volatilization process, care must be taken to select appropriate spectroscopic techniques for measuring the electronic transitions. Laser-induced fluorescence has proven to be a sensitive technique for monitoring the electronic absorption spectrum of jet-cooled molecules.[9] Here a molecule is excited from its ground electronic state to its first or possibly higher excited states and the fluorescence emitted back down to the ground state is monitored. Although this approach provides the easiest means of detecting transitions to an excited electronic state, the possibility of fragmentation during volatilization casts uncertainty as to the identity of the absorbing and emitting species. Another sensitive and convenient method for detecting the absorption of radiation is resonantly enhanced two-photon ionization (R2PI).[35,36] In this approach, one photon excites the molecule to its excited electronic state in the same manner as laser-induced fluorescence, however a second photon, either from the same dye laser or another one, further excites the neutral molecule into its ionization continuum, forming the molecular ion. The ion can then be mass analyzed and detected with almost unit efficiency. The ionization efficiency, and hence the ion signal at the parent mass of the sample, is greatly enhanced when the frequency of the first photon is tuned into resonance with an electronic transition, and a spectrum is obtained by monitoring the parent ion mass as a function of excitation laser frequency. Resonantly enhanced two-photon ionization has several advantages over fluorescence detection: (1) the mass identification of the absorbing molecule helps verify that the spectrum of the intended molecule is being observed and not simply an impurity or a fragmentation product formed during volatilization; and (2) the molecule need not have a very high fluorescence yield since the ionization process can compete with nonradiative processes much more efficiently than radiative decay. Using R2PI, the conditions of the molecular beam source that avoid fragmentation of the sample can be determined and then laser-induced fluorescence detection techniques can be applied with a reasonable degree of confidence in the identity of the emitting species.

The purpose of this chapter is to demonstrate the advantages that supersonic molecular beam techniques and mass spectrometry can bring to spectroscopic studies of biologically related molecules. In the following section we describe the details of the experimental scheme employed in our studies of aromatic amino acids with an emphasis on our molecular beam source for nonvolatile molecules and present data to compare this source with a conventional heated continuous nozzle. The next section describes our spectroscopic studies of the aromatic amino acid tryptophan and tryptophan analogs. Finally we summarize the types of data that are available from such studies and make projections as to the directions of this field in the future.

EXPERIMENTAL APPROACH

The experimental apparatus (Fig. 18.1a) has four basic components: the vacuum chamber with a linear time-of-flight mass spectrometer, the molecular beam source for "nonvolatile" molecules, the excitation laser, and the data collection electronics. The reader is referred to several recent publications that describe the details of this apparatus.[10,37] Since our molecular beam source for nonvolatile, thermally labile molecules is the unique aspect of our experimental approach, we will take the time to describe it in detail here. We describe the experimental parameters appropriate for producing molecular beams of tryptophan in particular, since our primary interest is in the spectroscopy of tryptophan and tryptophan-containing peptides.

Molecular Beam Source

Our pulsed molecular beam source combines a thermospray technique for sample transport with thermal desorption to introduce nonvolatile molecules into a molecular beam. The thermospray technique has gained fairly wide acceptance as one solution to the LC/MS coupling problem, and a detailed description can be found in the work of Vestal and co-workers.[38] As shown in Fig. 18.1b, our source consists of a 1-mm-i.d. pulsed gas valve that discharges the argon carrier gas at a nominal pressure of 2 atm into a 1-mm-i.d. $\times$ 9-mm-long cylindrical channel through a heated brass block. With the source outside the vacuum chamber, a dilute solution (typically $10^{-3} M$) of tryptophan in methanol is forced at 20 atm backing pressure through a heated 100-μm-i.d. stainless capillary and a 50-μm pinhole and sprayed into the channel, depositing the sample on the cylinder wall, which is kept at 170°C. After 15–20 min the solution flow is stopped, and the sample is dried at 170°C for 30 min before the source assembly is inserted into the vacuum chamber. The tem-

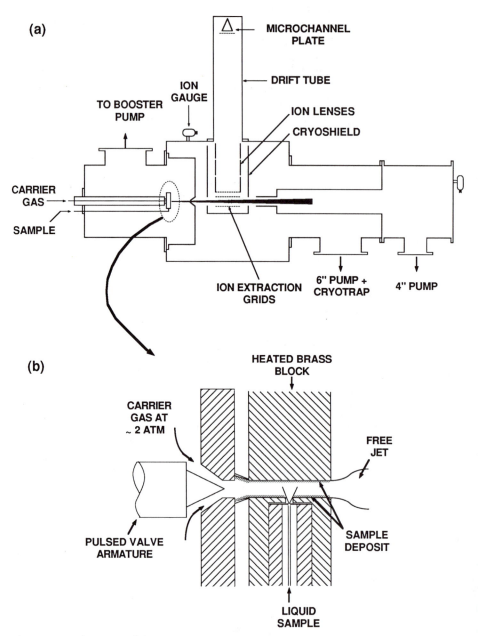

Fig. 18.1. Schematic of the apparatus used for spectroscopic studies of biological molecules. (a) Vacuum chamber with separately pumped regions for the molecular beam source, time-of-flight mass spectrometer, and beam catcher. The laser crosses the molecular beam between the first two extraction grids of the mass spectrometer. (b) Pulsed molecular beam source for thermally labile molecules (see text for description).

perature of the brass block is then raised to 230°C, and a free jet is produced by pulsing argon or helium over the tryptophan on the hot wall. The free jet is skimmed to form a molecular beam, and the resonantly enhanced two-photon ionization spectrum is taken in the time-of-flight mass spectrometer. Using typical liquid flow rates of 1–2 mL/min during sample deposition, the tryptophan sample typically lasts for 1–3 hr before needing to be replenished.

Digital delay generators determine the relative timing of the dye laser pulses and molecular beam pulses, and by scanning the 6-nsec laser pulse across the temporal profile of the gas pulse (0.3–1 msec duration) we can map out the relative concentration and temperature of our sample molecules. The concentration of tryptophan monotonically decreases from the front part of the helium pulse, whereas the relative temperature of the tryptophan molecules, as indicated by the rotational width of the electronic transitions, decreases rapidly at first, remains relatively constant, and then begins to increase again. We propose the following mechanism for the entrainment of the sample into the beam: while the valve is closed, tryptophan evaporates from the hot surface and fills the 1-mm channel with vapor (see Fig. 18.1b). When the valve opens, the shock front of the carrier gas pulse moves through the channel, and some tryptophan vapor manages to penetrate the shock structure. Tryptophan in the front part of the pulse remains warm due to collisions with the background gas, whereas that which penetrates farther back into the pulse is efficiently cooled. As the valve begins to close, the effective nozzle diameter decreases, and the cooling becomes less effective.

For some of our experiments on tryptophan analogs (e.g., tryptamine and indole-3-proprionic acid),[34] we found the samples to be thermally stable, and in this case we use a conventional heated continuous nozzle. The sample is mounted several centimeters behind a 100-μm orifice and the whole assembly is wrapped with heating tape and heated. A thermocouple monitors the temperature of the assembly. The amino acids tryptophan, tyrosine, and phenylalanine were not sufficiently stable in this continuous source, even at temperatures identical to those that were used in the pulsed source. A comparison of the mass spectra using the continuous and pulsed sources (shown below) dramatically emphasizes the difference between them.

Comparison of the Pulsed and Continuous Molecular Beam Sources

Figure 18.2a shows a time-of-flight mass spectrum of tryptophan that results from nonresonantly photoionizing the contents of the molecular beam that is formed using the pulsed source described in the previous section. The ion signal at the parent mass of tryptophan (204) dominates the mass spectrum, with a second smaller peak at mass 130 resulting from breakage of the $C_\alpha - C_\beta$ bond (see inset Fig. 18.2a). The mass 130 ion comes from photofragmentation of the tryptophan parent ion rather than from thermal decomposition in the beam source, since the mass 130/204 ratio is strongly dependent on the laser power and can be eliminated at sufficiently low power. In an earlier publication we described diagnostic experiments that use isotopic substitution to verify the identity of these ions.[33]

To demonstrate the unique performance of our pulsed molecular beam source, we show in Fig. 18.2b a time-of-flight mass spectrum of tryptophan using a con-

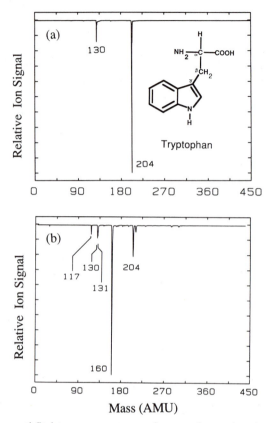

Fig. 18.2. (a) Time-of-flight mass spectrum of tryptophan using the pulsed molecular beam source of Fig. 18.1b. The molecules are nonresonantly photoionized at 290 nm. The source temperature is 230°C and the carrier gas is argon at a backing pressure of 2.3 atm. Mass 204 is the mass of the tryptophan parent ion and 130 is the mass of the fragment produced by breaking the $C_\alpha-C_\beta$ bond (see inset). (b) Time-of-flight mass spectrum of a tryptophan sample using a conventional heated continuous nozzle. The sample and nozzle temperatures are 245° and 265°C, respectively. The carrier gas is argon at a backing pressure of 1.5 atm. The sample is mounted approximately 5 cm upstream from the pinhole.

ventional heated continuous nozzle. A small amount of ion signal appears at the parent mass of tryptophan (204), but the majority of the signal appears at mass 160, the mass of decarboxylated tryptophan. In addition to the mass 130 photofragment, there are peaks at 131, the mass of 3-methylindole, and at 117, the mass of indole. These molecular fragments seem to be characteristic of the thermal decomposition of tryptophan, but none of them appears in the mass spectrum using our pulsed source. Using the continuous source, if we start at 25°C and slowly raise the temperature, we observe no significant ion signal until ~230°C at which time the masses of the thermal decomposition products appear along with a small amount of tryptophan. With time, the tryptophan peak disappears and only the thermal decomposition products are observed.

The significant difference between the pulsed source and a conventional heated nozzle is the time that the tryptophan remains at high temperature *in the gas phase.* In the continuous source, the flow velocity of the carrier gas over the sample is quite slow, and it can take up to a few seconds (assuming a 100-μm pinhole and a temperature of 230°C) for a tryptophan molecule to find its way to the pinhole and into the vacuum chamber. In the pulsed source, the high flow velocity of the carrier gas through the 1-mm orifice rapidly transports the tryptophan vapor into the vacuum chamber. On the basis of the mass spectra of Fig. 18.2 it appears that tryptophan decomposes much more rapidly in the vapor phase than it does in the solid. By minimizing the time that tryptophan remains in the gas phase before expansion into the vacuum chamber, we effectively avoid thermal decomposition. We have observed similar behavior with other aromatic amino acids in our beam source.

SPECTROSCOPIC STUDIES OF AROMATIC AMINO ACIDS

Motivation

Our primary interest in the spectroscopy of the aromatic amino acids, particularly tryptophan, has been stimulated by the important role that this chromophore plays in condensed-phase spectroscopic studies of proteins.[2,39] The large oscillator strength and high fluorescence quantum yield of the indole chromophore cause tryptophan to dominate the near-ultraviolet absorption and fluorescence of many proteins,[1] and its electronic spectroscopy and photophysics have long been the subject of intense investigation.[40] Studies have focused on the effect of conformation and solvent on the time- and wavelength-resolved fluorescence of the tryptophyl residue to characterize its use as an optical probe of protein structure and dynamics. However, there remain serious unresolved questions regarding the nature and location of the participating electronic states, the origin of the nonexponential fluorescence decay in aqueous solution, and the detailed role of solvent on tryptophan photophysics following excitation in the near ultraviolet. Many of the problems arise from the poor spectroscopic resolution in solution and from the complex nature of solvent interactions on dynamical processes that make it difficult to distinguish the intrinsic properties of the solute molecule from those that involve the solvent.

This seemed to us to be an ideal situation in which supersonic molecular beam techniques could elucidate some of the fundamental spectroscopy and photophysics of an important condensed-phase molecule. The cooling resulting from the supersonic expansion can greatly reduce spectral congestion in the electronic spectrum and facilitate the assignment of the remaining sharp structure, and the cold, isolated environment of a molecular beam allows the study of dynamical processes intrinsic to the molecule of interest. Furthermore, van der Waals complexes formed in a supersonic expansion permit the detailed investigation of microscopic solvent effects on the spectroscopy and dynamics of isolated molecules. Hence, in an effort to elucidate the properties of the tryptophan molecule in solution, we have studied its spectroscopy and dynamics in a supersonic molecular beam.

Resonantly Enhanced Two-Photon Ionization Spectrum of Tryptophan

Figure 18.3 displays a resonantly enhanced two-photon ionization spectrum of tryptophan obtained using our pulsed molecular beam source and collecting the ion signal at mass 204 as the excitation laser wavelength is scanned. The spectra show sharp, clearly resolved features whose residual widths are due to the rotational contour of the cold molecules. We have previously described diagnostic experiments that ensure that we are observing the spectrum of cold isolated tryptophan and not hot bands, fragmentation products, or clusters of tryptophan.[33] Reference 10 contains a complete list of the measured spectral transitions. For contrast, at the top of Fig. 18.3 we show the room temperature liquid phase UV absorption spectrum of tryptophan over the same spectral region. Clearly the cold environment of the supersonic expansion reveals information which simply is not available from the condensed phase spectrum.

We assign the prominent feature at 34,873 cm^{-1} to the origin of the transition to the lowest excited singlet state of neutral tryptophan. The indole chromophore of tryptophan has two low lying singlet states, designated ^1L_a and ^1L_b,[41-45] analogous to the two lowest singlet states of naphthalene. It is generally believed that the ^1L_a

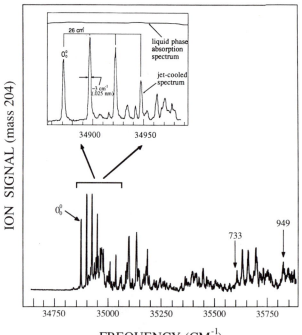

Fig. 18.3. Resonantly enhanced two-photon ionization spectrum of tryptophan in a molecular beam using the pulsed source of Fig. 18.1b. The source temperature is 230°C and the carrier gas is argon at a backing pressure is 1.4 atm. The inset shows the region of the spectrum near the origin on an expanded scale along with the liquid-phase absorption spectrum (10^{-3} M in methanol) over that same spectral region.

state is more sensitive to perturbations in the environment such as changes in solvent polarity.[46,47] Low-resolution studies of the absorption spectrum of indole and 3-methylindole by Strickland et al.[46] locate both the solvent insensitive features and those that vary strongly with the solvent, assigning them to the ^1L_b and ^1L_a bands, respectively. Hays et al[48] measure the fluorescence excitation spectrum of 3-methylindole cooled in a supersonic expansion and assign the major peaks in their spectrum to the ^1L_b band. The frequency of the 0–0 band of tryptophan that we measure is shifted only 2 cm^{-1} from the origin of the electronic transition of 3-methylindole (34,875 cm^{-1}) as measured by Hays et al.[48] On the basis of this small shift, if the assignment in 3-methylindole is correct, it appears that the upper state that we excite in tryptophan is ^1L_b.

An important question to address is whether in the gas phase tryptophan exists in the zwitterionic form, as it does in neutral solution, or in the molecular form. The small shift of the tryptophan origin from the origin of 3-methylindole as measured by Hays et al.[48] suggests that we are observing the molecular form of tryptophan rather than the zwitterion, which should show a substantially larger spectral shift. In the absence of solvent to stabilize the zwitterion, it is not surprising that the molecular form should be the predominant species in the molecular beam. This conclusion is consistent with both experimental[50] and theoretical[51] work on the amino acid glycine, which indicates that the molecular form is more stable than the zwitterion in the gas phase.

To the blue of the tryptophan origin (Fig. 18.3) there is a nearly harmonic 26 cm^{-1} vibrational progression followed by a few more irregularly spaced peaks. The intensity pattern of the low-frequency progression indicates that tryptophan undergoes a significant geometry change in the coordinate of this vibration upon electronic excitation. The absence of other extensive vibrational progressions indicates that the potential surface of the excited electronic state is similar to that of the ground state in the Franck–Condon region. None of the low-frequency structure in the tryptophan spectrum appears in the spectrum of indole[47,49] or 3-methylindole[48,49] indicating that these vibrations involve the amino acid part of the molecule. Higher frequency vibrational features characteristic of indole ring vibrations[46−49,52,53] do appear at 733 and 949 cm^{-1} from the origin as seen in Fig. 18.3, although in the tryptophan spectrum they show a distinctive progression of combination bands with the 26 cm^{-1} vibration that is absent in the spectra of the simpler indole derivatives.

Ground State Conformers of Tryptophan and Tryptophan Analogs

A prime motivation for studies of gas-phase tryptophan is to investigate the sensitivity of the electronic spectrum to different conformations of the molecule, because the identification of noninterconverting conformers of tryptophan is important for understanding its photophysics in solution. Individual tryptophan molecules in aqueous solution exhibit a nonexponential fluorescence decay[54−57] making time-resolved protein fluorescence, which is used as a probe of protein conformation and dynamics, difficult to interpret. Several models have been proposed to describe the fluorescence decay of tryptophan and other indole derivatives that presume that the excited molecule has a small number of stable conformers that do not interconvert during the fluorescence lifetime.[58] Until recently, however, there has been no direct evidence for the existence of such excited state conformers.

The use of a supersonic expansion to internally cool large molecules is a sensitive means of identifying the effects of conformation on their electronic spectra inasmuch as the sharpness of individual vibrational transitions permits detection of subtle spectral shifts.[9,27,59,60] In a study of the fluorescence excitation spectra of jet-cooled alkylbenzenes, Hopkins et al.[60] clearly observe effects of the conformation of the alkyl side chain on the electronic spectrum of the phenyl ring. They determine that conformations about certain bonds in the alkyl side chain that cause it to lie either toward or away from the phenyl ring exhibit a significant shift of the origin of the electronic transition, whereas conformations about other bonds have no such effect. They observe a simple doubling of the spectral features in those alkylbenzenes that have chains long enough to fold back on the ring (with little further splitting due to other conformations), and no splitting in molecules with side chains too short to fold back on the ring. Recently Bernstein and co-workers have used supersonic jet spectroscopy to make detailed studies of the conformations of several alkyl benzenes.[61]

In molecules that have a congested electronic spectrum, such as tryptophan, it is more difficult to assign features in the spectrum due to different conformations of the molecule, however in an earlier publication we described a method to make such assignments.[33] This method depends on the different response of spectral features due to different conformers to power saturation. If two lines in the spectrum have different intensities, there are two possible reasons for the intensity difference. The two lines could involve transitions from a common initial state, and, in this case, the intensity difference must be due to different transition strengths, (i.e., different Franck–Condon factors). If this is the case, as the laser power is increased the two transitions will begin to saturate, but since they have different transition strengths, the stronger line will saturate first, and the relative intensity of the weaker line will increase until the two lines have the same intensity. The second possibility is that the two lines involve transitions originating in different initial states. In this case the difference in intensity could arise either from a difference in transition strength or from a difference in population of the initial levels. If the transition strengths are different, an increase in laser power will again cause the weaker line to increase in relative intensity, but in the limit of total saturation the two features will still have unequal intensity because of the unequal populations in the two states. In the special case of two transitions with the same transition strength originating in two different initial levels, the relative intensity of the two features will not depend on laser power. Thus, power saturation measurements can be used to identify spectral features that have different initial states. If, having identified such features, the possibility of these features being due to vibrational hot bands can be ruled out, the features can then be assigned to different molecular conformers.

Figure 18.4b shows a multiphoton ionization spectrum of tryptophan at ~ 100 times the laser power density of the spectrum of Fig. 18.4a. Notice that certain peaks grow in and approach a common intensity as the power density is increased. Analysis of these peaks reveals that they form progressions of ~ 26 cm^{-1} built on either the lowest frequency feature (the electronic origin) or on other weak low-frequency vibrations. As the power density is increased and the stronger transitions

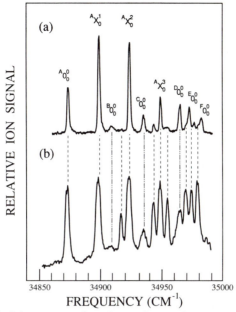

Fig. 18.4. A portion of the resonantly enhanced two-photon ionization spectrum of tryptophan taken at different laser power densities. The spectrum in (b) is taken at ~ 100 times the power density of that in (a). Other conditions are the same as those of Fig. 18.3. The origin of transitions due to different conformers are denoted as $^A0_0^0 - {}^F0_0^0$. (The notation for the spectral assignments is explained in the text.)

begin to saturate, the weaker features gain in relative intensity. However, a few of the small peaks in the spectra of Fig. 18.4 do not significantly change in intensity relative to the largest peak as we increase the power density. Because their relative intensity does not change at power densities that are sufficient to saturate the more intense transition, their Franck–Condon factors must be nearly the same and the intensity difference reflects a difference in initial state population. Having previously ruled out the possibility of hot bands,[34] the transitions must arise from two different conformers of the molecule (i.e., different initial states separated by a barrier). Careful analysis of the relative power dependence of the spectrum leads us to assign several of the features in Fig. 18.4 to transitions from different conformers of tryptophan.

In referring to spectral features due to different conformers we will use the following notation. The origin band of the most intense conformer will be labeled $^A0_0^0$, the origin of the next most intense conformer will be labeled $^B0_0^0$, etc. The members of the one identified low-frequency vibrational progression built on $^A0_0^0$ will be labeled $^AX_0^1$, $^AX_0^2$, etc. Unassigned bands or members of vibrational progressions with uncertain origins will be labeled 1,2, etc. in order of increasing frequency.

There seem to be six different conformers of tryptophan denoted as A–F in Fig. 18.4. Only the lowest energy conformer whose origin is the feature at 34,873 cm^{-1} shows a strong low-frequency vibrational progression, designated $^AX_0^1$–$^AX_0^3$. Although these assignments directly identify different conformers in the ground electronic state of tryptophan, the similarity of the ground and excited state poten-

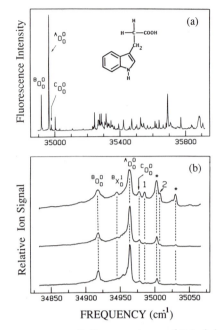

Fig. 18.5. (a) The fluorescence excitation spectrum of 3-indolepropionic acid cooled in a supersonic free jet. The carrier gas is helium at a stagnation pressure of 7.8 atm and the nozzle diameter is 0.1 mm. The origin of transitions due to different conformers are marked as $^A0_0^0$–$^C0_0^0$. (b) The two-photon ionization spectra of 3-indolepropionic acid near the origin taken at different laser power densities. The power densities were top: 6.7 mJ/cm^2/pulse; middle: 0.14 mJ/cm^2/pulse; and bottom: 0.052 mJ/cm^2/pulse.

tial surfaces (as evidenced by the lack of many extensive vibrational progressions) implies the existence of equivalent conformers in the excited state. We will come back to the issue of excited state conformers shortly.

The power dependence of the tryptophan spectrum provides no direct information about the specific configuration of the different conformers, however it is likely that the conformations that show significant spectral shifts are different rotamers about the C_α–C_β and C_β–C_γ bonds, since it is these coordinates that bring the acid or amine group closer or further from the indole ring. Using a continuous molecular beam source, we measured the electronic spectrum of the indole derivatives indole-3-propionic acid and tryptamine; the results of this work, as described below, have a bearing on the question of conformers of tryptophan.

3-Indolepropionic Acid

Figure 18.5a shows the fluorescence excitation spectrum of 3-indolepropionic acid in the spectral range 34,900–35,900 cm^{-1}.[34] The fluorescence excitation spectrum is essentially identical to the two-photon ionization spectrum except that the features are slightly better resolved. (This is a result of the greater cooling due to the increased backing pressure in the supersonic expansion that we can use when we take fluorescence spectra.) We can identify different conformers by observing the two-photon ionization spectrum at different laser power densities as shown in Fig.

18.5b. We assign the features at 34,964 and 34,917 cm^{-1} as origins of two different conformers of the molecule. The intensity of band $^C0_0^0$ varies only slightly with laser power, and we believe that it is the origin of a third conformer. All other features in this region have a large change in relative intensity as a function of laser power and are assigned to vibrational progressions. At high laser power the peaks marked with asterisks are more intense than the origin transition $^B0_0^0$, and these features must therefore be vibrational progressions built on the $^A0_0^0$ origin. Peak $^BX_0^1$ is at a lower frequency than the $^A0_0^0$ origin and must be a vibrational progression built on $^B0_0^0$. A complete table of spectral features can be found in Ref. 34.

Tryptamine

The fluorescence excitation spectrum of tryptamine cooled in a supersonic helium expansion is given in Fig. 18.6a.[34] This spectrum is dominated by an intense spectral feature at 34,918 cm^{-1}, which is assigned as the origin of one conformer of the molecule. The spectrum has extensive structure to the low-frequency side of this most intense feature, which cannot be due to vibrational progressions of low-frequency vibrational modes built on this feature since they are red shifted. The relative intensities of these lines do not change with the expansion conditions of the jet, ruling out the possibility of their being hot band transitions. The lack of laser

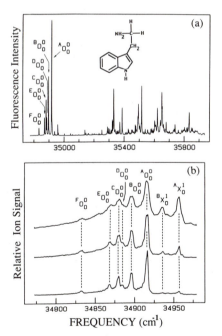

Fig. 18.6. (a) The fluorescence excitation spectrum of tryptamine cooled in a supersonic free jet. The conditions are the same as in the spectrum of Fig. 18.5. The origin of transitions due to different conformers are marked as $^A0_0^0$–$^F0_0^0$. (b) The two-photon ionization spectra of tryptamine near the origin taken at different laser power densities. The power densities were top: 6.7 mJ/cm^2/pulse; middle: 0.14 mJ/cm^2/pulse; and bottom: 0.052 mJ/cm^2/pulse.

power dependence of the relative intensities of these features, labeled $^A0_0^0$–$^F0_0^0$ in Fig. 18.6b, indicates that they are origins of different conformers of the molecule. Peaks labeled $^AX_0^1$ and $^BX_0^1$ do increase in intensity relative to $^A0_0^0$–$^F0_0^0$ as the laser power is increased, and these features may be assigned as vibrational progressions. The splittings between $^BX_0^1$ and $^B0_0^0$ and that between $^AX_0^1$ and $^A0_0^0$ are both 41 cm^{-1}, indicating this is a vibrational frequency that is common to these two conformers. A complete table of spectral features can be found in Ref. 34.

The spectra of both tryptamine and indole-3-propionic acid show evidence of several conformers, each of which is slightly shifted from the origin of 3-methylindole. It is likely that these shifts result from the direct interaction of the carboxylic acid or amine with the aromatic ring in a manner analogous to that seen in *n*-propylbenzene[60] rather than from inductive effects, since two C–C bonds insulate these groups from the indole ring. The magnitudes of the conformer shifts in the tryptophan spectrum identified by our power dependence technique are very close to the shifts of the indole-3-propionic acid and tryptamine conformers from the origin of 3-methylindole, supporting the notion that it is also the direct interaction of the acid or amine with the ring that causes the shifts in tryptophan. If the spectral shifts in the indole derivatives are indeed the result of direct interaction of the side chain substituents with the indole ring, different rotamers of tryptophan about the C_α–C_β and C_β–C_γ bonds should also show different spectral shifts, insofar as the proximity of the acid or amine group to the ring varies widely between these rotamers. It is therefore tempting to assign the conformer peaks in the tryptophan spectrum to different rotamers of the tryptophan molecule about the C_α–C_β and C_β–C_γ bonds, however there is some indication that more subtle conformational changes of the side chain substituents may be responsible for the conformer shifts in tryptamine.[62] The only way to unambiguously assign a particular spectral feature to a specific molecular conformation is to use a high-resolution laser and measure the spacing between individual rotational levels in the electronic transition.[19,20] This has been done for tryptamine,[20,62] and work is currently underway to extend these studies to tryptophan itself.[63]

Excited State Conformers

By examining the dispersed fluorescence subsequent to electronic excitation of a particular tryptophan conformer we can test our assignments of the features in the excitation spectrum, since the dispersed fluorescence of different conformers is likely to reveal different ground state vibrational frequencies and different emission intensity patterns. In addition, the observation of different emission spectra from different conformers would indicate that the conformers do not interconvert in the excited state during the fluorescence lifetime.

Figure 18.7 shows dispersed fluorescence spectra from exciting each of the four members of the 26 cm^{-1} excited state vibrational progression of conformer A.[11] The strong progression in the excitation spectrum of conformer A (Fig. 18.3) also appears in the emission spectra of Fig. 18.7. Because the vibrational progression results from a shift in the ground and excited potential surfaces, a similar pattern is expected both in the excitation spectrum and in emission. The spacings of the spectral features in each of the curves of Fig. 18.7 are the same and represent the

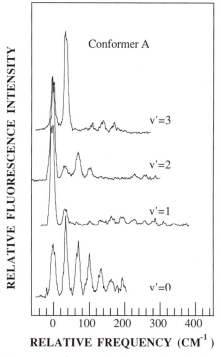

Fig. 18.7.   Dispersed fluorescence spectra of tryptophan conformer A subsequent to exciting each of the members of the 26 cm^{-1} vibrational progression in the excitation spectrum. The label v' indicates the number of vibrational quanta excited in this low-frequency vibration in the excited state. The monochromator slits are 0.25 mm producing a resolution of ~12 cm^{-1}.

vibrational frequencies of conformer A in the ground state. The spacing of the first member (~36 cm^{-1}) is consistent with the observation of the small hot band in the excitation spectrum.[10] The fact that exciting different members of the 26 cm^{-1} progression in the excitation spectrum reveals identical ground state vibrational frequencies confirms our original assignment of the excited state progression. The intensity patterns in the emission spectra of Fig. 18.7 are the result of variations in the Franck–Condon factors that occur in the transitions between two nearly harmonic potentials. Simple calculations of the Franck–Condon overlap integrals between two harmonic oscillators predict such fluctuations with remarkable accuracy.*

Figure 18.8 shows the dispersed emission spectra of conformers C–E following excitation of their respective electronic origins.[11] The emission from conformers C–E is clearly different from that of conformer A, both in the ground state vibrational

---

*The program for calculating Franck–Condon factors obtained from J. C. Light and R. B. Walker uses harmonic oscillators and calculates the overlap of the Hermite polynomials. The use of harmonic wavefunctions should be a good approximation in this case, because the 26 cm^{-1} vibration appears to be nearly harmonic. The dispersed fluorescence studies of Ref. 11 and the spacing of the small hot bands in the excitation spectra of Ref. 10 suggest that the corresponding vibration in the ground electronic state of tryptophan has a frequency of 36 cm^{-1} and is also nearly harmonic.

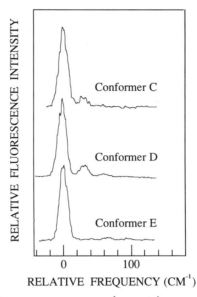

Fig. 18.8. Dispersed fluorescence spectra of tryptophan conformers C–E subsequent to exciting their respective electronic origins. The monochromator slits are 0.25 mm producing a resolution of ∼12 cm⁻¹.

frequencies and in the intensity patterns. The differences between conformers C–E are more subtle but are clearly discernible. The observation that the $\Delta v = 0$ transitions for conformers C–E carry almost all the transition intensity confirms our assignments of these features as origins of different conformers, since if they were higher members a vibrational progression in the excited state, the emission spectra would show strong $\Delta v \neq 0$ transitions. The consistency of the assignments from the dispersed fluorescence spectra with those made on the basis of the power dependence of the excitation spectrum substantiates the latter approach to revealing the spectral contributions from different conformers.[10] The stark contrast between the strong vibrational progression in the excitation spectrum and emission spectrum of conformer A (Figs. 18.3 and 18.7) and the absence of such a progression in the spectra of the other tryptophan conformers (Figs. 18.3 and 18.8) or in tryptamine or indole-3-propionic acid signifies a difference in the shape of the potential surface in this vibrational coordinate for this particular conformer. In addition, the observation of different dispersed fluorescence spectra from excitation of the origins of different tryptophan conformers indicates that the conformers do not interconvert in the excited state during the fluorescence lifetime.

The identification of noninterconverting rotamers of tryptophan in the excited electronic state that have different electronic properties is central to the rotamer model of Petrich et al.[58], which accounts for the nonexponential fluorescence decay of tryptophan and many indole derivatives in solution. They propose that upon excitation in the near ultraviolet, tryptophan undergoes charge transfer from the indole ring to an electrophilic group in the amino acid part of the molecule. The difference in the proximity of the electron acceptor to the ring accounts for the different nonradiative lifetimes of the rotamers. If an interaction with the amine or carboxylic acid of tryptophan is sufficient to shift the electronic spectrum of the indole chromophore as we have observed, it is likely that such an interaction can also account for different intramolecular charge transfer rates for different rotamers in solution. Our observations of different stable conformations of tryptophan in the gas phase and the spectral shifts caused by these conformations support the rotamer model of Petrich et al.[58] of tryptophan photophysics in solution.

Solvent Effects on Conformation: An Example

An important aspect of the study of molecular conformations is understanding the role that the solvent plays on stabilizing one particular conformer over another. One way to begin to achieve such an understanding is to observe the effect of forming van der Waals clusters between a molecule that can have several stable conformations and one or several solvent molecules in a molecular beam. We have studied the electronic spectroscopy of such a complex between tryptamine and one methanol molecule and have observed a particularly striking result.[34] Figure 18.9 shows the two-photon ionization spectrum taken by monitoring the mass of the tryptamine–methanol complex, which is produced in an expansion containing both tryptamine and methanol in a helium carrier gas. Unlike the spectrum of the parent molecule tryptamine, which shows a rich structure near the origin because of multiple conformers, the spectrum of the tryptamine–methanol complex has only a single line in the region near the origin. Presumably the addition of a single solvent molecule stablizes one conformer at the expense of the others. This emphasizes the importance of including solvent interactions in any attempt to model the behavior of these molecules in solution.

SUMMARY AND FUTURE DIRECTIONS

We have attempted to provide a clear example of how laser, molecular beam, and mass spectrometric techniques can be applied to spectroscopic studies of small biologically related molecules. We have demonstrated how the power saturation of resonantly enhanced two-photon ionization spectra can help distinguish features in

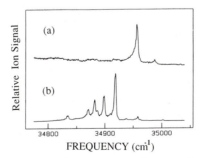

Fig. 18.9. The two-photon ionization spectrum of tryptamine seeded (a) in a helium–methanol mixture monitoring the mass of the tryptamine–methanol complex, and (b) in pure helium monitoring the mass of tryptamine.

the electronic spectrum arising from different conformations of a molecule, and combined with high-resolution studies underway at the University of Chicago,[19,20] this approach should yield precise geometrical information on small biologically important molecules. This type of data will provide important tests for calculations of molecular structure which are typically performed for isolated molecules. The identification of spectral features of specific molecular conformers that do not interconvert under our experimental conditions opens up the possibility of exciting a specific molecular conformers and exclusively studying their photophysical behavior. We have shown one example of this in the case of tryptophan, but the applications could potentially be much wider. For example, one might look at electron transfer processes in the gas phase, and study the effect of conformation on electron transfer rates. It will be quite informative to compare the photophysics in the gas phase with that observed in the condensed phase to help determine the detailed role of solvent. We have shown in the case of tryptamine how the addition of a single solvent molecule can change the distribution of stable conformations quite drastically. Similar effects may be observed on photophysical processes that depend strongly on molecular conformation.

The practical implications of being able to obtain highly resolved electronic spectra of biologically related molecules in supersonic molecular beams are intriguing. The sensitivity of the spectral features of a particular chromophore to small changes in its local environment may be exploited in providing nearest-neighbor information in small biopolymers and perhaps ever permit the selective photofragmentation of such molecules.

There are several directions for future research that we feel will be very fruitful. The successful design of molecular beam sources that are generally applicable to large classes of nonvolatile thermally labile molecules will clearly be the most important development in this field and will open the door even wider for imaginative applications of laser spectroscopy to study biological chromophores. Much of the progress may come in parallel to the development of successful LC/MS interfaces using techniques such as supercritical fluids, laser desorption, or electrospray. The continued application of ultrahigh resolution techniques to determining geometries of individual molecular conformers is important both for testing calculated geometries as well as for examining the geometry dependence of photophysical processes. Finally, studies of biologically related molecules in the gas phase with one or several solvent molecules attached should provide new insight into the role solvent plays in stabilizing specific molecular conformations as well as promoting or inhibiting certain photophysical processes.

ACKNOWLEDGMENT

The authors gratefully acknowledge the support of this work by the National Science Foundation through Grant No. CHE-8311971.

REFERENCES

1. C. R. Cantor and P. R. Schimmel, *Biophysical Chemistry,* Vol. II, Chaps. 7 and 8. Freeman, San Francisco, 1980.

2. I. Weinryb and R. F. Steiner, in *Excited States of Proteins and Nucleic Acids,* Chap. 5, edited by R. F. Steiner and I. Weinryb. Plenum Press, New York, 1971.

3. See, for example, *Excited States of Biological Molecules,* edited by J. B. Birks. Wiley-Interscience, London, 1976.

4. G. Herzberg, *Molecular Spectra and Molecular Structure, III. Electronic Spectra and Electronic Structure of Polyatomic Molecules.* Van Nostrand Reinhold, New York, 1966.

5. R. M. Hochstrasser, *Acc. Chem. Res.* **1,** 266 (1968).

6. D. S. McClure, in *Solid State Physics,* edited by F. Seitz and D. Turnbull. Academic Press, New York, 1959; R. M. Hochstrasser and G. J. Small, *J. Chem. Phys.* **48,** 3612 (1968).

7. T. J. Aartssma and D. S. Wiersma, *Phys. Rev. Lett.* **36,** 1360 (1976); T. E. Orlowski and A. H. Zewail, *J. Chem. Phys.* **70,** 1390 (1979).

8. D. Voet and A. Rich, *Prog. Nucl. Acid Res. Mol. Biol.* **10,** 183 (1970).

9. D. H. Levy, *Annu. Rev. Phys. Chem.* **31,** 197 (1980).

10. T. R. Rizzo, Y. D. Park, L. Peteanu, and D. H. Levy, *J. Chem.* **84,** 2534 (1986).

11. T. R. Rizzo, Y. D. Park, and D. H. Levy, *J. Chem. Phys.* **85,** 6945 (1986).

12. A. Kantrowitz and J. Grey, *Rev. Sci. Instrum.* **22,** 328 (1951).

13. G. B. Kistiakowsky and W. P. Slichter, *Rev. Sci. Instrum.* **22,** 333 (1951).

14. R. B. Bernstein, *Chemical Dynamics via Molecular Beam and Laser Techniques.* Oxford University Press, New York, 1982).

15. D. H. Levy, *Sci. Am.* **250,** 96 (1984).

16. See, for example, *Molecular and Quantum Pharmacology,* edited by E. D. Bergmann and B. Pullman. Reidel, Dordrecht, 1974.

17. See G. A. Segal, *Semiempirical Methods of Electronic Structure Calculations, Part A: Techniques.* Plenum, New York, 1977.

18. U. Burkert and N. L. Allinger, *Molecular Mechanics.* ACS Monograph 177, New York, 1982.

19. L. A. Phillips and D. H. Levy, *J. Chem. Phys.* **85,** 1327 (1986).

20. L. A. Phillips and D. H. Levy, *J. Phys. Chem.* **90,** 4921 (1986).

21. M. Gueron, J. Eisinger, and A. A. Lamola, in *Basic Principles in Nucleic Acid Chemistry,* Vol. 1, edited by P.O. P. Ts'o. Academic Press, New York, 1974.

22. E. M. Kosower and D. Huppert, *Annu. Rev. Phys. Chem.* **37,** 127 (1986).

23. N. P. Ernsting, *J. Phys. Chem.* **89,** 4932 (1985); M. Itoh and H. Kurokawa, *Chem. Phys. Lett.* **91,** 487 (1982); O. Cheshnovsky and S. Leutwyler, *Chem. Phys. Lett.* **121,** 1 (1985); L. Heimbrook, J. E. Kenny, B. E. Kohler, and G. W. Scott, *J. Phys. Chem.* **87,** 280 (1983).

24. D. Mauzerall and S. G. Ballard, *Annu. Rev. Phys. Chem.* **33,** 377 (1982).

25. M. Gordon and W. R. Ware (eds.), *The Exicplex.* Academic Press, New York, 1975.

26. R. Lumry and M. Hershberger, *Photochem. Photobiol.* **27,** 819 (1978).

27. R. E. Smalley, B. L. Ramakrishna, D. H. Levy, and L. Wharton, *J. Chem. Phys.* **61,** 4363 (1974); R. E. Smalley, L. Wharton, and D. H. Levy, *J. Chem. Phys.* **63,** 4977 (1975); D. H. Levy, L. Wharton, and R. E. Smalley, in *Chemical and Biochemical Applications of Lasers,* Vol. 2. Academic Press, New York, 1977, p. 1; D. H. Levy, L. Wharton, and R. E. Smalley, *Acc. Chem. Res.* **10,** 139 (1977); P.S.H. Fitch, L. Wharton, and D. H. Levy, *J. Chem. Phys.* **69,** 3424 (1978).

28. A. Amirav, U. Even, and J. Jortner, *Chem. Phys. Lett.* **67,** 9 (1979); A. Amirav, U. Even, and J. Jortner, *Opt. Commun.* **32,** 266 (1980); A. Amirav, U. Even, and J. Jortner, *Chem. Phys. Lett.* **67,** 14 (1979); A. Amirav, U. Even, and J. Jortner, *J. Chem. Phys.* **74,** 3145 (1981); A. Amirav, U. Even, and J. Jortner, *J. Chem. Phys.* **75,** 3771 (1981).

29. A. L. Burlingame, T. A. Baillie, and P. J. Derrick, *Anal. Chem.* **58,** 165R (1986).

30. A. L. Burlingame, J. O. Whitney, and D. H. Russell, *Anal. Chem.* **56,** 417R (1984).

31. P. J. Arpino and G. Guiochon, *Anal. Chem.* **51,** 682 (1979).

32. T. R. Covey, E. D. Lee, A. P. Bruins, and J. D. Henion, *Anal. Chem.* **58**, 1451A (1986).

33. T. R. Rizzo, Y. D. Park, L. Peteanu, and D. H. Levy, *J. Chem. Phys.* **83**, 4819 (1985).

34. Y. D. Park, T. R. Rizzo, L. A. Peteanu, and D. H. Levy, *J. Chem. Phys.* **84**, 6539 (1986).

35. For example, T. G. Dietz, M. A. Duncan, M. G. Liverman, and R. E. Smalley, *Chem. Phys. Lett.* **70**, 246 (1980); T. G. Dietz, M. A. Duncan, M. G. Liverman, and R. E. Smalley, *J. Chem. Phys.* **73**, 4816 (1980).

36. P. M. Johnson and C. E. Otis, *Annu. Rev. Phys. Chem.* **32**, 139 (1981); and references cited therein.

37. E. Carrasquillo M., T. Zweir, and D. H. Levy, *J. Chem. Phys.* **83**, 4990 (1985).

38. L. Yang, G. J. Fergusson, and M. L. Vestal, *Anal. Chem.* **56**, 2632 (1984); C. R. Blakely and M. L. Vestal, *Anal. Chem.* **55**, 750 (1983).

39. D. B. Wetlaufer, *Adv. Protein Chem.* **17**, 303 (1962).

40. See Refs. 1–16 of Ref. 10.

41. G. Weber, *Biochemistry* **75**, 335 (1960); B. Valeur and G. Weber, *Photochem. Photobiol.* **25**, 441 (1977).

42. M. Martinaud and A. Kadri, *Chem. Phys.* **28**, 473 (1978).

43. Y. Yamamoto and J. Tanaka, *Bull. Chem. Soc. Jpn.* **45**, 1362 (1972).

44. S. R. Meech, D. Phillips, and A. G. Lee, *Chem. Phys.* **80**, 317 (1983).

45. J. R. Platt, *J. Chem. Phys.* **19**, 101 (1951).

46. E. H. Strickland, J. Horowitz, and C. Billups, *Biochemistry* **9**, 4914 (1970); E. H. Strickland, C. Billups, and E. Kay, *Biochemistry* **11**, 3657 (1972).

47. J. Hager and S. C. Wallace, *J. Phys. Chem.* **87**, 2121 (1983).

48. T. R. Hays, W. E. Henke, H. L. Selzle, and E. W. Schlag, *Chem. Phys. Lett.* **97**, 347 (1983).

49. R. Bersohn, U. Even, and J. Jortner, *J. Chem. Phys.* **80**, 1050 (1984).

50. J. S. Gaffney, R. C. Pierce, and L. Friedman, *J. Am. Chem. Soc.* **99**, 4293 (1977); M. J. Locke and R. T. McIver, *J. Am. Chem. Soc.* **105**, 4226 (1983).

51. Y.-C. Tse, M. D. Newton, S. Vishveshwara, and J. A. Pople, *J. Am. Chem. Soc.* **100**, 4329 (1978).

52. J. M. Hollas, *Spectrochim. Acta* **19**, 752 (1963).

53. A. Mani and J. R. Lombardi, *J. Mol. Spectrosc.* **31**, 308 (1969).

54. G. R. Fleming, J. M. Morris, R. J. Robbins, G. J. Woolfe, P. J. Thistlethwaite, and G. W. Robinson, *Proc. Natl. Acad. Sci. U.S.A.* **75**, 4652 (1978); M. C. Chang, J. W. Petrich, D. B. McDonald, and G. R. Fleming, *J. Am. Chem. Soc.* **105**, 3819 (1983).

55. T. C. Werner and L. S. Forster, *Photochem. Photobiol.* **29**, 909 (1979).

56. A. G. Szabo and D. M. Rayner, *J. Am. Chem. Soc.* **102**, 554 (1980).

57. E. F. Gudgin-Templeton and W. R. Ware, *J. Phys. Chem.* **88**, 4626 (1984).

58. J. W. Petrich, M. C. Chang, D. B. McDonald, and G. R. Fleming, *J. Am. Chem. Soc.* **105**, 3824 (1983), and references cited therein.

59. D. H. Levy, *Adv. Chem. Phys.* **47**, 323 (1981).

60. J. B. Hopkins, D. E. Powers, and R. E. Smalley, *J. Chem. Phys.* **72**, 5039 (1980).

61. P. J. Breen, J. A. Warren, and E. R. Bernstein, *J. Am. Chem. Soc.* **109**, 3453 (1986); *J. Chem. Phys.* **87**, 1927 (1987); P. J. Breen, J. A. Warren, E. R. Bernstein, and J. I. Seeman, *J. Chem. Phys.* **87**, 1917 (1987); P. J. Breen, E. R. Bernstein, and J. I. Seeman, *J. Chem. Phys.* **87**, 3269 (1987).

62. L. A. Philips and D. H. Levy, *J. Chem. Phys.,* in press. **89**, 85 (1988).

63. D. H. Levy, unpublished results.

19

Two-Color Threshold Photoionization Spectroscopy of Polyatomic Molecules and Clusters

JAMES W. HAGER and STEPHEN C. WALLACE

In spite of the importance of ion chemistry in determining the outcome of micro-scopic and macroscopic molecular dynamics, much less is known about the prop-erties of ions in general, and polyatomic ions in particular, than their neutral coun-terparts. This is the result in large part of the experimental difficulties associated with preparing molecular ions in well-defined states, and subsequently probing the vibrational and electronic properties. Much of the available information regarding the electronic structure of ions has been obtained using conventional photoelectron spectroscopy.[1,2] In favorable cases, photoelectron spectra have provided details regarding electronic state ordering and spacings of a few vibrational progressions. However, complications arise for medium and large size polyatomic molecules because of the large number of vibrational degrees of freedom. The complexity of the vibrational profile and the limited resolution of photoelectron spectrometers make the assignment of the photoelectron band origin (the adiabatic ionization potential) possible for only a relatively few species. Furthermore, vibrational pro-gressions are often poorly resolved and frequency determinations have been lim-ited to an uncertainty of about 80 cm^{-1}. Consequently little detailed information can be obtained for intermediate and larger size polyatomic ions using conven-tional photoelectron spectroscopy.

When a neutral aromatic molecule in its ground electronic state absorbs a single vacuum ultraviolet photon of energy just sufficient to ionize the molecule, most often a bonding electron is liberated. This fact implies that the geometry of the ion will be considerably different from that of the neutral molecule.[1,2] This geometry change on ionization is manifested in the broad Franck–Condon envelope usually observed in the single-photon photoelectron spectra of intermediate and larger molecules. When the spectrum arises from a room temperature sample there can be a significant population of excited rotational and vibrational levels within the ground electronic state of the neutral molecule. This leads to the presence of hot band structure in the spectrum and compounds the already difficult task of spectral assignment.

The dilemma created by these problems has in large part been solved by the technique of two-color threshold photoionization spectroscopy, which has permitted the precise determination of adiabatic ionization potentials (IP_0) and vibrational frequencies of polyatomic ions.[3-5] Two-color photoionization spectroscopy is conceptually very simple and widely applicable. Laser light at a given energy ($h\nu_1$) is used to prepare the molecules in a specific vibrational level of an electronically excited state. At a prespecified time later the excited state molecules are ionized by the light from a second laser operating at $h\nu_2$. By holding $h\nu_1$ fixed and tuning $h\nu_2$ from below the ionization potential to deeper into the ionization continuum while collecting the produced ions, the absorption features of the ion can be mapped out. Alternatively, $h\nu_1$ can be tuned while $h\nu_2$ remains fixed such that $h\nu_1 + h\nu_2 > IP_0$. The resulting spectrum corresponds to the absorption spectrum of the electronically excited state of the neutral molecule.

There are several important advantages of this two-color approach compared with single-photon ionization. Two-color photoionization spectroscopy is inherently a high-resolution technique. The bandwidths of the two lasers are typically less than 0.5 cm^{-1}, so the ion absorption features are obtained with high precision and an accuracy determined by the wavelength calibration of the two lasers. Furthermore, the molecules are ionized from a selected vibrational level of an electronically excited state. The geometries of these excited electronic states are much more similar to that of the corresponding ion leading to Franck–Condon factors that strongly favor transitions with no change in vibration quantum numbers (i.e., $\Delta v = 0$ ionizing transitions).[3-5] Because of this strong propensity for $\Delta v = 0$ ionization in the two-color scheme, parent molecular ions are produced with virtually no fragmentation, with the excess energy of ionization being carried off as electron kinetic energy.[3-6]

A unique advantage of this technique is that ionization can be carried out from a variety of intermediate excited state vibrational levels, thus favoring transitions to a variety of ion vibrational levels. The ability to vary the Franck–Condon factors for ionization in this fashion provides a very powerful tool to explore regions of ion potential surfaces not accessible using conventional methods.

In spite of the fact that near threshold, two-color photoionization produces almost exclusively parent molecular ions, the mechanisms governing ion production in this energy regime can be quite different. Direct ionization and vibrational autoionization both lead to the formation of parent ions, but with much different vibrational state distributions. Two-color threshold photoionization can distinguish between these two mechanisms and provide details of the coupling between highly excited states of the neutral molecule and the ionization continuum.[7,8]

We have chosen to conduct these two-color photoionization investigations in the unique environment of a supersonic expansion. One of the most useful properties of a supersonic expansion is the cooling of the internal degrees of freedom of polyatomic molecules. Typical temperatures associated with continuous free jet expansions of intermediate size organic molecules seeded in a helium carrier gas are in the 1–5 K range for rotations and 20–50 K for vibrations.[9,10] The cooling of the molecular internal degrees of freedom concentrates population in the lowest rotational and vibrational levels of the ground electronic state, significantly simplifying the resulting electronic spectra and enhancing sensitivity. Supersonic cooling virtually eliminates vibrational hot bands and reduces rovibronic linewidths to 1–3 cm^{-1}. This can ease assignment tasks considerably.[10]

Particularly important to the work described here is the possibility of the formation of weakly bound van der Waals molecules in the expansion. Because of the very low temperatures associated with the various degrees of freedom, even van der Waals molecules between an organic molecule and a helium atom are stable.[10] Thus, it is possible to study isolated complexes between a particular chromophore and virtually any number of solvent species ranging from helium atoms to chemically relevant molecules such as water, alcohols, and amines.[11,12]

Neutral van der Waals molecules have also been shown to be excellent precursors for the preparation of solvated positive ions. If, after the formation of a weakly bound neutral complex, the chromophore is photoionized, the production of a charged solute–solvent cluster can be observed.[11,13] These ionic clusters are characterized by intermolecular forces such as charge-dipole and charge-induced dipole in addition to the dipole–dipole, dipole-induced dipole, and dispersion forces prevalent in the neutral precursors. These interactions are governed by intermolecular potentials of known functional form allowing accurate theoretical predictions of geometries, vibrational frequencies, and dissociation energies.[14] Thus, for the first time intermolecular interactions with characteristic strengths intermediate between those of neutral clusters and covalent bonds can be probed in a comprehensive fashion. Two-color photoionization studies of these species are of particular interest in understanding, on a molecular level, the effects of solvents on ionization potentials.

In this chapter, we will examine several of the ways in which two-color photoionization has contributed to the understanding of the structure and dynamics of neutrals, ions, and ionic clusters. First, the results of a detailed investigation of the aromatic molecule aniline will be presented. The photoionization spectra of aniline very clearly display the phenomena of direct ionization and vibrational autoionization. Second, the threshold photoionization of a group of molecular clusters is presented to illustrate the manifestations of van der Waals and hydrogen bonding interactions in photoionization spectra. Finally, a two-color photoionization spectroscopic exploration of several aromatic molecules with interactive side chains is discussed. This study demonstrates the way in which the side chain can act as a van der Waals partner and considerably complicate the interpretation of even jet-cooled spectra. Two-color photoionization, however, provides a tool with which to simplify and identify these intermolecular interactions.

EXPERIMENTAL DETAILS

The supersonic beam mass spectrometer used in the two-color photoionization studies has been described fully in previous publications.[7,8] Briefly, ions are produced within the isentropic core of a continuous free jet expansion in the region defined by the spatial overlap of two counterpropagating laser beams. The ions are accelerated into a differentially pumped quadrupole mass spectrometer chamber, focused, mass analyzed, and detected. An important feature of this photoionization spectrometer is that a very small accelerating voltage is applied, resulting in electric field strengths of <1 V/cm. This virtually eliminates Stark broadening of the ionization thresholds commonly observed at higher electric field strengths.

To obtain threshold photoionization spectra, the pump laser ($h\nu_1$) is fixed at the frequency of an individual electronic absorption feature and the ionizing laser ($h\nu_2$) is tuned from below the ionization onset to deeper in the ionization continuum with concurrent ion selection and detection. The resulting spectrum of $h\nu_1 + h\nu_2$ vs ion signal is referred to as a photoionization efficiency (PIE) spectrum.

TWO-COLOR PHOTOIONIZATION RESULTS AND DISCUSSION

Threshold Photoionization of Aniline

The aromatic molecule aniline has served as a prototypical test case for investigations of multiple photon photoionization of intermediate size molecules for several years. The ionization potential is at relatively low energies (~ 7.7 eV) and the first excited electronic state ($^1B_2$) is located at energies greater than IP/2. These facts alone make aniline a good candidate for studies employing one-color photoionization. Of even greater importance, however, is that the photoionization of aniline using the $^1B_2$ vibrational levels for resonant enhancement is a very high-efficiency process. Approximately 25% of the ground state neutral molecules that can absorb the first photon are photoionized at even moderate laser intensities.[15,16]

Whereas one-color multiple photon ionization yields almost exclusively information regarding excited states of the neutral molecule, two-color photoionization allows the investigator to obtain details pertaining to both the neutral excited states and the states of the ion. Figure 19.1 displays the one-color photoionization spectrum of the $^1B_2$ excited state of aniline. Here, the laser was tuned through the excited state and parent molecular ions were detected. The advantages of employing a supersonic expansion as the sample source are evident from this spectrum. The rotational envelopes of each vibronic band have collapsed to about 3 cm^{-1} and no vibrational hot bands or sequence structure can be seen. Consequently, spectral assignment of the vibronic transitions becomes an almost trivial matter even for a molecule with 36 vibrational degrees of freedom such as aniline.

One-color threshold photoionization of aniline can be accomplished by tuning a high power laser in the energy region around IP/2, in this case ~ 3200 Å. The threshold photoionization spectrum obtained under these experimental conditions is shown in Fig. 19.2. It is important to realize that at 3200 Å, there are no excited state levels available, so photoionization is nonresonant and inefficient.

The problems of a one-color approach to threshold studies are exemplified in the spectrum in Fig. 19.2. The ionization threshold is very broad (~ 350 cm^{-1}) compared with the resolution of the laser (<1 cm^{-1}). Since the experimental resolution is high, but the observed spectrum is broad, the problem is likely to be due to a significant molecular geometry change in going from the ground electronic state of the neutral to the low-energy region of the ion. These poor Franck–Condon factors make it very difficult to assign a true adiabatic ionization potential with confidence.

In the two-color threshold photoionization studies of aniline, the excitation

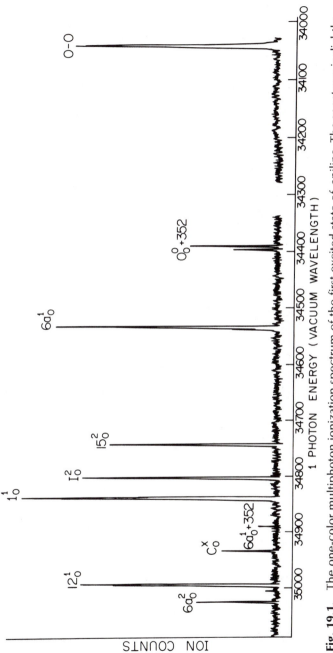

Fig. 19.1. The one-color multiphoton ionization spectrum of the first excited state of aniline. The spectrum is slightly saturated to more clearly display the weaker transitions.

laser was fixed at the energy corresponding to a specific $A(^1B_2)$ transition whereas the ionizing laser was tuned from below the ionization onset to deeper within the ionization continuum. Two spectra obtained in this way are displayed in Fig. 19.3. These spectra are presented to illustrate the advantages of the two-color approach, namely the excellent signal-to-noise ratio and the high resolution of the technique.

Both of the spectra in Fig. 19.3 are characterized by extremely sharp ionization onsets ($3-5$ cm^{-1}) in contrast to the spectrum in Fig. 19.2. In the case of ionization from the $^1B_2$ origin (Fig. 19.3a) the onset corresponds to the true adiabatic ionization potential ($IP_0 = 62,265 \pm 8$ cm^{-1}). The ionization threshold observed when the $6a^1$ level serves as the resonant state (Fig. 19.3b) corresponds to the position of the one quantum excitation of ν_{6a} in the ground electronic state of the ion. This observation is excellent confirmation of the propensity for $\Delta v = 0$ ionization of aromatic molecules when an excited state vibrational level is used as the intermediate state in the two-color photoionization process. This arises from the similar equilibrium geometries of the $A(^1B_2)$ and $X^+(^2B_1)$ electronic states of the neutral molecule and the ion, respectively. We have used the strong $\Delta v = 0$ propensity and the value of IP_0 to obtain the vibrational frequencies of the ground electronic state of the ion.[4] The frequency of vibration v is given by the relationship

$$E_v = IE_v - IP_0$$

where, E_v is the ion vibration frequency of mode v and IE_v is the energy of the ionization onset measured in the PIE spectrum when the same mode v serves as the resonant intermediate level in the two-color experiment.

A summary of the vibrational frequencies of the aniline radical cation is presented in Table 19.1. Comparison of the vibrational frequencies of the ground and excited states of the neutral and ground electronic states reveals several clear trends. With the exceptions of modes 15 and I, there are only small frequency changes upon photoionization. The energies of ν_{15} and ν_I increase considerably

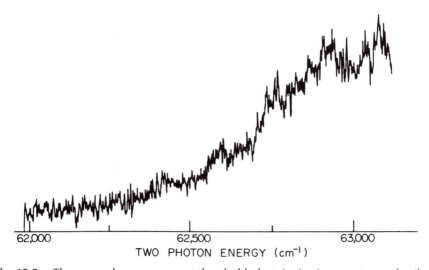

Fig. 19.2. The one-color nonresonant threshold photoionization spectrum of aniline.

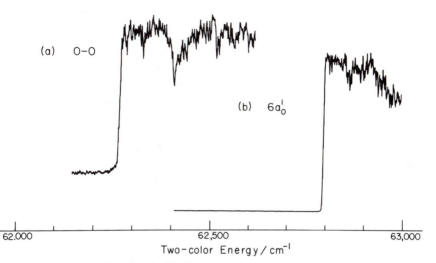

Fig. 19.3. The two-color threshold photoionization spectra of aniline obtained with $h\nu_1$ fixed at (a) the $^1B_2$ origin and (b) the $6a^1$ vibronic level. Note the excellent signal-to-noise ratio and the sharp thresholds as compared with the spectrum in Fig. 19.2.

going from the ground state to the excited state of the neutral and then to the ground electronic state of the cation. Both of these vibrations are localized in the region about the C−N bond, and the significant changes in the associated frequencies provide insight into the structural changes accompanying partial and complete removal of the bonding π-electron in the b_1 orbital.

In the neutral ground state, aniline is nonplanar about the nitrogen atom resulting in an inversion potential characterized by a double minimum and a barrier to planarity of about 500 cm^{-1}.[17] Excitation to the $^1B_2$ state leads to an inversion potential that is best fit by a quartic expression with no barrier.[18] From this it has been concluded that the excited state neutral molecule is "quasiplanar." The results we have obtained for the ionic ground state show a harmonic potential for ν_1 with

Table 19.1. Vibration energies for the neutral ground and excited states and ground ionic state of aniline (cm^{-1})

Mode		$E_v(X)$	$E_v(^1B_2)$	$E_v(^2B_1)$
6a	$v = 1$	529	492	521
	$v = 2$	—	984	1045
1	$v = 1$	820	798	815
12	$v = 1$	997	954	990
10b	$v = 2$	—	348	352
16a	$v = 2$	—	352	361
	$v = 4$	—	—	712
15	$v = 1$	—	—	724
	$v = 2$	—	702	1455
I	$v = 1$	41	340	655
	$v = 2$	423	760	1320
	$v = 3$	700	1221	1980

a significantly larger force constant than observed for the neutral. For ν_{15}, which involves tangential motion of the C—N bond, we observe an increase in the vibrational frequency of approximately a factor of two between the $A(^1B_2)$ and the $X^+(^2B_1)$ electronic states. The observed changes in the vibrational frequencies for modes 15 and I suggest a considerable shortening of the C—N bond upon photoionization and the production of an ion that is more rigidly planar than either the ground or excited states of the neutral. Recent multiphoton photoelectron results[6] and molecular orbital calculations[19] confirm this interpretation and also point to a considerable shortening and strengthening of this bond upon removal of an electron from the highest occupied molecular orbital.

These results demonstrate that considerable new insights into the structure and energetics of polyatomic ions can be gained using two-color threshold photoionization spectroscopy. Sufficient experimental information now exists to support sophisticated force field calculations of several aromatic positive ions. Consequently, it is now expected that much finer details regarding potential surfaces of these elusive species will be revealed.

Vibrational Autoionization of Aniline

When aniline is photoionized from the I^1, I^2, and I^3 $^1B_2$ first excited state vibrational levels, the PIE spectra displayed in Fig. 19.4 were recorded. In addition to the sharp $\Delta v = 0$ direct ionization onsets in these threshold spectra, regular band structure lying *above* the adiabatic ionization potential was observed. This band structure is due to vibrational autoionization of highly excited Rydberg states of the neutral molecule converging to the $\Delta v = 0$ direct ionization threshold. Assignments[7,8] of these progressions were carried out using the well known Rydberg equation,

$$E = IP - R(n - \delta)^{-2}$$

where the ionization potential of interest here is the direct ionization threshold displayed in each of the spectrum in Fig. 19.4. Rydberg series with quantum defects of $\delta = 0.78$, $\delta = 0.45$, $\delta = 0.41$, and $\delta = 0.22$ have been observed in the I^1 PIE spectrum. In each case these series converge to the IP associated with the aniline positive ion with the same number of quanta in the inversion mode as that of the intermediate level (i.e., $\Delta v = 0$ onset).

The spectra in Fig. 19.4 are interesting in several respects. All of the Rydberg bands lie within the ionization continuum and in each case the structure begins at approximately the same energy with respect to the direct ionization onset (~ -660 cm^{-1}). This value corresponds to the energy of one vibrational quantum of the inversion mode of the ground state ion as can be seen from Table 19.1. Thus, autoionization begins only at the energy corresponding to one quantum less vibrational excitation of the ion than that produced by direct ionization. This is a characteristic of a vibrational autoionization mechanism in which a single quantum of vibrational energy of the ion core is converted into additional electronic excitation of the Rydberg electron producing an ion with less vibrational energy than that characteristic of the Rydberg state. For vibrational autoionization of aniline, we have observed a strong propensity for the ion core to lose only a single quantum of vibrational energy (i.e., $\Delta v = -1$).

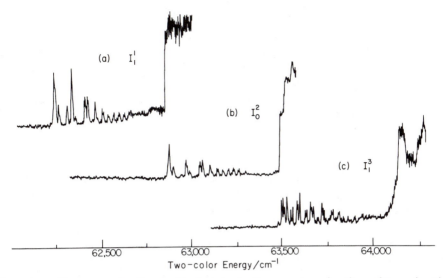

Fig. 19.4. The two-color threshold photoionization spectra of aniline obtained with $h\nu_1$ fixed at (a) the I^1 level, (b) the I^2 level, and (c) the I^3 level of the $^1B_2$ excited state. All of the structure lies above the adiabatic IP located at 62,263 cm^{-1}.

We have also observed vibrational autoionization in the two-color threshold photoionization spectra of the vibrational modes 10b and 15 as well as the combination vibrations of I, 10b, and 15 with the totally symmetric vibrations 6a and 1. In each case the autoionization band structure lies within the ionization continuum and exhibits the $\Delta v = -1$ propensity. In fact, only when the intermediate vibrational level contains a contribution from a nontotally symmetric mode is vibrational autoionization observed. This suggests that the symmetry of the vibrational mode plays a crucial role in the competition between direct ionization and vibrational autoionization.

Vibrational autoionization occurs through the coupling of electronic and vibrational motion and represents a breakdown of the Born–Oppenheimer (BO) approximation.[20] Within the BO limit, the nuclear momentum operator, $\partial/\partial Q$, leads to a dynamic electronic-vibration coupling. For coupling due to a single vibrational mode, Q, the matrix element of the major perturbation can be written as[20-22]

$$\langle \Psi_f(q,Q) | \langle \phi_f(Q) | \partial/\partial Q | \phi_i(Q) \rangle \partial/\partial Q | \Psi_i(q,Q) \rangle$$

where $\Psi(q,Q)$ and $\phi(Q)$ are the electronic and vibrational portions of the product BO wavefunction, respectively. The initial states are the highly excited Rydberg states, and the final states are levels of the ionic ground state. Upon making the Condon approximation, this expression can be rewritten as two matrix elements[20-22]

$$\langle \Psi_f(q,Q) | \partial/\partial Q | \Psi_i(q,Q) \rangle \langle \phi_f(Q) | \partial/\partial Q | \phi_i(Q) \rangle$$

These two matrix elements help explain the results for aniline. The vibrational matrix element leads to a $\Delta v = -1$ selection rule for the interaction of two harmonic potentials. In spite of the complicated vibrational motion of a polyatomic molecule such as aniline, our results show a strict adherence to this expected behavior.

The electronic matrix element is the type found in vibronic coupling problems.[23] This term couples electronic states that have symmetries connected by that of a normal coordinate. Consequently, for electronic states of different symmetries, the coupling will be effective for only nontotally symmetric vibrations. Again, this expected behavior is indeed observed in the two-color threshold photoionization investigations of aniline.

Further information regarding the aniline vibrational autoionization can be inferred from measurements of the spectral linewidths of the Rydberg bands. Typical linewidths (FWHM) are found to be approximately 9–10 cm^{-1}, placing an upper limit on the autoionization rate of 2×10^{12} sec^{-1}, or a minimum lifetime of the autoionizing state of about 0.5 psec. This compares with recently reported values of 0.3–0.9 psec for naphthalene[24] and approximately 0.05 psec for benzene autoionizing states.[3] The fact that aniline and naphthalene autoionizing levels exhibit such relatively long lifetimes points to the small vibrational-electronic coupling for these molecules and illustrates that the Born–Oppenheimer approximation is quite reasonable for these two aromatic molecules near the ionization limit.

Threshold Photoionization of Indole van der Waals and Molecular Clusters

Using supersonic expansion technology, it is a relatively simple matter to synthesize van der Waals and molecular clusters of virtually any size and composition.[10] When investigating the electronic properties of a species AX_m, in which A is the chromophore and X represents the "solvent" atoms or molecules, subtle intermolecular interactions serve to alter slightly the electronic transition energies of the chromophore. The intermolecular interactions of interest for neutral clusters are dispersion forces, dipole-induced dipole forces, and, when applicable, dipole-dipole interactions.[25] The direction and magnitude of this "spectral shift" observed in the electronic spectra of such complexes are actually a measure of the difference of the binding energy of the cluster in the two optically connected electronic states.[10] Investigations of these small "solute–solvent" species have provided unique insight into the nature of pairwise intermolecular forces and the early stages of solvation and nucleation phenomena.

Our approach has been to use these neutral clusters as precursors for the preparation of solvated positive ions. If, after the formation of a weakly bound neutral cluster, the parent molecule is photoionized, the opportunity exists to investigate the interactions between a cation and a neutral solvent molecule in the same isolated environment. The ionic complexes are more strongly bound than the neutral precursors, with the additional stabilization energy being directly given by the difference between the adiabatic ionization potential of the complex and that of the isolated monomer.[14] Thus, the threshold photoionization of van der Waals and molecular clusters allows the charge-dipole and charge-induced dipole forces characteristic of the ionic complexes to be probed.[11,12,26]

Indole was chosen for these ion-solvent studies because of its important relationship to protein photochemistry,[27] its use as a model compound for the amino acid tryptophan,[27] and its unusual condensed phase photoionization dynamics.[28] In solution, one photon-induced photoionization of indole from low lying excited states has been observed in several polar solvents.[28] The observed solution phase IPs are found to be as much as 3–4 eV below that of the gas-phase molecule. Observation of such a dramatic ionization potential decrease for a small cluster should not be expected due to the lack of a mechanism for stabilization of the ejected electron, such as a bulk polar solvent.[29] However, by varying the number and type of solvent partners in the complex, the "stepwise" solvation effects of the cation can be investigated. Within the context of this chapter, these effects are best illustrated by the two-color photoionization results of indole–$(CH_4)_{1,2}$ and indole–H_2O.

The two-color PIE spectra of indole–$(CH_4)_1$ and –$(CH_4)_2$ are shown in Fig. 19.5. These were obtained with $h\nu_1$ tuned to the corresponding cluster first excited state band origins, located at -36 and -74 cm^{-1}, respectively, relative to the indole $S_1(0-0)$ transition. The photoionization thresholds in these spectra yield adiabatic ionization potentials that are significantly lower in energy with respect to that of the uncomplexed molecule. For indole–$(CH_4)_1$, this IP_0 was measured to be -487 cm^{-1} and for indole–$(CH_4)_2$, -900 cm^{-1}. Thus, there is a nearly additive shift of approximately -450 cm^{-1} per added methane molecule. This is consistent with a picture in which the two methane molecules occupy geometrically equivalent positions above and below the plane of the indole molecule producing negligible charge screening effects.[11,12]

Investigations of other van der Waals complexes of indole and nonpolar partners have shown that the measured ΔIP is approximately a function of the polarizability of the complexing partner.[11,12,26] This is to be expected since the primary contribution to the stabilization of such complexes is charge-induced dipole intermolecular interactions. The result is an ion-neutral complex with a characteristic binding energy governed by the electrostatic interaction[14]

$$V = -|F(r)|^2\alpha/2e^2$$

where α is the polarizability of the neutral nonpolar molecule and F is the electrostatic force exerted by the charge distribution of the positive ion. Of course, the dipole-induced dipole and dispersion forces that stabilize the neutral van der Waals complex also make a contribution to the binding energy of the ionic species, but the dominant term is that previously given. It is the charge-induced dipole interaction that leads to the deeper potential wells associated with the ionic clusters compared with the neutral species. Furthermore, it is this difference in binding energies that produces some of the qualitative differences between the threshold photoionization behavior of monomers and clusters.

Earlier it was demonstrated that two-color photoionization of an aromatic molecule from an excited electronic state usually results in a preponderance of $\Delta v = 0$ ionizing transitions. This is because the geometry of the parent ion is often very similar to that of the electronically excited neutral molecule. However, threshold photoionization studies of neutral van der Waals complexes and molecular clusters are somewhat more complicated. Ionic complexes are often much more tightly bound than the excited state neutral species due to the presence of a full positive charge. Consequently, the Franck–Condon factors for photoionization will often favor large changes in the degree of vibrational excitation. This, in turn, can give rise to lower ionization cross sections near threshold and broader, more diffuse

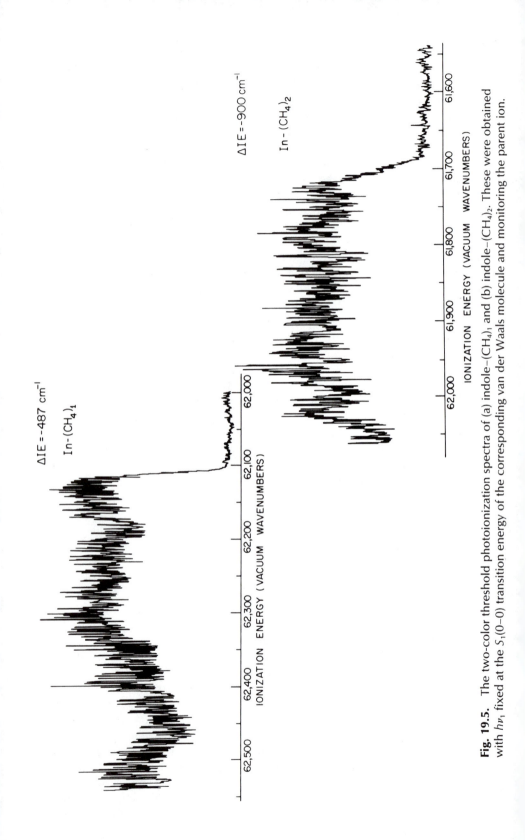

Fig. 19.5. The two-color threshold photoionization spectra of (a) indole–$(CH_4)_1$ and (b) indole–$(CH_4)_2$. These were obtained with $h\nu_1$ fixed at the $S_1(0-0)$ transition energy of the corresponding van der Waals molecule and monitoring the parent ion.

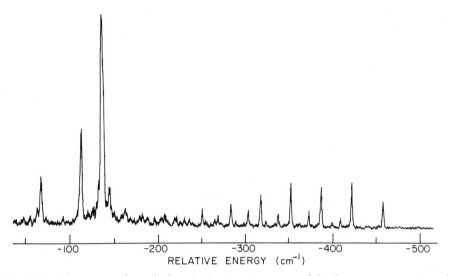

RELATIVE ENERGY (cm⁻¹)

Fig. 19.6. The mass-selected photoionization spectrum of the low-energy portion of the first excited state of indole–$(H_2O)_1$. The energy units are relative to the indole $S_1(0-0)$ band.

onsets.[11,12,26] An example of such phenomena can be seen in Fig. 19.5. The threshold for indole–$(CH_4)_1$ was found to be 20 cm^{-1} in breadth and that for indole–$(CH_4)_2$ is even broader, 38 cm^{-1}. It should be recalled at this point that the ionization onsets observed for aniline were typically 3 cm^{-1}.

To examine the effects of more complicated intermolecular interactions on threshold photoionization of molecular clusters, consider the indole–water hydrogen bonded system.[11,12] The lowest energy portion of the electronic spectrum of indole–$(H_2O)_1$ is displayed in Fig. 19.6. There are two distinct sets of absorption features in this spectrum: one set that begins at about −450 cm$_{-1}$ relative to the indole $S_1(0-0)$ band and another that starts at −135 cm^{-1}. Both sets of bands can be assigned to a complex wtih a 1:1 stoichiometry.

Figure 19.7 presents the two-color photoionization spectra obtained from a representative band from each set. When the feature at −135 cm^{-1} serves as the resonant intermediate level, the spectrum in Fig. 19.7a was obtained. This PIE spectrum exhibits multiple thresholds separated by approximately 180 cm^{-1}. The lowest energy of these onsets is located 3027 cm^{-1} lower in energy than the adiabatic IP of the indole monomer.

The large change in ionization potential can be attributed to the very strong intermolecular interactions operative for ion-neutral complexes involving a polar "solvent" molecule. The combination of charge-induced dipole and charge-dipole forces leads to anionic complex that is substantially more tightly bound than the neutral. The experimental manifestation of this are large Franck–Condon factors for $\Delta v \neq 0$ ionizing transitions as can be seen by the multiple thresholds shown in Fig. 19.7a. We have assigned this 180 cm^{-1} interval to excitation of the hydrogen bond stretching frequency in the ionic ground state of the cluster. This is approximately a twofold increase over the value of 90 cm^{-1} for the analogous vibration of the excited state neutral species.

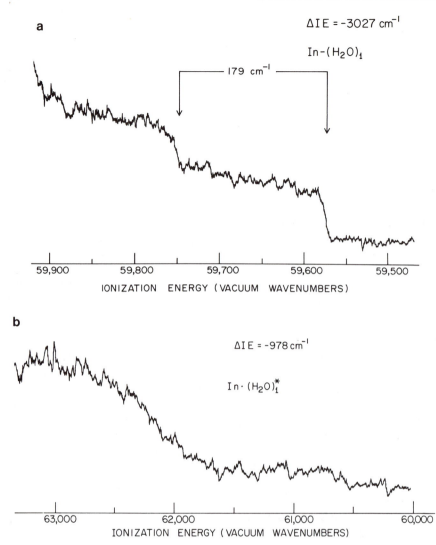

Fig. 19.7. The two-color threshold photoionization spectra of indole–$(H_2O)_1$. These spectra were obtained with $h\nu_1$ fixed at (a) the -135 cm^{-1} cluster band and (b) the -452 cm^{-1} feature. The ΔIE values are relative to the adiabatic IP of indole located at 62,598 cm^{-1}.

Figure 19.7b displays the threshold photoionization spectrum obtained using the indole–$(H_2O)_1$ band at -452 cm^{-1} as the resonant intermediate level. The ionization threshold is quite broad in this case (~ 130 cm^{-1}) and is shifted by less than -1000 cm^{-1} from the indole IP$_0$. This suggests that we have an extreme example of poor Franck–Condon factors between the excited state neutral and the lowest energy levels of the ground state of this ionic complex.

The fact that the two-color threshold photoionization spectra of two closely lying excited state levels are so radically different suggests that there are considerably different intermolecular interaction for the 1:1 molecular complexes that give rise to these electronic absorption bands. We have proposed[12] that there are two distinct conformations of indole–$(H_2O)_1$ "frozen out" in the supersonic expansion. Electronic excitation of one conformation leads to the bands shown between -50 and -150 cm^{-1} in Fig. 19.6 whereas the other conformation is responsible for the absorption features at lower energies. Using the two-color technique it is a simple matter to identify conformations of molecular clusters since the differences in intermolecular interactions are magnified when one of the partners takes on a positive charge. This illustrates the potential of such studies for investigation and elucidation of site-specific intermolecular interactions.

Conformation-Selective Spectroscopy: Tryptophan Analogs

As previously discussed, indole has been used by many investigators as a model for the amino acid tryptophan. Although certain insights into the electronic spectroscopy and photophysics of tryptophan have been gained through these pursuits,[30-34] more representative model systems have been sought. This has led to the use of "tryptophan analogs,"[35] which are actually just 3-substituted indole derivatives as illustrated in Fig. 19.8. From structural considerations, the excited state energetics of these tryptophan analogs are expected to be similar to indole itself with deviations being attributable to the extent of the interaction between the side chain and the rigid indole nucleus. These deviations from "indole-like" behavior can be examined from the perspective of "pseudointermolecular" interactions. The tryptophan analogs can be thought of as being comprised of two subunits: the indole nucleus and the side chain. The interactions between the two are limited because the side chain–nucleus relative geometries are limited by the basic molecular geometry of the tryptophan analog. In this respect the problem is somewhat simpler than that of the intermolecular interactions between separate molecules. Of course, it should be kept in mind that this is strictly a zero-order approach that is aimed at providing insight into the small side chain perturbations to electronic properties.

The lowest energy portions of electronic spectra of several tryptophan analogs are displayed in Fig. 19.8. Each molecule is characterized by strong electronic transitions in the region around 2865 Å. The relatively small differences in transition energies certainly suggest that the perturbation of the side chain to the indole nucleus is largely determined by the methylene unit attached at the 3-position. Conformation of this is provided by the fact that the $S_1(0-0)$ band of 3-methylindole is located at 2866.7 Å. It is quite surprising that a substitution as radical as exchanging a $-COOH$ group for $-NH_2$ does little to alter the energetics of the first excited singlet states of these molecules.

The spectra in Fig. 19.8 are interesting in another respect. In each case, the presence of a side chain leads to the appearance of multiple, intense bands in the lowest energy regions of the spectra. Levy and co-workers[35,36] have assigned these features to the 0–0 transitions of multiple molecular conformations of the side chains relative to the indole nucleus. The rapid cooling of the supersonic expansion effectively freezes out a set of conformations that is stable in this ultracold environment. Because of the slightly different side chain-nucleus orientations, the interactions between the subunits are slightly different leading to the observed distribution of

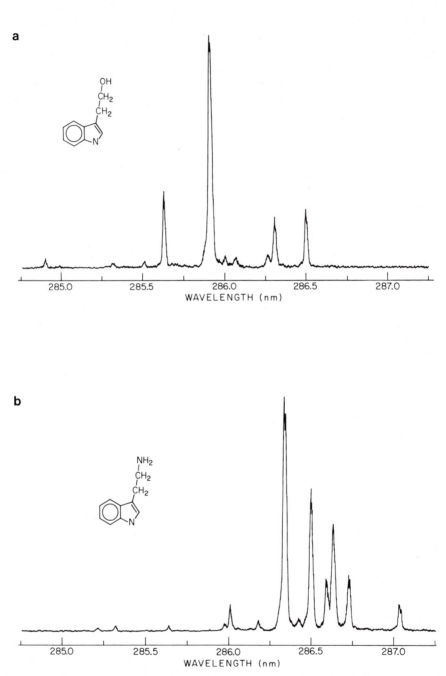

Fig. 19.8. The mass-selected photoionization spectra of the tryptophan analogs (a) tryptophol, (b) tryptamine.

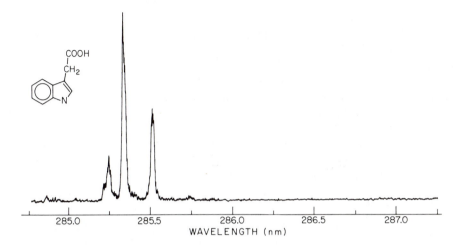

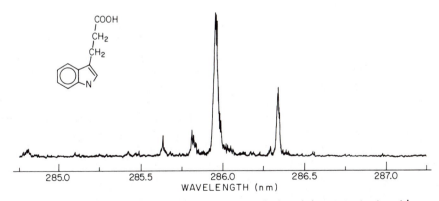

Fig. 19.8. (*Continued*) (c) indole-3-acetic acid, and (d) indole-3-propionic acid.

transition energies for each tryptophan analog. Thus, the excellent cooling of the supersonic jet not only allows the formation of these conformers, but also allows the properties of each one to be studied individually. We have conducted[37] a comprehensive investigation of the excited state and photoionization behavior of these tryptophan analogs. As an example, the results for the molecule tryptophol are presented below.

The electronic spectrum of tryptophol in Fig. 19.8a exhibits four strong bands separated by 10 Å. The small energy differences indicates that the side chain–nucleus interactions are relatively small in the first excited state. It would be expected, however, that the pseudointermolecular interactions, and hence the resulting energy separations, will be significantly greater for the corresponding cations because of the strong charge-dipole and charge-induced dipole contributions. This should be reflected by the positions of respective ionization thresholds.

Figure 19.9 displays the two-color threshold photoionization spectra associated with these four excited state levels. There are several points regarding these spectra that are worthy of comment. The very different ionization energies obtained from the four closely spaced electronic bands provide direct confirmation of the presence of multiple tryptophol conformers. Furthermore, from the considerable energy spread between the four ionization onsets, it can be seen that the side chain–nucleus interactions are significantly greater for the charged molecule.

One should also notice that the early portions of the ionization onsets are quite different. The lowest energy threshold in Fig. 19.9 is the sharpest (~ 35 cm^{-1}), but is still significantly broader than that commonly observed for rigid aromatic molecules (~ 3–5 cm^{-1}). The almost 400 cm^{-1} broad onset shown in Fig. 19.9d is very similar to those observed for strongly bound hydrogen bonded complexes that undergo radical equilibrium geometry changes upon photoionization. As discussed earlier, the shape of an ionization onset is a manifestation of the Franck–Condon factors for ionization of the electronically excited molecule. Thus, the degree of threshold broadening observed in these spectra is a measure of the differences of side chain-nucleus equilibrium geometries between the excited state and ground ionic species.

Finally, two-color photoionization provides a multidimensional method of investigating contributions of an individual conformer to the total electronic spectrum. We have obtained electronic spectra of tryptophol under several ionization conditions. For these studies, the ionizing laser was fixed at different wavelengths near the ionization thresholds, and the excitation laser was tuned through the electronically excited state. The results are presented in Fig. 19.10.

The necessary conditions for production of ions under these circumstances are that $h\nu_1$ be resonant with a vibronic transition and $h\nu_1 + h\nu_2$ be greater than the minimum energy required to ionize the absorbing species. Since the conformer with its $S_1(0-0)$ level at 2863.5 Å has the lowest ionization energy, the contributions of the other conformers in the supersonic expansion to the electronic spectrum can be stripped away gradually by reducing the energy of the ionizing laser as is shown in Fig. 19.10. In this way it is possible to obtain an uncontaminated view of individual conformers unavailable with other techniques. Such a conformation-selective approach may soon provide details regarding small side chain perturbations

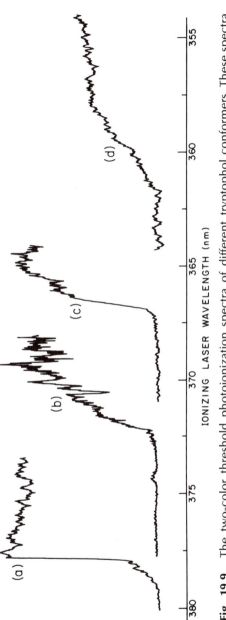

Fig. 19.9. The two-color threshold photoionization spectra of different tryptophol conformers. These spectra were obtained with $h\nu_1$ fixed at each of the four strong low energy electronic transitions shown in Fig. 19.8a: (a) 2863.5 Å, (b) 2865 Å, (c) 2857 Å, and (d) 2859 Å.

IONIZING LASER WAVELENGTH (nm)

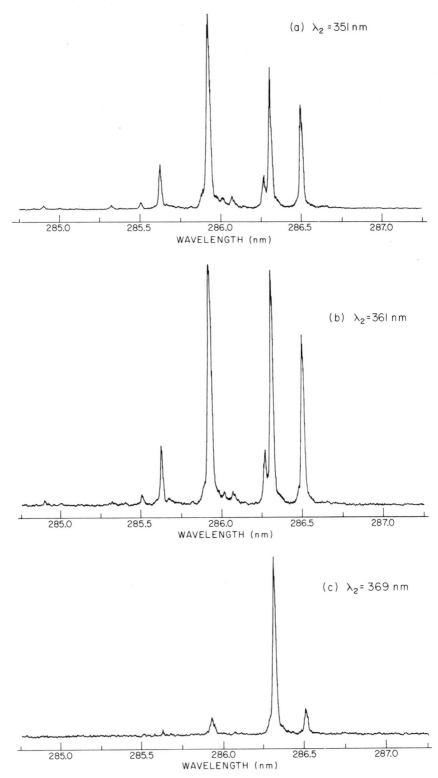

Fig. 19.10. The mass-selected two-color photoionization spectra of tryptophol at different ionizing laser wavelengths. These spectra were obtained by tuning the excitation laser ($h\nu_1$) while the wavelength of the ionizing laser remained fixed at (a) $\lambda = 351$ nm, (b) $\lambda = 361$ nm, and (c) $\lambda = 369$ nm.

on electronic and vibrational energetics and energy dissipation dynamics in large nonrigid molecules.

CONCLUSIONS

This chapter has touched only briefly on a few of the many applications of two-color threshold photoionization to investigations of spectroscopic and dynamic phenomena of neutrals and ions. On the simplest level, this technique provides a method of obtaining high-resolution spectra at vacuum ultraviolet energies without a VUV laser. Ionization potentials and ion vibrational frequencies can be determined with unprecedented precision and accuracy; as illustrated for aniline, such studies provide access to regions of ionic potential surfaces not previously accessible and have identified a new autoionization pathway for this much studied molecule. From a molecular dynamics point of view, these latter results help enlarge our understanding of bound state–continuum coupling interactions in intermediate size molecules.

One of the potentially most useful applications of two-color photoionization is for the production of vibrational state-selected molecular ions that can be used as reactants in ion–molecule dynamics investigations. The ability to probe ion–molecule chemistry as a function of the degree of vibrational excitation and even the nature of the vibrational motion itself has already had a considerable impact on this fundamental area.[38]

Two-color photoionization is also an excellent technique with which to probe the excited state dynamics of the resonant intermediate state of the neutral molecule. The rate of ionization, which can to a large extent be controlled, provides a "built-in clock" with which to measure the rates of other excited state pathways.[5,39] Competing processes such as internal conversion, intersystem crossing, and predissociation must have rates comparable to ionization to proceed to any measurable extent. Photoionization serves to "freeze" the excited state dynamics at a particular time and "project" the resulting products onto the ionization continuum. Often these products will have ionization thresholds at significantly different energies compared with the initially excited species. By monitoring time-resolved ion production at a variety of threshold energies, a direct picture of the excited state dynamics can be obtained.[5,39]

With the development of supersonic expansion technology, great interest has been sparked in the physics and chemistry of neutral and ionic clusters.[40-44] Much of the emphasis in this rapidly growing field has been directed at elucidating factors that influence variations in electronic properties, spectroscopic transitions, and photoionization dynamics during the change from the gaseous to the condensed phase. Fundamental research in cluster chemistry has implications for areas such as gas phase ion-neutral reactions of atmospheric and astronomical importance, aerosol particle formation, as well as nucleation and solvation phenomena.[40-44] Two-color threshold photoionization has had a significant impact on understanding the sometimes very complicated structure, energetics, and dynamics of small complexes involving aromatic chromophores.[44] The results of the indole studies have aided in identification of the excited electronic state ordering[33] in addition to providing insights[11,12] into the early stages of solvation of this biologically important molecule.

Finally, two-color photoionization combined with supersonic cooling has recently been demonstrated[45] to be a sensitive and very selective analytical technique. The simplified electronic spectra of aromatic molecules under expansion-cooled conditions have been exploited as "molecular fingerprints" for the analysis of multicomponent samples, most notably by Lubman.[46] Two-color photoionization provides several extra dimensions for resonant ionization detection.[45] The ability to tune the ionization laser often provides a more efficient method of photoionization as well as allowing additional selectivity based on small differences in ionization potentials. An excellent illustration of this can be seen in the conformer-selective photoionization spectra of tryptophol previously discussed. The fact that individual conformers of tryptophol can be easily isolated and detected demonstrates the extreme selectivity of this technique and opens the door to analysis of multicomponent samples with no prior separation.

ACKNOWLEDGMENTS

We would like to acknowlege Dr. Mark Smith, Dr. Michael Ivanco, Dr. David Demmer, Gary Leach, and Al Outhouse for their significant contributions to these studies. Much of the work described here has been supported by the National Sciences and Engineering Research Council of Canada.

REFERENCES

1. J. Berkowitz, *Photoabsorption, Photoionization, and Photoelectron Spectroscopy.* Academic Press, New York, 1979.

2. J. W. Rabalais, *Principles of Ultraviolet Photoelectron Spectroscopy.* John Wiley, New York, 1977.

3. M. A. Duncan, T. G. Dietz, and R. E. Smalley, *J. Chem. Phys.* **75**, 2118 (1981).

4. M. A. Smith, J. W. Hager, and S. C. Wallace, *J. Chem. Phys.* **80**, 3097 (1984).

5. M. A. Smith, J. W. Hager, and S. C. Wallace, *J. Phys. Chem.* **88**, 2250 (1984).

6. J. T. Meek, E. Sekreta, W. Wilson, K. S. Viswanathan, and J. P. Reilly, *J. Chem. Phys.* **82**, 1741 (1985).

7. J. W. Hager, M. A. Smith, and S. C. Wallace, *J. Chem. Phys.* **83**, 4820 (1986).

8. J. W. Hager, M. A. Smith, and S. C. Wallace, *J. Chem. Phys.* **84**, 6771 (1986).

9. J. B. Fenn, *Adv. Chem. Phys.* **10**, 275 (1966).

10. D. H. Levy, *Adv. Chem. Phys.* **47**, 323 (1982).

11. J. W. Hager, M. Ivanco, M. A. Smith, and S. C. Wallace, *Chem. Phys. Lett.* **113**, 503 (1985).

12. J. W. Hager, M. Ivanco, M. A. Smith, and S. C. Wallace, *Chem. Phys.* **105**, 397 (1986).

13. C. Y. Ng, *Adv. Chem. Phys.* **52**, 263 (1983).

14. J. Jortner, U. Even, S. Leutwyler, and Z. Berkovitch-Yellin, *J. Chem. Phys.* **78**, 309 (1983).

15. U. Boesl, H. Neusser, and E. W. Schlag, *Chem. Phys.* **55**, 193 (1981).

16. J. H. Brophy and C. T. Rettner, *Chem. Phys. Lett.* **67**, 351 (1979).

17. R. A. Kydd and P. J. Kreuger, *Chem. Phys. Lett.* **49,** 539 (1971).

18. J. M. Hollas, M. R. Howson, T. Ridley, and L. Haloren, *Chem. Phys. Lett.* **74,** 531 (1983).

19. G. W. King and A.A.G. van Putten, *J. Mol. Spectrosc.* **44,** 286 (1972).

20. R. S. Berry, *J. Chem. Phys.* **45,** 1228 (1966).

21. S. E. Neilsen and R. S. Berry, *Chem. Phys. Lett.* **2,** 503 (1968).

22. J. N. Bardsley, *Chem. Phys. Lett.* **1,** 229 (1967).

23. G. Fischer, *Vibronic Coupling.* Academic Press, New York, 1984.

24. J. A. Syage and J. E. Wessel, *J. Chem. Phys.* **87,** 6207 (1987).

25. J. O. Hirschfelder, C. F. Curtiss, and R. B. Bird, *Molecular Theory of Gases and Liquids,* Chap. 13. John Wiley, New York, 1964.

26. J. W. Hager and S. C. Wallace, *J. Phys. Chem.* **89,** 3833 (1985).

27. D. Creed, *Photochem. Photobiol.* **39,** 537 (1984).

28. J. Lee and G. W. Robinson, *J. Chem. Phys.* **81,** 1203 (1984).

29. G. A. Kenney-Wallace, in *Photoselective Chemistry,* Pt. 2, edited by J. Jortner. John Wiley, New York, 1981, pp. 535–577.

30. J. W. Hager and S. C. Wallace, *J. Phys. Chem.* **87,** 2121 (1983).

31. J. W. Hager and S. C. Wallace, *J. Phys. Chem.* **88,** 5513 (1984).

32. J. W. Hager and S. C. Wallace, *Can. J. Chem.* **63,** 1502 (1985).

33. J. W. Hager, D. R. Demmer, and S. C. Wallace, *J. Phys. Chem.* **91,** 1375 (1987).

34. G. A. Bickel, G. W. Leach, D. R. Demmer, J. W. Hager, and S. C. Wallace, *J. Chem. Phys,.* **88,** 1 (1988).

35. Y. D. Park, T. R. Rizzo, L. Peteanu, and D. H. Levy, *J. Chem. Phys.* **84,** 6539 (1986).

36. T. R. Rizzo, Y. D. Park, L. Peteanu, and D. H. Levy, *J. Chem. Phys.* **84,** 2534 (1986).

37. J. W. Hager, E. A. Outhouse, and S. C. Wallace, *J. Chem. Phys.,* in press, 1989.

38. W. A. Conway, R.J.S. Morrison, and R. N. Zare, *Chem. Phys. Lett.* **113,** 429 (1985).

39. J. W. Hager, G. W. Leach, D. R. Demmer, and S. C. Wallace, *J. Phys. Chem.* **91,** 3750 (1987).

40. A. W. Castleman, Jr., B. D. Kay, V. Hermann, P. M. Holland, and T. D. Mark, *Surf. Sci.* **106,** 179 (1981).

41. J. Jortner, *Ber. Bunsenges. Phys. Chem.* **88,** 188 (1984).

42. T. D. Mark and A. W. Castleman, Jr., *Adv. At. Mol. Phys.* **20,** 65 (1985).

43. R. G. Keese and A. W. Castleman, Jr., *Annu. Rev. Phys. Chem.* **37,** 525 (1986).

44. J. W. Hager and S. C. Wallace, *Cmt. Mod. Phys., Pt. D* **20,** 63 (1987).

45. J. W. Hager and S. C. Wallace, *Anal. Chem.* **60,** 5 (1988).

46. D. M. Lubman, *Prog. Anal. Spectrosc.* **10,** 529 (1987), and references therein.

20

Combustion Species Detection by Resonance-Enhanced Ionization

TERRILL A. COOL

Methods for efficient, controlled combustion, without contamination of the environment from harmful by-products, are urgently needed to meet increasing demands for energy and for the safe disposal of hazardous wastes. In the past several years laser-based diagnostic methods have been under development for *in situ* detection and monitoring of flame species. Such laser-based methods are used in studies of combustion reaction mechanisms in laboratory flames to supplement information gained by direct sampling and mass spectrometric species detection.

An understanding of the mechanisms by which hazardous by-products are formed requires an accurate description of the formation and consumption of many highly reactive radical intermediates.[1,2] Sophisticated combustion models are required that specify the rate constants for considerably more reactions than the minimal set required to account for the profiles of the major species formed in the overall sequence from reagents through stable intermediates to products.[3] Important examples have been cited in studies of hydrocarbon combustion in which the measurement of radical species profiles has led to significant revisions in kinetic models to account for previously unsuspected reaction mechanisms.[3]

Although much of our present knowledge of combustion mechanisms is based on experimental observations of species profiles with conventional flame sampling and molecular beam mass spectrometry,[2,4,5] flame perturbations and sampling difficulties inherent in these techniques limit their reliability for the monitoring of small concentrations of highly reactive intermediates.[4] Methods for the detection of fluorescence and resonance-enhanced ionization initiated by laser absorption offer comprehensive monitoring of the concentrations of radical intermediates with high sensitivity, excellent spatial resolution, and negligible perturbation of the flame environment.

Newly developed laser-based methods may also find application in the ultrasensitive detection of hazardous contaminants and as monitors of combustion efficiency. Adequate diagnostic techniques are presently unavailable for the direct continuous monitoring of the concentrations of the principal hazardous contaminants in the flue gases of municipal and hazardous waste incinerators. Continuous monitoring of the highly toxic polychlorinated dibenzo-*p*-dioxin (PCDD) and polychlorinated dibenzofuran (PCDF) compounds in real time would require a technique capable of detection limits at the part-per-billion level.[6-8] Detection limits this low are obtained in current practice only by methods unsuitable for continuous monitoring. Large samples must be trapped, concentrated, separated chromatographically, and often mass analyzed for accurate performance evaluation.[9] Such evaluations are expensive and generally require several days to complete. It is therefore not possible, at present, to directly detect variations in the concentrations of toxic compounds for continuous control of incinerator operation.[6,10-12] The absence of such real-time monitoring may result in the undetected release of unacceptable levels of toxic effluents caused by fluctuations in the amount and composition of the waste feed or degradation in the performance of system components.[6,10,11]

Recent research indicates that the resonance-enhanced multiphoton ionization (REMPI) technique, when combined with molecular beam mass spectrometry, is capable of detecting a wide class of substituted aromatic hydrocarbons,[13,14] which may include PCDDs and PCDFs at part-per-billion levels. The REMPI technique has also recently been employed for the detection of several radical intermediates including the species H,[15-17] O,[17,18] CH_3,[19-24] HCO,[24-27] CCl,[28] ClO,[29] and $CHCl_2$,[30] which are of importance in the combustion of chlorinated hydrocarbons.

DESCRIPTION OF THE REMPI APPROACH TO SPECIES MONITORING

Progress has been made in the implementation of REMPI as a laser-based diagnostic method for monitoring the concentrations of combustion species at the part-per-million level.[15-18,22,23,27,31-37] The REMPI approach has a sensitivity rivaling that of the well-established laser induced fluorescence (LIF) method. Both techniques possess the sensitivity for *in situ* monitoring of trace concentrations of reaction intermediates, a capability not found in two other popular laser-based diagnostics: Raman scattering and coherent anti-Stokes Raman scattering (CARS). The REMPI and LIF methods are complementary, with the REMPI approach well suited to detect some weakly fluorescing species that cannot be easily monitored by LIF. Both techniques offer excellent spatial resolution and minimal perturbation of the combustion environment.

The REMPI approach to flame species detection and monitoring, in its simplest form, consists of the collection of electrons on a thin ($\leq$1-mm-diameter) platinum or tungsten wire placed within the flame. The REMPI electrons are produced with a tunable dye laser focused within the flame a millimeter or so from the wire. The REMPI process refers to the resonant excitation of a Rydberg electronic state of the species of interest by multiphoton absorption, followed by ionization produced when one (usually) or more photons are incoherently absorbed following the initial resonant excitation step. Useful variations, accomplished with a second dye laser, may include the use of a different wavelength for the ionization step, or a doubly resonant excitation via two intermediate states.

An ionization probe, designed for the detection of REMPI electrons from flame species,[38] is illustrated in Fig. 20.1. The probe is designed for use in flames with the laser focused about 1-2 mm in front of the anode surface. The probe has a small cross-sectional area normal to the combustion gas flow direction to minimize flow disturbances at the laser focus. The cathode for the ionization probe can either be part of the probe assembly, as illustrated in Fig. 20.1, or, in a simpler version, the grounded burner head itself may be used as the cathode (or anode).[17,22]

At low probe voltages, many electrons are lost by recombination before reaching the anode. As the voltage is increased, the ionization signal rises until a "plateau" region is reached at which the electron collection is saturated; essentially all of the electrons produced are collected. Higher voltages lead to avalanche ionization in the "proportional counting" regime. Quantitative measurements require that the probe be operated at a voltage within the plateau region. A discussion of the procedures required for quantitative density measurements has been given.[38] The *in situ* REMPI detection of flame radicals requires that the spectral signature of REMPI ionization unique to a given species be resolvable and identifiable against a background of resonant and nonresonant ionization of other flame species. The spectrum associated with the detected REMPI electrons, obtained by tuning the dye laser through the wavelengths of resonantly excited transitions, provides species identification and a quantitative *in situ* measure of species density. Absolute densities are not obtained directly; the response of the probe must be calibrated for each species of interest. The sensitivity of the method is ultimately limited by the levels of the laser-induced background ionization.

The primary limitation to species detection with REMPI electrons is the scar-

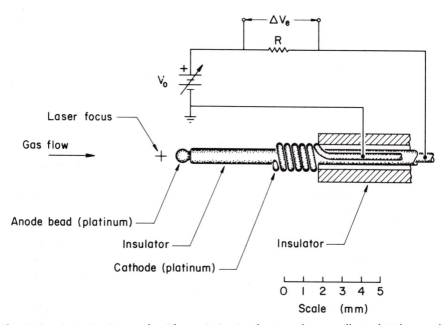

Fig. 20.1. An ionization probe. The variation in electron charge collected at the anode of the probe is recorded as a function of dye laser wavelength for observations of REMPI electron spectra from flame species. A comparison between the observed REMPI electron spectrum for a given species with its spectral signature obtained by other means, for example, a mass-resolved REMPI ion spectrum, provides species identification and discrimination against background ionization interferences.

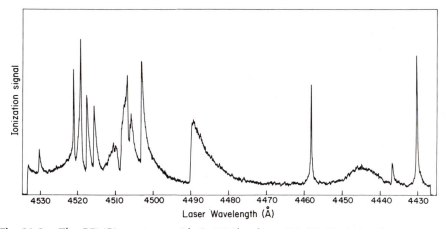

Fig. 20.2. The REMPI spectrum with C-450 dye for a $CH_4/O_2/Ar$ (1/2.6/0.1) $\phi = 0.8$ flame at a pressure of 31 torr.[36] The laser focus was located 8 mm above the burner surface and 1 mm below the probe anode surface. A laser pulse energy of 0.8 mJ at 4500 Å was employed; the variation in pulse energy with wavelength is indicated by the solid points. The identification of C_2O as the carrier is still tentative, pending mass-selective detection of the REMPI ions associated with this spectrum.

city of spectroscopic data for the Rydberg states of most combustion radicals of interest. In the absence of a priori knowledge of the spectral signature of the REMPI electrons produced from a given species, it is very difficult to unambiguously assign spectral features of an observed REMPI spectrum to the carrier species. Indeed, over the past several years we have accumulated a rich library of unidentified REMPI electron spectra for a variety of hydrocarbon flames operated over wide ranges of conditions. These spectra are beautifully resolved from the nonresonant background and have spatial profiles within the flame exhibiting the growth and decay characteristic of intermediate radical species. An example is the spectrum of Fig. 20.2, tentatively believed to originate from the C_2O radical.[36]

The association of the spectrum of Fig. 20.2 with C_2O is based on circumstantial evidence. Comparisons of spectra obtained from several hydrocarbon flames with spectra obtained for their deuterated analogs demonstrate the absence of hydrogen in the carrier species.[36] The most likely candidate species, containing only carbon and oxygen, for the carrier of the spectrum of Fig. 2 is the C_2O radical. Confirmation of this identification awaits further experiments designed to sample and mass analyze C_2O^+ REMPI ions formed within the flame zone; an apparatus for this purpose is described later in the chapter.

REMPI SPECTROSCOPY OF COMBUSTION RADICALS

REMPI spectroscopy has provided a remarkable body of new spectroscopic data on previously unobserved electronic states of many molecules. The REMPI spectroscopy of transient radicals has been recently reviewed.[39] The CH_3[19-21] and HCO[25,26] radicals are examples of reaction intermediates of prime importance in hydrocarbon combustion for which the recent identification of new electronic states with REMPI spectroscopy now permits flame diagnostics to be performed with this sensitive and unique laser-based method. The growing list of species that have been detected in flames by REMPI includes: C,[22,31] H,[15-17] O,[17,18] CH,[31] CO,[36] O_2,[31] NO,[32,33] PO,[34] HCO,[24,27] C_2O,[36] CH_3,[22-24] and C_4H_6.[35,37]

At present, the primary limitation to the widespread application of REMPI spectroscopy for the detection of flame radicals is a lack of spectral information for the Rydberg states of species of interest. The assignment of the carrier species of REMPI electron spectra observed in hydrocarbon flames has proven to be quite difficult, as previously discussed. Mass-selected REMPI spectroscopy by which radicals, formed by laser photolysis, by pyrolysis, or in flow reactors and electrical discharges, are resonantly ionized and then detected mass spectrometrically, has provided spectral signatures for many new electronic states of numerous combustion-related radicals.[39,40]

Recent studies of HCO[26] and C_2[41] illustrate the now commonplace use of a pulsed valve and skimmer to form a cooled supersonic molecular beam, convenient for mass-selected REMPI spectroscopy.[40,42] Suitable parent molecules, CH_3CHO and C_2H_2, were photolyzed in the molecular beam with an excimer laser to yield HCO and C_2 radicals, respectively. Mass selected REMPI ions, formed with an excimer-pumped tunable dye laser focused within the ionization region of either a quadrupole or time-of-flight mass spectrometer (TOF-MS), were detected with an electron multiplier.

HCO

The HCO radical is an important intermediate in hydrocarbon combustion[43,44] and atmospheric photochemical reactions.[45,46] It is also a primary product in the photolysis of aldehydes[47-49] and has astrophysical importance.[50-52] The need for reliable laser-based diagnostic methods for the *in situ* monitoring of HCO in reaction environments has spurred recent spectroscopic measurements.[25,26]

High-resolution REMPI spectra of HCO and DCO have been obtained for wavelengths from 373 to 460 nm.[25,26] Two-photon resonant, three-photon ionization (2 + 1) REMPI of HCO via the $3p^2\Pi$ Rydberg state, studied in this work, is illustrated in Fig. 20.3. The linear $3p^2\Pi$ state is resonantly excited by two-photon absorption from the bent $\tilde{X}^2\Pi$ ground state to form long progressions in the bending mode of the upper state. The mass selection of HCO^+ ions removes background ionization interferences that would be present in the REMPI electron spectrum. The excellent rotational resolution, thus afforded, is illustrated in Fig. 20.4 for the $(K',K'') = (0,2)$ subband of the $(0,7,0) \leftarrow (0,0,0)$ vibronic band of the $3p^2\Pi(A'') \leftarrow \tilde{X}^2\Pi(A')$ transition.[26] A comparison of HCO REMPI electron spectra obtained in a stoichiometric methane/oxygen flame[27] and a 300 K photolysis cell[25] with the higher resolution, mass-selected REMPI ion spectrum[26] is presented in Fig. 20.5. The feature at 50,328 cm^{-1} that appears prominently in the spectra of Fig. 5A and C is a P-branch head of the $(K',K'') = (3,1)$ subband of the $(0,6,0) \leftarrow (0,0,0)$ $3p^2\Pi(A'') \leftarrow \tilde{X}^2\Pi(A')$ vibronic transition of HCO. This peak occurs at a dye laser wavelength of 397.4 nm, and has been used for the relative density profile measurements of HCO presented in the following section.

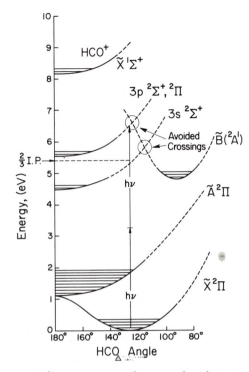

Fig. 20.3. Angular potential energy curves for some low-lying states of HCO, based on *ab initio* calculations [P. J. Bruna, R. J. Buenker, and S. D. Peyerimhoff, *J. Mol. Struct.* **32**, 217 (1976)]. Two-photon resonant, three-photon ionization (2 + 1) REMPI to the $3p\,^2\Pi$ Rydberg state is illustrated; resonances below the limit, $\frac{2}{3}$ IP, are not accessible to a single-photon ionization step in the absence of collisions.

C_2

The electronic states of the C_2 radical have been extensively studied because of the importance of C_2 in astrophysics and combustion chemistry. *Ab initio* calculations of potential energy curves are available for 62 valence states; of these, 29 are possible bound sates for which molecular parameters have been calculated.[53–55] Spectroscopic constants, obtained with older methods, are known for 14 of these bound states[56]; values have been recently given for the lowest lying $^1\Delta_u$ state, predicted by *ab initio* calculations, but experimentally unknown until the application of REMPI spectroscopy.[41] Multiphoton selection rules and the high sensitivity of REMPI spectroscopy make possible the observation of transitions to electronic states inaccessible with older methods.

For the REMPI studies, C_2 radicals were prepared in the $A\,^1\Pi_u$ state under collisionless conditions via the two-step photolysis of C_2H_2 at 193 nm:[57]

$$C_2H_2 + h\nu\ (193\ \text{nm}) \rightarrow C_2H + H \tag{20.1}$$

$$C_2H + h\nu\ (193\ \text{nm}) \rightarrow C_2(A\,^1\Pi_u) + H \tag{20.2}$$

The radiative lifetime of the $C_2(A\,^1\Pi_u)$ state is long enough (10^{-5} sec)[58] for REMPI

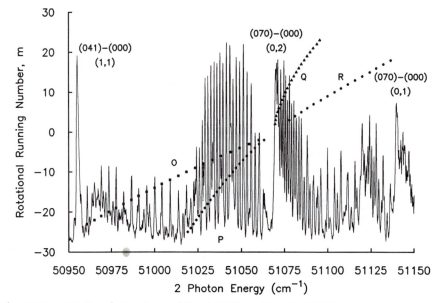

Fig. 20.4. Rotational structure of the $(0,7,0) \leftarrow (0,0,0)$, $(K',K'') = (0,2)$ $3p^2\Pi(A'')\leftarrow$ $\tilde{X}^2\Pi(A')$ vibronic subband of HCO near 51,069 cm^{-1}. The accompanying solid points lie on Fortrat parabolas articulated by the rotational running number m, where $m = N + 1$ for the R branch, $m = N$ for the Q branch, and $m = -N$ for the P and O branches.[26]

ionization from the $A\,^1\Pi_u$ state, via two-photon resonant excitation of the $1\,^1\Delta_u$ state, to be produced with a tunable dye laser timed to fire about 2 μsec after the photolysis laser pulse. REMPI ions produced with the tunable dye laser were mass analyzed with a time-of-flight mass spectrometer.

REMPI ion spectra in the mass 24 (C_2^+) channel were recorded for dye laser wavelengths ranging from 370 to 480 nm. Fragmentary vibronic progressions were observed for $v' = 0$, 1, and 2 with energy spacings, based on the assumption of a two-photon resonant absorption, that conformed with the known ω_e and $\omega_e x_e$ values fo the $A\,^1\Pi_u$ state. Ionization to the $^2\Pi_u$ ground state of C_2^+ (IP = 12.15 eV) for each of the observed two-photon resonances requires the absorption of two additional photons [i.e., (2 + 2) REMPI]. Spectroscopic constants for the $A\,^1\Pi_u$ state are accurately known.[56] It was thus possible to analyze the rotational structure of 13 vibronic bands with $0 \leq v' \leq 2$ and $0 \leq v'' \leq 4$ of the two-photon $1\,^1\Delta_u \leftarrow A\,^1\Pi_u$ transition to obtain spectroscopic constants for the upper $1\,^1\Delta_u$ state. The bands exhibit well developed O, P, Q, R, and S branches, characteristic of two-photon absorptions, as indicated for the $(v',v'') = (0,4)$ band shown in Fig. 20.6. The spectral resolution was limited by the dye laser linewidth and power broadening to about 0.3 cm^{-1}. Selection rules for two-photon resonant REMPI permit excitation of $^1\Sigma_u$, $^1\Pi_u$, $^1\Delta_u$, and $^1\Phi_u$ states from the $A\,^1\Pi_u$ state. Confirmation of the assignment of the upper level to a $^1\Delta_u$ state is provided by the portion of the (0,4) band spectrum illustrated in Fig. 20.7. The line assignments of Fig. 20.7 indicate that the Q(2) transition is present, but that no Q(1) transition appears; moreover, the P branches

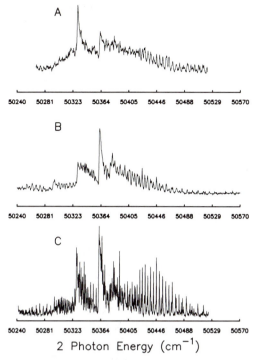

Fig. 20.5. (2 + 1) REMPI spectra of HCO near 397 nm. The $(K',K'') = (3,1)$ and $(1,1)$ subbands of the $(0,6,0) \leftarrow (0,0,0)$ $3p^2\Pi(A'') \leftarrow \tilde{X}^2\Pi(A')$ vibronic transition[25,26] are illustrated. (A) Spectrum observed from a $CH_4/O_2/Ar$ flame at 25 torr with flow rates (slm) of 0.75/1.5/0.6; a laser pulse energy of 2.5 mJ and probe bias voltage of 120 V were used.[24] (B) Spectrum of HCO produced from the 308 nm photolysis of CH_3CHO in a 300 K photolysis cell at 10 torr.[25] (C) Spectrum of HCO photofragments observed under collisionless conditions.[26]

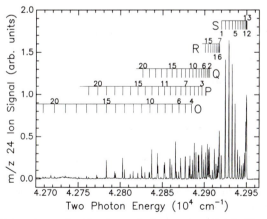

Fig. 20.6. REMPI spectrum for the two-photon resonant $(v',v'') = (0,4)$ rovibronic band of the $C_2(1^1\Delta_u \leftarrow A^1\Pi_u)$ transition near 466 nm. The ordinate is the mass 24 ion signal in arbitrary units.

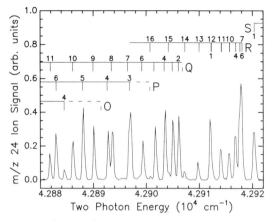

Fig. 20.7. A region near the band origin of the REMPI spectrum of Fig. 20.6, shown on an expanded scale.

terminate on the P(3) transition and the O branches terminate on the O(4) transition. The upper state can thus only be a $^1\Delta_u$ state.

The values of the molecular constants ω_e, $\omega_e x_e$, B_e and α_e for the $1^1\Delta_u$ state are given in Table 20.1. Morse potential curves are drawn for the presently known singlet states of C_2 in Fig. 20.8. For clarity, only the states of importance in the (2 + 2) REMPI ionization of C_2 via the $1^1\Delta_u$ state are labeled. The dashed curve of Fig. 20.8 is a Morse representation of the unobserved $2^1\Delta_u$ state, predicted by *ab initio* calculations[53,55] to undergo an avoided crossing with the $1^1\Delta_u$ state. The $1^1\Delta_u$ and $2^1\Delta_u$ states correlate, respectively, with the $C(2p^2\ ^1D) + C(2p^2\ ^1D)$ and $C(2p^2\ ^1D) + C(2p^2\ ^1S)$ separated atom states.[53,55] The $1^1\Delta_u$ state has not been observed with conventional absorption spectroscopy as it is not accessible from the two lowest lying singlet states, $X^1\Sigma_g^+$ and $A^1\Pi_u$, by one-photon absorption.

The spectroscopy of HCO and C_2, just summarized, and the extensive efforts devoted to the REMPI spectroscopy of CH_3 discussed elsewhere[19–21] serve as useful examples of the spectroscopic characterization of Rydberg states yielding the REMPI spectral signatures required for flame species density measurements. REMPI measurements of relative density profiles of HCO and CH_3 in hydrocarbon flames are discussed in the following section.

USE OF REMPI FOR FLAME PROFILE MEASUREMENTS OF HCO AND CH_3

As illustrations of the state of the art in the use of REMPI for measurements of the relative density profiles of key intermediates, we have recently measured profiles of CH_3 and HCO for methane, acetylene, ethylene, ethane, and propane flames.[24,59] The excellent spatial resolution and high signal-to-noise typical of REMPI measurements in flames are exhibited by these data. Measurements of the concentrations and profiles of these important intermediate species can provide sensitive checks on the rate constants and mechanisms built into models of hydrocarbon combustion.

Table 20.1. Spectroscopic constants for the $1^1\Delta_u$ state of C_2

$D_0 = 12{,}990 \text{ cm}^{-1a}$	$B_e = 1.361 \pm 0.002 \text{ cm}^{-1}$
$T_e = 57{,}720 \pm 2 \text{ cm}^{-1}$	$\alpha_e = 0.026 \pm 0.002 \text{ cm}^{-1}$
$\omega_e = 1150 \pm 1 \text{ cm}^{-1}$	$D_e = 7.7 \times 10^{-6} \text{ cm}^{-1}$
$\omega_e x_e = 21.3 \pm .5 \text{ cm}^{-1}$	$r_e = 1.437 \pm 0.002 \text{ Å}$
$\nu_{00} = 49097 \pm 1 \text{ cm}^{-1}$ $(1^1\Delta_u \leftarrow A^1\Pi_u)$	

^{a}Calculated using $D_0^0 = 6.21$ eV.

The methane/oxygen flame has been extensively studied. Measurements of the concentration profiles for all major species and many radical intermediates have been obtained with conventional molecular beam mass spectrometry.[4,60–63] These measurements have provided the basis for comprehensive computer models that include all major reaction mechanisms involved in the oxidation of methane.[3,64–68] The methane/oxygen flame thus provides a useful comparison with less precisely characterized hydrocarbon flames. Differences revealed by such comparisons are helpful in delineating the mechanisms for formation and consumption of key reactive intermediates. An example is provided by relative density profile measurements of CH_3 and HCO in low-pressure, premixed, stoichiometric $CH_4/O_2/Ar$ and $C_2H_4/O_2/Ar$ flames.

The P-branch head of the $(0,6,0) \leftarrow (0,0,0)$, $(K',K'') = (3,1)$ subband of the $(2 + 1)$ REMPI spectrum of HCO, shown in Fig. 20.5A at a two-photon energy of 50,328 cm^{-1}, was used for HCO profile measurements at a laser wavelength of 397.4 nm. The Q-branch head of the $3p^2A_2'' \leftarrow \tilde{X}^2A_2''$ band of CH_3, shown in Fig. 20.9 at a 59,970 cm^{-1} two-photon energy corresponding to a dye laser wavelength of 333.5 nm, was convenient for flame profile measurements of CH_3, as shown by previous investigators.[22,23]

Measurements of spatial profiles of the relative densities of flame species

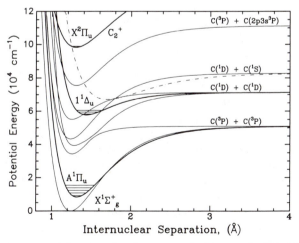

Fig. 20.8. Morse potential curves for the presently known singlet states of C_2 (solid curves).[41,54] An avoided crossing predicted by *ab initio* calculations[51,53] is shown between the $2^1\Delta_u$ state (dashed curve) and the $1^1\Delta_u$ state.

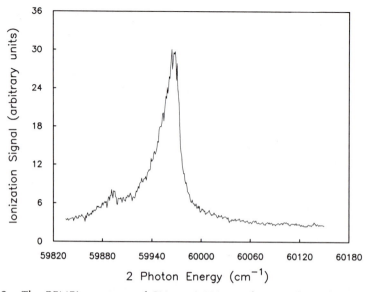

Fig. 20.9. The REMPI spectrum of CH_3 near 333 nm showing the Q-branch head of the $3p^2A_2'' \leftarrow \tilde{X}^2A_2''$ band. Taken in a stoichiometric $C_2H_4/O_2/Ar$ flame.[24]

require that the ionization probe be operated at a bias voltage corresponding to saturated collection of the REMPI electrons associated with the species of interest. The appropriate voltage is determined from observations of the variation in ionization signal with probe voltage. These variations in probe response with probe voltage are graphed in Fig. 20.10 for CH_3 ionized at 333.5 nm and HCO ionized at 397.4 nm, respectively. Both curves exhibit a well defined plateau region of nearly constant probe response, in which saturated electron collection occurs, followed by a "proportional counting" region in which the ionization signal increases quasilinearly with probe voltage.

A series of measurements was made in which the dye laser was tuned to monitor either the HCO or CH_3 radical and the probe was given the appropriate bias voltage for saturated collection of REMPI electrons. REMPI ionization profiles were recorded for both CH_4/O_2 and C_2H_4/O_2 flames with the dye laser and probe parameters fixed for CH_3 detection. The measurements were then repeated in the two flames with the laser and probe parameters appropriate for the HCO radical. Care was taken to ensure that measurements for the two flames were performed in identical fashion so that the profiles in both flames could be compared for each radical. A flame pressure of 18 torr and flow rates (slm) of 0.75/1.5/3.5 (cold flow velocity = 156 cm/sec) and 0.37/1.11/3.0 (cold flow velocity = 122 cm/sec) were used, respectively, for the $CH_4/O_2/Ar$ and $C_2H_4/O_2/Ar$ flames. These flow conditions give equal (per carbon atom) fuel flow rates to facilitate comparisons of CH_3 and HCO densities between the methane and ethylene flames.

The CH_3 radical profiles for the two flames are compared in Fig. 20.11. The CH_3 density is highest in the methane flame. The maximum for the ethylene flame is reached 1.9 mm from the burner head, whereas the maximum CH_3 density for the methane flame is not reached until 2.9 mm from the burner head.

Figure 20.12 illustrates the formyl radical profiles for the two flames. The two HCO profiles are similar in shape and magnitude. Like the CH_3 profiles, the maximum HCO density is reached closest to the burner for the ethylene flame. The lower edge of the visible flame zone is located about 1.4 mm above the HCO peak, at about 5.2 and 5.5 mm, respectively, for the ethylene and methane flames. The HCO profiles peak much further from the burner than the corresponding CH_3 profiles; the separations between the HCO and CH_3 peaks are about 1.2 and 2.0 mm for the methane and ethylene flames, respectively.

The excellent spatial resolution and high sensitivity of the REMPI method are exemplified by data for HCO and CH_3 in Figs. 20.11 and 20.12. The SNR at the peaks of the density profiles for HCO exceeded 20 with a laser pulse energy of 2.5 mJ at 397.4 nm; for CH_3, with a pulse energy of 5 mJ at 333.5 nm, the SNR exceeded 10^3 at the density peaks.

The reproducibility and good precision of these data may warrant comparisons of the qualitative features of the density profiles with existing flame models, even in the absence of absolute density calibrations. The significant lag in HCO forma-

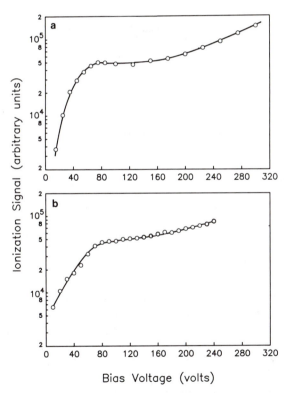

Fig. 20.10. Operating characteristics of the ionization probe of Fig. 20.1 for the REMPI detection of CH_3 (upper) and HCO (lower) in a stoichiometric $CH_4/O_2/Ar$ flame.[24] The ionization probe signal (collected electron charge) is plotted as a function of the probe bias voltage. Saturated collection of REMPI electrons from CH_3 (upper) occurs on the plateau region extending from about 70 to 190 V. The plateau for the detection of REMPI electrons from HCO (lower) extends from about 80 to 150 V.

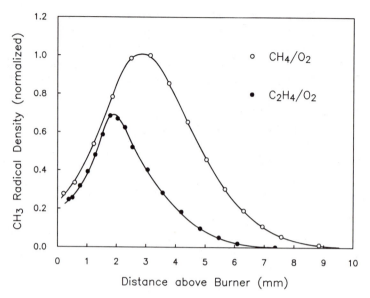

Fig. 20.11. A comparison of the CH_3 radical profiles obtained for stoichiometric $CH_4/O_2/Ar$ and $C_2H_4/O_2/Ar$ flames at a pressure of 18 torr.

tion with respect to CH_3 formation for the two flames has been mentioned. The flames exhibit a second kinetic similarity; comparable amounts of HCO are present in the methane and ethylene flames with equal (per carbon) fuel flow rates.

HCO is produced in the methane flame via CH_3 and H_2CO intermediates through the well-documented[64] reactions

$$CH_3 + O \rightarrow H_2CO + H \tag{20.3}$$

and

$$H_2CO + R \rightarrow HCO + RH \tag{20.4}$$

where R = O, H, and OH, following CH_3 production via H-atom abstraction from methane by the reactions

$$CH_4 + R \rightarrow CH_3 + RH \tag{20.5}$$

where R = O, H, and OH. Thus the methane flame reactions (20.3)–(20.5) provide an efficient route to HCO formation via intermediate formaldehyde; since HCO is formed with the consumption of CH_3, the peaking of the HCO profile follows the decay of the CH_3 profile in agreement with the data of Figs. 20.11 and 20.12.

The kinetic similarities between the ethylene and methane flames suggest a comparably efficient mechanism for HCO formation in the ethylene flame. The product channels for the oxidation of C_2H_4 by O and OH have been much discussed.[2,43,64,69,70,73–75] The kinetics of the formation of CH_3 and HCO in the ethylene flame is complicated by the low flame pressure, which lies within the fall-off region for adduct stabilization, and by the wide variation of temperature in the preheat zone between the burner head and the luminous flame zone.

Methyl radicals may be produced in the ethylene flame by the reactions[43,64,71,72]

$$C_2H_4 + OH \rightarrow CH_3 + H_2CO \tag{20.6}$$

and

$$C_2H_4 + O \rightarrow CH_3 + HCO \tag{20.7}$$

or

$$C_2H_4 + O \rightarrow C_2H_3O + H \tag{20.8}$$

followed by

$$C_2H_3O \rightarrow CH_3 + CO \tag{20.9}$$

and

$$C_2H_3O + O \rightarrow CH_3 + CO_2 \tag{20.10}$$

The branching ratio between reactions (20.7) and (20.8) has been controversial. Buss, et al.[69] found no evidence for the $CH_3 + HCO$ product channel of reaction (20.7) under single collision conditions. Other studies indicate that reaction (20.7), which requires an intersystem crossing between triplet and singlet energy surfaces,[69] does indeed occur under typical flame conditions.[43,64]

The equal per carbon atom production of HCO for the methane and ethylene flames would be consistent with strong contributions from reactions (20.6) or (20.7) followed by the formaldehyde route of reactions (20.3) and (20.4). Figures 20.8 and 20.9 exhibit little evidence for the direct formation of HCO coincident with the formation of CH_3 as a result of reaction (20.7) in the ethylene flame. The delayed peaking of HCO would be more consistent with reaction (20.6), which is known to

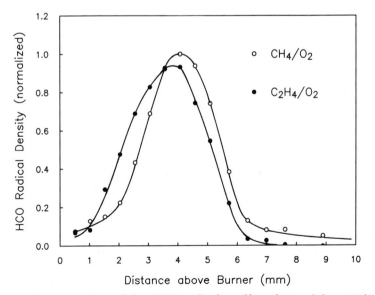

Fig. 20.12. A comparison of the HCO radical profiles obtained for stoichiometric $CH_4/O_2/Ar$ and $C_2H_4/O_2/Ar$ flames at a pressure of 18 torr.

occur at low temperatures[73] and has been included in recent ethylene flame models.[71,72] Several studies,[74] most recently those of Tully,[75] have shown, however, that the addition reaction

$$C_2H_4 + OH \rightarrow C_2H_4OH^* \tag{20.11}$$

followed by adduct stabilization

$$C_2H_4OH^* + M \rightarrow C_2H_4OH + M \tag{20.12}$$

dominates other OH reaction channels [including reaction (20.6)] for temperatures below 600–700 K and pressures above a few torr.

At higher temperatures the H-atom abstraction

$$C_2H_4 + OH \rightarrow C_2H_3 + H_2O \tag{20.13}$$

reaction is expected to predominate.[71,72,74,75] The studies of Baldwin and Walker[76] and of Gutman and coworkers[77] indicate that the subsequent reaction of vinyl radicals with O_2,

$$C_2H_3 + O_2 \rightarrow H_2CO + HCO \tag{20.14}$$

may be quite important under the present flame conditions. Reactions (20.13), (20.14), and (20.4) provide an efficient mechanism for HCO formation in the ethylene flame consistent with the equal per carbon atom formation of HCO observed for the methane and ethylene flames.

The kinetics of CH_3 and HCO formation in the preheat zone of the low pressure stoichiometric ethylene flame is thus apparently more complex than for the comparable methane flame. A comparison of the CH_3 profiles of Fig. 20.11 and the HCO profiles of Fig. 20.12 with flame model calculations, for the present flame conditions, may help resolve present uncertainties in the mechanisms of CH_3 and HCO production and consumption for the two flames. An effort is in progress in our laboratory to place the REMPI density measurements of HCO on an absolute scale with a calibration based on a measurement of laser absorption on the HCO $\tilde{A}^2\Pi(A'') \leftarrow \tilde{X}^2\Pi(A')$ band system.

FUTURE APPLICATIONS OF REMPI SPECTROSCOPY FOR FLAME SPECIES DETECTION

The Laser Ionization Mass Spectrometer

The REMPI method provides a useful laser-based alternative to conventional flame sampling and mass spectrometric techniques[4,5] that have provided much of the experimental justification for existing flame models. The REMPI approach provides an essential complement to the LIF method (the method of choice for strongly fluorescing species) and to direct mass spectrometric sampling for studies of combustion mechanisms. Unique advantages of REMPI detection, compared with conventional mass spectrometric sampling methods, include: a very high sensitivity, good spatial resolution, a capacity for discrimination between isomers, negligible perturbation of the flame environment, and a capability for direct *in situ* measurements of species concentrations. The REMPI method is better suited for monitoring highly reactive intermediates, since reactions occurring during the sampling process in conventional molecular beam mass spectrometry may change the composition of the sample before it is ionized and mass analyzed. The danger exists, therefore, that the mass-analyzed ion spectrum is not representative of the actual flame composition; reliable corrections for these changes may be difficult or impossible to make.

The principal disadvantages of REMPI detection in flames are a current lack of spectral information requisite for unambiguous species identification, the possible complications of photolytically induced changes in flame composition,[78,79] and the absence of reliable calibration methods for converting measured ionization signals to absolute species densities.

Because, as already mentioned, REMPI spectra for many radical species formed in a variety of sources have been obtained via mass-selected REMPI spectroscopy, it should be feasible to perform mass-selected REMPI spectroscopy directly on the radicals formed within flames. A proposed apparatus, designed to yield the spectral signatures needed for monitoring flame radicals with a REMPI electron probe is schematically illustrated in Fig. 20.13. The flame sampling and TOF-MS apparatus are of conventional design, well documented in previous mass spectrometric sampling studies of species profiles in hydrocarbon flames.[5] The novel aspect of the proposed apparatus is the use of REMPI as an ionization source. The laser ionization may be accomplished either prior to ion extraction from the flame (option 1), or just as the extracted neutral radicals reach the acceleration region of the TOF-MS (option 2). When laser ionization occurs within the flame, the sampled ions are focused with an electrostatic lens through the orifice of a skimmer into the TOF source region. The REMPI ion packet, spatially elongated in transit from the flame to the acceleration plates of the TOF-MS, is introduced to the flight tube by pulse-gating the ion extraction plates. Electrostatic focusing and pulsed ion extraction are not used with option (2). With either technique, partial discrimination of the REMPI ions from background ions within the flame is accomplished by signal detection with a boxcar signal averager, gated with a precisely controlled time-delay in synchronization with the ionizing laser pulses.

Laser ionization within the flame (option 1) provides several advantages over conventional mass spectrometric flame sampling methods:

1. The ionization distinguishes species located at the laser focus within the flame itself rather than species that exist after extraction from the flame. Because the ionization is species specific, the number of possible species that can contribute to the observed mass spectrum is very small even when rapid reactions scramble the sample composition during the gas-dynamic sampling process. Thus, the identity of the parent species may be assigned with a high degree of certainty from the observed mass spectrum. In many cases the identity of the parent REMPI ion may also be established from a knowledge of the dependence of the ionization signal on laser wavelength.
2. Laser resonance ionization provides a unique selectivity. Conventional molecular beam mass spectrometry cannot distinguish between isomers such as the methoxy (CH_3O) and hydromethoxy (CH_2OH) radicals that are distinguishable[80,81] with the REMPI approach. The REMPI method can also,

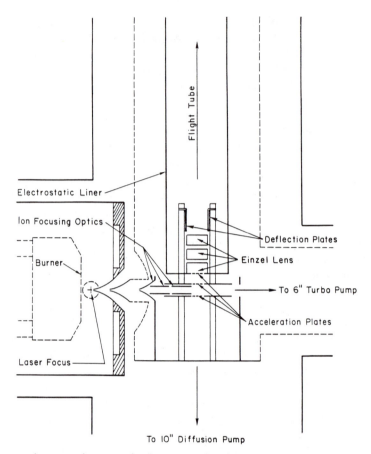

Fig. 20.13. Schematic diagram of a flame sampling laser ionization time-of-flight mass spectrometer for REMPI monitoring the flame radicals. The apparatus may be used in two modes. With option (1) REMPI ions are produced by focusing a tunable dye laser within the flame zone of a low-pressure flat flame burner. Laser-produced ions, sampled via a two-stage differentially pumped quartz sampling cone and skimmer, are accelerated along an evacuated (10^{-6} to 10^{-7} torr) flight tube and recorded by an electron multiplier. With option (2) neutral radical species are sampled; laser ionization is produced with the laser focused within the draw-out region between the acceleration grids. The apparatus is contained within a demountable stainless-steel vacuum housing equipped with optical and flow pumping ports.

in principle, distinguish between different states of the same species such as singlet and triplet CH_2, which play different kinetic roles in combustion processes.[82] These spectral signatures also avoid confusion between different species with similar masses such as C_2O and C_3H_4, or HCO and C_2H_5, for example.

3. Detection within flames at the sub-part-per-million level is expected to be routinely achievable for many species of importance in combustion chemistry, observable with the REMPI technique.

With the acquisition of sufficient spectral data for key radical species using options (1) and (2), diagnostic measurements may ultimately be made with the electron collection probe, eliminating the need for complex flame sampling and mass spectrometry.

The Ultrasensitive Detection of Toxic Effluents from Incinerators

In recent years, widespread public awareness of the need for safe disposal of hazardous wastes has resulted in mandated changes in state and federal waste management practices. A critical issue in public acceptance of the technology of hazardous waste control is the proper and accurate monitoring of the stack emissions from municipal and hazardous waste incinerators. Present regulations require that hazardous waste incinerator facilities meet performance standards based on trial burn procedures.[6,9–11] The current technology for sampling and analysis of flue gas emissions is highly developed. The destruction and removal efficiency (DRE) of an incinerator facility for selected principal organic hazardous constituents (POHCs) is determined by expensive and time-consuming testing, requiring sophisticated sampling procedures and analysis via chromatography and mass spectrometry.[6,7,9] Unfortunately, the day-to-day process monitoring that follows successful burn trials is based solely on indirect indicators of the DRE, as there are no reliable methods presently available for continuous real-time monitoring of the efficiency for the removal of specific POHCs.[6,10] Monitoring of the levels of O_2, CO, NO_x and total unburned hydrocarbons in the stack gases along with measurements of temperatures, flow rates, and scrubber solution pH are relied upon as indirect indicators of the DRE of toxic pollutants.[6,10] The successful development of an ultrasensitive technique for the accurate continuous *in situ* monitoring of the levels of POHCs within stack effluents would permit real-time corrections to be made for complex transient variations in the amount and composition of waste feed to the incinerator.[6,10,12] Moreover, a demonstrated capability for accurate continuous monitoring of potentially harmful emissions may provide an answer to present public health concerns regarding the management of municipal and hazardous wastes.

An on-line continuous monitoring device based on the mass-resolved REMPI detection of the parent or photofragment ions of either the POHCs or their suitable surrogates is envisioned as a direct extension of equipment currently in widespread use for REMPI spectroscopy.[13,14,40] It would consist of a sampling system similar to those used for the monitoring of CO and hydrocarbons in stack gases,[10] followed by a supersonic molecular beam, resonance-enhanced multiphoton ionization, time-of-flight mass spectrometer apparatus (MB/REMPI/MS).[13,14,83] Pulsed molecular beam nozzles are commonly used to permit sampling of high-pressure gases (≥ 1 atm) with vacuum systems of modest capacity.[84]

Prototypical MB/REMPI/MS apparatus capable of ultrasensitive detection limits for chlorinated aromatic hydrocarbons at the ppb level[13] and for a variety of organophosphonates and sulfide molecules as low as 300 ppt[14] have recently been reported. The REMPI spectra for molecules at the low rotational temperatures (typically <10 K) of the expanded molecular beam exhibit a considerably simplified and better resolved structure than at room temperature.[42] The selectivity afforded by the combined effects of rotational cooling and mass discrimination has been used to distinguish between isomers of chlorinated aromatic hydrocarbons.[13] Selectivities of 10^4 have been demonstrated for the REMPI detection of target organophosphonates against background interferants.[14,83]

These studies demonstrate the potential use of mass-resolved REMPI spectroscopy for the real-time detection of trace concentrations of hazardous compounds and their surrogates at the ppb level. Considerable additional research is required on the REMPI spectroscopy of both selected POHCs and potential surrogates to establish the selectivity, detection limits, and feasibility of this promising new approach to the problem of continuous monitoring and control of the combustion of hazardous wastes.

There is also a need to develop appropriate diagnostic methods for the detection of important radical intermediates in the combustion of chlorinated hydrocarbons. Relatively little is known concerning the mechanisms of the oxidation of chlorinated hydrocarbons; the present design of thermal destruction facilities is based primarily on general models for hydrocarbon combustion,[3,44] without a detailed knowledge of the mechanisms by which toxic chlorinated by-products are formed. The presence of chlorine in the reagent mix greatly alters the composition of the pool of combustion radicals leading to profound effects on ignition limits, flame speeds, sooting characteristics, temperatures, and trace species formation rates and compositions.[1] The traditional use of flame sampling and mass spectrometry for concentration measurements of the stable intermediates and products has proved useful in the modeling of combustion mechanisms for some representative chlorinated hydrocarbons.[85-88] The highly sensitive laser-based methods of laser-induced fluorescence and resonance ionization have yet to be used for studies of the growth and decay of radical intermediates such as C_2Cl_3, CCl_3, ClO, CCl, $CHCl_2$, $COCl$, and $CHOCl$, which may play key roles in the formation of stable toxic byproducts. The use of laser-based methods for studies of such intermediates may be helpful in selecting appropriate surrogate species[6,10-12,88,89] that can be monitored in real-time to gauge the destruction efficiencies accurately for such highly toxic compounds as the polychlorinated dibenzo-p-dioxins and dibenzofurans.

ACKNOWLEDGMENTS

The experiments discussed in this paper were performed by the author's graduate students: Jeffrey S. Bernstein, Peter M. Goodwin, Xiao-Mei Song, and Paul J. H. Tjossem. This work was supported by the U.S. Army Research Office under Grant DAAL03-87-G-0053 and Contract DAAL-87-K-0066 and the U.S. Department of Energy, Office of Basic Energy Sciences, Division of Chemical Sciences, under Grant DE-FG02-86-ER13508.

REFERENCES

1. S. M. Senkan, in *Detoxication of Hazardous Waste,* Chap. 3, edited by J. H. Exner. Ann Arbor Science, Ann Arbor, 1982.

2. J. A. Miller and G. A. Fisk, *Chem. Engr. News* **65**, 22, 1987.

3. C. K. Westbrook and F. L. Dryer, *Prog. Energy Combust. Sci.* **10**, 1 (1984).

4. J. C. Biordi, *Prog. Energy Combust. Sci.* **3**, 151 (1977).

5. See, for example, J. C. Biordi, in *Experimental Diagnostics in Gas Phase Combustion Systems*, edited by B. T. Zinn, *Progress in Astronautics and Aeronautics*, Vol. 53. American Institute of Aeronautics and Astronautics, New York, 1977; R. M. Fristrom, *Int. J. Mass Spectrom. Ion Phys.* **16**, 15 (1975); J. W. Hastie, *Int. J. Mass Spectrom. Ion Phys.* **16**, 89 (1975); R. V. Serauskas, G. R. Brown, and R. Pertel, *Int. J. Mass Spectrom. Ion Phys.* **16**, 69 (1975); J. D. Bittner, *A Molecular Beam Mass Spectrometer Study of Fuel-Rich and Sooting Benzene-Oxygen Flames*, Sc.D. Thesis, Department of Chemical Engineering, M.I.T., Cambridge, MA, 1981; J. D. Bittner and J. B. Howard, *Eighteenth Symposium (International) on Combustion*. The Combustion Institute, Pittsburgh, 1981, p. 1105.

6. E. T. Oppelt, *J. Air Pollut. Control Assoc.* **37**, 558 (1987).

7. J. W. A. Lustenhouwer, K. Olie, and O. Hutzinger, *Chemosphere* **9**, 501 (1980).

8. R. R. Bumb et al., *Science* **210**, 385 (1960).

9. R. E. Adams, R. H. James, L. B. Farr, M. M. Thomason, H. C. Miller, and L. D. Johnson, *Environ. Sci. Technol.* **20**, 711 (1986).

10. R. K. La Fond, J. C. Kramlich, W. M. Seeker, and G. S. Samuelsen, *J. Air Pollut. Control Assoc.* **35**, 658 (1985).

11. B. Dellinger and D. L. Hall, *J. Air Pollut. Control Assoc.* **36**, 179 (1986).

12. W. Tsang, and W. Shaub, in *Detoxication of Hazardous Waste*, Chap. 2, edited by J. H. Exner. Ann Arbor Science, Ann Arbor, 1982.

13. E. A. Rohlfing and D. W. Chandler, in *Advances in Laser Science II, Optical and Engineering Sciences 8*, Edited by M. Lapp, W. C. Stwalley, and G. A. Kinney-Wallace, The American Institute of Physics, New York, 1987, p. 618.

14. J. A. Syage, J. E. Pollard, and R. B. Cohen, *Appl. Opt.* **26**, 3516 (1987).

15. J.E.M. Goldsmith, *Opt. Lett.* **7**, 437 (1982).

16. P.J.H. Tjossem and T. A. Cool, *Chem. Phys. Lett.* **100**, 479 (1983).

17. J.E.M. Goldsmith, *Twentieth Symposium (International) on Combustion*. The Combustion Institute, Pittsburgh, 1984, p. 1331.

18. J.E.M. Goldsmith, *J. Chem. Phys.* **78**, 1610 (1983).

19. T. G. DiGiuseppe, J. W. Hudgens, and M. C. Lin, *J. Phys. Chem.* **86**, 36 (1982); *Chem. Phys. Lett.* **82**, 267 (1981); *J. Chem. Phys.* **76**, 3338 (1982).

20. J. Danon, H. Zacharias, H. Rottke, and K. H. Welge, *J. Chem. Phys.* **76**, 2399 (1982).

21. J. W. Hudgens, T. G. DiGuiseppe, and M. C. Lin, *J. Chem. Phys.* **79**, 571 (1983).

22. K. C. Smyth and P. H. Taylor, *Chem. Phys. Lett.* **122**, 518 (1985).

23. U. Meier and K. Kohse-Höinghaus, *Chem. Phys. Lett.* **142**, 498 (1987).

24. J. S. Bernstein, X.-M. Song, P. M. Goodwin and T. A. Cool, *Twenty-Second Symposium (International) on Combustion*. The University of Washington, Seattle, August 14–19, 1988, in press.

25. P.J.H. Tjossem, P. M. Goodwin, and T. A. Cool, *J. Chem. Phys.* **84**, 5334 (1986).

26. P.J.H. Tjossem, T. A. Cool, D. A. Webb, and E. R. Grant, *J. Chem. Phys.* **88**, 617 (1988).

27. J. S. Bernstein, X.-M. Song, and T. A. Cool, *Chem. Phys. Lett.* **145**, 188 (1988).

28. S. Sharpe and J. W. Hudgens, *Chem. Phys. Lett.* **107**, 35 (1984).

29. M. T. Duignan and J. W. Hudgens, *J. Chem. Phys.* **82**, 4426 (1985).

30. G. R. Long and J. W. Hudgens, *J. Phys. Chem.* **91**, 5870 (1987).

31. P.J.H. Tjossem and K. C. Smyth, *Chem. Phys. Lett.* **144**, 51, (1988).

32. W. G. Mallard, J. H. Miller, and K. C. Smyth, *J. Chem. Phys.* **76**, 3483 (1982).

33. B. H. Rockney, T. A. Cool, and E. R. Grant, *Chem. Phys. Lett.* **87**, 141 (1982).

34. K. C. Smyth and W. G. Mallard, *J. Chem. Phys.* **77**, 1779 (1982).

35. W. G. Mallard, J. H. Miller, and K. C. Smyth, *J. Chem. Phys.* **79**, 5900 (1983).

36. P.J.H. Tjossem and T. A. Cool, *Twentieth Symposium (International) on Combustion*. The Combustion Institute, Pittsburgh, 1984, p. 1321.

37. P. H. Taylor, W. G. Mallard, and K. C. Smyth, *J. Chem. Phys.* **84,** 1053 (1986).

38. T. A. Cool, *Appl. Opt.* **23,** 1571 (1984).

39. J. W. Hudgens, *Advances in Multi-photon Processes and Spectroscopy,* Vol. 4 edited by S. H. Lin. World Scientific, Singapore, 1988, p. 171.

40. D. H. Parker, in *Ultrasensitive Laser Spectroscopy,* Chap. 4, edited by D. S. Kliger. Academic Press, New York, 1983.

41. P. M. Goowdin and T. A. Cool, *J. Chem. Phys.* **88,** 4548 (1988).

42. D. H. Levy, L. Wharton, and R. E. Smalley, in *Chemical and Biological Applications of Lasers,* Vol II. Academic Press, New York, 1977, p. 1.

43. D. J. Hucknall, *Chemistry of Hydrocarbon Combustion.* Chapman and Hall, London, 1985.

44. I. Glassman, *Combustion.* Academic Press, Orlando, FL, 1987.

45. J. S. Levine, *The Photochemistry of Atmospheres.* Academic Press, New York, 1985.

46. J. Weaver, J. Meager, and J. Heicklen, *J. Photochem.* **6,** 111 (1976/77).

47. J. G. Calvert and J. N. Pitts, *Photochemistry.* John Wiley, New York, 1966.

48. A. Horowitz, C. J. Kershner, and J. G. Calvert, *J. Phys. Chem.* **86,** 3094 (1982).

49. C. B. Moore and J. C. Weisshaar, *Annu. Rev. Phys. Chem.* **34,** 525 (1983).

50. L. E. Snyder, J. M. Hollis, and B. L. Ulrich, *Astrophys. J.* **208,** L91 (1976).

51. J. M. Hollis and E. Churchwell, *Astrophys. J.* **271,** 170 (1983).

52. L. J. Allamandola, *J. Mol. Struct.* **157,** 255 (1987).

53. P. F. Fougere and R. K. Nesbet, *J. Chem. Phys.* **44,** 285 (1966).

54. J. Barsuhn, *Z. Naturforsch. Teil A* **27,** 1031 (1972).

55. K. Kirby and B. Liu, *J. Chem. Phys.* **70,** 893 (1979).

56. K. P. Huber and G. Herzberg, *Molecular Spectra and Molecular Structure IV. Constants of Diatomic Molecules.* Van Nostrand Reinhold, New York, 1979; contains a tabulation of the spectroscopic constants for the known electronic states of C_2.

57. A. M. Wodtke and Y. T. Lee, *J. Phys. Chem.* **89,** 4744 (1985).

58. J. R. McDonald, A. P. Baronavski, and V. M. Donnelly, *Chem. Phys.* **33,** 161 (1978); S. V. ONeil, P. Rosmus, and H.-J. Werner, *J. Chem. Phys.* **87,** 2847 (1988).

59. J. S. Bernstein, X.-M. Song, and T. A. Cool, unpublished.

60. J. Peeters and G. Mahnen, *Fourteenth Symposium (International) on Combustion.* The Combustion Institute, Pittsburgh, 1972, p. 133.

61. J. Peeters and C. Vinckier, *Fifteenth Symposium (International) on Combustion.* The Combustion Institute, Pittsburgh, 1974, p. 969.

62. J. C. Biordi, C. P. Lazzara, and J. F. Papp, *Fifteenth Symposium (International) on Combustion.* The Combustion Institute, Pittsburgh, 1974, p. 917; *Fourteenth Symposium (International) on Combustion.* The Combustion Institute, Pittsburgh, 1972, p. 367; C. P. Lazarra, J. C. Biordi, and J. F. Papp, *Combust. Flame* **21,** 371 (1973).

63. R. Harvey and A. Maccoll, *Seventeenth Symposium (International) on Combustion.* The Combustion Institute, Pittsburgh, 1978, p. 857.

64. J. Warnatz, *Combustion Chemistry,* Chap. 5, edited by W. C. Gardiner, Jr. Springer-Verlag, New York, 1984.

65. J. O. Olsson and L. L. Andersson, *Combust. Flame* **67,** 99 (1987).

66. T. P. Coffee, *Combust. Flame* **55,** 161 (1984).

67. J. H. Bechtel, R. J. Blint, C. J. Dasch, and D. A. Weinberger, *Combust. Flame* **42,** 197 (1981).

68. J. Warnatz, *Eighteenth Symposium (International) on Combustion.* The Combustion Institute, Pittsburgh, 1980, p. 369.

69. R. J. Buss, R. J. Baseman, H. Guozhong, and Y. T. Lee, *J. Photochem.* **17,** 389 (1981).

70. A. R. Clemo, G. L. Duncan, and R. Grice, *J. Chem. Soc., Faraday Trans. II* **78,** 1231 (1982).

71. C. K. Westbrook, F. L. Dryer, and K. P. Schug, *Nineteenth Symposium (International) on Combustion.* The Combustion Institute, Pittsburgh, 1982, p. 153.

72. C. K. Westbrook, F. L. Dryer, and K. P. Schug., *Combust. Flame* **52,** 299 (1983).

73. M. Bartels, K. Hoyermann, and R. Sievert, *Nineteenth Symposium (International) on Combustion.* The Combustion Institute, Pittsburgh, 1982, p. 61.

74. A.-D. Liu, W. A. Mulac, and C. D. Jonah, *Int. J. Chem. Kinet.* **19,** 25 (1987), and references therein.

75. F. P. Tully, *Chem. Phys. Lett.* **96,** 148 (1983); *Chem. Phys. Lett.* **143,** 510 (1988).

76. R. R. Baldwin and R. W. Walker, *Eighteenth Symposium (International) on Combustion.* The Combustion Institute, 1980, p. 819.

77. I. R. Slagle, J.-Y., Park, M. C. Heaven, and D. Gutman, *J. Am. Chem. Soc.* **106,** 4356 (1984).

78. J.E.M. Goldsmith, *Opt. Lett.* **11,** 416 (1986).

79. J.E.M. Goldsmith, *Opt. Lett.* **11,** 67 (1986).

80. C. S. Dulcey and J. E. Hudgens, *J. Chem. Phys.* **84,** 5262 (1986); *J. Phys. Chem.* **87,** 2296 (1983).

81. G. R. Long, R. D. Johnson, and J. W. Hudgens, *J. Phys. Chem.* **90,** 4901 (1986).

82. C. E. Canosa-Mas, H. M. Frey, and R. Walsh, *J. Chem. Soc. Faraday Trans. II* **80,** 561 (1984).

83. J. A. Syage, J. E. Pollard, and R. B. Cohen, *Ultratrace Detection of Chemical Warfare Agent Simulants Using Supersonic Molecular Beam, Resonance Enhanced Multiphoton Ionization, Time-of-Flight Mass Spectroscopy.* Rept. SD-TR-88-13, The Aerospace Corporation, El Segundo, CA, Feb. 15, 1988.

84. W. R. Gentry and C. F. Giese, *Rev. Sci. Instrum.* **49,** 595 (1978).

85. W. D. Chang, S. B. Karra, and S. M. Senkan, *Combust. Sci. Technol.* **49,** 107 (1986).

86. D. Bose, D. and S. M. Senkan, *Combust. Sci. Technol.* **35,** 187 (1983).

87. S. M. Senkan, *Combust. Sci. Technol.* **38,** 197 (1984).

88. W. D. Chang, S. B. Karra, and S. M. Senkan, *Environ. Sci. Technol.* **20,** 1243 (1986).

89. J. L. Graham, D. L. Hall, and B. Dellinger, *Environ. Sci. Technol.* **20,** 703 (1986).

21

New Developments in Molecular Detection by Supersonic Molecular Beam, Laser Mass Spectrometry

JACK A. SYAGE

Sensitive methods of detection serve a wide variety of applications.[1] Chemical analysis, for instance, often involves determining the composition of a given sample (i.e., assays) or the extent of contamination. Particular requirements might include the separation of isomers or isotopic substituents in molecules. These applications usually require high levels of sensitivity and selectivity, but not necessarily rapid analysis times (at least on a real-time basis). Another important area requiring trace detection methods is atmospheric monitoring. Applications are great in situations that have high environmental impact, such as toxic waste sites, space launch facilities, and nuclear power plants. Here, rapid analysis is more important. This statement is particularly true in situations involving toxic and hazardous compounds that might endanger life (i.e., chemical warfare agents).

As sensitivity continues to improve, the problem of interference from more abundant compounds becomes a greater concern. This awareness has motivated work to develop methods capable of realizing greater selectivity. Our approach to this problem has centered on the use of multiphoton ionization (MPI). This process relies on the sequential absorption of photons until the total energy of the molecule exceeds the ionization potential. If the photon wavelength is chosen to match an absorbing state of the molecule, the process is called resonance enhanced (i.e., REMPI). The resonance step not only improves sensitivity substantially, but it also provides a means for selectively ionizing molecules. In some of the earlier MPI applications, additional selectivity was introduced by tandem arrangements that included stages such as gas chromatography (GC) to separate the constituents of a mixed gas sample before ionization. Optical selectivity, though not adequate by itself for thermal samples, was used to augment the GC selectivity by taking advantage of the different absorption properties and ionization potentials. Klimcak and Wessel demonstrated high levels of sensitivity and selectivity using such an arrangement for the isomers phenanthrene and anthracene.[2] More recently, Reilly and co-workers demonstrated selective detection using an arrangement based on the combination of GC and time-of-flight (TOF) mass spectrometry.[3] This method provides mass selectivity; however, they opted for a fixed wavelength excimer laser for ionization, thus sacrificing some optical selectivity for simplicity of operation.

Some form of mass spectrometry is now used by most detection methods based on MPI. Preparation of gaseous samples by supersonic expansion offers enormous benefits with regard to selectivity and will be discussed at length shortly. However, it can be argued that this improvement comes at the expense of sensitivity since the gaseous sample, on expansion in the vacuum chamber, is at a significantly lower density in the region of detection. In response to this concern, Kolaitis and Lubman have demonstrated the advantages of atmospheric pressure MPI mass spectrometry in conjunction with an ion mobility drift tube apparatus.[4] Exceptional levels of sensitivity are possible at atmospheric pressures, in which part-per-billion (ppb) concentrations correspond to greater than 10^{10} molecules/cm^3. The optical selectivity of REMPI, however, is reduced in atmospheric samples, and the separability of the ion drift times is limited by diffusion-controlled broadening; hence the major source of selectivity is provided by the quadrupole mass spectrometer.[4]

Another approach to improving sensitivity involves accumulating trace molecules on a cooled substrate. After adequate coverage has been achieved, the molecules can then be desorbed by a pulsed laser and photoionized above the surface.[5] Pulsed desorption and MPI has also been applied extensively to surface analysis of materials[6] and is motivated to a large extent by the semiconductor community. The technique is likewise gaining popularity as a means for introducing nonvolatile or thermally unstable molecules into the gas phase or vacuum[7].

Efforts to improve selectivity have often centered on advancing the mass spectrometer performance. Mass resolution can be improved by simply operating two spectrometers in tandem (MS/MS).[8] However, the most important development for obtaining ultrahigh mass resolution is Fourier transform mass spectrometry (FTMS), which has recently been reviewed by Marshall.[9] The basis for this technique is the use of an ion trap cell that derives from ion cyclotron resonance (ICR) spectrometers, hence the technique is often referred to as FTICR. The resolution depends on the residence time of the ions in the ion trap cell, and can reach levels exceeding 100,000. This capability permits separation of compounds of the same nominal molecular weight, but of different molecular structure.

Detection methods that improve optical selectivity are understandably based on techniques originally developed for high-resolution spectroscopy. Foremost among these is the use of supersonic molecular beams. For detection applications, supersonic cooling accounts for excellent sensitivity and selectivity because (1) the cold molecules are in virtually the same energy state and can therefore be nearly completely ionized by resonance excitation using a single wavelength of light, and (2) each molecule typically has its own characteristic set of narrow excitation wavelengths (i.e., spectral fingerprints) that is distinct from that of other molecules. In the following sections, recent work will be reviewed that demonstrates the enormous enhancement in optical selectivity that becomes possible by supersonic molecular beam detection. However, it is emphasized that sensitivity is likewise considerable. The loss in sensitivity due to the dilution in molecular density that occurs upon expansion is largely compensated by the benefit that results from the very narrow distribution of initial energy states.

EXPERIMENTAL TECHNIQUE

Supersonic Molecular Beam

The expansion of a high pressure of gas into a vacuum through an aperture is termed a supersonic expansion and can lead to substantial cooling of molecular internal energy. The method is well established and extensively reviewed.[10] The very narrow distribution of initial rovibrational states in conjunction with laser excitation has many valuable benefits. From a detection and sensitivity standpoint, nearly complete molecular excitation can be achieved using a single wavelength of light. For selective excitation, it is possible to populate single vibronic levels in excited electronic states and to form rovibrationally cold ions.

Most molecular beam photoionization experiments employ some form of mass resolved detection, usually using either a quadrupole or TOF mass spectrometer. A schematic of our molecular beam system[11] is illustrated in Fig. 21.1. The compact apparatus consists of two differentially pumped chambers, the first containing the molecular beam source and the second the excitation and detection components. A 1-mm-diameter skimmer, situated at the interface of the two chambers, allows a narrow collimated portion of the supersonic jet to enter the second chamber. This arrangement directs the majority of the gas pumping load to the first (source) chamber, allowing the second (detection) chamber to operate at higher vacuum. The source chamber is pumped by a liquid nitrogen trapped 2000 L/sec diffusion pump. The detection chamber is pumped by a 1200 L/sec turbomolecular pump that is mounted on the side to catch the direct output of the molecular beam. An all metal conflat design was chosen for the overall system to allow high temperature (up to 200°C) bakeouts to remove internal chamber deposits that can contribute to background signals. Ambient base pressure in the detection chamber is about 1×10^{-8} torr, rising to about 2×10^{-7} torr during operation of the pulsed molecular beam source under typical conditions (e.g., 10 Hz, 50 psi He, 500-μm-diameter nozzle, 400-μsec gas pulse width, 2.5-cm nozzle–skimmer distance). The detection chamber pumpdown time of 10 msec, however, ensures that the instantaneous pressure in the chamber at the time of a molecular beam pulse is close to the base pressure.

The use of pulsed nozzles for supersonic expansions is now well established with several designs available to meet specific requirements. However, the lack of a commercial high-repetition rate, high-temperature pulsed nozzle prompted us to construct an in-house pulsed source using a design based on fast solenoid activation.[12] This design offers a unique combination of features important to the detection method: (1) high repetition rates (up to 100 Hz), (2) temperature control (up to 200°C), (3) continuous gas sample flow capabilities, and (4) internally mountable sample holder.

The TOF mass spectrometer employs a configuration of three parallel grids,[13] spaced 1 cm apart, to direct the ions toward the detector. This is accomplished by applying different voltages to the grids (the final grid is held at ground), thereby creating the necessary electric fields for accelerating the ions. A relatively small field of about 100–200 V/cm is applied across the first two grids. This region encom-

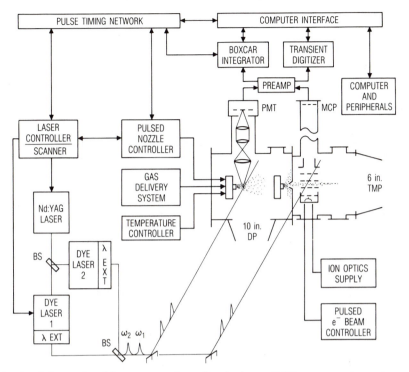

Fig. 21.1. Schematic of the supersonic molecular beam TOF mass spectrometer apparatus and associated components: PMT, photomultiplier tube; MCP, microchannel plate; DP, diffusion pump; TMP, turbomolecular pump; BS, beamsplitter; λ EXT, wavelength extension components.

passes the volume where the molecular beam and the laser intersect. The ions that are formed then move into the region between the second and third grids where they encounter a much greater field (about 1500–2000 V/cm). The ions are accelerated to high energy in this region. They then enter a 1-m field-free TOF drift region that allows the slower heavy ions to separate from the faster light ions. A time-resolved microchannel plate detector monitors the intensity and arrival time of each ion mass. The recorded TOF signals are then routinely converted to a mass spectrum. Most of the experiments described here use MPI to form ions. A fourth grid and a tungsten filament are used to generate a pulsed electron beam for applications requiring electron impact ionization.[14]

Laser Excitation and Detection

Laser excitation is provided by a tunable dye laser pumped by an Nd:YAG laser operating at 532 nm (500 mJ/pulse). By frequency doubling the visible dye output or by frequency mixing the visible with the Nd:YAG 1064 nm output, we obtain a tuning range of 275–440 nm (10–20 mJ/pulse). Frequency doubling followed by mixing with 1064 nm extends the tuning range to 222 nm (1–3 mJ/pulse). Shorter wavelengths to 195 nm (10–100 μJ) are generated using anti-Stokes Raman shifting in an H_2 pressure cell. A spherical lens is used to focus the light into the molecular beam.

TOF mass spectra are recorded using a 200-MHz transient digitizer. Ionization-detected excitation spectra are recorded by scanning the ionizing laser wavelength and detecting a specific mass, using a time-gated boxcar integrator (channel A). A second boxcar integrator (channel B) is used to record the laser pulse intensity. The recorded spectrum is normalized approximately by storing the ratio A/B^2 or, more precisely, by storing the pulse-to-pulse values of A and B for subsequent analysis using an appropriate model for the MPI.

Sample Preparation

The trace concentrations reported in this review were prepared by evacuating a 2-L stainless-steel cylinder, introducing the target molecules to low pressure (typically 1 torr as measured using a digital capacitance manometer), and backfilling to high pressure (typically 400 psi) with the carrier gas. The resulting mixture ($\sim$50 ppm) was diluted further by releasing the pressure and backfilling to attain the desired concentration. For concentrations below 1 ppm, successive dilutions were prepared and the signal levels monitored to ensure a linear dependence of signal with concentration. Dilution of dimethyl sulfide followed this dependence reasonably well to 60 ppb, below which the signal decreased faster than the concentration (presumably due to wall adsorption on the polyethylene and stainless-steel delivery line).

TANDEM OPTICAL/MASS DETECTION

Sensitive detection relies on finding photoionization processes that are efficient. Aromatic molecules are ideally suited as they involve strong absorption cross sections at convenient wavelengths, require just two photons to ionize, and form relatively stable parent ions. Aromatic molecules have therefore been employed extensively as model compounds for evaluating detection capabilities by REMPI. Other classes of molecules pose more serious detection challenges. Ionization efficiency may be reduced by competitive neutral pathways (e.g., dissociation, internal conversion), inconvenient lasing wavelengths, or large ionization potentials (IP) requiring more than two photons to ionize. Mass selectivity may also be compromised by extensive ion fragmentation. Finally, many molecules of interest are nonvolatile (i.e., biological molecules), thereby making the introduction of sufficient material into a photoionization source difficult.

The following sections focus primarily on recent work from our laboratories, with reference to related work by other researchers. We shall refer to specific excitation pathways as $m + n$ REMPI, where m refers to the number of photons absorbed to excite a resonance state, and n the additional absorbed photons necessary to ionize the molecule. An REMPI excitation spectrum can be recorded by measuring the ionization signal for the parent ion or some characteristic fragment ion as a function of excitation wavelength. This provides a sensitive analog to a one-photon absorption spectrum.

Aromatic Molecules

Lubman and co-workers have made contributions in demonstrating the detection capabilities of supersonic molecular beam MPI mass spectometry.[15-18] In a series of experiments, they determined the extent of separability attainable by molecular beam MPI mass spectrometry for (1) structural isomers,[16] (2) molecules of similar mass,[17] and (3) isotopic substituents.[18] For instance, they demonstrated that the 1 + 1 REMPI (sometimes referred to a R2PI) excitation spectra for structural isomers of dichlorobenzene are dramatically different, allowing a sensitive optical means for separation that is not available by mass detection alone. A similar study was conducted for the o-, m-, and p-cresol isomers.[18] These experiments were notable as they actually considered the separability of the isomers in supersonic expansions of air samples, in which cooling is not as effective. Nonetheless, discrimination limits between 10^2 and 10^3 were achieved, and a theoretical detection limit of 20 ppb was reported.[16]

Other groups have demonstrated optical selectivity for isomers without the aid of molecular beams. These include the selective ionization of either anthracene or phenanthrene using excitation at either 285 or 310 nm.[2] Another example is the selective ionization of chrysene over triphenylene using XeCl excimer excitation at 308 nm in which the two-photon energy exceeds the IP of chrysene (7.8 eV), but not triphenylene (8.1 eV).[3] Although optically selective excitation has been emphasized in the above studies of isomers, it is also possible to obtain mass selectivity by identifying different fragmentation patterns as a function of ionizing wavelength. The related challenge of separating molecules of similar or of the same nominal mass is less demanding than the isomer problem, since the parent masses can often be separated, and the optical and fragmentation properties tend to be more different for less similar molecules.

We have used aromatic molecules as models for establishing enhanced two-stage optical selectivity by using double resonance excitation in the detection process. This work is described later in the chapter.

Challenging Applications

Many molecules present special challenges or sensitive detection. Often, the most hazardous molecules, which demand the greatest levels of detection sensitivity, are poorly understood spectroscopically. We have been engaged in studies to establish the detectability of more "troublesome" classes of molecules by the supersonic molecular beam MPI mass spectrometry technique. Good examples are the organophosphates and phosphonates that represent groups of insecticides and pesticides, as well as chemical nerve agents. Vacuum UV absorption spectra indicate that strong single-photon absorptions occur only at relatively high energies (i.e., wavelengths of about 200 nm and shorter).[19] This situation also characterizes many organosulfide molecules.[11,19-21] The spectroscopy of these compounds poses problems for sensitive detection, as adequate pulse energies are difficult to generate at these wavelengths using conventional pulsed dye lasers and nonlinear frequency generation methods. Hydrazines are an important class of toxic molecules that are used as rocket fuels, but whose physical properties are a challenge to laser-based detection methods. We consider each of these groups of compounds in the following sections.

Sensitive and selective laser detection requires specific excitation processes that uniquely probe the molecule of interest. The high-resolution spectroscopic data necessary for this task, however, were not available for the classes of compounds previously mentioned. Therefore significant experimental effort was directed toward fundamental spectroscopic studies to provide the so-called spectral data base or signatures. This was accomplished by first recording room temperature vapor phase absorption spectra using a home-built UV and vacuum UV spectrometer. Once absorptive wavelength regions were defined, high-resolution REMPI excitation spectra were recorded using a scanning laser in the supersonic molecular beam mass spectrometer apparatus.

Organophosphonates

A variety of excitation processes through different electronic states were investigated for diisopropyl methylphosphonate (DIMP). These excitation schemes, which are diagramed in Fig. 21.2, invariably lead to ion fragmentation with no evidence for a parent ion at mass 180. However, characteristic fragment ions are evident in the mass spectra in Fig. 21.3 that afford mass selective detection unique for DIMP. The 2 + 2 REMPI excitation spectrum of DIMP in Fig. 21.4 shows a highly resolved vibrational structure for the resonance electronic state transition. In this case, the absorptions were detected by recording the predominant isopropyl ion fragment at mass 43 (see Fig. 21.3c). The spectrum in Fig. 21.4 was recorded using room air as the carrier gas to demonstrate the potential of supersonic molecular beam spectroscopy for atmospheric monitoring. The extent of molecular cooling in terms of reduced rotational broadening and hot band intensity (due to rotational and vibrational energy, respectively, in the nascent molecule) compares favorably to that obtained using 50 psi He as a carrier gas. Hence, considerable optical selectivity is made possible by supersonic expansion of atmospheric samples. An additional stage of selectivity is of course provided by the TOF mass spectrometer.

Because sufficient tunable pulse energy at 200 nm was not available, it was necessary to drive the formally allowed one-photon transition by using two photons at half the transition energy. Although this is not a favorable excitation mechanism for sensitivity purposes, a detection limit on the order to 50 ppb was nonetheless obtained. The other excitation mechanisms shown in Fig. 21.2 were examined in an attempt to improve the detection efficiency. A 1 + 2 REMPI excitation via a lower energy state at about 266 nm was also studied. However, the resonance transition was very weak and diffuse and, hence, sensitivity was not improved by this route. Interestingly, there was no resemblance in the mass spectra when exciting by these two different routes, as illustrated in Fig. 21.3. There is some evidence that the 200-nm state may dissociate, thereby altering the fragmentation pattern in addition to reducing the ionization efficiency.[11,19]

The molecule dimethyl methylphosphonate (DMMP) exhibits rather different spectroscopic and photochemical behavior than the structurally related molecule DIMP. Broadband UV/VUV absorption spectra of room temperature vapor-phase samples showed no evidence for any strong absorption bands ($\epsilon > 500$) at wave-

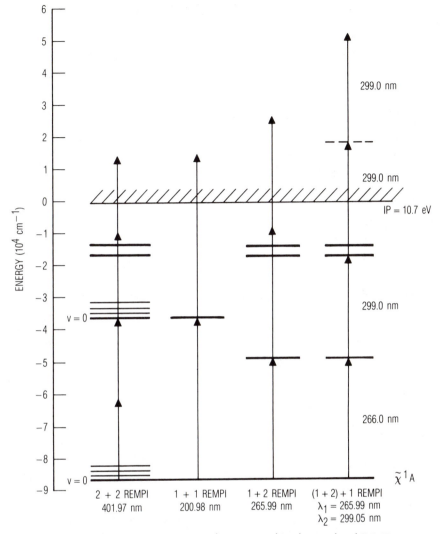

Fig. 21.2. REMPI excitation schemes used in the study of DIMP.

lengths longer than 140 nm, nor were any vibrationally resolved electronic state features revealed.[19] Since optical selectivity by discrete resonance excitation was apparently not available for DMMP, we sought to obtain optical selectivity by REMPI excitation of a major photofragment of DIMP. Other investigators have reported the efficient production of PO radical from DMMP by a variety of excitation sources that include ArF and KrF excimer laser excitation,[22] CO_2 laser excitation,[23] and microwave discharge.[24] We therefore used the narrow PO vibrationless transition $\tilde{A}^2\Sigma^+ \leftarrow \tilde{X}^2\Pi$ at 247.7 nm to ionize the fragment by 1 + 1 REMPI. Fortuitously, this wavelength also proved effective for photodissociating DMMP to give the characteristic fragment PO, thereby avoiding the need for a second wavelength of light. A more selective example of the use of photofragment spectroscopy that used two-color excitation is described later in this chapter.

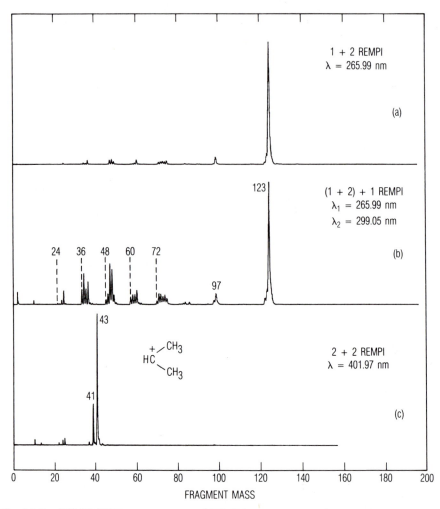

Fig. 21.3. REMPI/TOF mass spectra of DIMP in a supersonic beam: (a) 1 + 2 REMPI using 265.99 nm; (b) 1 + 2 REMPI using resonant 265.99 nm, followed by 299.05 nm to ionize and further fragment the ions; (c) 2 + 2 REMPI at 401.97 nm.

Organosulfides

The REMPI detection efficiency of dimethyl sulfide (DMS) was, likewise, hindered by the necessity of driving an otherwise single-photon transition (at 195 nm) using two photons in a 2 + 2 REMPI excitation. The molecular beam excitation spectrum, for an electronic transition assigned as a $3p$ Rydberg state, is shown in Fig. 21.5. Although vibrationally well resolved, the linewidths (~ 15 cm^{-1}) are considerably broader than those observed for DIMP (~ 7 cm^{-1}) in Fig. 21.4. This property is not unusual for transitions to higher excited vibronic levels due to the greater density of states, as well as interactions with many nearby electronic states (see, for example, the discussion under Excitation to Higher Electronic and Rydberg States). However, predissociation, which can compete with ionization, also seemed possi-

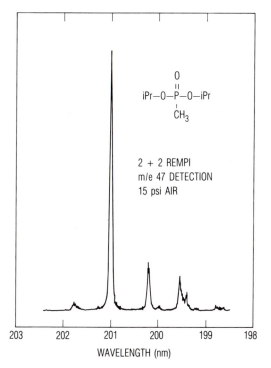

Fig. 21.4. 2 + 2 REMPI molecular beam excitation spectrum of DIMP using atmospheric room air as the carrier gas (0.5% DIMP in 15 psi air). The isopropyl fragment ion at mass 43 was monitored in recording the spectrum.

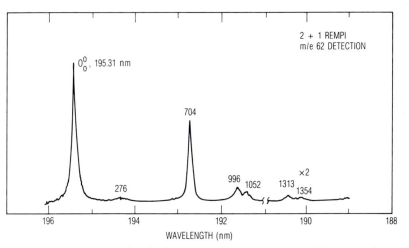

Fig. 21.5. 2 + 1 REMPI molecular beam excitation spectrum of the 3*p* Rydberg state of DMS (0.1% DMS in 50 psi He). The spectrum was recorded by detecting the parent ion at mass 62. Vibrational intervals are reported in cm^{-1}.

ble. For instance, DMS has been reported to undergo dissociation to form $CH_3S\cdot$ radicals following 193-nm excitation.[20]

We investigated the extent of photodissociation of DMS following excitation into the $3p$ Rydberg state, as well as into a lower energy Rydberg and valence state. In each case, the major signal was the parent ion.[11] This result is evidence not only for stable electronically excited neutral states (at least over the duration of the 5-nsec laser pulse), but for stable ions to energies as high as 16000 cm^{-1} (corresponding to the excess absorbed photon energy above the IP of DMS). The stability of DMS made possible a detection limit on the order of 1 ppb (see the section on Detection Levels) for the $2 + 1$ REMPI excitation through the $3p$ state in spite of the two-photon absorption step. Unfortunately, the lower energy $3s$ transition at 227.9 nm that allowed a more favorable $1 + 1$ REMPI excitation did not improve on this figure because of a weak absorption cross section.[11]

The molecular stability, and hence detection sensitivity, can be seriously compromised by the introduction of a weakly bound substituent. The substituted analog chloro-DMS demonstrates this quite well. The VUV absorption spectrum shows a very long Franck–Condon progression[19] in what is presumed to be the corresponding $3p$ state previously discussed for DMS. The vibrational frequency of

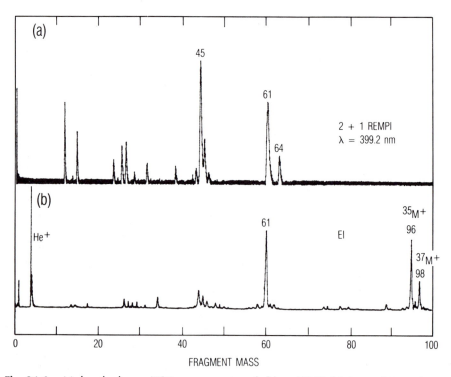

Fig. 21.6. Molecular beam TOF mass spectra of chloro-DMS: (a) $2 + 1$ REMPI (399.2 nm) excitation showing the absence of parent ion and poor sensitivity (low signal-to-noise ratio); (b) 70-eV EI excitation showing the presence of parent ion and much better signal-to-noise ratio.

405 cm^{-1} is consistent with an S$-$C$-$Cl deformation or bending motion. The 2 + 1 REMPI molecular beam mass spectrum via the $v = 4$ overtone vibration at 199.6 nm is shown in Fig. 21.6. This excitation was chosen so that the three-photon energy exceeded the IP. The REMPI mass spectrum is compared to the spectrum recorded, using 70-eV electron impact (EI) ionization. EI excitation leads to direct ionization, thereby avoiding dissociative intermediate states in the neutral molecule. EI ionization of chloro-DMS proved to be efficient, and the resulting mass spectrum shown in Fig. 21.6 exhibits significant parent ion formation. This is in marked contrast to the REMPI results, which suffer from extremely weak ionization efficiency and the absence of parent ion. Instead, the prominent ion signal occurs at fragment mass 61, resulting from loss of Cl. These results are suggestive of a rapid C$-$Cl dissociation in the neutral excited state, followed by nonresonant MPI of the neutral mass 61 fragment.[19] For detection purposes, the method of photofragment excitation, which we employed for DMMP detection, would be more appropriate for monitoring dissociative molecules such as chloro-DMS.

The cases described for the organophosphonates and organosulfides represent detection difficulties as a result of unfavorable spectroscopic properties. These limitations will mostly be overcome by new developments in laser and crystal technology (e.g., broadly tunable ultrafast dye lasers, solid-state lasers, β-barium borate crystals) as well as additional spectroscopic studies to identify more efficient excitation methods. Nonetheless, the detection efficiency (see the section on Detection Levels) by the the molecular beam MPI method for these rather unfavorable molecules compares well with other techniques. Also, the present technique offers real-time detection.

Hydrazine

The molecule hydrazine (N_2H_4) and the methylated analogs, methyl hydrazine and 1,1'-dimethyl hydrazine, are most commonly noted for their use as propellant fuels for launch vehicles and satellite thrusters. They are also quite toxic, and hence methods are sought to detect their presence in environments in which the fuel is handled. The initial phase of our laser/molecular beam studies on hydrazine detection required basic electronic state spectroscopic information to be obtained. Published spectra were limited to a single, vibrationally unresolved ambient vapor-phase UV absorption spectrum, which showed an onset to absorption at about 260 nm.[25] The source of the broadening was at first believed to be due to either very rapid photodissociation or thermal congestion. However, preliminary work in our laboratory using molecular beam expansions rules out the latter possibility. Other possible sources of broadening include large geometry changes in the excitation process, as well as strong interactions of the excited electronic state with the dense vibrational level spacing in a lower electronic state.[26] We have been pursuing this problem along two directions. Experimentally, direct absorption spectra are being recorded in a supersonic expansion using a monochromatized broad band light source for medium resolution, but wide tunability extending into the VUV. Theoretically, *ab initio* calculations are in progress to calculate lower electronic state energies and absorption strengths.

Detection Levels

A major objective in our molecular detection effort is to establish the limits of sensitivity and selectivity by molecular beam MPI mass spectrometry. To judge the performance levels for different classes of molecules, we have adopted the following simple model for detection levels.

A Working Model

For simplicity, we define sensitivity as the concentration of a target molecule that gives a detectable signal. A detectable signal, in this case, is one that exceeds the fluctuation of the background signal. The detectable concentration C_0 is determined by recording the signal strength S_{ref} for a calibrated reference sample at concentration C_{ref}, and comparing it to the background fluctuation level S_0. This leads to a simple relation for detection limit given by

$$C_0 = C_{ref} \frac{S_0}{S_{ref}} \tag{21.1}$$

The only assumption in Eq. (21.1) is that the ion count rate does not vanish at the extrapolated detection limit.

We define two response times: (1) τ_e being the time required for the signal to grow to within $1/e$ of its final value, where we assume exponential growth of the form

$$S(t) = S(\tau_e)[1 - e^{-t/\tau e}] \tag{21.2}$$

and (2) τ_0 being the time required for the detected signal to exceed the background fluctuation level S_0. We may consider τ_0 to be the response time for detection and τ_e the response time for measurement of concentration. They can be related to each other using Eq. (21.2) to give

$$S(\tau_0) = S(\tau_e)[1 - e^{-\tau_0/\tau e}] \tag{21.3}$$

Defining $S(\tau_0) = S_0$ and rearranging then gives

$$\frac{\tau_0}{\tau_e} = -\ln\left[1 - \frac{S_0}{S(\tau_e)}\right] \tag{21.4}$$

It is not difficult to see that the response time for detection, τ_0, can be considerably shorter than τ_e for target molecule concentrations well above the detection limit.

Sensitivity

The response of our detection system to a 60 ppb sample of DMS in water-saturated air (i.e., 100% relative humidity, 17,000 ppm) is presented in Fig. 21.7. An enlargement of the baseline region, along with a detectability scale, is included for convenience. The ratio of the reference sample signal to background signal S_{ref}/S_0 is greater than 200. This corresponds to a detection limit for DMS of less than 300 ppt. The measurement response time τ_e, based on the signal strength vs time in Fig. 21.7, is 8 sec; however, the detection response time τ_0 at 60 ppb is on the order of 100 msec. Naturally, other factors (i.e., transit time of sample through delivery

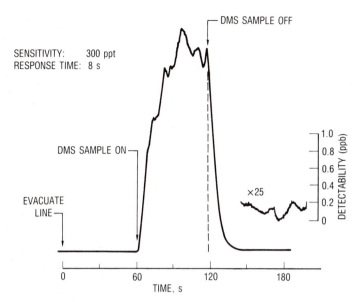

Fig. 21.7. Sensitivity determination for a 60 ppb sample of dimethyl sulfide (DMS) in water-saturated air. The baseline signal represents the background level of our molecular beam apparatus. The sample signal represents the response to turning the sample flow on and off at specified times.

lines, laser repetition rate, etc.) can increase the actual "system" response time; however, the intrinsic value of τ_0 can be approached with careful design. Finally, it is noted that by varying experimental conditions (e.g., smoothing factors), tradeoffs between measurement response time τ_e and detection level C_0 can be made to enhance one parameter at the expense of the other.

The excitation scheme leading to the signal in Fig. 21.7 involved $2 + 1$ REMPI at a wavelength of 390.62 nm. This corresponds to resonance excitation through the vibrationless level of the $3p$ Rydberg-like state at 195.31 nm. As discussed earlier, the transition to the $3p$ state is formally allowed by one-photon excitation, but symmetry forbidden by two-photon excitation. Hence the two-photon absorption cross section is very small and not favorable from the standpoint of ionization efficiency. New developments are now making possible the generation of high-power far-UV laser pulses. This will soon enable one-photon resonance excitation of the organosulfides; thus, an appreciable improvement in sensitivity can be expected.

Selectivity

The usable sensitivity (i.e., detectability) requires a high level of discrimination against contaminant molecules. A test of the selectivity of the MB/REMPI/MS technique against chemically similar compounds was performed by mixing a flowing 10 ppm DIMP sample (in N_2) with a 1300 ppm DMMP sample. Excitation of DIMP was by $2 + 2$ REMPI via the 200.98-nm electronic state (Figs. 21.2–21.4). Detection was for the $m/e = 43$ fragment ion (Fig. 21.3c). The selectivity was determined by recording the signal intensity in the absence and presence of the target

DIMP sample and taking account of the ratio of the DIMP and DMMP concentrations. This determination resulted in a selectivity of about 10^4. A similar level of selectivity was observed using gasoline as an interferent. In a related experiment the response of the DIMP sample was compared to a room air sample saturated with water. The results, which were obtained in the same manner as previously described, indicate that the selectivity of DIMP to water is greater than 10^5. However, more importantly, it demonstrates that water vapor does not form molecular complexes with DIMP (or DMS) that would degrade the signal strength due to non-resonant excitation and loss of mass identification.

Atmospheric Monitoring

The value of supersonic molecular beam mass spectroscopy for trace detection was recognized from the start. The molecular cooling accounts for narrow excitation linewidths, which contributes to an appreciable optical selectivity. For MPI excitation, the selectivity is further augmented by single mass detection. Sensitivity is considerable, largely because of the ability to detect single ions using modern ion lenses and detectors. Detection is very rapid in the sense that the total trip, from the time the neutral molecules exit the molecular beam source to the time the ionized particles reach the detector, is well under a millisecond. For chemical analysis, which allows the preparation of samples in convenient carrier gases, impressive detection capabilities have been reported for a variety of applications. The more challenging application is in atmospheric monitoring in which the carrier gas is 1 atm of air. Several potential concerns were investigated in our laboratories, including the following questions: (1) Does the supersonic expansion of atmospheric air provide adequate molecular cooling? (2) Do the target molecules form complexes with water vapor, thus undermining detection efforts? (3) Do the target molecules form clusters among themselves? (4) Can the gas inlet system to the detection apparatus be designed to avoid delivery lines that can adsorb trace amounts of target molecules?

The extent of molecular cooling that occurs in a supersonic expansion is strongly dependent on the carrier gas and pressure. Without adequate cooling, the spectral lines of a sample will become rotationally broadened, leading to a decrease in selectivity. The cooling efficiency of atmospheric air is usually less than ideal and, hence, a reduction in the realizable selectivity may be expected. For the organophosphonates, considerable supersonic cooling was achieved for atmospheric expansions, as evident by the REMPI excitation spectrum of DIMP in Fig. 21.4.[11] The extent of cooling was judged to be comparable to that observed using 50 psi of He as a carrier gas. This is a rigorous test, as large "floppy" molecules with many low-frequency vibrations are difficult to cool using diatomic carrier gases.[12]

A particular concern in detection systems that rely on the introduction of seeded gas samples into a vacuum system is the complexation of the target molecule with other molecules through hydrogen bonding or van der Waals forces.[10] This can be particularly troublesome when water vapor is present, as much higher concentrations of water compared to trace target molecules are likely to be encountered in atmospheric monitoring. The possibility for complexation between water and the target molecule under humid conditions can cause a loss of wavelength and mass identity. Problems with complexation are often encountered in atmospheric pressure ionization (API),[6] in which molecules are ionized by protonation, thus increasing their propensity to undergo hydrogen bonding. This problem is amplified since the ionization occurs at atmospheric pressures and not under the collisionless conditions of a molecular beam. Our ionization approach differs in that the neutral, unprotonated molecules are allowed to expand into the vacuum, forming a molecular beam. The molecules are collision free when ionization occurs, so that no complexation with ions is possible. The immunity to water complexation was demonstrated for DIMP and DMS, in which it was found that the excitation/detection signal described earlier was insensitive to whether the carrier gas was dry or entrained with room temperature water vapor.

The question of complexation of target molecules in a supersonic expansion to form homologous clusters was also investigated using electron-impact ionization. DMS was particularly prone to complexation at high concentrations (10,000 ppm). Because cluster (or at least dimer) formation is a bimolecular event, the extent of clustering can be expected to decrease at lower concentration. Clustering of DMS became insignificant at concentrations below 1000 ppm and should, therefore, be negligible at trace concentration.

During our work it became obvious (to no great surprise) that for very low premixed concentrations, a large proportion of the target molecules were adhering to the delivery lines, thereby decreasing the number of molecules that reached the pulsed molecular beam valve. Our pulsed valve has two characteristics that can alleviate this problem. First, the design incorporates continuous flow capabilities that ensure that a fresh atmospheric sample is always present at the injection aperture. Second, the valve may alternatively be constructed using an open structure that allows the atmosphere to be in direct contact with the pulsed aperture. Both adaptations would minimize the surface interactions that can degrade detection capabilities.

DOUBLE RESONANCE ENHANCED SELECTIVITY

As detection limits improve, the requirements for selectivity become greater. We have been exploring the potential for an additional level of optical selectivity by using two-color double resonance excitation augmented by TOF mass detection.[27-31] Several variations have been examined involving a second resonance excitation from the intermediate neutral state to discrete vibronic states below[27,30] and above[28,30] the ionization potential (IP). The method of ion dip detection[32,33] figures prominently in these experiments. In this technique, the first dye laser generates a constant ion signal by 1 + 1 REMPI through an intermediate state at the one-photon level. In one scheme, a second dye laser is used to excite from the intermediate state to another state, which competes with the ionization step and depletes the ion signal (see Fig. 21.8a). In another scheme, the second dye laser is used to photofragment, and therefore deplete, the parent ion signal, which in turn causes a concomitant increase in the fragment ion signal.[29]

The appeal of the two-color detection schemes to be described is that they do

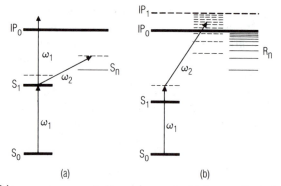

Fig. 21.8. Double resonance excitation scheme for higher excited electronic state and Rydberg detection: (a) excitation to final states below the IP, using ion dip detection; (b) excitation to final states above IP that undergo autoionization.

not require both colors for successful detection. As discussed in the preceding sections, the one-color REMPI method provides exceptional sensitivity and selectivity of its own accord when used in conjunction with molecular beam cooling and mass detection. The role of the second color is to serve as a desirable step for obtaining a greater level of selectivity. The use of two-color excitation may be viewed as providing detection by the first color and verification by the second color.

Excitation to Higher Electronic and Rydberg States

Discrete higher excited electronic states can offer very specific conditions for molecular detection by double resonance excitation. This is particularly true of Rydberg states, which tend to have narrow excitation linewidths. We consider excitation to final resonant states below and above the first ionization threshold. The excitation schemes, which involve ion dip and autoionization detection, respectively, are portrayed in Fig. 21.8.

Below Threshold Excitation—Ion Dip Detection

Detection of a higher excited state, S_n, in aniline and aniline-d_5 is illustrated in Fig. 21.9 for an ion dip excitation/detection scheme similar to Fig. 21.8a. The aniline S_n spectra were recorded by first generating an ion signal by $1 + 1$ REMPI via the vibrationless ($v = 0$) $S_1(\tilde{A}^1B_2) \leftarrow S_0(\tilde{X}^1A_1)$ transition using the first tunable laser pulse ω_1. The second tunable pulse ω_2, overlapped spatially and temporally with the first pulse, was tuned through the $S_n(v) \leftarrow S_1(0)$ resonances.[27,30] These absorptions compete with ionization, thereby creating the ion dip signals. The excitation spectra were mass-selected by detecting only the aniline parent ion signal M^+, using the boxcar gated integrator.

Spectroscopically, the two-color ion dip technique is important for detecting higher excited electronic states that are too short lived to be detected by conventional ionization or fluorescence detection methods. In fact, the lifetime of the final state determines in part the strength of the ion dip signal and, hence, the sensitivity for that particular excitation. For example, saturation of a long-lived state under intense laser power would depopulate the intermediate state by no more than 50%, hence the ion dip signal could not exceed 50%. However, if the upper state decays at a rate greater than the ω_2 pumping rate (i.e., the up-pumping process is irreversible), then the lower state can be nearly completely depleted, leading to ion dip signals greater than 50%. This holds the potential for very sensitive detection. Ion dip signals exceeding 50% could be obtained for nearly all the observed $S_n(v) \leftarrow S_1(0)$ transitions in aniline and aniline-d_5 by decreasing the ω_1 and increasing the ω_2 intensity. Apparently, the S_n state undergoes rapid decay,[27,34] either by dissociation or radiationless decay, to a set of states that is not easily ionized by ω_1 or ω_2.

It is interesting to note in Fig. 21.9 that the spectral linewidths for aniline-d_5 are significantly narrower than for the protiated molecule. This behavior is due to the reduction in vibrational frequencies that occurs upon deuteration. Vibrational interactions between initially excited states and underlying vibrations from lower electronic states are made weaker because of the larger differences in vibrational quantum number. This Franck–Condon effect outweighs the effects of an increased density of states. Finally, we have recently obtained evidence for narrower electronic transitions at higher energy, which may prove promising from the viewpoint of selective detection.

Above Threshold Excitation—Autoionization

Rydberg series, which are excited electronic states, are commonly known to converge to ionization thresholds. However, in molecules in which internal energy

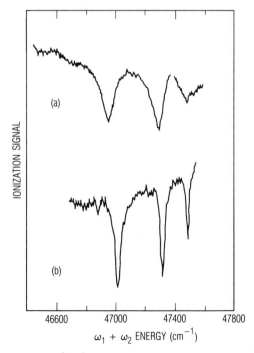

Fig. 21.9. Ion dip spectrum for the $S_n(v) \leftarrow S_1(v = 0)$ absorption of (a) aniline and (b) aniline-d_5. Ion dip intensities are plotted relative to the ion signal level in the absence of the ω_2 beam.

(i.e., vibrations and rotations) is possible, the total molecular energy in excited Rydberg states can exceed the IP. Excitation to vibrationally excited Rydberg states represents narrow transitions, which can spontaneously ionize for facile detection by mass spectrometry. An important mechanism for ionization is called vibrational autoionization, and is due to the interaction of a vibration with the ionization continua. The interactions are often sufficiently weak that transitions to these states can be rather sharp. In our experiments,[28] ω_1 is used to excite a specific vibrational level in S_1; ω_2 is then used to excite to a Rydberg state in an efficient step that preserves the vibrational excitation (Fig. 21.8b). Often many members of an autoionizing Rydberg series can be identified,[28,30] providing additional possibilities for selective detection. The observed linewidths for the autoionizing transitions are significantly narrower upon deuteration of naphthalene, a trend also observed for the ion dip transitions in aniline previously discussed.

Resonance Ion Dissociation in Naphthalene Ions

We have been investigating new methods for obtaining double mass selectivity along with the usual double optical selectivity made possible by two-color excitation. A promising excitation technique, which we refer to as resonance ion dissociation (RID), is illustrated in Fig. 21.10a. In this scheme, the first laser populates a single vibrational level of the ion by $1 + 1$ REMPI and the second scanning laser dissociates the ion by resonance-enhanced multiphoton dissociation. The dissociation process is evident by observing a decrease in the parent ion M^+ signal (i.e., ion dip) and a corresponding increase in the fragment ion, $[M - m]^+$ signal. Detection of both the parent ion and the characteristic fragment ion, therefore, may hold promise as a very selective detection method.

Singly charged cations of aromatics have been studied in a variety of low-temperature matrices.[35] The vibrationally resolved ion spectra are distinctive, showing strong red absorption bands corresponding to transitions between normally filled molecular orbitals in the neutral. Low-pressure vapor phase studies of the spec-

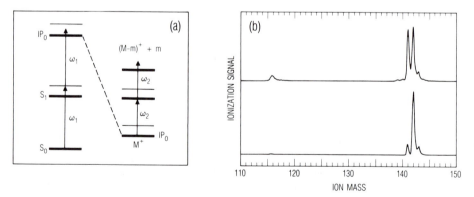

Fig. 21.10. (a) Excitation and detection sequence for resonance ion dissociation. (b) Mass spectrum of 2-methylnaphthalene showing resonance ion dissociation: (lower trace) $1 + 1$ REMPI by ω_1 only at 308.3 nm; (upper trace) $1 + 1$ REMPI by ω_1 and RID by ω_2 at 662 nm, revealing an efficient and specific H-atom dissociation.

troscopy of ions formed by electron impact in an ICR spectrometer[36] and by MPI in a mass spectrometer,[37] however, failed to reveal the vibrational structure that was evident in the low-temperature matrix spectra.[35] This may be attributed to the thermal energy content of the precursor neutral molecules. Cooling the precursor molecules by supersonic expansion may lead to the formation of cold ions by 1 + 1 REMPI.

We have observed vibrationally resolved resonances in naphthalene and 2-methylnaphthalene (2-MeN) by recording RID spectra following selective vibronic preparation of the molecular cation.[29,30] These studies relied on the detection of the low-energy H-atom photodissociation channel of naphthalene and 2-MeN cations as ω_2 was scanned through resonant absorptions of the molecular ion (Fig. 21.10a). For the 2-MeN cation, vibrational structure was observed for a visible and a UV state; for naphthalene, only a UV state was observed. The UV vibrationless electronic transition was observed to occur at 304.7 nm (naphthalene) and 308.9 nm (2-MeN). The origin of the 2-MeN cation absorption in the visible region occurred at 662 nm. The ω_1 1 + 1 REMPI mass spectrum of 2-MeN in the absence and presence of the ω_2 RID pulse is presented in Fig. 21.10b. The H-atom dissociation at the ω_2 wavelength of 662 nm is very distinctive and specific. Only a small yield of the next lowest energy dissociation channel, loss of C_2H_2, is evident.

The linewidths in the RID spectra of the cations are rather broad for both the visible ($\sim$200 cm^{-1}) and the UV resonances ($\sim$150 cm^{-1}), although they are comparable to the matrix absorption spectra.[25] However, a pronounced power broadening was evident in which the visible transitions of the cooled 2-MeN ions were saturated, even relatively far from resonance. This behavior can be explained[29] by the presence of a strong continuum background absorption to longer wavelengths, which is believed to be associated with a lower excited electronic state.

For selective detection, the broadness of the ion transitions is somewhat discouraging. However, the decreased optical selectivity in the ω_2 excitation step is compensated by the increased mass selectivity resulting from parent/fragment detection. Furthermore, the visible absorptions in the aromatic ions are very selective in that they are typically inactive in neutral molecules. Finally, the dissociation process is very efficient (Fig. 21.10); consequently, the detection efficiency is favorable by this scheme. The important point to consider is that molecular beam excitation and detection using a single color has proved to be a very powerful means for sensitive and selective detection; hence the second color should be viewed as having a nonessential, but still a very enhancing, role in sensitive and selective detection. In this context, the ω_2 RID pulse aptly serves that purpose.

SUMMARY AND CONCLUSIONS

Multiphoton ionization in molecular beam systems has advanced into a remarkably diverse tool since its infancy about a decade ago. Creative adaptations of the MPI/molecular beam method have contributed enormously to our fundamental understanding of the spectroscopy and dynamics of excited state molecules. It is only natural that the benefits of high resolution and sensitive detection would be recognized and applied toward molecular detection. The nature of this chapter limits us to reporting on recent and ongoing efforts in our laboratory; however, this represents only a small part of a very active field. We hope that our reference to other work will present the reader with a broader view of the subject matter.

The molecular detection work reviewed here should provide a glimpse of the potential performance obtainable by molecular beam MPI mass spectrometry in terms of speed, sensitivity, and selectivity of detection. There are many classes of compounds that are favorably disposed to detection by laser techniques such as REMPI. Our attempt here is to show that it is possible to overcome the challenges of spectroscopically and physically more demanding classes of molecules. It seems reasonable to expect further refinements of experimental methods in this area. The maturing of molecular beam, laser mass spectrometry as a viable broad-based analytical tool only awaits the imminent development of rugged, compact, and widely tunable sources of laser pulses.

ACKNOWLEDGMENTS

The work presented in this review represents valuable contributions from a number of collaborators. I would like to thank Drs. J. E. Pollard and R. B. Cohen for their extensive assistance in the organophosphonate and sulfide work. The efforts of Mr. R. A. Hertz and Ms. M. L. Homer toward the hydrazine studies are also appreciated. Finally, the double-resonance work owes its success to a fruitful collaboration with Dr. J. E. Wessel, who originally developed the ion dip technique, and to the experimental expertise of Mr. L. J. Pruitt. Funds for this work were provided by the USAF Space Division, the Brooks AFB School of Aerospace Medicine, and the Department of Energy.

REFERENCES

1. R. N. Zare, *Science* **226,** 298 (1984); D. M. Lubman, *Anal. Chem.* **59,** 31A (1987).

2. C. Klimcak and J. Wessel, *Anal. Chem.* **52,** 1233 (1980).

3. G. Rhodes, R. B. Opsal, J. T. Meek, and J. P. Reilly, *Anal. Chem.* **55,** 280 (1983); R. B. Opsal and J. P. Reilly, *Opt. News* **12,** 18 (1986); R. B. Opsal and J. P. Reilly, *Anal. Chem.* **58,** 2919 (1986).

4. L. Kolaitis and D. M. Lubman, *Anal. Chem.* **58,** 1993 (1986).

5. S. E. Egorov, V. S. Letokhov, and A. N. Shibanov, *Chem. Phys.* **85,** 349 (1984); F. Engelke, J. H. Hahn, W. Henke, and R. N. Zare, *Anal. Chem.* **59,** 909 (1987).

6. F. M. Klimock, J. P. Baxter, and N. Winograd, *Surf. Sci. Lett.* **124,** L41 (1983); C. H. Becker and K. T. Gillen, *Anal. Chem.* **56,** 1671 (1984).

7. R. Tembreull and D. M. Lubman, *Anal. Chem.* **58,** 1299 (1986); M. A. Posthumus, P. G. Kistemaker, H.L.C. Meuzelaar, and M. C. Ten Noever de Brauw, *Anal. Chem.* **50,** 985 (1978); D. A. McCrery, E. B. Ledford, and M. L. Gross, *Anal. Chem.* **54,** 1435 (1982); C. L. Wilkins, D. A. Weil, C.L.C. Yang, and C. F. Ijames, *Anal. Chem.* **57,** 520 (1985); R. Tembreull and D. M. Lubman, *Anal. Chem.* **59,** 1003 (1987); R. Tembreull and D. M. Lubman, *Anal. Chem.* **59,** 1082 (1987).

8. F. T. McLafferty (ed.), *Tandem Mass Spectrometry.* John Wiley, New York, 1983.

9. A. G. Marshall, *Acc. Chem. Res.* **18,** 316 (1985).

10. D. H. Levy, L. Wharton, and R. E. Smalley, in *Chemical and Biological Applications*

of Lasers, Vol. II, edited by C. B. Moore. Academic Press, New York, 1977, p. 1; R. E. Smalley, L. Wharton, and D. H. Levy, *Acc. Chem. Res.* **10,** 139 (1977).

11. J. A. Syage, J. E. Pollard, and R. B. Cohen, *Appl. Opt.* **26,** 3516 (1987).

12. J. A. Syage, P. M. Felker, and A. H. Zewail, *J. Chem. Phys.* **81,** 2233 (1984); J. A. Syage, P. M. Felker, and A. H. Zewail, *J. Chem. Phys.* **81,** 4685 (1984).

13. W. C. Wiley and I. H. McClaren, *Rev. Sci. Instrum.* **26,** 1150 (1955).

14. J. A. Syage, *Chem. Phys, Lett.* **143,** 19 (1988); J. E. Pollard and R. B. Cohen, *Rev. Sci. Instrum.* **58,** 32 (1987).

15. D. M. Lubman, *Anal. Chem.* **59,** 31A (1987).

16. R. Tembreull, C. H. Sin, P. Li, H. M. Pang, and D. M. Lubman, *Anal. Chem.* **57,** 1186 (1985); R. Tembreull and D. M. Lubman, *Anal. Chem.* **56,** 1962 (1984).

17. D. M. Lubman, R. Tembreull, and C. H. Sin, *Anal. Chem.* **57,** 1084 (1985).

18. C. H. Sin, R. Tembreull, and D. M. Lubman, *Anal. Chem.* **56,** 2776 (1984).

19. J. A. Syage, J. E. Pollard, and R. B. Cohen, unpublished work.

20. M. Suzuki, G. Inoue, and H. Akimoto, *J. Chem. Phys.* **81,** 5405 (1984).

21. J. C. Scott, G. C. Causley, and B. R. Russell, *J. Chem. Phys.* **59,** 6577 (1973).

22. S. R. Long, R. C. Sausa, and A. W. Miziolek, *Chem. Phys. Lett.* **117,** 505 (1985).

23. J. S. Chou, D. S. Sumida, and C. Wittig, *J. Chem. Phys.* **82,** 1376 (1985).

24. K. N. Wong, W. R. Anderson, A. J. Kotlar, M. A. DeWilde, and L. J. Decker, *J. Chem. Phys,* **84,** 81 (1986).

25. C. Willis and R. A. Back, *Can. J. Chem.* **51,** 3605 (1973).

26. J. A. Syage, *J. Phys. Chem.,* **93,** 170 (1989).

27. J. A. Syage and J. E. Wessel, *J. Chem. Phys.* **85,** 6806 (1986).

28. J. A. Syage and J. E. Wessel, *J. Chem. Phys.* **87,** 6207 (1987).

29. J. A. Syage and J. E. Wessel, *J. Chem. Phys.* **87,** 3313 (1987).

30. J. A. Syage and J. E. Wessel, *Appl. Opt.* **26,** 3573 (1987).

31. J. A. Syage and J. E. Wessel, *Appl. Spectrosc. Rev.,* **24,** 1 (1988).

32. D. E. Cooper, C. M. Klimcak, and J. E. Wessel, *Phys. Rev. Lett.* **46,** 324 (1981); D. E. Cooper and J. E. Wessel, *J. Chem. Phys.* **76,** 2155 (1982).

33. T. Suzuki, N. Mikami, and M. Ito, *Chem. Phys. Lett.* **120,** 333 (1985); J. Xie, G. Sha, X. Zhang, and C. Zhang, *Chem. Phys, Lett.* **124,** 99 (1986).

34. D. J. Moll, G. R. Parker, and A. Kuppermann, *J. Chem. Phys.* **80,** 4808 (1984).

35. L. Andrews, B. Kelsall, and T. A. Blankenship, *J. Phys. Chem.* **86,** 2916 (1982); L. Andrews, R. S. Friedman, and B. J. Kelsall, *J. Phys. Chem.* **89,** 4016 (1985); L. Andrews, R. S. Friedman, and B. J. Kelsall, *J. Phys. Chem.* **89,** 4550 (1985).

36. J. P. Honovich, J. Segall, and R. C. Dunbar, *J. Phys. Chem.* **89,** 3617 (1985); M. S. Kim and R. C. Dunbar, *J. Chem. Phys.* **72,** 4405 (1980); R. C. Dunbar and R. Klein, *J. Am. Chem. Soc.* **98,** 7994 (1976).

37. D. Mordaunt, G. Loper, and J. Wessel, *J. Phys. Chem.* **88,** 5197 (1984).

22

Resolution and Sensitivity in Laser Ionization Mass Spectrometry

RICHARD B. OPSAL, STEVEN M. COLBY,
CHARLES W. WILKERSON, Jr., and JAMES P. REILLY

It has been more than a decade since the successful demonstration of single atom detection by two-step laser ionization of alkali atoms.[1] This pioneering physics experiment did more than elegantly demonstrate the simplicity and power of resonantly enhanced multiphoton ionization; it captured the imagination of analytical chemists who for years had considered nanogram to picogram detection sensitivity for most techniques as state of the art. Since a single atom having a nominal mass of 60 amu corresponds to 10^{-22} g of material, a whole new definition of and appreciation for sensitivity became possible.

As spectacular as these early experiments were, their connection with analytical reality was somewhat distant. For example, no handling of ultrasmall quantities of material was actually required or performed in this work. Exceedingly low alkali atom concentrations were obtained simply by controlling the temperature and thus the vapor pressure of the metal in its sample cell. Compared to other atoms and molecules, alkali metals have particularly low-lying excited electronic states and ionization potentials. It was therefore possible to ionize alkali atoms selectively in an essentially background-free manner. Furthermore, the photons required to ionize alkali atoms could be produced with easily tuned and conveniently generated visible dye laser radiation. In contrast, most other substances require the use of less powerful, and harder to tune, frequency-doubled dye laser light.

Background ionization can become a significant problem with the use of UV light. Even high-quality vacuum pump oil is usually present at concentrations of 10^5 molecules/cm^3 or more, and ultraviolet laser-induced ionization of these molecules has been amply demonstrated.[2] In an experimental environment in which background can play a role, it is essential that the sample signal be distinguishable from the background. Since the ions associated with each normally differ in mass, this is straightforwardly accomplished with a mass spectrometer. In principle, any type of mass analyzer [magnetic, quadrupole, Fourier transform ion cyclotron resonance (FT/MS), time-of-flight (TOF), etc.] can be utilized. However, the relatively slow repetition rates of pulsed lasers, compared, for example, with the high duty cycles of electron impact ion sources, make it almost imperative for the mass analyzer to be able to detect ions of all masses produced during each laser pulse. FT/MS and time-of-flight instruments uniquely share this advantage. The short duration (~10 nsec) of the output pulses from UV excimer and Q-switched Nd:YAG lasers facilitate the operation of time-of-flight laser ionization mass spectrometers. The simplicity, small size, low cost, and ease of construction of TOF instruments have made them almost the exclusive choice of researchers involved in laser ionization experiments. Their principal disadvantage, compared with FT/MS instruments, is their limited mass resolution.

In the next section the limits on resolution in TOF mass spectrometers will be explored in detail. As will be seen, considerable progress has been made in this area. Mass resolution has become respectable and the prospects for attaining further dramatic improvements in the future are good. In the second part of the chapter, we return to the general topic of laser ionization as a sensitive detection method for molecules and consider some of the complexities that distinguish atoms from molecules and one compound from another.

RESOLUTION OF TIME-OF-FLIGHT MASS SPECTROMETERS

Resolution Limiting Factors and Approaches to Dealing with Them

The two-step acceleration linear TOF mass spectrometer originally described by Wiley and McLaren[3] has served as the prototype instrument in this field for over three decades. This design has a high ion transmission, a theoretically unlimited mass range, and all of the other TOF advantages previously enumerated. Its resolution, defined as $m/\Delta m$, is typically a few hundred and is almost independent of whether ionization is induced by electron impact or laser excitation. (The reason for this will be apparent from the analysis to be presented.)

A schematic diagram of a two-step acceleration TOF mass spectrometer is displayed in Fig. 22.1. Ideally, all ions are formed at a single position $(+)$ and then accelerated to the same kinetic energy by extraction fields that are produced by applying voltages to the parallel acceleration grids G_0, G_1, and G_2. It is also assumed that all ions are produced instantaneously or, if they are generated over a finite period, they experience a drawout field that turns on at t_0 with zero risetime. Following their acceleration, ions enter the field-free drift region. Since they all have the same kinetic energy, those with different masses fly through the drift region with different velocities and arrive at the detector at different times. Lighter ions arrive earliest, and drift time is directly proportional to the square root of ion mass.

If an ion's flight time were a function only of its mass-to-charge ratio, the time-of-flight mass spectrometer would have infinite resolution. Unfortunately, several other factors are involved. These include space charge effects, nonuniform electric fields, the finite risetime of the detector and signal averaging electronics, the temporal width of the action that determines t_0, the spatial width of the region over which ions are generated, and the nascent velocity distribution of the ions. To the extent that these factors are not controlled, mass resolution is impaired. Fortunately, some of them can be handled in a straightforward manner. Space charge effects result from coulombic repulsions among the nascent ions. Too many ions

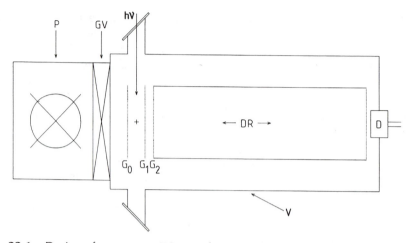

Fig. 22.1. Design of a two-step TOFMS showing grids G_0, G_1, and G_2, the field free drift region (DR), detector (D), the ionization position (+), laser light (hν), vacuum chamber (V), gate valve (GV), and pump (P).

generated too close together will noticeably perturb each other's flight trajectories. With electron impact ionization the incident electron flux is sufficiently small that this is not a major problem. However, the high photon fluxes available from pulsed lasers can lead to significant space charge effects. These can be avoided by using low laser powers or sample pressures, thereby producing a low density of ions during each light pulse. To obtain highly homogeneous acceleration fields, the grid material must have a rather fine mesh to minimize electric field punch-through between the wires. The grids must also be mounted flat and be large enough so that edge effects are not important. The use of particle timing devices such as time-to-amplitude or time-to-digital converters that nuclear and high-energy physicists have employed for years enables the measurement of flight times with a precision of better than 100 psec. In electron impact TOF sources the filament current is normally continuously on. Therefore, pulsed ion drawout fields are required to initiate each timing cycle. With laser ionization the light pulse itself is sufficiently short that it represents a convenient and accurate t_0 for each cycle and all grid potentials are usually static. Although most work has been performed with nanosecond duration light pulses, picosecond light pulses from mode-locked lasers are now conveniently available. In this case the ionizing light pulse can easily be made shorter than the limiting resolution of the timing electronics. Therefore, under realistic experimental conditions, the variation in ion flight time is mainly determined by the initial ion spatial and velocity distributions.

Ions with a particular mass-to-charge ratio are formed in the source with initial velocities that reflect those of the neutral precursors. In gas-phase samples this distribution is normally Maxwellian. Since it is the velocity component along the axis of the ion flight tube that critically affects mass spectral peak widths, the distribution of interest is a one-dimensional Gaussian function with a mean of zero and a standard deviation of $(kT/m)^{1/2}$ where m is the mass of the molecule, T is the absolute temperature, and k is Boltzmann's constant. For benzene at 300 K the standard deviation of the velocity is 178 m/sec corresponding to an FWHM spread of 419 m/sec.

The initial ion spatial distribution results from the fact that neither electrons nor laser light can be focused down to a waist of zero breadth. Since the electrostatic potential varies linearly with position between the first two grids of our mass spectrometer, ions formed at different positions will be accelerated to different final kinetic energies. This leads to a spread in ion flight times whose magnitude depends on the size of the laser or electron beam focal spot.

Over the past few decades many methods have been proposed to increase the mass resolution in the TOF-MS by compensating for these initial variations in ion velocity and position. Wiley and McLaren[3] introduced a space-focusing condition that makes flight time independent of initial position to first order. In practice, this is accomplished by judicious choice of the acceleration field strengths. Mass resolution is then primarily limited by the initial velocity distribution. In the same paper they also presented a technique called time-lag focusing. In this case, ions are produced under field-free conditions using a pulsed electron source. After a time lag τ, a positive acceleration pulse is applied to the backing plate and the ions are ejected into the mass spectrometer. The effect of the lag period τ is to allow the ions to move to new positions in the ionization region. The amount of kinetic energy received by each of the ions upon application of the acceleration pulse then depends upon its initial velocity either toward or away from the direction of flight. Unfortunately, the conditions required to both space focus and energy focus using the time-lag technique are mutually exclusive. Although resolution is increased, it also becomes a function of mass.

Marable and Sanzone[4] have also described the use of a time-dependent drawout field as a means of increasing resolution in the TOF-MS. A high drawout field E_r is applied for a time τ after which the field reverts to a value that satisfies the space-focusing condition described by Wiley and McLaren.[3] This technique has become known as impulse field focusing. In effect it reduces the impact of the initial ion velocity distribution by reducing the turnaround time of those ions formed with initial velocities in the direction opposite to the direction of flight. Although theory predicts that the mass resolution can be increased to several thousand in this way, resolution of only about 600 has been experimentally demonstrated.[5]

Muga[6] has recently patented a mass spectrometer design that incorporates linear drift regions with acceleration regions to which are applied exponentially increasing or decreasing potentials. This allows the instrument to perform both space and velocity "compaction." It is claimed that resolution can be increased dramatically by cascading several drift/compaction regions together. Resolution of 2000 has thus far been demonstrated with a single-stage instrument.[7]

Time-of-flight instruments incorporating magnetic and electric sectors were described theoretically in two papers by Poschenreider.[8,9] In these, velocity focusing is achieved by combining linear drift regions with one or more sectors in various geometries. Magnetic sectors compensate for differences in ion momentum. Müller and Krishnaswamy[10] have incorporated a 163.2° toroidal deflector into their time-of-flight atom probe spectrometer and report an FWHM resolution of 4800 for Rh^{2+} ions having an estimated initial energy spread of 300 eV.

A reflection TOF-MS was introduced in the early 1970s by Mamyrin et al.[11] and has more recently been used with laser ionization sources.[12] This type of mass spectrometer can effectively reduce the degradation in mass resolution caused by both the finite ionization volume and the spread in ion velocities. The reflectron mass spectrometer accomplishes first-order space focusing as in a conventional linear TOF-MS. To compensate for the ion velocity distribution present in the drift region, ions enter an electrostatic reflector in which those having greater velocities must travel longer distances than their slower moving counterparts. In this way slower ions catch up with faster ones and narrow temporal profiles are observed at the detector for ions with a particular m/z ratio. Resolution of about 4000 FWHM appears to be routine with this type of mass spectrometer, and recent results of 10,000 to 11,000 have been reported.[13-16] Calculations suggest that this may be increased to tens of thousands by using a single-step ionization source and higher electric fields.[17]

An alternative to compensating for the initial velocity or spatial spread of the ions is to reduce or eliminate the distributions. This can be achieved in at least two ways. Ionization from a surface has the effect of limiting the production of ions to a very well-defined spatial region.[18,19] Furthermore, in this case initial ion velocities cannot be negative. This eliminates contributions to peak widths from the turnaround time of ions. The alternative to drastically reducing the spatial distribution is to truncate the initial ion velocity distribution. Both Lubman's[20] and our own group have reported using a supersonic molecular beam source to accomplish this.[21]

In the next section we quantitatively examine the two-step TOF-MS in detail. The results of computer simulations of ion flight times and mass spectral peak profiles as a function of spectrometer operating parameters enable us to estimate practical resolution limits under various experimental conditions. In these calculations it is assumed that mass resolution is limited only by the initial spread in ion kinetic energies, the finite volume in which ionization occurs, and the temporal profile of the ionization source, and that other factors leading to decreased resolution can be made negligible. A comparison of experimental results with theoretical predictions for the supersonic beam and surface ionization source cases is then presented.

Ion TOF Equations and Focusing

The physical parameters of the mass spectrometer are the principal factors that determine an ion's time of flight. They include the distances and electric fields between grids and the length of the drift tube. The two-step TOF-MS consists of three distinct regions, as shown in Fig. 22.1. Potentials V_0, V_1, and V_2 are applied to grids G_0, G_1, and G_2, respectively, which are separated by distances d_1 and d_2. A three-step instrument has one additional acceleration region.

Each region of the mass spectrometer is bounded by an ideal metal grid or screen. An ideal grid is perfectly flat, and has high ion transmission and holes of negligible size. It is also large enough that edge effects can be ignored. With such grids, homogeneous electric fields can be produced in the two acceleration regions of the TOF-MS. Punching of electric fields through such a fine mesh can be shown by calculations to be quite minimal.[22]

In our model, an atom or molecule with a velocity component v_0 along the main axis of the mass spectrometer is ionized at a distance x_0 from grid 1 forming an ion of mass m and charge q. The ion is accelerated out of the ionization region under the influence of the electric field E_1. It then experiences a second electric field in region 2, after which it reaches the drift region of the mass spectrometer. The latter is a tube whose walls and end grids are at the same potential. In this region the ions travel with a constant velocity until they impinge on the ion detector.

The magnitudes of the electric fields (E_1 and E_2) in regions 1 and 2 are given by

$$E_1 = (V_0 - V_1)/d_1 \tag{22.1a}$$

$$E_2 = (V_1 - V_2)/d_2 \tag{22.1b}$$

The acceleration experienced by an ion in an electric field is proportional to both the electric field strength and the ion's mass to charge ratio:

$$a_1 = qE_1/m \tag{22.2a}$$

$$a_2 = qE_2/m \tag{22.2b}$$

The velocities of ions leaving regions 1 and 2 are given by v_1 and v_2, respectively:

$$v_1 = [2a_1x_0 + (v_0)^2]^{1/2} \tag{22.3a}$$

$$v_2 = [2a_1x_0 + 2a_2d_2 + (v_0)^2]^{1/2} \tag{22.3b}$$

From these, the time that an ion spends in each of the regions is then calculated:

$$t_1 = (v_1 - v_0)/a_1 \tag{22.4a}$$

$$t_2 = (v_2 - v_1)/a_2 \tag{22.4b}$$

$$t_3 = L/v_2 \tag{22.4c}$$

The total ion flight time T is the sum of the times that an ion spends in each region of the TOF-MS:

$$T(x_0,v_0) = t_1 + t_2 + t_3 \tag{22.5}$$

Through optimal choice of ion drawout and acceleration fields, it should be possible to minimize the degradation in resolution brought about by the initial and velocity distributions of the ions. The dependence of T on x_0 is given by

$$\delta T/\delta x_0 = 1/v_1 + a_1(1/v_2 - 1/v_1)/a_2 - La_1/(v_2)^3 \tag{22.6}$$

By setting this derivative equal to zero the space-focusing condition described by Wiley and McLaren[3] is obtained. The value of x_0 at which $\delta T/\delta x_0 = 0$ is denoted x_0^f. Physically, the variation in ion time of flight caused by a change in initial position is zero to first order, that is, T is independent of the initial ion position at least for small deviations from x_0^f. If x_0, d_1, d_2, and L are held fixed, the space focusing condition is realized by adjusting the potentials V_0, V_1, and V_2 (i.e., the electric fields E_1 and E_2). The first-order space-focusing condition is satisfied by a unique value of E_1/E_2 that depends only on the geometry of the instrument.

A second-order space-focusing condition can be obtained by solving for $\delta^2 T/\delta x_0^2$ and setting it equal to zero. This derivative is given by

$$\delta^2 T/\delta x_0^2 = -a_1/(v_1)^3 + [1/(v_1)^3 - 1/(v_2)^3](a_1)^3/a_2 + 3(a_1)^2 L/(v_2)^5 \qquad (22.7)$$

Since the first-order space-focusing condition yields a unique value for the ratio E_1/E_2, the second-order space-focusing condition should lead to unique values of E_1 and E_2. Unfortunately, Eq. (22.7) yields the trivial solution that $\delta^2 T/\delta x_0^2$ approaches zero as E_1 (and hence E_2) approaches infinity. With $\delta T/\delta x_0$ equal to zero and finite values of E_1 and E_2, the value of $\delta^2 T/\delta x_0^2$ at x_0^f is negative, indicating that ion time of flight is a maximum at this position.

Analogously, a velocity-focusing expression can be derived by taking the partial derivative of T with respect to initial velocity v_0:

$$\delta T/\delta v_0 = v_0[1/a_1 v_1 - 1/a_1 v_0 + 1/a_2 v_2 - 1/a_2 v_1 - L/v_2^3] \qquad (22.8)$$

By setting this equal to zero, the velocity-focusing condition is obtained. Since under normal conditions the mean of the velocity distribution is equal to zero, the value of this expression approaches zero as the acceleration a_1 approaches infinity. This makes intuitive sense, since by making a_1 large, the ratio of the initial spread in ion kinetic energy to the energy the ion receives in region 1 becomes small. The result is that as E_1 is increased, the initial ion velocity distribution has a diminishing effect on the ion's time of flight.

To minimize the dependence of ion flight time on the initial ion spatial and velocity distributions, it is necessary to simultaneously satisfy both the space- and velocity-focusing conditions given above. From Eq. (22.8) it is apparent that to minimize the effect of the velocity distribution, it is necessary to make the acceleration a_1 as large as possible. Since Eq. (22.6) indicates that space focusing does not rely on the absolute magnitudes of a_1 and a_2 but on their ratio, it is possible to use large values of a_1 and a_2 and still maintain the space-focusing condition.

Ion Peak Widths and Shapes

The above equations make it possible to readily calculate how large of an effect the spatial and velocity contributions have on mass spectral peak broadening. Figures 22.2 and 22.3 are plots of ion flight time as a function of ionizing position and initial ion velocity for ions of mass 78 and 39 amu. Parts A, B, C, and D in each of these figures correspond to different drawout voltages applied to the acceleration grids of a three-step instrument. These voltages are listed in Table 22.1. Basically, A corresponds to space-focusing conditions, whereas B, C, and D correspond to larger and somewhat arbitrarily chosen accelerating fields. It is quite apparent from Fig. 22.2 that low-voltage, space-focusing conditions limit the variation in ion flight time caused by differences in ionizing position of ±1 mm to a few nanoseconds or less. High voltages increase these temporal variations dramatically. On the other hand, Fig. 22.3A demonstrates that low-voltage space-focusing conditions are not optimal for sources in which ions are initially formed with a broad thermal velocity distribution. In that case, much higher voltages should be applied to reduce the dependence of ion flight time on initial ion velocity. (At higher potentials, the ini-

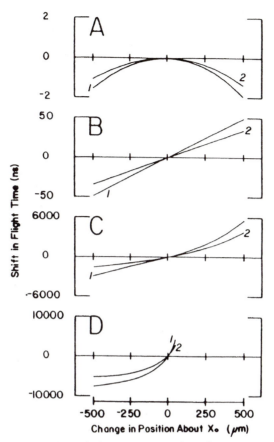

Fig. 22.2. Dependence of ion flight time on initial ion formation position for four different sets of conditions. Negative shifts in flight time indicate a shorter ion TOF relative to an ion generated at x_0^i. Curves 1 and 2 are for ions of mass 78 and 39 amu, respectively. Spectrometer conditions are listed in Table 22.1.

tial ion kinetic energy spread becomes a less significant fraction of the ion drift energy.)

It is clear from these simple examples that for a gas-phase sample that is characterized by a Boltzmann velocity distribution and that is ionized by a 1-mm-wide electron or light beam, selection of accelerating grid voltages requires that a compromise be made. *Moreover, if either the spatial or the velocity distribution of the ions can be drastically narrowed then grid voltages optimized for reducing the effect of the other distribution can be employed.* As previously mentioned, when sample molecules are ionized on or extremely close to a surface, the narrow axial spatial distribution of the nascent ions enables high drawout voltages to be employed without encountering the problems displayed in Fig. 22.2B, C, and D. In this case, even if the nascent ion velocity distribution is wide, its effect on TOF peak broadening will be small. Figure 22.4 demonstrates this point. Here a mass spectrum recorded by ionizing aniline molecules adsorbed to a metal film is displayed. High drawout

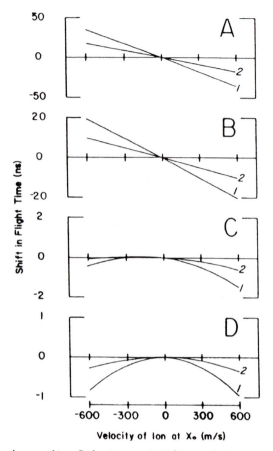

Fig. 22.3. Dependence of ion flight time on initial ion velocity. Positive values correspond to initial velocities toward G_1. These curves are for the same conditions as those listed in Fig. 22.2.

voltage conditions ensured that the mass spectral peak width was limited by the temporal duration of the ionizing light pulse (see Ref. 19 for further details).

If molecules in a molecular beam are irradiated with laser light such that only those traveling orthogonal to the TOF flight tube axis are ionized, then the gas-phase Boltzmann velocity distribution along the axial direction can easily be narrowed by more than two orders of magnitude. In this case the velocity distribution contributes little to mass spectral peak broadening, and low-voltage space-focusing conditions enable the attainment of resolution of up to 4600.[23]

Although the above method of considering the magnitudes of individual spatial and velocity contributions leads to a correct general picture of how to optimize TOF mass spectral operating parameters, quantitatively accurate peak profile simulations require a general formalism to convolute these distributions with the temporal profile of the ionizing source. This can be accomplished in at least two ways. The first is a largely analytical approach in which the TOF mass spectrometer is treated as an operator that transforms distributions for ion formation time, posi-

Table 22.1. Conditions used for the results of the simulations displayed in Figs. 22.2 and 22.3

	V_0	V_1
A	34.275 V	−34.275 V
B	60.48 V	−60.48 V
C	2500.00 V	−2500.00 V
D	2500.00 V	−2500.00 V

$V_2 = -500$ V $\qquad V_3 = -1000$ V

$d_1 = 0.005$ m $\qquad L = 0.60$ m
$d_2 = 0.010$ m $\qquad X_0 = 0.0025$ m
$d_3 = 0.010$ m $\qquad v_0 = 0.0$ m/sec

tion, and nascent velocity into a marginal or overall probability density function that represents a mass spectral peak profile. This was discussed thoroughly in Ref. 23. A second approach, which leads to equivalent results and is somewhat more straightforward to execute, involves directly using the ion flight time equations previously listed. The expected flight time for a particular m/z ion generated at one specific position at a certain time and with a certain initial velocity is calculated. Then this calculation is repeated for a range of assumed ionizing positions, times, and initial velocities. The resulting computed flight times are sorted into bins in a manner analogous to the operation of a multichannel analyzer. If appropriate weighting functions are applied to the choice of initial ion position, formation time, and velocity, then the composite peak profile accurately reflects the convolution of these three distributions. By using these calculational approaches, we have concluded that for room temperature gas-phase samples in a typical Wylie–McLaren linear TOF instrument, mass resolution is predominantly determined by the effect of the velocity distribution. The spatial contribution due to a 2-mm-wide ionizing beam is presented in Fig. 22.5A, whereas the temporal contribution due to a 2-nsec

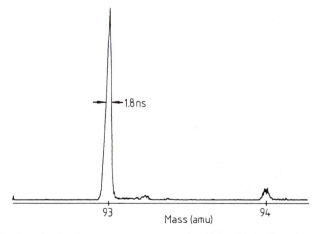

Fig. 22.4. Surface ionization mass spectrum of aniline. Note that the peak width is limited by the temporal width of the laser pulse (∼1.5 nsec).

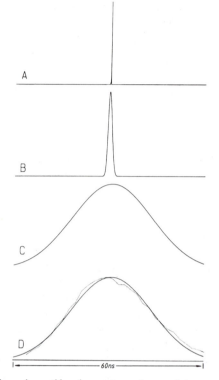

Fig. 22.5. Calculated peak profiles for an ion of mass/charge ratio of 78. Spatial (A), temporal (B), and velocity (C) components of the total ion peak profile (D) are shown. The conditions used in the calculations assume that room temperature molecules are ionized with a laser beam of diameter (FWHM) = 750 μm and temporal duration (FWHM) = 1.5 nsec. An experimental peak (dotted line) is shown for comparison.

ionizing pulse is displayed in Fig. 22.5B. The slight asymmetry of the former profile results from the fact that under space-focusing conditions there is a maximum ion flight time that occurs for ions formed at a certain position in the source region. [This results from the condition established by Eq. (22.6).] Those ions generated in other positions, at which the electrostatic potential is higher or lower, all have smaller flight times. The velocity contribution is displayed in Fig. 22.5C. Its similarity to the overall predicted and experimentally observed peak shapes displayed in Fig. 22.5D indicates just how important is its effect. For further discussion of the effect of reducing the width of the velocity distribution, see Ref. 23.

Perhaps the most exciting possibilities for future improvement in TOF mass spectrometer resolution will involve a conjunction of some of the above techniques. Schlag and co-workers have obtained mass resolution of $m/\Delta m = 10,000$ by combining a molecular beam source with the reflectron geometry.[13] Benninghoven and co-workers have obtained $m/\Delta m = 12,000$ by combining a SIMS surface ionization source with a reflectron.[14] Similarly, we have obtained resolution of 11,000 using laser-induced surface ionization in our reflectron instrument.[15,16] For aniline ions, mass spectral peak widths of as little as 7 nsec in flight times of 150 μsec have been obtained. Nevertheless, our calculations suggest that the peaks should be about three times narrower than this. The reason for this discrepancy probably involves uncontrolled electric or magnetic field fluctuations in our instrument. If this problem can be overcome, another quantum leap in TOF mass spectrometer technology may be possible.

MOLECULAR DETECTION SENSITIVITY

Having concluded from the earlier discussion that mass spectral analysis of laser-generated ions is essential for distinguishing signal from background in ultrasensitive applications, and that a time-of-flight instrument is the most convenient and cost-effective method of mass analysis, the next step is arranging for a simple and reproducible method of sample introduction. Gas chromatography not only serves this purpose well, but also provides added selectivity, which, as we will see, can be helpful for distinguishing sample and impurity ionization. Because several compounds can be simultaneously injected into the gas chromatograph, their relative ionization efficiencies can be measured with a high precision that is independent of the reproducibility of sample injection volumes. In our laboratory we use a Varian Model 3700 capillary column gas chromatograph with nitrogen, argon, or helium carrier gas. The slow flow rate of approximately 1 mL/min can easily be pumped away by a 6-in. diffusion pump, enabling vacuum in the realm of 10^{-5} torr to be maintained in the mass spectrometer. At times we have used a cryopump of equivalent speed. This reduces the background due to pump oil, but it is not able to pump helium. The last 30 cm of the fused silica bonded phase capillary column is surrounded by a heated interface that protrudes into the vacuum system of our mass spectrometer.

Effluent from the column sprays into the vacuum system continuously and is irradiated by pulsed ultraviolet laser light emitted at the repetition rate of approximately 20 Hz. The wavelengths that we commonly use in our experiments include 193, 248, and 308 nm from various excimer lasers and 266 nm, which is the fourth harmonic of an Nd:YAG laser. These diverse wavelengths allow us to resonantly excite a variety of molecules or different electronic states within some particular molecule. The latter can be advantageous since electronic transitions to certain states are stronger than transitions to other states, and since different electronic states display varying relaxation propensities. Although frequency-doubled dye laser radiation can also be employed to enhance the selectivity of the two-step ionization process, it is substantially less convenient to generate this light, and the output pulse energy is much lower than that available from the UV sources just mentioned. For molecules with sharp spectral features and particularly for those cooled in supersonic beams, light source tunability is a necessity. However, large aromatic compounds eluting from a 250°C chromatography column exhibit rather broad absorption spectra and the quasitunability of excimer lasers is sufficient to guarantee they can be resonantly excited.

Following their drift through the field-free flight tube, ions are detected with a tandem pair of microchannel plates, the output of which is digitized by a fast waveform recorder. Information is then transferred to an IBM/PC-AT computer where signal averaging takes place. Most of our chromatograms are plots of mass-integrated ion yields as a function of retention time. For highest sensitivity we employ selected ion monitoring.

Various mixtures of related compounds have been injected into our laser ionization GC/MS apparatus. These include polyaromatic hydrocarbons,[24] alkyl-benzenes,[25,26] halogenated benzenes, phthalates, nitro- and nitroso-containing compounds,[27] and phenols.[28] Compounds were selected for study either because of their significance as targets for trace environmental analysis or because of their interesting photophysical or photochemical properties.

Polyaromatic hydrocarbons have certainly received the most attention by workers in this field,[24,29] and their characteristics typify those of other molecules. Different compounds are found to ionize with efficiencies that vary by orders of magnitude. Understanding of the factors that determine these efficiencies leads to new insights about molecular properties and improvements in the analytical method. A particularly interesting example is the comparatively inefficient ionization of biphenyl relative to other polyaromatic hydrocarbons. We found that at 248 nm, the relative naphthalene:biphenyl ion yield is 22:1. At 266 nm, Gross and co-workers measured a somewhat smaller ratio of 8:1.[29] Liquid-phase ultraviolet absorption data indicate that biphenyl absorbs about 10 and 1.5 times more strongly than naphthalene at these two wavelengths.[30] Although for molecules with structured ultraviolet spectra gas-phase absorption cross sections would be more relevant, the above numbers suggest that the relative ionization yields of naphthalene and biphenyl should be the converse of that found experimentally. On the other hand, it is well known to physical organic chemists that molecules having significant torsional flexibility exhibit small quantum yields of fluorescence indicating that their radiationless relaxation is very fast. The picture that develops is that the electronic excitation in biphenyl is partially or completely converted to vibrational excitation by an internal conversion or intersystem crossing process. In the 10^{-5} torr environment of our laser ionization mass spectrometer source, molecules do not undergo collisions on a submicrosecond time scale. Therefore, any vibrational excitation generated by either of these relaxation processes must be retained in the molecule. Because of the Franck–Condon effect, this conversion of electronic to vibrational excitation tends to raise the molecular vertical ionization potential, lowering the probability of photoionization by the next photon that is absorbed. The overall process is schematically illustrated in Fig. 22.6. It is worth pointing out that molecules that have absorbed two photons without ionizing will most likely dissociate on an extremely rapid time scale as 10 eV of excitation is much larger than the strengths of single chemical bonds. Dissociation products may or may not in turn ionize. It is noteworthy that rapid radiationless relaxation in excited electronic states of molecules is not uncommon, but is in fact rather typical behavior. Its rate tends to be enhanced by the presence of electron-withdrawing groups and heavy atoms. Dissociation and radiationless relaxations are two ionization-limiting phenomena that occur only in molecules. They represent just two more reasons why atoms can, in general, be laser ionized more easily than molecules. Nevertheless, for compounds that ionize efficiently, such as naphthalene, we have been able to inject as little as 500 fg of sample and still see a signal that was five times as large as our background. The signal increased linearly over a four to five order of mag-

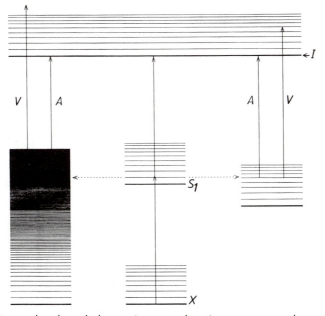

Fig. 22.6. Energy levels and electronic state relaxation processes relevant to the laser ionization of polyatomic molecules. After excitation from the ground state (X), electronic energy is converted into vibrational energy. This leaves the molecule in a highly excited vibrational state. A transition to a vibrationless state in the ion (A) is now very unlikely because of the poor Franck–Condon overlap between the vastly different vibrational levels. For there to be a reasonable probability of generating an ion, additional energy (V) has to be supplied to reach a higher ion vibrational level.

nitude dynamic range. This provides some indication of the technique's analytical potential.

From this discussion it would appear that the way to improve the ionization efficiencies of molecules that absorb strongly but do not ionize well would be to successfully compete with the excited state relaxation. We have used two approaches to do this. The simplest one is to increase the intensity of the light irradiating the chromatographic effluent. Since this raises the rate of photoionization, in principle this enables us to compete with excited state relaxation. Furthermore, it makes it possible to saturate weak transitions so that the ionization efficiency of a compound is not strictly limited by its single-photon absorption cross section. A typical example of how varying the light intensity can affect a laser ionization chromatogram is displayed in Fig. 22.7. By varying the 248-nm KrF light energy by a factor of 20, it is clear that peak area ratios can change by almost one order of magnitude and that they tend to approach unity. Even so, it is evident from this figure that biphenyl and fluoranthene ionize inefficiently at even the highest light intensity employed. High laser light intensities can under some circumstances lead to confusion and added complications. In studies of the 193-nm laser ionization of benzene, we demonstrate that at low light intensities parent ions are generated but at higher intensities C^+, C^{2+}, and C^{3+} ions dominate the mass spec-

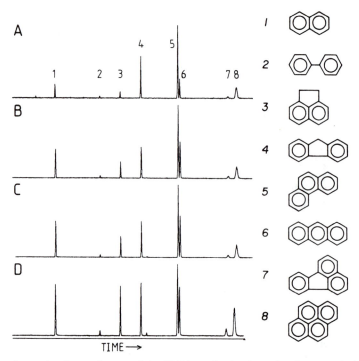

Fig. 22.7. Intensity dependence of the KrF laser ionization of polyaromatic hydrocarbons. Although ion yield increases monotonically with light intensity, each chromatogram is plotted such that the largest peak (phenanthrene) has the same height. The sample contained 5 ng of each component. (A) 1.0 mJ/pulse. (B) 3.1 mJ/pulse. (C) 6.4 mJ/pulse. (D) 19.5 mJ/pulse.

trum.[31] By measuring the kinetic energy distribution of the photoelectrons generated during the ionization process, we concluded that neutral carbon atoms are produced in abundance and that their two-step ionization dominates the ionization yield. From this we know that any attempt to model the C^+ ion yield quantitatively as a two-step ionization of benzene followed by fragmentation of the ions is destined to fail!

These complications suggest that a better way to try to compete with excited state relaxation is to use light pulses whose duration is so short that relaxation does not take place during the pulse itself. In this case, relaxation cannot affect the ionization process. We have recently undertaken such a study using the fourth harmonic of an Nd:YAG laser. A Quantel Model 571 nano/pico laser produced light pulses of either 10 nsec or 18 psec nominal duration at a 20-Hz repetition rate. Pulse energies of up to 2.4 and 0.4 mJ at 266 nm were used in the nano- and pico-modes, respectively. Laser ionization gas chromatograms recorded in two cases along with a flame ionization chromatogram for comparison are displayed in Fig. 22.8. In Fig. 22.8B the nanosecond results are seen to be similar, though not identical, to the data in Fig. 22.7D. Note that the biphenyl/naphthalene signal ratio is larger at 266 than at 248 nm, as previously observed by Sack et al.[29] In Fig. 22.8C it is clear that the biphenyl/naphthalene signal ratio increases, as expected, as the light pulse duration is shortened. The fact that biphenyl still ionizes less efficiently means that either its resolved vibronic structure gives naphthalene a larger gas-phase absorption coefficient at 266 nm or that at least some of the biphenyl molecules are relaxing or dissociating on a time scale faster than our laser pulse width. Future gas-phase UV absorption experiments should resolve this issue and establish whether ionization pulses of even shorter duration are required to enable biphenyl to ionize as efficiently as other polyaromatic hydrocarbons. The varying ionization efficiencies exhibited by different polyaromatic hydrocarbons exemplify how ion yields can be dramatically affected by molecules' physical properties. Although all polyaromatic compounds ionize to at least some degree, halogenated substances are particularly capricious. Figure 22.9 exhibits chromatograms of a series of chlorinated compounds recorded with flame ionization, ArF excimer, and KrF excimer laser-induced ionization. The striking variation in peak areas seen in Fig. 22.9A occurs because the flame ionization detector is mainly responsive to

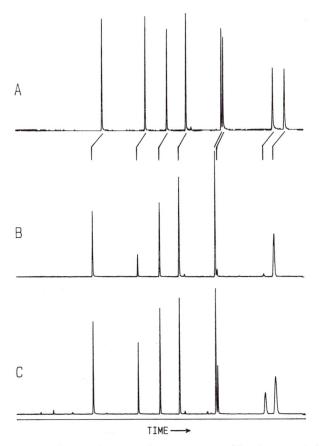

Fig. 22.8. The dependence of 266-nm laser ionization of polyaromatic hydrocarbons on laser temporal width. The compounds in this sample are the same as those in Fig. 22.7. (A) An FID chromatogram. (B) 2.4–2.8 mJ/pulse of 10-nsec pulses. (C) 400–500 μJ/pulse of 18-psec pulses. Note that the retention times for the FID chromatogram are slightly shifted due to the change in pressure at the outlet of the column.

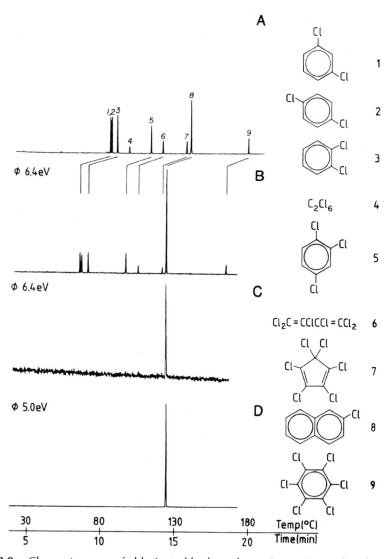

Fig. 22.9. Chromatograms of chlorinated hydrocarbon mixture (5 ng of each component). (A) Flame ionization detection. (B) ArF laser ionization: 5.5–4.3 mJ/pulse, ions
summed from m/z 10 to 27 and 29 to 590. Mass 28 was skipped in order to remove
the N_2 carrier gas ionization signal. (C) ArF laser ionization: 2.4–1.6 mJ/pulse, ions
summed from m/z 10 to 590. (D) KrF laser ionization: 6.9 mJ/pulse, ions summed from
m/z 10 to 590.

organic carbon.[32] KrF laser light and low-intensity ArF light appear to ionize
monochloronaphthalene only. This compound absorbs at least 10 times more
strongly than the others at 248 nm. At high ArF light intensities the compounds all
ionize to some extent. In fact, with the exception of C_2Cl_6, which does not absorb
because it does not have the ability to undergo a π–π^* transition, the different com-

pounds ionize with efficiencies remarkably similar to the flame-generated results. With KrF or low-intensity ArF radiation only monochloronaphthalene appears in the chromatogram. None of the single-ring aromatics ionizes efficiently under either of these conditions. Tembreull and Lubman have examined the laser ionization of a number of halogenated compounds in supersonic expansions with tunable UV laser radiation.[33] They were able to record two-photon ionization spectra in the 260–320 nm range for cold dichlorobenzenes. The fact that their sample molecules were internally cold and were excited at longer wavelengths probably enhanced their ion yields. Although this is still an area of intense study, it appears that reducing the rotational and vibrational state of excitation of a molecule by supersonic cooling can slow down certain radiationless relaxation processes.[34]

Future efforts in laser ionization GC/MS will be directed toward improving detection sensitivity by increasing the ionization efficiency of molecules through picosecond and tunable laser excitation and by increasing the number of molecules interacting with the UV radiation. The latter can be accomplished in a few ways. Since molecules elute from our GC column continuously but the UV laser light is irradiating the ionization zone only intermittently, it would obviously help to use a higher repetition rate laser. Similarly, it would be advantageous to connect a pulsed valve to the end of the column and release molecules only when there is light available to ionize them. This approach has already been employed to limit the throughput of gas into an FT/MS system.[35] A second advance that may be helpful is the use of gas dynamic focusing to limit the angular spread of the analyte as it sprays out of the column and into the vacuum chamber. Keller and co-workers originally demonstrated that this general method works,[36] and Stiller and Johnson showed that it can be applied to capillary column effluent slowly flowing into a vacuum chamber.[37] With improvements such as these, the prospects for improving on our previously demonstrated multifemtogram limit of detection appear to be excellent, and sensitivity in the attogram realm should be attainable.

Before closing the discussion of the sensitivity of laser ionization gas chromatography, it is perhaps worthwhile to cite an example of its utility in solving a scientific problem. A few years ago we were investigating the laser ionization characteristics of benzaldehyde.[38] We found that with 259-nm radiation, ions of mass 78, not mass 106, were produced. Three different scientific hypotheses could explain this observation. First, $C_6H_5COH^+$ ions could be initially generated and subsequently photofragmented to $C_6H_6^+$ ions. Second, neutral excited C_6H_5COH could dissociate into C_6H_6 and CO and the benzene then ionize. Third, benzene impurity might be present in the benzaldehyde, and only this might be ionizing. Photoelectron kinetic energy distribution measurements indicated that benzene and not benzaldehyde was ionizing. To distinguish between the second and third mechanisms we used laser GC/MS. Although a very small benzene impurity was indeed found in the benzaldehyde, the gas chromatograph conveniently separated this from the rest of the sample and enabled us to record a laser ionization mass spectrum of benzene-free benzaldehyde. Under these conditions only $C_6H_6^+$ ions were observed, indicating that the second proposed mechanism is correct. El Sayed and co-workers have investigated this same phenomenon using picosecond light pulses and have come to similar conclusions about the mechanism.[39–41] It is evident that laser ionization GC/MS can be very effective in simplifying complex problems. With continued advances in coherent ultraviolet light sources, we expect that the number of these applications will rapidly proliferate.

REFERENCES

1. G. S. Hurst, M. H. Nayleh, and J. P. Young, *Appl. Phys. Lett.* **30,** 229 (1977).
2. S. Rockwood, J. P. Reilly, K. Hohla, and K. L. Kompa, *Opt. Commun.* **28,** 175 (1979).
3. W. W. Wiley and I. H. McLaren, *Rev. Sci. Instrum.* **26,** 1150 (1955).
4. N. L. Marable and G. Sanzone, *Int. J. Mass Spectrom. Ion. Phys.* **13,** 185 (1974).
5. J. A. Browder, R. L. Miller, W. A. Thomas, and G. Sanzone, *Int. J. Mass Spectrom. Ion Phys.* **37,** 99, (1981).
6. M. L. Muga, U.S. Patent #4458149, July 3, 1984.
7. M. L. Muga, presented at the 1984 Pittsburgh Conference and Exhibition on Analytical Chemistry and Applied Spectroscopy, March 5–9, 1984, Abstract #296.
8. W. P. Poschenreider, *Int. J. Mass Spectrom. Ion Phys.* **6,** 413 (1971).
9. W. P. Poschenreider, *Int. J. Mass Spectrom. Ion Phys.* **9,** 357 (1972).
10. E. W. Müller and S. V. Krishnaswamy, *Rev. Sci. Instrum.* **45,** 1053 (1974).
11. B. A. Mamyrin, V. I. Karataev, D. V. Shmikk, and B. A. Zagulin, *Sov. Phys.–JETP* **37,** 45 (1973).
12. U. Boesl, H. J. Neusser, R. Weinkauf, and E. W. Schlag, *J. Phys. Chem.* **86,** 4857 (1982).
13. J. Grotemeyer, U. Boesl, K. Walter, and E. W. Schlag, *Org. Mass Spectrom.* **21,** 645 (1986).
14. E. Nielus, T. Heller, H. Feld, and A. Benninghoven (eds.), *Ion Formation from Organic Solids,* IFOS III. Springer, Berlin, 1986, p. 198.
15. M. Yang and J. P. Reilly, *Int. J. Mass Spectrom. Ion Process.* **75,** 209 (1987).
16. M. Yang and J. P. Reilly, *Anal. Inst.* 16, **1,** 133 (1987).
17. R. B. Opsal and J. P. Reilly, results to be published.
18. R. B. Opsal and J. P. Reilly, *Chem. Phys. Lett.* **99,** 461 (1983).
19. M. Yang, J. R. Millard, and J. P. Reilly, *Opt. Commun.* **55,** 41 (1985).
20. D. M. Lubman, *Laser Focus/Electro-Optics* **20,** 110 (1984).
21. K. G. Owens and J. P. Reilly, *J. Opt. Soc. Am. B.* **2,** 1589 (1985).
22. J. L. Verster, *Philips Res. Rep.* **18,** 465 (1963).
23. R. B. Opsal, K. G. Owens, and J. P. Reilly, *Anal. Chem.* **57,** 1884 (1985).
24. G. Rhodes, R. B. Opsal, J. T. Meek, and J. P. Reilly, *Anal. Chem.* **55,** 280 (1983).
25. R. B. Opsal, G. Rhodes, and J. P. Reilly, *Proceedings of the 1982 Scientific Conference on Chemical Defense Research,* Aberdeen Proving Ground, 321 (1982).
26. R. B.Opsal and J. P. Reilly, *Anal. Chem.* **60,** 1060 (1988).
27. R. B. Opsal and J. P. Reilly, *Anal. Chem.* **58,** 2919 (1986).
28. R. B. Opsal, Ph.D. Thesis, Indiana University, 1985.
29. T. M. Sack, D. A. McCrery, and M. L. Gross, *Anal. Chem.* **57,** 1290 (1985).
30. I. B. Berlman, *Handbook of Fluorescence Spectra of Aromatic Molecules.* Academic Press, New York, 1971.
31. E. Sekreta, K. G. Owens, and J. P. Reilly, *Chem. Phys. Lett.* **132** (4,5), 450 (1986).
32. H. H. Willard, L. L. Merritt, Jr., J. A. Dean, and F. A. Settle, Jr., *Instrumental Methods of Analysis.* Wadsworth, Belmont, CA 1981, p. 468.
33. R. D. Tembreull and D. M. Lubman, *AIP Conf. Proc.* **146** (Adv. Laser Sci.-1), 620 (1986).

34. C. S. Parmenter, *J. Phys. Chem.* **86,** 1735 (1982).

35. T. M. Sack and M. L. Gross, *Anal. Chem.* **55,** 2419 (1983).

36. R. A. Keller and N. S. Nogar, *Appl. Opt.* **23,** 2146 (1984).

37. S. W. Stiller and M. V. Johnston, *Anal. Chem.* **59,** 567 (1987).

38. S. R. Long, J. T. Meek, P. J. Harrington, and J. P. Reilly, *J. Chem. Phys.* **78,** 3741 (1983).

39. J. J. Yang, D. A. Bogel, R. S. Pandolfi, and M. A. El-Sayed, *J. Phys. Chem.* **87,** 2255 (1983).

40. J. J. Yang, M. A. El-Sayed, and F. Rebetrost, *Chem. Phys.* **96,** 1 (1985).

41. J. J. Yang, D. A. Gobeli, and M. A. El Sayed, *J. Phys. Chem.* **89,** 3426 (1985).

23

Multidimensional, Resonance-Enhanced Multiphoton Ionization Time-of-Flight Mass Spectrometry for Characterizing Environmental and Biological Samples for Polycyclic Aromatic Compounds

STEPHAN WEEKS, ARTHUR P. D'SILVA,
and ROY L. M. DOBSON

Analysis of environmental samples for polycyclic aromatic compounds (PACs) is the proverbial tip of the iceberg. These compounds are ubiquitous. The potential human health hazard of some PACs, such as the infamous benzo[a]pyrene (B[a]P), as mutagens and carcinogens is well established.[1] Common wisdom dictates the development of analytical methods that allow studies determining the sources of these potential carcinogenic compounds, their transport in the environment, and, finally, the metabolism to their ultimate carcinogenic form.

Analytical techniques must be versatile, precise, and accurate. Hundreds, and perhaps thousands of PACs may need to be determined in a variety of complex environmental and biological samples. Analyses of engine exhausts, air particulates, and industrial stack gases can aid in determining ways to reduce source emissions. Analyses of urine, blood, and DNA samples will answer the question "what is the real level of human exposure to potentially carcinogenic PACs?" Because of a potentially widespread need, analyses of samples for PACs must be routine, and because of the human health risks, and the economic and social impact, the analyst must use techniques and methods that allow efficient reporting of analyses with confidence.

Difficulties abound, however, which place stringent requirements on analytical methodology. Some interest exists in on-line, real-time monitoring. However, thousands of PACs are typically present at trace levels in complex sample matrices, generally requiring sampling, extraction, and separation prior to analysis. Some PACs can present solubility and storage problems, or may photodegrade while only small quantities (milligram) of biological samples may be available. In addition, PACs have many isomers, may be adsorbed on particles or adducted to large molecules, and are known to exhibit carcinogenic synergistic effects. The need is evident for a multidimensional technique capable of providing the necessary selectivity and detectability for PACs, as well as providing independent measurement checks on interferences, precision, and accuracy. Appropriate data acquisition, reduction, and statistical analysis will provide an essential part of such a multidimensional instrument responsible for the accurate measurement of hundreds of PACs in real time at trace levels. Unfortunately, most techniques either target a few selected carcinogens or group isomers together (e.g., benzofluoranthenes or methyl derivatives of B[a]P) rather than characterizing samples as completely as possible. Chemometric data treatment would enhance the significance of the analyst's report allowing synergistic effects and structure–activity relationships to be identified and measurement confidence levels to be determined.

Work at this laboratory is aimed at developing this total analysis methodology. The emphasis has been placed on developing versatile instrumentation that could completely characterize samples for PAC content. Understanding the importance of sampling, extraction techniques, and chemometrics, we have developed instrumentation that combines the powers of (1) chromatography, (2) laser spectroscopy, (3) mass spectrometry, and (4) fluorescence spectrometry. This is best shown by a flow chart (Fig. 23.1). Polycyclic aromatic compound separation can be accomplished using capillary gas chromatography (CGC) or supercritical fluid chromatography (SFC). High-resolution chromatography (HRC) simplifies the "working" matrix by ideally introducing PACs one at a time through either an effusive or supersonic expansion for PAC–selective laser excitation and ionization. When the resonance-enhanced multiphoton ionization (REMPI) involves only two photons, it is often termed R2PI or (1 + 1) REMPI. The products of the laser–analyte interaction are cations, electrons, and photons. Time-of-flight mass spectrometry (TOF-MS), total electron current detection (TECD), and laser-induced fluorescence (LIF) are the respective methods of detection. Mass and fluorescence spectrometry provide independent and highly selective and sensitive measurement capabilities. The system we developed for the characterization of complex samples for PACs was CGC–REMPI–TOF-MS–TECD–LIF.[2]

ENVIRONMENTAL AND BIOLOGICAL SAMPLES

Approximately 70–90% of all cancers in humans have been attributed to environmental causes.[3] The main source of polycyclic aromatic compounds is the result of incomplete combustion processes. These sources range from industrial burner stack gas emission to your backyard barbecue. Because they include energy-generating processes, automobile exhaust, and cigarette smoke, PACs exist everywhere and exposure is unavoidable. Extensive analyses of PACs in a variety of samples have been performed.[1] These include coal, oil, shale oil, solvent-refined coal (SRC), effluent water stream, marine sediment, and diesel and urban air particulate samples.

Recent emphasis has been placed on determining the mechanism of carcinogenesis. A wide variety of PACs have been studied for their mutagenic/carcinogenic effects.[1] Analyses of urine, blood, hemoglobin, tissue, and DNA have been per-

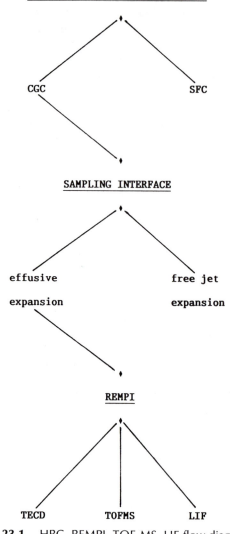

Fig. 23.1.   HRC–REMPI–TOF–MS–LIF flow diagram.

formed.[1,4,5] Carcinogens are known to undergo metabolic activation prior to adducting to DNA in the ultimate carcinogenic form. Techniques that can determine the procarcinogen, its metabolic products, and the intact DNA–carcinogen adduct are being sought. The need is to measure the adducted DNA at levels of $1:10^8$ adducts:base pair, or lower.[5]

PACs AND CARCINOGENESIS

Over the past decade there has been a flurry of activity combining many scientific disciplines aimed at understanding the chemistry and carcinogenic activity of

PACs.[1] Current thought is that many PACs are procarcinogens and must be metabolized to a form (ultimate carcinogens) that can then adduct to DNA. For B[a]P, the major adduct is believed to occur at the N-2 position of guanine through the C-10 position of B[a]P, after it has been metabolized to the diol epoxide (BPDE).[6] The adducted DNA is then believed to hinder (perhaps by cleavage of the adducted base from the DNA) the normal DNA replication process, resulting in (unregulated) cancerous cell growth. The "Bay Region" theory[7] and the one-electron theory[8] have attempted to explain the metabolic activation pathways. What triggers cancerous cell growth and what synergistic or antagonistic effects exist in complex chemical systems are largely unknown.

Analytically, carcinogenic and mutagenic studies are themselves hindered by trace level concentrations, small sample size, and resolution of components in complex mixtures. This last item is particularly poignant considering the fact that isomers and derivatives of PACs vary greatly in their carcinogenic or mutagenic activity. For example, B[a]P is known to be quite mutagenic whereas its isomer B[e]P is considered almost innocuous.[9] Isomeric heterocyclic compounds and substitutional derivatives of parent PACs, such as dimethyl derivatives of benz[a]anthracene, are also known to exhibit a wide range of mutagenic and carcinogenic effects.[9]

HYBRIDIZED ANALYTICAL APPROACH

The approach taken was to develop analytical instrumentation capable of gaining as much useful information as necessary to solve the problem in the shortest possible time. For targeting specific PAC isomers, this may mean utilizing the high spectral resolution of rotationally cooled jet spectroscopy. For characterizing environmental and biological samples for hundreds of PACs, however, this means taking a hybridized or hyphenated analytical instrumentation approach. This approach is essential to gain the necessary selectivity and detectabilities while maintaining the analyte throughput common to chromatographic techniques. Sufficient information would then be available to determine parent PACs, reactant products, metabolite products, and adducts and to study synergistic effects and structure–activity relationships.

In the instrumental system developed (Fig. 23.2) for PAC determinations in complex matrices, the sample characterization utilizes three basic steps: (1) *matrix simplification* by HRC, (2) *selective excitation and ionization* with tunable laser radiation, and (3) *selective detection* of the laser–analyte interaction products using mass spectrometry and fluorescence spectrometry. The ion source region, "gas dynamic cell" shown in Fig. 23.3 depicts the multidimensional nature of the detection of the physical products created with each laser pulse.

Briefly, microliter size aliquots of extracted samples are injected on-column into a capillary column gas chromatograph (CGC). In addition to providing chromatographic selectivity, the CGC sample introduction system *simplifies* the matrix for real-time spectrometric analysis. The analytes undergo an effusive expansion into the source region of a time-of-flight mass spectrometer (TOF-MS). In the dynamic gas sampling cell further selectivity for PACs over other types of compounds is achieved because only those compounds that absorb and are ionized by laser irradiation are interrogated. [Our laser provides tunable, narrow band, high spectral irradiance that would allow the high-resolution absorption spectra available from the supersonic jet expansion to be probed. The main advantage of supersonic jet fluorescence and ionization measurements is the ability to resolve spectrally isomers that cannot be chromatographically resolved, or distinguished by MS or ambient-temperature fluorescence spectrometry.] The excitation wavelength

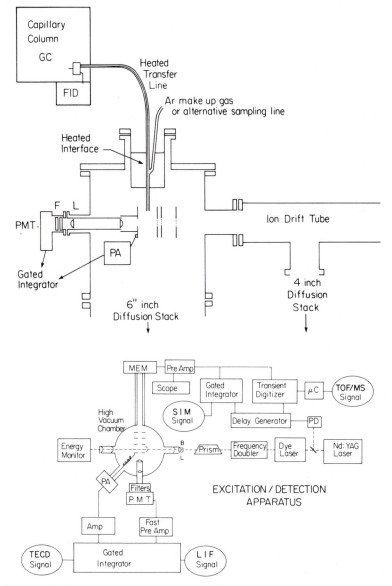

Fig. 23.2. Multidimensional, REMPI-TOF-MS based instrumentation (two part schematic diagram described in text).

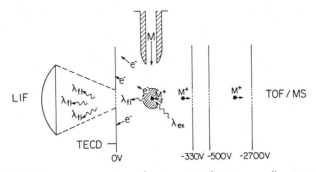

Fig. 23.3. TOF-MS ion source region showing simultaneous collection of laser–analyte interaction products.

and the laser power density controls the degree of ionization and fragmentation obtained during laser–analyte interaction. The TOF-MS provides the heart of the instrument, ideally, collecting and mass analyzing all of the ions created by each laser pulse, giving high selectivity and high sensitivity measurements. Additionally, we have added a novel electron detector and fluorescence measurement capability allowing us to simultaneously collect electrons and photons. All the analytically useful products of the laser–analyte interaction are therefore simultaneously collected from the same sample volume, making this a truly multidimensional technique in which information is separated in time, mass, and potentially fluorescent wavelength and lifetime.

Additional information about the sample can also be achieved by splitting the CGC effluent and using conventional detectors such as the flame ionization detector (FID) or electron capture detector (ECD).

ANALYTICAL FIGURES OF MERIT

The importance of figures of merit must be emphasized in light of the needs of biological and environmental measurements. Promising techniques are evaluated by their analytical figures of merit. These are defined only within the complete analytical procedure for a particular analysis.[10,11] They include functional and statistical descriptive and prognostic values. Functional figures of merit include accuracy and sensitivity, where sensitivity refers to the slope of the analytical calibration curve (i.e., a plot of concentration vs the measured value for the analytical signal). When interferences are present, the functional figures of merit include selectivity and specificity. A selective procedure is one in which the analyte can be detected in a manner such that there is no interference from other species. A specific method targets only one analyte, therefore the informing power of the analysis is low. The varying degrees of selectivity available from the simultaneous detection modes of the CGC-REMPI-TOF-MS-TECD-LIF-FID system have been demonstrated.

Statistical descriptive figures of merit generally used in quantitative analysis are the mean, $\bar{x}$, the standard deviation, s, and the relative standard deviation, RSD $= s/\bar{x}$. Statistical prognostic figures of merit, reported later, are used to appraise an analytical procedure. These include the limit of detection either as relative concentration (e.g., parts per million) or an absolute amount (e.g., picograms), the (linear) dynamic range, and, also quite useful, a precision curve (a plot of concentration vs relative standard deviation). Generally, the limit of detection is stated for an analyte signal that is three times the blank (matrix without analyte) noise level. The practical level of quantitation is generally considered to be two to three times the detection limit. To improve figures of merit within a given procedure random errors and system noise must be reduced. Figures of merit translate into the confidence an analyst has in using a particular instrument, method, and ultimately reporting analyte determinations. Functional figures of merit for multicomponent analysis of complex samples can be dramatically enhanced via the multidimensionality of a REMPI-MS-LIF instrumental system.

SIGNIFICANCE OF THE MULTIDIMENSIONAL APPROACH

Confidence in analytical measurements is of utmost importance when the reported results are to be used to make decisions relating to human health, cancer research, energy-producing and industrial operations, and possible waste site cleanup. Chemometrics involves applying mathematical and statistical methods to chemical measurements to extract useful chemical information to solve chemical problems. Multidimensional techniques are perhaps custom made for applying chemometric methods to the analysis. The independent measurement parameters inherent in our system increase the confidence in the analytical determinations.[11] These methods provide direct, efficient checks on possible interferences. Known synergistic effects could be flagged. Structure–activity relationships could also be efficiently studied or predicted. Dramatic improvements in multicomponent resolution have been shown to be possible by combining multivariate data analysis with multichannel detection (e.g., mass spectrometer or photodiode array) and parallel-column chromatography.[12] Chemometrics is useful in design, control, evaluation, and validation of the analytical measurement process. In essence, there is improved efficiency, as well as confidence in reporting analytical results.

RESONANCE-ENHANCED MULTIPHOTON IONIZATION TIME-OF-FLIGHT MASS SPECTROMETRY

Principles and Approach

Resonance-enhanced multiphoton ionization (REMPI) is analytically useful largely because it is a two-step process: first, the absorption of a photon into the upper vibronic level of the molecule (S_i in Fig. 23.4) and then the absorption of another photon to ionize the molecule.

Resonance Excitation

Resonance excitation gives the REMPI method its power of detectability. The absorption cross section for a nonresonant event through a virtual level is orders of magnitude smaller than for a resonant event. Additionally, the lifetime of the

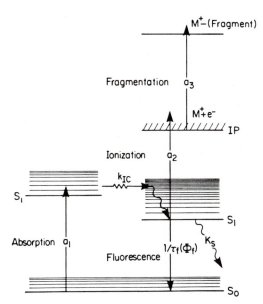

Fig. 23.4. Simplified Jablonski diagram representing possible energy transfer pathways of laser–analyte interaction. Absorption cross sections, a_1, a_2, a_3; fluorescence decay lifetime, τ_f; fluorescence quantum efficiency, Φ_f; (intramolecular) internal conversion rate, k_{ic}; and total nonradiative S_1 relaxation rate, K_s, are characteristic of the analyte.[17]

virtual state is $\sim 10^{-15}$ sec, whereas most polycyclic aromatic compounds have $\sim 10^{-9}$ sec lifetimes of their excited resonant levels. Obviously, the power density or photon flux of the laser would need to be considerably higher to increase the probability of absorbing at least one more photon to produce ionization in the non-resonant case. In addition to higher power density laser systems needed for non-resonant multiphoton ionization (MPI) methods, only limited selectivity, based on the ionization potential, is available. Typically a high degree of fragmentation also occurs in MPI experiments due to the high photon flux.

The larger absorption cross section, a_1, the longer lifetimes of the molecules in an excited state (e.g., S_1), and the saturation effects all lower the requirements of the necessary photon flux needed to optimize the fluorescence signal and, therefore, in part, the ionization efficiency. In fact, these factors limit the maximum power density of the first laser photon (λ_i) needed to optimize ion production and/or fluorescence signal. Power saturation effects can cause spectral broadening.[13] Saturation occurs when the lower energy level is significantly depleted so that any increase in laser energy density no longer provides the corresponding increase in excited level population. Saturation effects can also increase fragmentation and degrade the mass spectra and are known to exhibit temporal, spatial, and line profile effects.[13]

For most aromatic molecules, the absorption cross section to the first excited state, [a $\{S_0 \rightarrow S_1\}$], is large and approximately one to two orders of magnitude greater than that from the excited state to the ground ionic state [a $\{S_0 \rightarrow I_0\}$]. Two-color ionization experiments have been employed to avoid saturating the first excited state by limiting the photon energy density of the first laser pulse to ~ 2 μJ/mm^2 for an approximately 15-nsec pulsewidth.[13] For most PACs the first excited singlet state is less than half way to the ionization potential (IP). The highest selectivity and typically the best detectability for ionization exists if one can (1) cool the molecules in a supersonic expansion (which, in essence, highly populates the ground state and produces sharp line absorption spectra), (2) "just" saturate the $S_0 \rightarrow S_1$ transition in which the absorption cross sections are significantly higher nearer the band origin and the spectral region of excitation is generally in a less congested region for PACs, and (3) "just" saturate the ionizing S_1-I_0 transition.

When a polynuclear aromatic molecule is ionized from the zero vibrational level of the first excited state, the ionization cross section is highest slightly above the threshold for ionization (i.e., the ionization cross section is energy dependent due to Franck–Condon factors for ionization from an excited electronic state).[13] Since (1) the spectral irradiance needed to produce saturation for the first transition is typically significantly less than that needed for saturating the ionizing transition, and (2) the energies for these transitions are different (typically the ionizing transition energy > resonance excitation energy) for $1 + 1$ REMPI, then it is obvious that the use of two laser beams at different wavelengths, spatially and temporally overlapped in the analyte sample cell, is required to optimize the ionization signal. The use of two lasers, in addition to providing more selectivity and optimized ion production for individual molecular species, can also enhance selectivity based on differences in ionization potential and excited state lifetimes when compared to one-color REMPI or R2PI. Ionization thresholds are very sensitive to subtle structural factors. Isomers whose ionization potentials differ by more than 0.5 meV can be discriminated in a supersonic expansion.[13] Analogous to time-resolved fluorescence spectrometry, varying the delay time between the excitation and ionization laser pulses can selectively discriminate molecules with longer lifetimes from those with shorter lifetimes (e.g., phenanthrene, $\tau \sim 63$ nsec and anthracene, $\tau \sim 3$ nsec).[13]

However, when the analytical problem is to completely characterize unknown PACs in a complex sample matrix, initially, at least, compromises must be made. If these measurements are to be performed on a routine basis, cost and sample throughput are important. Obviously, two independent tunable laser systems increases the cost and the complexity (e.g., temporal and spatial overlap, power density considerations, I_1 and I_2 wavelength choices with appropriate data bases) of the instrumental system. Also, without returning to elaborate sample preparation and separation schemes, the complexity of the sample matrix may prove to be overwhelming. Even in a supersonic jet, as the energy of excitation or ionization increases (or λ_{ex} decreases) the electronic spectra of a polyatomic molecule becomes very congested due to the many closely spaced rovibronic bands.[13] Therefore, for example, it would be impossible to determine trace levels of a PAC whose excited state lifetime is shorter, first excited electronic level higher, and ionization potential greater relative to the interfering specie(s), and have the same molecular weight. In this example, time resolution will not be of use and spectral resolution is at best hindered from the interferent signal. When the ionizing laser radiation, which is typically 10 to 100 times more intense than the excitation radiation for optimal ion production, has sufficient energy ($h\nu_1 < h\nu_2$) for ionization (if $h\nu_1 < \frac{1}{2}$ IP, then $h\nu_2 > \frac{1}{2}$ IP), then with typical PACs and typical ionizing laser energy densities ($I_2 \sim 100 \times I_1 \simeq 200 \ \mu J/mm^2 \simeq 2 \times 10^6 \ W/cm^2$) the upper rovibronic levels of the molecule will be excited and ionized as in a simple one-color R2PI experiment. Because molecules cannot a priori distinguish between the excitation beam, (I_1), or ionizing radiation (I_2), significant interference in complex matrices containing hundreds of PACs can be expected from the absorption of two I_2 photons in what is often termed nonresonant background ionization (somewhat of a misnomer since the excitation is definitely occurring to real upper rovibronic levels). Although the selectivity is greatly diminished in real samples, selectivity for PACs over many other matrix components (e.g., alkanes, alcohols) does exist because of the significantly greater energy needed for resonance absorption of these species.

As we typically apply REMPI, a molecule is resonantly excited into an upper vibronic state, then it undergoes immediate intramolecular conversion to the congested upper vibronic levels of the lowest excited electronic singlet followed by relaxation to the zero point vibrational level.[14] After the excited molecules settle into the S_1 level, the relative probabilities for ionization, spontaneous emission, and nonradiative decay for a given photon energy density are determined by a_2 (photoionization cross section from S_1), τ_f (fluorescence decay lifetime), and Φ_f (fluorescence quantum yield), and the nonradiative decay probability, respectively. The molecule may then absorb more photons to become ionized and perhaps undergo subsequent fragmentation. Nonresonant MPI contributions are negligible when the laser energy density is low. Intermolecular deactivation is essentially nonexistent because the laser–analyte interaction occurs at "collision-free" pressures of 10^{-4} torr or less.

Instrumentation

To gain the most versatility, provide trace level detection, and minimize interferences, the high selectivities of capillary gas chromatography (CGC), laser absorption (excitation and ionization), mass spectrometry, and fluorescence spectrometry were combined. This is diagrammed in Fig. 23.1 and the instrumental system is shown in Fig. 23.2 To demonstrate the potential informing power of this instrumentation, compromise operation conditions were selected to characterize a 64-component mixture containing 24 polycyclic aromatic compounds. It is important to note that REMPI signals are wavelength and pulse energy dependent since each molecule is defined by its own unique set of physical and spectroscopic constants. Although enhancements in selectivity and detectabilities are possible, compromise conditions are necessary for real-time characterization of components in complex real-sample matrices.[11] Obviously, the instrumental operating conditions can be adjusted to attack a wide variety of analytical problems.

High-Resolution Gas Chromatographic Sample Introduction

Capillary gas chromatography (CGC) was selected as the preferred method for sample introduction of complex mixtures of PACs. First, CGC sample introduction of reference compounds and real samples is *reproducible,* making quantitation straightforward.[15] CGC separates and concentrates microliter sized samples and delivers the matrix simplified molecules as approximately 1-sec transient peaks into the ion source region of the mass spectrometer. This provides gas-phase molecules required for MS and, quite importantly, means only small amounts of the compounds enter the MS vacuum chamber reducing the chance of causing residual background signals that might severely limit detectabilities. Additionally, advances made in the areas of gas (CGC), supercritical fluid (SFC), and high-performance liquid (HPLC) chromatography can be readily adapted. The interface may be modified to operate as a rotationally cooled supersonic expansion (as opposed to an ambient temperature effusive expansion) by installing an appropriately sized restriction orifice (nozzle).

A temperature programmable gas chromatograph (Tracor Model 560) with a capillary inlet system and an FID detector was used in most of these studies. Because the emphasis was on sample throughput and not on high-resolution chromatography, a modest 15 m × 0.32-mm DB-5 bond stationary-phase fused silica capillary column (J&W Scientific) was used. This column phase is commonly used in characterization studies[16] in which samples contain high boiling point PACs.

The CGC effluent was split (30:70) for parallel detection by universal FID and the selective laser-based detectors. A 2 m × 0.20-mm-i.d. deactivated fused silica capillary was fed through a laboratory-constructed temperature-controlled transfer line and a vacuum flange mounted within an alumina-sleeved envelope used to prevent excessive heating of the vacuum chamber. The transfer line and mantle were heated to a temperature (typically 350°C) greater than the final GC over temperature to prevent condensation. Typically, 1-μL samples were injected into the GC with a 50:1 split ratio. Later studies employed an on-column injector. Helium was used as the carrier gas with a 50 cm/sec linear flow rate.

Laser Excitation and Ionization

The significance of laser excitation/ionization has been previously discussed. Narrow band radiation was obtained from an Nd:YAG laser-pumped tunable dye laser (Quantel YG 481 and TDL III). The fundamental output of the dye laser was collimated to a 2-mm beam diameter and frequency doubled (INRAD) to provide UV excitation and ionization. The fundamental radiation was separated from the UV radiation by a Pellin–Broca prism. The UV radiation had a wavelength range from 282 to 350 nm, pulse energies of 0.1–2.6 mJ, pulsewidth of ~15 nsec, repetition rate of 10 Hz, and spectral bandwidth of ~0.2 cm^{-1} at 300 nm. A portion of the 532-nm Nd:YAG pump laser beam illuminated a fast photodiode (PD). The resulting signal was used to trigger the detection electronics.

Dynamic Gas Sampling Cell

The multidimensional nature of the instrumentation is best shown in the dynamic gas sampling cell (Fig. 23.3). The sample was delivered from the CGC through a heated probe in an effusive expansion between two properly biased grids 15 mm apart. Tunable pulsed laser irradiation was absorbed by the analyte molecules producing fluorescence and ionization. All products of this laser–analyte interaction (i.e., cations, electrons, and photons) were collected and measured. The photons were imaged by a quartz lens through an appropriate optical cut-off filter (F) onto a photomultiplier tube (PMT). The electrons strike the backing grid that was electrically isolated from the other nude ion source components and grounded across a 10-kΩ resistor. The ions created in the laser–analyte interaction were accelerated into a time-of-flight mass spectrometer (TOF-MS).

Multidimensional Detection

The detection systems have been described in detail elsewhere.[2,17] Only the details of the TOF-MS system will be recapped here. Parallel detection was performed by splitting the chromatographic eluent. A conventional flame ionization detector served as a universal detector to give general information about the sample matrix. The laser-based detectors included laser-induced fluorescence (LIF), total electron current detection (TECD), and the TOF-MS. The total fluorescence imaged onto the PMT was amplified and measured by a gated integrator with a 1 sec time constant. Since PACs exhibit varying lifetimes (typically 1–500 nsec), a compromise gate delay (20 nsec) and gate width (40 nsec) were selected. The electron current was converted to a voltage signal by a fast operational amplifier mounted directly on the backing grid. The output signal was amplified and processed by a gated integrator with a 1 sec time constant and a 200 nsec gate width. These settings provided a compromise condition between chromatographic peak distortion and baseline noise. The TECD signal provided total ionization chromatograms and also could serve as a signal to begin TOF-MS data collection. The response of this detector was, of course, independent of compound mass and therefore, not subject to m/z biases observed in MS detection schemes.

The time-of-flight mass spectrometer was chosen for use in these pulsed laser excitation studies because of its inherently high ion detection efficiency. In principle, all ions created for each laser pulse may be simultaneously extracted and mass analyzed. The spacing between the backing and first ion drawout grids was enlarged from the standard 3 mm to a 15 mm spacing. This enlargement was primarily done to allow for incorporation of a supersonic jet nozzle/rotational cooling sample system at a later date. Also, the wider spacing somewhat relaxed the ionizing beam diameter restrictions of the instrument allowing the laser beam diameter to be varied without significant degradation in mass spectral resolution. The first and second ion drawout grids were held at dc potentials that were empirically optimized to yield maximum ion transmission and highest mass resolution through the 2-m flight tube. The mass spectral resolution achieved, under the experimental parameters employed, was greater than 300.

A variable-gain, magnetic electron multiplier (MEM) provided an output signal that was first preamplified and then processed by either a gated integrator, for selective ion monitoring (SIM) or a 200-MHz (5-nsec resolution) transient digitizer and microcomputer system (LeCroy 3500 SA 200) for collection, processing, and storage of any mass spectral window of interest during the elution of selected chromatographic peaks. Proper synchronization of all TOF-MS detection electronics was achieved through the incorporation of a 1- to 100-μsec variable trigger delay generator, in conjunction with the photodiode trigger system.

While the analytical system was in operation, a typical pressure of 6×10^{-6} torr was maintained in the main vacuum chamber (ion source region) by a 6-in. (15.2-cm) diffusion stack backed by a 500 L/min mechanical pump. A pressure of approximately 3×10^{-7} torr was maintained in the 2-m flight tube by an additional 4-in. (10.2-cm) diffusion stack backed by a 160 L/min mechanical pump.

Analytical Characterization of the Instrumentation

An idea of the functional and statistical figures of merit is necessary to select a particular technique for performing analyses. Because the primary emphasis was multidimensional, multicomponent sample characterization, many of these fundamental and practical studies were performed under compromise conditions. Typical conditions within the ionization source region were laser irradiation of 1 mJ/pulse at 285 nm in a 2-mm-diameter laser beam intersecting an effusive expansion ~2 cm from the end of the GC capillary.

Laser Power Dependency

The TECD response vs laser pulse energy (0.1–2.6 mJ) was linear over the power density range (0.2–5.5 $\times$ 10^6 W/cm^2). Power density plots for acenaphthene and carbazole, for example, resulted in linear correlation coefficients of 0.996 and 0.977, respectively. This study indicated that the primary ionization mechanism was a stepwise one-photon limited process for this range of power densities. Quite likely the resonance excitation step was saturated. To improve precision all REMPI signals should be normalized with respect to laser pulse energy (at a given laser beam diameter) to compensate for any drift in laser power.

Fragmentation

Complete TOF-MS fragmentation patterns were collected for each of four representative PAC (acenaphthene, dibenzofuran, fluorene, and carbazole) chromatographic eluents over a range of UV pulse energies. For the same power density range and wavelength used in the linearity study, the extent of fragmentation increased slightly, particularly for the heterocyclic PACs, dibenzofuran and carbazole. The parent ion still dominated even at 2.6 mJ/pulse. This indicated that the PACs largely experience "soft ionization," absorbing just enough photons (two:R2PI) to lose an electron.

By simply placing a lens in front of the vacuum chamber window to increase the laser power density even further, a significant increase in fragmentation can be obtained. In Fig. 23.5, a comparison of the TECD and TOF-MS data obtained for condensed vs unfocused laser beam at 2 mJ/pulse is shown. The laser beam diameter was reduced from 2.0 to 0.5 mm within the laser–analyte interaction region. Note that the respective TECD peak heights are approximately the same at both power densities even though the sample volume interrogated decreased by the same factor as the power density increased. This implies linearity to even greater power densities than discussed above.

Although the extent of fragmentation increased significantly, the parent ion peak continued to dominate. These studies indicate controllable fragmentation is possible with this system. The option of obtaining a complete fragmentation pattern for well-resolved chromatographic eluents is a valuable tool for identification of unknown polycyclic aromatic compounds. However, the "soft ionization" alternative is more practical when detecting coeluting photoionizable PAC. The general

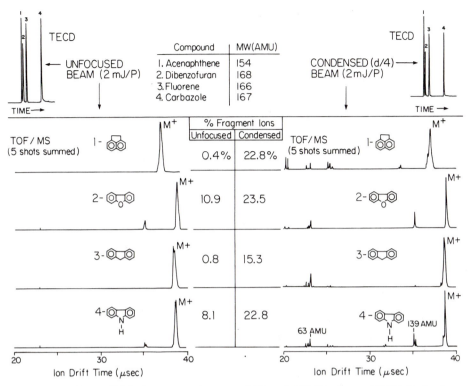

Fig. 23.5. Controllable fragmentation capability of REMPI shows TOF-MS for low (unfocused) and higher (condensed) laser power density. TECD signals remained constant.

conditions selected for analytical sample characterization were, therefore, a 2-mm-diameter laser beam at 1 mJ/pulse and 285 nm.

Wavelength Selective Ionization

Although conventional mass spectrometry cannot distinguish between the two closely eluting geometric isomers anthracene and phenanthrene, the REMPI spectra suggest highly selective photoionization of anthracene in the presence of phenanthrene at 300-nm laser excitation. This prediction was verified using both TECD and TOF-MS to monitor photoionization products of a chromatographic eluent consisting of a 1:1 mixture of these two polycyclic aromatic compounds (Fig. 23.6). In this study, deuterated phenanthrene was substituted to make obvious the identity of the ionized species. Each mass spectrum represents the summation of all ions produced, in a 42 amu mass window, during the elution of both species. As predicted by the REMPI spectra only anthracene was ionized at 300 nm, with a calculated selectivity factor of greater than 400. Spectral selectivity factors (SF) were determined by ratioing mass spectral peak areas of the protonated species to those of the phenanthrene-d_{10}. For comparison, both isomers were ionized at 285 nm with deuterated phenanthrene being favored, as expected. Also, a 1:1 protonated:deuterated phenanthrene standard was run under identical conditions. A

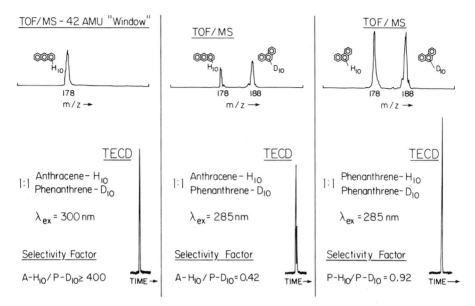

Fig. 23.6. Wavelength selective ionization for anthracene over phenanthrene at λ_{ex} = 300 nm vs interference effects at λ_{ex} = 285 nm for the two closely eluting isomers. Use of deuterated analog phenanthrene-d_{10} as internal reference standard.

selectivity factor of 0.92 demonstrated similar responses for both species, which in part justified the substitution of the deuterated analog for these studies.

Internal Reference Compounds

A further point to note concerning deuterated analogs is that they should prove to be ideal as internal reference compounds for REMPI–TOF-MS detection because they meet all of the important criteria: mainly, they have similar chemical and spectroscopic properties and they are absent in the sample. When response factors are known, these compounds provide a straightforward means of quantitation.

Linear Dynamic Range

The log–log plot of the TECD chromatographic peak heights as a function of the mass of PAC delivered to the gas sample cell was linear over four orders of magnitude change in concentration for the compounds examined. All analytical signals were normalized for laser pulse energy. The linear correlation coefficients for the analytical curves were all >0.995 with slopes equal to one. The upper concentration limits of ~50 ng was caused by overloading of the TECD amplifiers. The limiting noise in the instrumental system was from residual background and radio frequency interference (RFI).

Limits of Detection

The absolute limits of detection (LOD for S/N = 3) for TECD and TOF-MS detection are given in Table 23.1 for 21 PACs. While the TOF-MS detection scheme provides lower LODs, both detection methods operating under compromise con-

Table 23.1. CGC–REMPI–TECD–TOF-MS Limits of Detection (LOD)

Peak number	Compound	Structure	MW	LOD (pg at 285 nm)	
				TECD	TOF-MS
1	p-Cresol		108	6	1
2	Naphthalene		128	13	2
3	Biphenyl		154	a	a
4	Acenaphthene		154	1	0.1
5	Dibenzofuran		168	2	0.2
6	Fluorene		166	2	0.2
7	p-Phenylphenol		178	6	1
8	Dibenzothiophene		184	26	4
9	Phenanthrene		178	5	1
10	Anthracene		178	27	4
11	Carbazole		167	1	0.1
12	Pyrene		202	4	0.5
13	p-Terphenyl		230	2	0.2
14	Benz[a]anthracene		228	2	0.2
15	Chrysene		228	3	0.3
16	Triphenylene		228	5	0.5
17	Benzo[b]fluoranthene		252	50	7
18	Benzo[k]fluoranthene		252	4	0.5
19	Benzo[e]pyrene		252	2	0.3
20	Benzo[a]pyrene		252	2	0.2
21	Perylene		252	a	a
22	Dibenz[a,h]anthracene		278	a	a
23	Dibenz[a,c]anthracene		278	4	0.5
24	Benzo[ghi]perylene		276	4	0.4

[a]Detected but not quantitated.

ditions (1 mJ/pulse at 285 nm) offer low picogram detectabilities. It is important to remember that the REMPI signals are wavelength and pulse energy density dependent.

The detection limits reported in Table 23.1 were improved by one order of magnitude by using an on-column injection technique. On-column injection systems also provide improved accuracy and precision of sample introduction. Additionally, they eliminate discrimination against the less volatile solutes and allow some thermally labile species to be introduced into the analytical system.

Simplification of Chromatograms

The various degrees of chromatogram simplification available from this multidimensional analytical system were qualitatively assessed by studying a 64-compo-

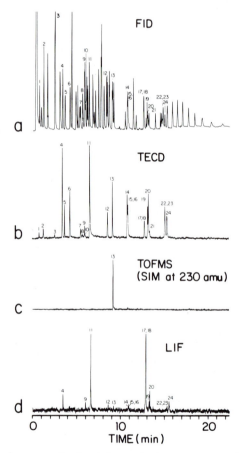

Fig. 23.7. Varying degrees of selectivity and chromatogram simplification demonstrated for a 64-component mixture by simultaneous (a) FID, (b) TECD, (c) TOF-MS, SIM at 230 amu (1 amu window), λ_{ex} = 285 nm, and (d) LIF, with WG 320 and WG 335 optical filters. Temperature programmed from 100° to 320°C at 14°C/min with 1 min initial hold and 6 min final hold.

nent mixture containing a variety of PACs and other aromatics, as well as aliphatics, alcohols, sulfones, and esters. Those species, mostly PACs, that were photoionized at 285 nm, are listed in Table 23.1, in the order of chromatographic elution. Figure 23.7 clearly demonstrates the degree of chromatogram simplification provided by the laser-based detection systems. The four chromatograms of this 64-component mixture were obtained simultaneously for a single sample injection. The parallel FID chromatogram is very complex with many unresolved peaks and includes the solvent peak. The laser-based portion of the detection system provided selective excitation–ionization based largely on characteristic molecular absorption coefficients and ionization cross sections at the excitation wavelength employed.

The TECD responded only to 24 compounds, mostly PACs, that were ionized. By utilizing the TOF-MS in the selective ion monitoring (SIM) mode, at 230 amu, only *p*-terphenyl was detected. This, in part, illustrates the additional selectivity available by employing a mass spectrometer in the detection system. When the transient digitizer data processing system is utilized, full advantage of the TOF-MS detector may be realized, as an entire mass spectrum may then be collected for each ionizable chromatographic effluent.

The LIF chromatogram shows detector response for only those fluorescing compounds that absorb at 285 nm. The LIF chromatogram was obtained under compromise detector gate width/delay and emission filtering conditions, all of which affected selectivity and detectability for each compound. The considerable variation in LIF relative peak heights as compared to TECD peak heights illustrates the complementary nature of the LIF information, as well as the potential utility of multivariate analysis for chromatogram peak deconvolution.

All of these chromatograms (FID, TECD, LIF, and TOF-MS–SIM or full mass spectra) are available simultaneously for each sample injection.

Applications

As noted in the introduction, a wide variety of environmental and biological samples contain trace levels of PACs. The samples investigated in our laboratory have included diesel and air particulate extracts, feed coal extracts, solvent-refined coal, and crude and shale oil fractions. Many real samples are far more complex than the above 64-component mixture. The neutral PAC fraction of a crude shale oil provides a qualitative indication of the complexity and the potential interferences that can occur in a real sample. The chromatograms, shown in Fig. 23.8, show the degree of selectivity simultaneously available using the current operating conditions. The FID is extremely complex, consisting mainly of unresolved peaks and including the solvent peak. The TECD chromatogram is still too complex using these chromatographic and REMPI conditions. This is not surprising as this fraction typically contains hundreds of PACs. However, the potential of TOF-MS detection is shown in Fig. 23.8c by the selective ion monitoring (184 ± 1 amu window) chromatogram. TOF-MS provided selective detection of dibenzothiophene in this real and extremely complex sample. The identification was verified by mass spectral data and a match of chromatographic retention time.

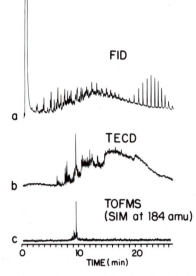

Fig. 23.8. Simultaneous chromatograms of a Paraho shale oil fraction: (a) FID, (b) TECD, and TOF-MS, SIM at 184 amu (±1 amu window). Temperature was programmed from 60° to 340°C at 12°C/min with 1 min initial hold and 2 min final hold. (Reprinted with permission from *Anal. Chem.* **58,** 2129 (1986). Copyright 1986 American Chemical Society.]

SUMMARY

Motivated to increase the potential informing power and efficiency of instrumental systems, a hybridized analytical approach was developed to determine PACs in complex sample matrices. The system described allows the analyst to take full advantage of developments in the fields of high-resolution chromatography, laser spectroscopy, fluorescence spectrometry, mass spectrometry, and chemometrics. HRC–REMPI–TOF-MS–LIF instrumentation provides, perhaps, the most promising technique for trace molecular sample characterization. REMPI–TOF-MS-based instrumentation will provide a versatile tool for developing methods for determining or monitoring potential PAC carcinogens in environmental and biological samples. The degree of selectivity and multidimensionality, combined with deuterated analogs as internal references and multivariate data reduction, will lead to shorter total analysis times—*and* give the analyst greater confidence in the analysis of complex, real samples.

REFERENCES

1. M. Cooke and A. J. Dennis (eds.), *Polynuclear Aromatic Hydrocarbons: A Decade of Progress.* Battelle Press, Columbus, 1988.
2. R.L.M. Dobson, A. P. D'Silva, S. J. Weeks, and V. A. Fassel, *Anal. Chem.* **58,** 2129 (1986).

3. J. H. Weisburger and G. M. Williams, in *Chemical Carcinogens,* Vol. 1, edited by C. E. Searle. American Chemical Society, Washington, D.C., 1984, p. 1323.

4. L. Shugart, *Anal. Biochem.* **152,** 365 (1986).

5. R. Jankowiak, R. S. Cooper, D. Zamzow, G. J. Small, G. Doskocil, and A. M. Jeffrey, *Chem. Res. Toxicol.* **1,** 60 (1988).

6. A. Dipple, R. C. Moschel, and C.A.H. Bigger, in *Chemical Carcinogens,* edited by C. E. Searle. American Chemical Society, Washington, D.C., 1984.

7. R. G. Harvey, *Acct. Chem. Res.* **14,** 218 (1981).

8. E. Cavalieri and E. Rogan, *EHP, Environ. Health Perspect.* **64,** 69 (1985).

9. M. L. Lee, M. V. Novotny, and K. D. Bartle, *Analytical Chemistry of Polycyclic Aromatic Compounds.* Academic Press, New York, 1981.

10. J. D. Winefordner (ed.), *Trace Analysis.* John Wiley, New York, 1976.

11. L. A. Currie (ed.), *Detection in Analytical Chemistry, Importance, Theory, and Practice.* American Chemical Society, Washington, D.C., 1988.

12. L. S. Ramos, J. E. Burger, and B. R. Kowalski, *Anal. Chem.* **57,** 2620 (1985).

13. J. W. Hager and S. C. Wallace, *Anal. Chem.* **60,** 5 (1988).

14. W. Siebrand, in *Dynamics of Molecular Collisions,* Part A, Chap. 6, edited by W. Miller. Plenum Press, New York, 1976.

15. J. M. Hayes and G. J. Small, *Anal. Chem.* **54,** 1202 (1982).

16. H. Y. Tong and F. W. Karasek, *Anal. Chem.* **56,** 2129 (1984).

17. R.L.M. Dobson, Ph.D. Dissertation, Iowa State University, Ames, IA, 1986.

Index

The italic letter *f* following an arabic numeral denotes a figure. Italic letter *t* following a numeral denotes a table.